THE
ORGANOMETALLIC
CHEMISTRY OF THE
TRANSITION METALS

THE ORGANOMETALLIC CHEMISTRY OF THE TRANSITION METALS

Sixth Edition

ROBERT H. CRABTREE
Yale University, New Haven, Connecticut

WILEY

Library of Congress Cataloging-in-Publication Data:

Crabtree, Robert H., 1948–
  The organometallic chemistry of the transition metals / by Robert H. Crabtree.—
Sixth edition.
    pages cm
  Includes bibliographical references and index.
  ISBN 978-1-118-13807-6 (cloth)
  1. Organometallic chemistry. 2. Organotransition metal compounds. I. Title.
  QD411.8.T73C73 2014
  547′.056–dc23
                                        2013046043

ISBN: 9781118138076

10 9 8 7 6 5 4 3 2 1

CONTENTS

Preface   xi

List of Abbreviations   xiii

1 Introduction   1

    1.1   Why Study Organometallic Chemistry?, 1
    1.2   Coordination Chemistry, 3
    1.3   Werner Complexes, 4
    1.4   The Trans Effect, 9
    1.5   Soft versus Hard Ligands, 10
    1.6   The Crystal Field, 11
    1.7   The Ligand Field, 19
    1.8   The sd^n Model and Hypervalency, 21
    1.9   Back Bonding, 23
    1.10 Electroneutrality, 27
    1.11 Types of Ligand, 29
    References, 37
    Problems, 38

2 Making Sense of Organometallic Complexes   40

    2.1   The 18-Electron Rule, 40
    2.2   Limitations of the 18-Electron Rule, 48
    2.3   Electron Counting in Reactions, 50
    2.4   Oxidation State, 51

v

2.5   Coordination Number and Geometry, 57
2.6   Effects of Complexation, 60
2.7   Differences between Metals, 63
References, 66
Problems, 67

3   Alkyls and Hydrides   **69**

3.1   Alkyls and Aryls, 69
3.2   Other σ-Bonded Ligands, 84
3.3   Metal Hydrides, 86
3.4   Sigma Complexes, 89
3.5   Bond Strengths, 92
References, 95
Problems, 96

4   Carbonyls, Phosphines, and Substitution   **98**

4.1   Metal Carbonyls, 98
4.2   Phosphines, 109
4.3   N-Heterocyclic Carbenes (NHCs), 113
4.4   Dissociative Substitution, 115
4.5   Associative Substitution, 120
4.6   Redox Effects and Interchange Substitution, 122
4.7   Photochemical Substitution, 124
4.8   Counterions and Solvents in Substitution, 127
References, 129
Problems, 131

5   Pi-Complexes   **134**

5.1   Alkene and Alkyne Complexes, 134
5.2   Allyls, 140
5.3   Diene Complexes, 144
5.4   Cyclopentadienyl Complexes, 147
5.5   Arenes and Other Alicyclic Ligands, 154
5.6   Isolobal Replacement and Metalacycles, 158
5.7   Stability of Polyene and Polyenyl Complexes, 159
References, 160
Problems, 161

6   Oxidative Addition and Reductive Elimination   **163**

6.1   Introduction, 163
6.2   Concerted Additions, 166

6.3 S_N2 Pathways, 168
6.4   Radical Mechanisms, 170
6.5   Ionic Mechanisms, 172
6.6   Reductive Elimination, 173
6.7   σ-Bond Metathesis, 179
6.8   Oxidative Coupling and Reductive Fragmentation, 180
References, 182
Problems, 182

7   Insertion and Elimination   **185**

7.1   Introduction, 185
7.2   CO Insertion, 187
7.3   Alkene Insertion, 192
7.4   Outer Sphere Insertions, 197
7.5   α, β, γ, and δ Elimination, 198
References, 201
Problems, 201

8   Addition and Abstraction   **204**

8.1   Introduction, 204
8.2   Nucleophilic Addition to CO, 207
8.3   Nucleophilic Addition to Polyenes and Polyenyls, 208
8.4   Nucleophilic Abstraction in Hydrides, Alkyls, and Acyls, 215
8.5   Electrophilic Addition and Abstraction, 216
8.6   Single-Electron Transfer and Radical Reactions, 219
References, 221
Problems, 222

9   Homogeneous Catalysis   **224**

9.1   Catalytic Cycles, 224
9.2   Alkene Isomerization, 231
9.3   Hydrogenation, 233
9.4   Alkene Hydroformylation, 242
9.5   Alkene Hydrocyanation, 245
9.6   Alkene Hydrosilylation and Hydroboration, 246
9.7   Coupling Reactions, 248
9.8   Organometallic Oxidation Catalysis, 250
9.9   Surface, Supported, and Cooperative Catalysis, 251
References, 253
Problems, 256

10   Physical Methods   259

    10.1   Isolation, 259
    10.2 ^{1}H NMR Spectroscopy, 260
    10.3 ^{13}C NMR Spectroscopy, 264
    10.4 ^{31}P NMR Spectroscopy, 266
    10.5   Dynamic NMR, 268
    10.6   Spin Saturation Transfer, 271
    10.7 T_1 and the Nuclear Overhauser Effect, 272
    10.8   IR Spectroscopy, 276
    10.9   Crystallography, 279
    10.10 Electrochemistry and EPR, 281
    10.11 Computation, 283
    10.12 Other Methods, 285
    References, 287
    Problems, 288

11   M–L Multiple Bonds   290

    11.1   Carbenes, 290
    11.2   Carbynes, 302
    11.3   Bridging Carbenes and Carbynes, 305
    11.4   *N*-Heterocyclic Carbenes, 306
    11.5   Multiple Bonds to Heteroatoms, 310
    References, 313
    Problems, 315

12   Applications   317

    12.1   Alkene Metathesis, 317
    12.2   Dimerization, Oligomerization, and Polymerization of
           Alkenes, 324
    12.3   Activation of CO and CO_2, 332
    12.4   C–H Activation, 336
    12.5   Green Chemistry, 343
    12.6   Energy Chemistry, 344
    References, 347
    Problems, 349

13   Clusters, Nanoparticles, Materials, and Surfaces   353

    13.1   Cluster Structures, 354
    13.2   The Isolobal Analogy, 364
    13.3   Nanoparticles, 368

13.4   Organometallic Materials, 371
References, 379
Problems, 381

14   Organic Applications   **383**

14.1   Carbon–Carbon Coupling, 384
14.2   Metathesis, 391
14.3   Cyclopropanation and C–H Insertion, 393
14.4   Hydrogenation, 394
14.5   Carbonylation, 396
14.6   Oxidation, 399
14.7   C–H Activation, 401
14.8   Click Chemistry, 405
References, 406
Problems, 408

15   Paramagnetic and High Oxidation-State Complexes   **411**

15.1   Magnetism and Spin States, 413
15.2   Polyalkyls and Polyhydrides, 420
15.3   Cyclopentadienyl Complexes, 425
15.4 f-Block Complexes, 426
References, 433
Problems, 435

16   Bioorganometallic Chemistry   **436**

16.1   Introduction, 437
16.2   Coenzyme B_{12}, 442
16.3   Nitrogen Fixation, 449
16.4   Nickel Enzymes, 457
16.5   Biomedical and Biocatalytic Applications, 463
References, 465
Problems, 467

Appendix A: Useful Texts on Allied Topics   **469**

**Appendix B: Major Reaction Types and Hints on
                        Problem Solving**   **472**

Solutions to Problems   **475**

Index   **493**

PREFACE

This book is a study of the logic of organometallic chemistry as well as some of its leading applications. It should give starting scholars everything they need to set out on this field and develop their own approaches and ideas. I would again like to thank the many colleagues and readers who kindly pointed out errors in the fifth edition or otherwise contributed: Professors Pat Holland, Jack Faller, Yao Fu, Lin Pu, Samuel Johnson, Odile Eisenstein, Ann Valentine, Gary Brudvig, Alan Goldman, Ulrich Hintermair, and Nilay Hazari, as well as students Liam Sharninghausen, Nathan Schley, Jason Rowley, William Howard, Joshua Hummel, Meng Zhou, Jonathan Graeupner, Oana Luca, Alexandra Schatz, and Kari Young. I also thank the Department of Energy for funding our work in this area.

Robert H. Crabtree

New Haven, Connecticut
August 2013

LIST OF ABBREVIATIONS

[ ]	Encloses complex molecules or ions
□	Vacant site or labile ligand
°	Degrees Celsius
$1°, 2°, 3° \ldots$	Primary, secondary, tertiary
A	Associative substitution (Section 4.5)
acac	Acetylacetonate
AO	Atomic orbital
at.	Pressure in atmospheres
bipy	2,2′-Bipyridyl
Bu	Butyl
ca.	about
cata	Catalyst
CIDNP	Chemically induced dynamic nuclear polarization (Section 6.4)
CN	Coordination number
cod	1,5-Cyclooctadiene
coe	Cyclooctene
cot	Cyclooctatetraene
Cp, Cp*	C_5H_5, C_5Me_5
Cy	Cyclohexyl
D	Dissociative substitution mechanism (Section 4.4)
D-C	Dewar–Chatt model of M(C=C) bonding involving weak back donation (Section 5.1)
d^n	Electron configuration (Section 1.4)
d_σ, d_π	σ-Acceptor and π-donor metal orbitals (see Section 1.4)
diars	$Me_2AsCH_2CH_2AsMe_2$
dmf	Dimethylformamide
dmg	Dimethyl glyoximate
dmpe	$Me_2PCH_2CH_2PMe_2$
DMSO	Dimethyl sulfoxide
dpe or dppe	$Ph_2PCH_2CH_2PPh_2$

e	Electron, as in 18e rule
E, E^+	Generalized electrophile such as H^+
e.e.	Enantiomeric excess (Section 9.3)
en	$H_2NCH_2CH_2NH_2$
EPR	Electron paramagnetic resonance
eq	Equivalent or equatorial
Et	Ethyl
eu	Entropy units
eV	Electron volt (1 eV = 23 kcal/mol)
fac	Facial (stereochemistry)
Fp	$(C_5H_5)(CO)_2Fe$
Hal	Halogen
$HBpz_3$	Tris(pyrazolyl)borate
HOMO	Highest occupied molecular orbital
hs	high spin
I	Nuclear spin
I	Intermediate substitution mechanism
IR	Infrared
L	Generalized ligand, most often a 2e ligand (the L model for ligand binding is discussed in Section 2.1)
L_nM	Metal fragment with n generalized ligands
lin	Linear
lp	Lone pair
ls	Low spin
LUMO	Lowest unoccupied molecular orbital
m-	Meta
m_r	Reduced mass
MCP	metalacyclopropane model of M(C=C) bonding involving strong back donation (Section 5.1)
Me	Methyl
mer	Meridional (stereochemistry)
MO	Molecular orbital
N	Group number of M (=number of valence e in the neutral atom)
nbd	Norbornadiene
NHC	N-heterocyclic carbene (Section 4.3)
NMR	Nuclear magnetic resonance (Sections 10.2–10.7)
NOE	Nuclear Overhauser effect (Section 10.7)
Np	Neopentyl
Nu, Nu^-	Generalized nucleophile, such as H^-
o-	Ortho
OA	Oxidative addition
OAc	Acetate
oct	Octahedral (p. 5)

OS	Oxidation state (Section 2.4)
oz.	Ounce (28.35 g)
p-	Para
Ph	Phenyl
pin	Pinacolate
py	Pyridine
RE	Reductive elimination
RF	Radio frequency
SET	Single-electron transfer (Section 8.6)
solv	Solvent
sq. pl.	Square planar
sq. py.	Square pyramidal (Figure 1.5)
T	A structure with three of the ligands disposed as in the letter T
T_1	Spin-lattice relaxation time
tacn	1,4,7-triazacyclononane
tacn*	N, N',N''-trimethyl-1,4,7-triazacyclononane
tbe	t-BuCH=CH$_2$
TBP or trig. bipy	Trigonal bipyramidal (Figure 4.4)
tet	Tetrahedral
thf	Tetrahydrofuran
TMEDA	Me$_2$NCH$_2$CH$_2$NMe$_2$
TMS	Trimethylsilyl
Tp	Tris(pyrazolyl)borate (**5.26**)
triphos	MeC(CH$_2$PPh$_2$)$_3$
Ts	p-tolylSO$_2$
TTP	Tricapped trigonal prism (Figure 2.1)
VB	Valence bond
X	Generalized 1e anionic ligand (see Section 2.1)
Y	A structure with three of the ligands disposed as in the letter Y
∂^+	Partial positive charge
δ	Chemical shift (NMR)
Δ	Crystal field splitting (Section 1.6)
Δ_{EN}	Electronegativity difference
$\Delta G^\ddagger \ \Delta H^\ddagger \ \Delta S^\ddagger$	Free energy, enthalpy and entropy of activation needed to reach the transition state for a reaction
η	Hapticity in ligands with contiguous donor atoms (e.g., C$_2$H$_4$. See Section 2.1)
κ	Hapticity in ligands with noncontiguous donor atoms (e.g., H$_2$NCH$_2$CH$_2$NH$_2$; see Section 2.1)
μ	Descriptor for bridging with a superscript for the number of metals bridged, as in M$_3$(μ^3-CO)
ν	Frequency

Periodic Table of the Elements

GROUP 1	2	3	4	5	6	7	8	9	10	11	12	13	14	15	16	17	18
1 **H** 1.0079																	2 **He** 4.0026
3 **Li** 6.941	4 **Be** 9.0122											5 **B** 10.811	6 **C** 12.011	7 **N** 14.007	8 **O** 15.999	9 **F** 18.998	10 **Ne** 20.180
11 **Na** 22.990	12 **Mg** 24.305											13 **Al** 26.982	14 **Si** 28.086	15 **P** 30.974	16 **S** 32.066	17 **Cl** 35.453	18 **Ar** 39.948
19 **K** 39.088	20 **Ca** 40.078	21 **Sc** 44.956	22 **Ti** 47.867	23 **V** 50.942	24 **Cr** 51.996	25 **Mn** 54.938	26 **Fe** 55.845	27 **Co** 58.933	28 **Ni** 58.693	29 **Cu** 63.546	30 **Zn** 65.39	31 **Ga** 69.723	32 **Ge** 72.61	33 **As** 74.922	34 **Se** 78.96	35 **Br** 79.904	36 **Kr** 83.80
37 **Rb** 85.468	38 **Sr** 87.62	39 **Y** 88.906	40 **Zr** 91.224	41 **Nb** 92.906	42 **Mo** 95.94	43 **Tc** 98.906*	44 **Ru** 101.07	45 **Rh** 102.91	46 **Pd** 106.42	47 **Ag** 107.87	48 **Cd** 112.41	49 **In** 114.82	50 **Sn** 118.71	51 **Sb** 121.76	52 **Te** 127.60	53 **I** 126.90	54 **Xe** 131.29
55 **Cs** 132.91	56 **Ba** 137.33	57 **La** ★ 138.91	72 **Hf** 178.49	73 **Ta** 180.95	74 **W** 183.84	75 **Re** 186.21	76 **Os** 190.23	77 **Ir** 192.22	78 **Pt** 195.08	79 **Au** 196.97	80 **Hg** 200.59	81 **Tl** 204.38	82 **Pb** 207.2	83 **Bi** 208.98	84 **Po** 209.98*	85 **At** 209.99*	86 **Rn** 222.02*
87 **Fr** 223.02	88 **Ra** 226.03*	89 **Ac** ▲ 227.03	104 **Rf** (261)	105 **Db** (262)	106 **Sg** (266)	107 **Bh** (262)	108 **Hs** (269)	109 **Mt** (266)	110 (273)	111 (272)	112 (294)						

★ Lanthanide series

58 **Ce** 140.12	59 **Pr** 140.91	60 **Nd** 144.24	61 **Pm** 146.92*	62 **Sm** 150.36	63 **Eu** 151.96	64 **Gd** 157.25	65 **Tb** 158.93	66 **Dy** 162.50	67 **Ho** 164.93	68 **Er** 167.26	69 **Tm** 168.93	70 **Yb** 173.04	71 **Lu** 174.97

▲ Actinide series

90 **Th** 232.04	91 **Pa** 231.04*	92 **U** 238.03	93 **Np** 237.05*	94 **Pu** 239.05*	95 **Am** 241.06*	96 **Cm** 244.06*	97 **Bk** 1249.08*	98 **Cf** 252.08*	99 **Es** 252.08*	100 **Fm** 257.10*	101 **Md** 258.10*	102 **No** 259.10*	103 **Lr** 262.11*

Note: Atomic masses shown here are the 1993 IUPAC values with a maxium of five significant figures (T. B. Coplen et al., Inorg. Chim. Acta 1994, 217, 217).
An asterisk indicates the mass of a commonly known radioisotope. Numbers in parentheses are the mass numbers of the corresponding longer-lived isotope.

1

INTRODUCTION

1.1 WHY STUDY ORGANOMETALLIC CHEMISTRY?

Organometallic chemists try to understand how organic molecules or groups interact with compounds of the inorganic elements, chiefly metals. These elements can be divided into the main group, consisting of the s and p blocks of the periodic table, and the transition elements of the d and f blocks. Main-group organometallics, such as n-BuLi and $PhB(OH)_2$, have proved so useful for organic synthesis that their leading characteristics are usually extensively covered in organic chemistry courses. Here, we look instead at the transition metals because their chemistry involves the intervention of d and f orbitals that bring into play reaction pathways not readily accessible elsewhere in the periodic table. While main-group organometallics are typically stoichiometric reagents, many of their transition metal analogs are most effective when they act as catalysts. Indeed, the expanding range of applications of catalysis is a major reason for the continued rising interest in organometallics. As late as 1975, the majority of organic syntheses had no recourse to transition metals at any stage; in contrast, they now very often appear, almost always as catalysts. Catalysis is also a central principle of Green Chemistry[1] because it helps avoid the waste formation,

The Organometallic Chemistry of the Transition Metals, Sixth Edition.
Robert H. Crabtree.
© 2014 John Wiley & Sons, Inc. Published 2014 by John Wiley & Sons, Inc.

for example, of Mg salts from Grignard reactions, that tends to accompany the use of stoichiometric reagents. The field thus occupies the borderland between organic and inorganic chemistry.

The noted organic chemist and Associate Editor of the *Journal of Organic Chemistry*, Carsten Bolm,[2] has published a ringing endorsement of organometallic methods as applied to organic synthesis:

> In 1989, OMCOS-VI [the 6th International Conference on Organometallic Chemistry Directed Toward Organic Synthesis] took place in Florence and . . . left me with the impression that all important transformations could—now or in the future—be performed with the aid of adequately fine-tuned metal catalysts. Today, it is safe to say that those early findings were key discoveries for a conceptual revolution that occurred in organic chemistry in recent years. Metal catalysts can be found everywhere, and many synthetic advances are directly linked to . . . developments in catalytic chemistry.

Organometallic catalysts have a long industrial history in the production of organic compounds and polymers. Organometallic chemistry was applied to nickel refining as early as the 1880s, when Ludwig Mond showed how crude Ni can be purified with CO to volatilize the Ni in the form of $Ni(CO)_4$ as a vapor that can subsequently be heated to deposit pure Ni. In a catalytic application dating from the 1930s, $Co_2(CO)_8$ brings about hydroformylation, in which H_2 and CO add to an olefin, such as 1- or 2-butene, to give n-pentanal or n-pentanol, depending on the conditions.

A whole series of industrial processes has been developed based on transition metal organometallic catalysts. For example, there is intense activity today in the production of homochiral molecules, in which racemic reagents can be transformed into single pure enantiomers of the product by an asymmetric catalyst. This application is of most significance in the pharmaceutical industry where only one enantiomer of a drug is typically active but the other may even be harmful. Other examples include polymerization of alkenes to give polyethylene and polypropylene, hydrocyanation of butadiene for nylon manufacture, acetic acid manufacture from MeOH and CO, and hydrosilylation to produce silicones and related materials.

Beyond the multitude of applications to organic chemistry in industry and academia, organometallics are beginning to find applications elsewhere. For example, several of the organic light-emitting diode (OLED) materials recently introduced into cell phone displays rely on organometallic iridium compounds. They are also useful in solid-state light-emitting electrochemical cells (LECs).[3] Samsung has a plant that has been producing OLED screens since 2008 that use a cyclometallated

Ir complex as the red emitter. Cyclometallated Ru complexes may have potential as photosensitizers for solar cells.[4] Organometallic drugs are also on the horizon.

Bioinorganic chemistry has traditionally been concerned with classical coordination chemistry—the chemistry of metal ions surrounded by N- or O-donor ligands, such as imidazole or acetate—because metalloenzymes typically bind metals via such N or O donors. Recent work has identified a small but growing class of metalloenzymes with organometallic ligands such as CO and CN^- in hydrogenases or the remarkable central carbide bound to six Fe atoms in the active site MoFe cluster of nitrogenase. Medicinally useful organometallics, such as the ferrocene-based antimalarial, ferroquine, are also emerging, together with a variety of diagnostic imaging agents.[5]

The scientific community is increasingly being urged to tackle problems of practical interest.[6] In this context, alternative energy research, driven by climate change concerns,[7] and green chemistry, driven by environmental concerns, are rising areas that should also benefit from organometallic catalysis.[8] Solar and wind energy being intermittent, conversion of the resulting electrical energy into a storable fuel is proposed. Splitting water into H_2 and O_2 is the most popular suggestion for converting this electrical energy into chemical energy in the form of H–H bonds, and organometallics are currently being applied as catalyst precursors for water splitting.[9] Storage of the resulting hydrogen fuel in a convenient form has attracted much attention and will probably require catalysis for the storage and release steps. The recent extreme volatility in rare metal prices has led to "earth-abundant" metals being eagerly sought[10] as replacements for the precious metal catalysts that are most often used today for these and other practically important reactions.

1.2 COORDINATION CHEMISTRY

Even in organometallic compounds, N- or O-donor coligands typical of coordination chemistry are very often present along with C donors. With the rise of such mixed ligand sets, the distinction between coordination and organometallic chemistry is becoming blurred, an added reason to look at the principles of coordination chemistry that also underlie the organometallic area. The fundamentals of metal–ligand bonding were first established for coordination compounds by the founder of the field, Alfred Werner (1866–1919). He was able to identify the octahedral geometric preference of CoL_6 complexes without any of the standard spectroscopic or crystallographic techniques.[11]

Central to our modern understanding of both coordination and organometallic compounds are d orbitals. Main-group compounds either have a filled d level that is too stable (e.g., Sn) or an empty d level that is too unstable (e.g., C) to participate significantly in bonding. Partial filling of the d orbitals imparts the characteristic properties of the transition metals. Some early-transition metal ions with no d electrons (e.g., group 4 Ti^{4+}) and some late metals with a filled set of 10 (e.g., group 12 Zn^{2+}) more closely resemble main-group elements.

Transition metal ions can bind *ligands* (L) to give a coordination compound, or *complex* ML_n, as in the familiar aqua ions $[M(OH_2)_6]^{2+}$ (M = V, Cr, Mn, Fe, Co, or Ni). Together with being a subfield of organic chemistry, organometallic chemistry can thus also be seen as a subfield of coordination chemistry in which the complex contains an M–C bond (e.g., $Mo(CO)_6$). In addition to M–C bonds, we include M–L bonds, where L is more electropositive than O, N, and halide (e.g., $M–SiR_3$ and M–H). These organometallic species tend to be more covalent, and the metal more reduced, than in classical coordination compounds. Typical ligands that usually bind to metal ions in their more reduced, low valent forms are CO, alkenes, and arenes, as in $Mo(CO)_6$, $Pt(C_2H_4)_3$, and $(C_6H_6)Cr(CO)_3$. Higher valent states are beginning to play a more important role, however, as in hexavalent WMe_6 and pentavalent $O=Ir(mesityl)_3$ (Chapter 15).

1.3   WERNER COMPLEXES

In classical *Werner complexes*, such as $[Co(NH_3)_6]^{3+}$, a relatively high valent metal ion binds to the lone pairs of electronegative donor atoms, typically, O, N, or halide. The M–L bond has a marked polar covalent character, as in $L_nM–NH_3$, where L_n represents the other ligands present. The M–NH_3 bond consists of the two electrons present in lone pair of free NH_3, but now donated to the metal to form the complex.

Stereochemistry

The most common type of complex, octahedral ML_6, adopts a geometry (**1.1**) based on the Pythagorean octahedron. By occupying the six vertices of an octahedron, the ligands can establish appropriate M–L bonding distances, while maximizing their L$\cdots$L nonbonding distances. For the coordination chemist, it is unfortunate that Pythagoras decided to name his solids after the number of faces rather than the number of vertices. The solid and dashed wedges in **1.1** indicate bonds that point toward or away from us, respectively:

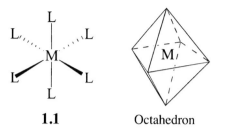

1.1   Octahedron

The assembly of metal and ligands that we call a *complex* may have a net ionic charge, in which case it is a complex ion (e.g., $[PtCl_4]^{2-}$). Together with the counterions, we have a complex salt (e.g., $K_2[PtCl_4]$). In some cases, both cation and anion may be complex, as in the picturesquely named *Magnus' green salt* $[Pt(NH_3)_4][PtCl_4]$, where the square brackets enclose the individual ions.

Ligands that have a donor atom with more than one lone pair can often donate one pair to each of two or more metal ions to give polynuclear complexes, such as **1.2** ($L = PR_3$). The bridging group is represented by the Greek letter μ (mu) as in $[Ru_2(\mu\text{-}Cl)_3(PR_3)_6]^+$. Dinuclear **1.2** consists of two octahedra sharing a face containing three chloride bridges.

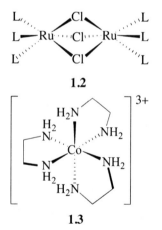

1.2

1.3

Chelate Effect

Ligands with more than one donor atom, such as ethylenediamine ($NH_2CH_2CH_2NH_2$, or "en"), can donate both lone pairs to form a *chelate* ring (**1.3**). The most favorable ring size is five, but six is often seen. Chelating ligands are much less easily displaced from a complex than are comparable monodentate ligands for the reason illustrated in Eq. 1.1:

$$[M(NH_3)_6]^{n+} + 3en \rightarrow [M(en)_3]^{n+} + 6NH_3 \tag{1.1}$$

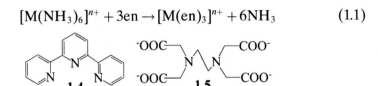

When the reactants release six NH_3 molecules in Eq. 1.1, the total number of particles increases from four to seven. This creates entropy and so favors the chelate. Each chelate ring usually leads to an additional factor of about 10^5 in the equilibrium constant for the reaction. Equilibrium constants for complex formation are usually called *formation constants*; the higher the value, the more stable the complex.

Chelate ligands can also be polydentate, as in tridentate **1.4** and hexadentate **1.5**. As a tridentate ligand, **1.4** is termed a *pincer* ligand, a type attracting much recent attention.[12] Ethylenediaminetetracetic acid, (EDTA, **1.5**) can take up all six sites of an octahedron and thus completely wrap up many different metal ions. As a common food preservative, EDTA binds free metal ions so that they can no longer catalyze aerial oxidation of the foodstuff. Reactivity in metal complexes usually requires the availability of open sites or at least labile sites at the metal.

Werner's Coordination Theory

Alfred Werner developed the modern picture of coordination complexes in the 20 years that followed 1893, when, as a young scientist, he proposed that the well-known cobalt ammines (ammonia complexes) have an octahedral structure as in **1.3** and **1.6**.

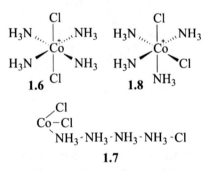

In doing so, he opposed the standard view that the ligands were bound in chains with the metal at one end (e.g., **1.7**), as held by everyone else in the field. Naturally, he was opposed by supporters of the standard model, who only went so far as adjusting their model to take account of new data. Jørgensen, who led the traditionalists against the Werner insurgency, was not willing to accept that a trivalent metal, Co^{3+}, could form bonds to more than three groups and so held to the chain theory. At first, as each new "proof" came from Werner, Jørgensen was able to point to problems or reinterpret the chain theory to fit the new facts. For example, coordination theory calls for two isomers of $[Co(NH_3)_4Cl_2]^+$ (**1.6** and **1.8**). Up to that time, only a green one had ever been found, now called the *trans* isomer (**1.6**) because the two Cl ligands occupy opposite vertices of the octahedron. According to Werner, a second isomer, **1.8** (cis), then unknown, should have had the Cl ligands in adjacent vertices—he therefore needed to find this isomer. Changing the chloride to nitrite, Werner was indeed able to obtain both green cis and purple trans isomers of $[Co(NH_3)_4(NO_2)_2]^+$ (**1.9** and **1.10**). Jørgensen quite reasonably—but wrongly—countered this finding by saying that the nitrite ligands in the two isomers were simply bound in a different way (*linkage isomers*), via N in one case ($Co-NO_2$) and O ($Co-ONO$) in the other (**1.11** and **1.12**). Undismayed, Werner then found the green and purple isomers, **1.13** and **1.14**, of $[Co(en)_2Cl_2]^+$, in a case where no linkage isomerism was possible. Jørgensen brushed this observation aside by invoking different chain arrangements, as in **1.15** and **1.16**:

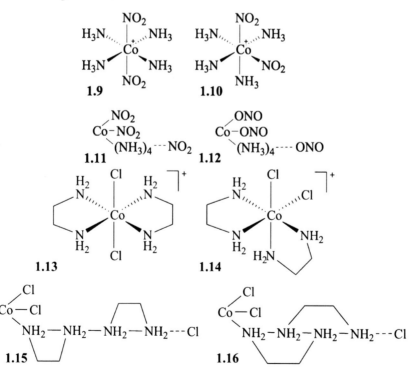

In 1907, Werner finally made the elusive purple isomer of $[Co(NH_3)_4Cl_2]^+$ by an ingenious route (Eq. 1.2) via the necessarily cis carbonate $[Co(NH_3)_4(O_2CO)]$. Treatment with HCl in the solid state at 0°C liberates CO_2 and gives the elusive cis dichloride. Jørgensen, receiving a sample of this purple complex by mail, finally conceded defeat.

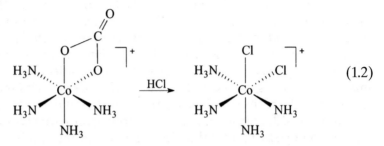

(1.2)

Werner later resolved optical isomers of the halides $[Co(en)_2X_2]^+$ (**1.17** and **1.18**), where the isomerism can arise from an octahedral array, but not from a chain. Even this point was challenged on the grounds that only organic compounds could be optically active, and so this activity must come from the organic ligands in some undefined way. Werner responded by resolving a complex (**1.19**) containing only inorganic elements. This has the extraordinarily high specific rotation of 36,000° and required 1000 recrystallizations to resolve.

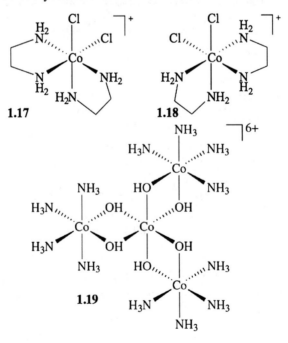

This episode provides general conclusions of importance: some of our current ideas are likely to be wrong—we just do not know which ones. The literature must thus be read critically with an eye for possible flaws in the results, inferences, or arguments. Nugent has reviewed a series of ideas, once generally held, that subsequently fell from grace.[13] Another lesson from Werner is that we must take objections seriously and devise critical experiments that distinguish between possible theories, not merely ones that confirm our own ideas.

1.4   THE TRANS EFFECT

We now move from complexes of Co^{3+}, or "Co(III)," to the case of Pt(II), where the II and III refer to the +2 or +3 oxidation states (Section 2.4) of the metal. Pt(II) is four coordinate and adopts a square planar geometry, as in **1.20**. These complexes can react with incoming ligands, L^i, to replace an existing ligand L in a *substitution* reaction. Where a choice exists between two possible geometries of the product, as in Eq. 1.3 and Eq. 1.4, the outcome is governed by the trans effect. For example, in the second step of Eq. 1.3, the NH_3 does not replace the Cl trans to NH_3, but only the Cl trans to Cl. This observation means that Cl is a higher trans effect ligand than NH_3. Once again, in Eq. 1.4, the NH_3 trans to Cl is displaced, not the one trans to NH_3.

1.20

Ligands, L^i, that are more effective at labilizing a ligand trans to themselves have a higher trans effect. We see the reason in Sections 4.3–4.4, but for the moment, only note that the effect is very marked for Pt(II), and that the highest trans effect ligands either (i) form strong σ bonds, such as $L^i = H^-$ or Me^-, or (ii) are strong π acceptors, such as $L^i = CO$, C_2H_4, or (iii) have polarizable period 3 or higher p block elements as donor as in S-bound thiourea, {$(NH_2)_2CS$ or "tu"}. One of the highest trans effect ligands of all, CF_3^-,[14] falls into classes (i) and (ii).

High trans effect L^i ligands also lengthen and weaken trans M–L bonds, as shown in X-ray crystallography by an increase in the M–L distance or in nuclear magnetic resonance (NMR) spectroscopy by a decrease in the M,L coupling (Section 10.4), or in the IR (infrared) spectrum (Section 10.8) by a decrease in the ν(M–L) stretching frequency. When L^i changes the ground-state thermodynamic properties of a complex in one of these ways, we use the term *trans influence* to distinguish the situation from the trans effect proper in which L^t accelerates the rate of substitution, a kinetic effect.

An important application of the trans effect is the synthesis of specific isomers of coordination compounds. Equation 1.3 and Equation1.4 show how the cis and trans isomers of $Pt(NH_3)_2Cl_2$ can be prepared selectively by taking advantage of the trans effect order $Cl > NH_3$, where $L^t = Cl$. This example is also of practical interest because the cis isomer is a very important antitumor drug (Section 16.5), but the trans isomer is toxic.

$$\tag{1.3}$$

$$\tag{1.4}$$

A trans effect series for a typical Pt(II) system is given below. The order can change somewhat for different metals and oxidation states.

$$OH^- < NH_3 < Cl^- < Br^- < CN^- \approx CO \approx C_2H_4 \approx CH_3^- < I^- < PR_3 < H^- < CF_3^-$$

$\longleftarrow$ low trans effect   high trans effect $\longrightarrow$

1.5   SOFT VERSUS HARD LIGANDS

Ligands may be *hard* or *soft* depending on their propensity for ionic (hard) or covalent (soft) bonding. Likewise, metals can also be hard or soft. The favored, well-matched combinations are a hard ligand with a hard metal and a soft ligand with a soft metal; hard–soft combinations are disfavored because of the mismatch of bonding preferences.[15]

Table 1.1 shows formation constants for different metal ion–halide ligand combinations,[15] where large positive numbers reflect strong binding. The hardest halide is F^- because it is small, difficult to polarize, and forms predominantly ionic bonds. It binds best to a hard cation, H^+, also small and difficult to polarize. This hard–hard combination therefore leads to strong bonding and HF is a weak acid (pK_a +3).

Iodide is the softest halide because it is large, easy to polarize, and forms predominantly covalent bonds. It binds best to a soft cation, Hg^{2+}, also large and easy to polarize. In this context, high polarizability means that electrons from each partner readily engage in covalent bonding.

TABLE 1.1 Hard and Soft Acids and Bases: Some Formation Constants[a]

Metal Ion (Acid)	Ligand (Base)			
	F^- (Hard)	Cl^-	Br^-	I^- (Soft)
H^+ (hard)	3	−7	−9	−9.5
Zn^{2+}	0.7	−0.2	−0.6	−1.3
Cu^{2+}	0.05	0.05	−0.03	–
Hg^{2+} (soft)	1.03	6.74	8.94	12.87

[a]The values are the negative logarithms of the equilibrium constant for $[M.aq]^{n+} + X^- \rightleftharpoons [MX.aq]^{(n-1)+}$ and show how H^+ and Zn^{2+} are hard acids, forming stronger complexes with F^- than with Cl^-, Br^-, or I^-. Cu^{2+} is a borderline case, and Hg^{2+} is a very soft acid, forming much stronger complexes with the more polarizable halide ions.

The Hg^{2+}/I^- soft–soft combination is therefore a very good one — by far the best in the table — and dominated by covalent bonding. HI, a mismatched pairing, produces a strong acid (pK_a −9.5).

Soft bases either have lone pairs on atoms of the second or later row of the periodic table (e.g., Cl^-, Br^-, and PPh_3) or have double or triple bonds (e.g., CN^-, C_2H_4, and benzene) directly adjacent to the donor atom. Soft acids can come from the second or later row of the periodic table (e.g., Hg^{2+}) or contain atoms that are relatively electropositive (e.g., BH_3) or are metals in a low (≤ 2) oxidation state (e.g., Ni(0), Re(I), Pt(II), and Ti(II)). Organometallic chemistry is dominated by soft–soft interactions, as in metal carbonyl, alkene, and arene chemistry, while traditional coordination chemistry involves harder metals and ligands.

1.6 THE CRYSTAL FIELD

An important advance in understanding the spectra, structure, and magnetism of transition metal complexes is provided by *crystal field theory* (CFT) which shows how the *d* orbitals of the transition metal are affected by the ligands. CFT is based on the very simple model that these ligands act as negative charges, hence crystal *field*. For Cl^-, this is the negative charge on the ion, and for NH_3, it is the N lone pair, a local concentration of negative charge. If the metal ion is isolated in space, then the five *d* orbitals are *degenerate* (have the same energy). As the six ligands approach from the octahedral directions $\pm x$, $\pm y$, and $\pm z$, the *d* orbitals take the form shown in Fig. 1.1. The *d* orbitals that point along the axes toward the incoming L groups ($d_{(x^2-y^2)}$ and d_{z^2}) are destabilized by the negative charge of the ligands and move to higher energy. Those that point away from L (d_{xy}, d_{yz}, and d_{xz}) are less destabilized.

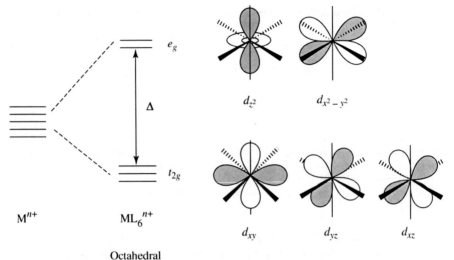

$$M^{n+} \qquad ML_6{}^{n+}$$

Octahedral

FIGURE 1.1 Effect on the d orbitals of bringing up six ligands along the $\pm x$, $\pm y$, and $\pm z$ directions. In this figure, shading represents the symmetry (not the occupation) of the d orbitals; shaded parts have the same sign of ψ. For convenience, energies are shown relative to the average d-orbital energy.

The most strongly destabilized pair of orbitals are labeled e_g, from their symmetry, or more simply as d_σ, because they point directly along the M–L directions. The set of three more stable orbitals has the label t_{2g}, or simply d_π—they point between the ligand directions but can still form π bonds with suitable ligands. The energy difference between the d_σ and d_π set, the *crystal field splitting*, is labeled Δ (or sometimes $10Dq$) and depends on the value of the effective negative charges and therefore on the nature of the ligands. A higher Δ means we have stronger M–L bonds.

High Spin versus Low Spin

In group 9 cobalt, the nine valence electrons have the configuration $[Ar]4s^2 3d^7$, but only in the free atom. Once a complex forms, however, the $3d$ orbitals become more stable than the $4s$ as a result of M–L bonding, and the effective electron configuration becomes $[Ar]4s^0 3d^9$ for a Co(0) complex, or $[Ar]3s^0 4d^6$ for Co(III), usually shortened to d^9 and d^6, respectively. The $4s$ orbital is now less stable than $3d$ because, pointing as it does in all directions, the $4s$ suffers CFT repulsion from all the ligands in any Co complex, while the $3d$ orbitals only interact with a subset of the ligands in the case of the d_σ set or, even less destabilizing, point between the ligands in the case of the d_π set.

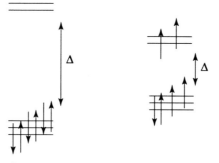

FIGURE 1.2 In a d^6 metal ion, both low- and high-spin complexes are possible depending on the value of Δ. A high Δ leads to the low-spin form (*left*).

This crystal field picture explains why Werner's d^6 Co^{3+} has such a strong octahedral preference. Its six electrons just fill the three low-lying d_π orbitals of the octahedral crystal field diagram and leave the higher energy d_σ orbitals empty. Stabilizing the electrons in a molecule is equivalent to stabilizing the molecule itself. Octahedral d^6 is by far the commonest type of metal complex in all of organometallic chemistry, as in Mo(0), Re(I), Fe(II), Ir(III), and Pt(IV) complexes. In spite of the high tendency to spin-pair the electrons in the d^6 configuration (to give the common *low-spin* form $t_{2g}^6 e_g^0$), if the ligand field splitting is small enough, the electrons may rearrange to give the rare *high-spin* form $t_{2g}^4 e_g^2$. In high spin (h.s.), all the unpaired spins are aligned (Fig. 1.2), as called for in the free ion by Hund's rule. Two spin-paired ($\uparrow\downarrow$) electrons in the same orbital suffer increased electron–electron repulsion than if they each occupied a separate orbital ($\uparrow$)($\uparrow$). The h.s. form thus benefits from having fewer electrons paired up in this way. Unless Δ is very small, however, the energy gained by dropping from the e_g to the t_{2g} level to go from h.s. to l.s. is sufficient to overcome the $e^- - e^-$ repulsion from spin pairing, resulting in an l.s. state.

The spin state is found from the magnetic moment, determined by comparing the apparent weight of a sample of the complex in the presence and absence of a magnetic field gradient. In l.s. d^6, the complex is *diamagnetic* and very weakly repelled by the field, as is found for most organic compounds, also spin paired. On the other hand, the h.s. form is *paramagnetic*, in which case it is attracted into the field because of the magnetic field of the unpaired electrons. The complex does not itself form a permanent magnet as can a piece of iron or nickel—this is *ferromagnetism*—because the spins are not aligned in the crystal in the absence of an external field, but they do respond to the external field by aligning against the applied field when we put them in a magnetic field to measure the magnetic moment.

With their high-field ligands, even d^n configurations and high Δ, the majority of organometallic complexes are diamagnetic, but interest in paramagnetic organometallics (Chapter 15) is on the rise. Mononuclear complexes with an uneven number of electrons, such as d^5 V(CO)$_6$, cannot avoid paramagnetism even in the low-spin case. For even d^n configurations, high spin is more often seen for the first row metals, where Δ tends to be smaller than in the later rows. Sometimes, the low- and high-spin isomers have almost exactly the same energy. Each state can now be populated in a temperature-dependent ratio, as in Fe(dpe)$_2$Cl$_2$. Different spin states have different structures and reactivity and, unlike resonance forms, may have a separate existence.

Inert versus Labile Coordination

In octahedral d^7, one electron has to go into the higher energy, less stable e_g level to give the low-spin $t_{2g}^6 e_g^1$ configuration and make the complex paramagnetic (Fig. 1.3). The *crystal field stabilization energy* (CFSE) of such a system is therefore less than for low-spin d^6, where we can put all the electrons into the more stable t_{2g} level. This is reflected in the chemistry of octahedral d^7 ions, such as Co(II), that are orders of magnitude more reactive in ligand dissociation than their d^6 analogs because the e_g or d_σ levels are M–L σ-antibonding (Section 1.7). Werner studied Co(III) precisely because the ligands tend to stay put. This is why Co(III) and other low-spin, octahedral d^6 ions are considered *coordinatively inert*. A half-filled t_{2g} level is also stable, so octahedral d^3 is also coordination inert, as seen for Cr(III). On the other hand,

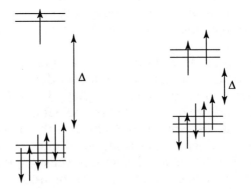

FIGURE 1.3 A d^7 octahedral ion is paramagnetic in both the low-spin (*left*) and high-spin (*right*) forms.

Co(II), Cr(II) and all other non-d^6 low-spin and non-d^3 ions are considered *coordinatively labile*. Second- and third-row transition metals form much more inert complexes than the first-row because of their higher Δ and CFSE.

Jahn–Teller Distortion

The lability of some coordination-labile ions, such as d^7 low spin, is aided by a geometrical distortion. This Jahn–Teller (J–T) distortion occurs whenever the individual orbitals in a set of orbitals of the same energy—degenerate orbitals—are unequally occupied. For a pair of degenerate e_g orbitals, this requires occupation by one or three electrons. Such is the case for low-spin d^7, where only one of the e_g orbitals is half-filled (Fig. 1.4). In such a case, a pair of ligands that lie along one axis—call this the z axis—either shows an elongation or a contraction

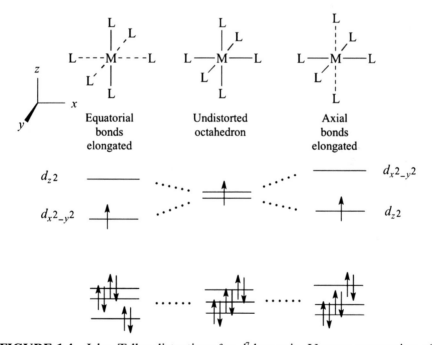

FIGURE 1.4 Jahn–Teller distortions for d^7 low-spin. Uneven occupation of the d_σ orbitals leads to a distortion in which either the xy ML_4 ligand set (left) or the z ML_2 ligand set (right) shows an M–L elongation because of electron–electron repulsions. Minor splitting also occurs in the d_π set. These types of diagrams do not show absolute energies—instead, the "center of gravity" of the orbital pattern is artificially kept the same for clarity of exposition.

of the M–L distances relative to those in the xy plane, depending on whether the $(d_{(x^2-y^2)}$ or $d_{z^2})$ orbital is half-occupied. On crystal field ideas, the electron in the half-filled d_{z^2} orbital repels the ligands that lie on the z axis, making these M–L bonds longer; if the $d_{(x^2-y^2)}$ orbital is half occupied, the bonds in the xy plane are longer. This distortion promotes ligand dissociation because two or four of the M–L distances are already elongated and weakened relative to the d^6 low-spin comparison case. A J-T distortion also occurs if the t_{2g} set of three orbitals are unevenly occupied with 1, 2, 4, or 5 electrons in t_{2g}, as in d^6 high spin (Fig. 1.2, right), but the distortion is now smaller because these t_{2g} orbitals do not point directly at the ligands. The J-T distortion splits the d orbitals to give a net electron stabilization relative to the pure octahedron. This is seen in Fig. 1.4, where the seventh electron is stabilized whichever of the two distortions, axial or equatorial, is favored.

Low- versus High-Field Ligands

Light absorption at an energy that corresponds to the d_π–d_σ splitting, Δ, leads to temporary promotion of a d_π electron to the d_σ level, typically giving d block ions their bright colors. The UV-visible spectrum of the complex can then give a direct measure of Δ and therefore of the crystal field strength of the ligands. *High-field* ligands, such as CO and C_2H_4, lead to a large Δ. *Low-field* ligands, such as F^- or H_2O, can give such a low Δ that even the d^6 configuration can become high spin and thus paramagnetic (Fig. 1.2, right side).

The *spectrochemical series* ranks ligands in order of increasing Δ. The range extends from weak-field π-donor ligands, such as halide and H_2O with low Δ, to strong-field π-acceptor ligands, such as CO that give high Δ (Section 1.6). These π effects are not the whole story,[16] however, because H, although not a π-bonding ligand, nevertheless is a very strong-field ligand from its very strong M–H σ bonds (Section 1.8).

$$I^- < Br^- < Cl^- < F^- < H_2O < NH_3 < PPh_3 < CO, H < SnCl_3^-$$

$\leftarrow$ low Δ high Δ $\rightarrow$

$\leftarrow$ π donor π acceptor/strong σ donor $\rightarrow$

Hydrides and carbonyls, with their strong M–L bonds (L = H, CO) and high Δ, are most often diamagnetic. High-field ligands resemble high trans-effect ligands in forming strong σ and/or π bonds, but the precise order differs a little in the two series and for different sets of complexes.

Magnetism and Nuclearity

A d^n configuration where n is odd, such as in d^7 [Re(CO)$_3$(PCy$_3$)$_2$], leads to paramagnetism in a mononuclear complex. In a dinuclear complex, however, the odd electron on each metal can now pair up in forming the M–M bond, as in the diamagnetic d^7–d^7 dimer, [(OC)$_5$Re–Re(CO)$_5$]. Mononuclear complexes with an even d^n configuration can be diamagnetic or paramagnetic depending on whether they are low or high spin. The practical difficulties of working with paramagnetic complexes, such as the complexity of analyzing their NMR spectra—if indeed any NMR spectrum is detectable at all (Section 10.2)—has slowed research in the area. Paramagnetism is more common in the first row because their smaller Δ favors high-spin species. The rising cost of the precious metals and the influence of green chemistry has made us take much more recent interest in the cheaper first-row metals.

Other Geometries

After octahedral, the next most common geometries are three types of 4- or 5-coordination: tetrahedral, square pyramidal and square planar. Tetrahedral is seen for d^0, d^5 (h.s.), and d^{10}, where we have symmetrical occupation of all the d orbitals, each having zero, one, or two electrons as in Ti(IV), Mn(II), and Pt(0). Since ligand field effects require *un*symmetrical d orbital occupation, such effects no longer apply and a tetrahedral geometry is adopted on purely steric grounds. The orbital pattern—three up, two down (Fig. 1.5, top)—is the opposite of that for octahedral geometry, and Δ_{tet} is smaller than Δ_{oct}, all else being equal, because we now only have four ligands rather than six to split the d orbitals. Tetrahedral geometry is typical for d^4 (low spin), as in Re(III), where only the low-lying pair of d orbitals is occupied.

The important square planar geometry, formally derived from an octahedron by removing a pair of trans ligands along the $\pm z$ axis, has a more complex splitting pattern (Fig. 1.5, lower). This derives from the octahedral pattern by pushing the distortion of Fig. 1.4 (right) to the limit. The big splitting, Δ in Fig. 1.5 (right), separates the two highest-energy orbitals. The square planar geometry is most often seen for d^8 (l.s.), as in Pd(II), where only the highest energy orbital remains unoccupied. It is also common for paramagnetic d^9, as in Cu(II). In square pyramidal geometry, only one axial L is removed from octahedral.

Holding the geometry and ligand set fixed, different metal ions can have very different values of Δ. For example, first-row metals and metals in a low oxidation state tend to have low Δ, while second- and third-row metals and metals in a high oxidation state tend to have high

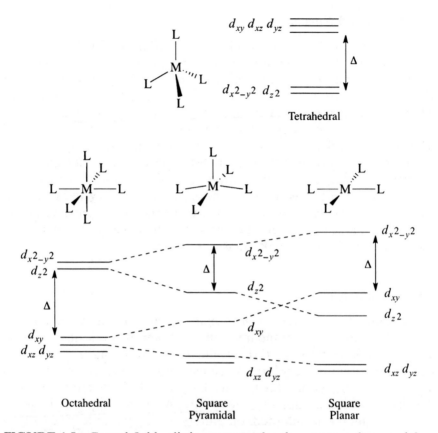

FIGURE 1.5 Crystal field splitting patterns for the common four- and five-coordinate geometries: tetrahedral, square pyramidal, and square planar. For the square pyramidal and square planar arrangements, the z axis is conventionally taken to be perpendicular to the L_4 plane. Octahedral geometry is expected for d^6 while square planar and square pyramidal are preferred in d^8; the Δ HOMO–LUMO splittings shown apply to those d^n configurations.

Δ. The trend is illustrated by the *spectrochemical series* of metal ions in order of increasing Δ:

$$I^- < Br^- < Cl^- < F^- < H_2O < NH_3 < PPh_3^- < CO, H^- < SnCl_3^-$$

low Δ,	high Δ,
← π donor/	π acceptor/ →
weak σ donor	strong σ donor

Second- and particularly third-row metals tend to have a higher Δ than first-row metals thus have stronger M–L bonds, give more thermally stable complexes that are also more likely to be diamagnetic. Higher oxidation states of a given metal also tend to produce higher Δ, enhancing these trends, but for a fair comparison, we would need to keep the same M and L_n in different oxidation states. This is rarely the case, because low oxidation state metals are usually found with strong-field ligands that tend to give a high Δ (see the spectrochemical series of ligands earlier) and high oxidation state metals are usually most accessible with weak-field ligands that tend to give a low Δ. The oxidation state trend is therefore partially counteracted by the change in ligand preferences.

Isoconfigurational Ions

Ions of the same d^n configuration show important similarities independent of the identity of the element. This means that d^6 Co(III) is closer in many properties to d^6 Fe(II) than to d^7 Co(II). The variable valency of the transition metals leads to many cases of isoconfigurational ions, and this idea helps us predict new complexes from the existence of isoconfigurational analogs. Numerous analogies of this type have been established for the pair Ir(III) and Ru(II), for example.

1.7   THE LIGAND FIELD

The crystal field picture gives a useful qualitative understanding, but for a more complete picture, we turn to the more sophisticated *ligand field theory* (LFT), really a conventional molecular orbital, or MO, picture. In this model (Fig. 1.6), we consider the s, the three p, and the five d orbitals of the valence shell of the isolated ion, as well as the six lone-pair orbitals of a set of pure σ-donor ligands in an octahedron around the metal. Six of the metal orbitals, the s, the three p, and the two d_σ, the dsp_σ set, find symmetry matches in the six ligand lone-pair orbitals. In combining the six metal orbitals with the six ligand orbitals, we make a bonding set of six (the M–L σ bonds) that are stabilized, and an antibonding set of six (the M–L σ^* levels) that are destabilized. The remaining three d orbitals, the d_π set, do not overlap with the ligand orbitals and remain nonbonding, somewhat resembling lone pairs in p block compounds. In a d^6 ion, we have 6e from Co^{3+} and 12e from the six :NH_3 ligands, giving 18e in all. This means that all the levels up to and including the d_π set are filled, and the M–L σ^* levels remain unfilled—the most favorable situation for high stability. Note that we can identify the familiar crystal field d orbital splitting pattern in the d_π set and two of the M–L σ^* levels. The Δ splitting increases as the strength of the M–L σ bonds increases, so bond strength is analogous to the effective charge in the crystal field model. In the ligand field

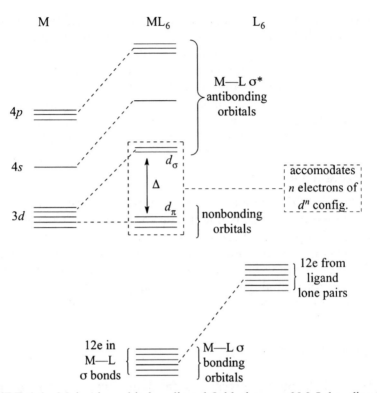

FIGURE 1.6 Molecular orbital, or ligand field picture, of M–L bonding in an octahedral ML_6 complex. The box contains the d orbitals that are filled with n electrons to give the d^n electron configuration. The star denotes antibonding.

picture, one class of high-field ligands form strong σ bonds, for example, H or CH_3. We can now see that the d_σ orbital of the crystal field picture becomes an M–L σ-antibonding orbital in the ligand field model.

The L lone pairs in the free ligand become bonding pairs shared between L and M when the M–L σ bonds are formed; these are the six lowest orbitals in Fig. 1.6 and are always completely filled with 12e. Each M–L σ-bonding MO is formed by the combination of the ligand lone pair, L(σ), with M(d_σ), and has both M and L character, but L(σ) predominates. Any MO more closely resembles the parent atomic orbital that lies closest to it in energy, and L(σ) almost always lies below M(d_σ) and therefore closer to the M–L σ-bonding orbitals. Electrons that were purely L lone pairs in free L now gain some metal character in ML_6; in other words, the L(σ) lone pairs are partially transferred to the metal. As L becomes more basic, the energy of the L(σ) orbital increases together with the extent of lone pair transfer. An orbital that moves to higher energy moves higher in the MO diagram and tends to occupy a larger volume of space; any electrons it contains become less stable and more available for chemical bonding or removal by electron loss in any oxidation.

Ligands are generally *nucleophilic* because they have high-lying lone pair electrons available, while a metal ion is *electrophilic* because it has low-lying empty *d* orbitals available. A nucleophilic ligand, a lone-pair donor, can thus attack an electrophilic metal, a lone pair acceptor, to give a metal complex. Metal ions can accept multiple lone pairs so that the complex formed is ML_n ($n = 2$–9).

1.8   THE sd^n MODEL AND HYPERVALENCY

The ligand field model is currently being challenged by the sd^n model.[17] This considers the np orbital as being ineffective in M–L bonding owing to poor overlap and mismatched energies and proposes that only the ns and five $(n-1)d$ orbitals contribute, n being 4, 5, and 6 for the first-, second-, and third-row *d* block metals. For example, photoelectron spectroscopy shows that Me_2TiCl_2 has sd^3 hybridization, not the familiar sp^3 hybridization as in Me_2CCl_2.[18] If so, one might expect d^6 metal complexes to prefer a 12-valence electron count, not 18e, since 12e would entirely fill the sd^5 set. This would, however, wrongly lead us to expect $Mo(CO)_3$ rather than the observed $Mo(CO)_6$. To account for the additional bonding power of $Mo(CO)_3$, hypervalency is invoked.

Hypervalency, the ability of an element to exceed the valence electron count normally appropriate for the orbitals that are available, is best established in the main-group elements, such as sulfur, where an octet of eight valence electrons is appropriate for its single *s* and three *p* orbitals. In hypervalent SF_6, for example, six electrons come from S and one each from the six F atoms for a total of 12 valence electrons, greatly exceeding the expected octet. The modern theory of hypervalency avoids the earlier idea, now exploded, that empty *d* orbitals (3*d* orbitals for S) allow the atom to house the excess electrons.

Hypervalent bonding is most simply illustrated for [FHF]⁻ anion, where H has four valence electrons, exceeding its normal maximum of 2e. In [FHF]⁻, the zero electron H⁺ receives 2e from each of the lone pairs of the two F⁻ anions coordinated to it, thus resembling an ML_2 complex. The bonding pattern, shown in Fig. 1.7, allows the 4e from the two F⁻ to occupy two lower-lying orbitals each having predominant F character—one bonding, one nonbonding—while leaving the highest energy orbital empty. In effect, one 2e bond is spread over two H–F bonds, and the remaining 2e in the nonbonding orbital are predominantly located on F. The resulting 4 electron–3 center (4e–3c) bonding

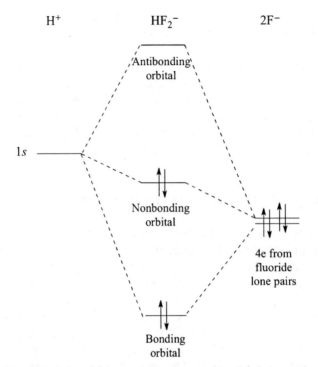

FIGURE 1.7 The four electron–three center (4e–3c) hypervalent bonding model for [FHF]⁻ anion in which the fluoride ions are considered ligands for the central H⁺. The bonding and nonbonding orbitals are occupied and the antibonding orbital left vacant.

leads to half-order bonding between H and each F, resulting in somewhat longer bonds (1.15 Å) than in the corresponding nonhypervalent species, HF (0.92 Å). [FHF]⁻ anion, normally considered as a strong hydrogen-bonded adduct of HF and F⁻, is here seen as hypervalent. Moving to the heavier p block elements, hypervalent octahedral SF_6, for example, can be considered as having three trans F–S–F units, each bonded via 4e–3c bonds.

Main-group hypervalency requires an electronegative ligand, often F or O, that can stabilize the bonding and nonbonding orbitals of Fig. 1.7. This results in the accumulation of negative charge on the terminal F atoms that are best able to accommodate it. In coordination complexes, the ligands are indeed almost always more electronegative than the metal even when we expand the ligand choice beyond F and O to N, P and C donors. To return to $Mo(CO)_6$, the bonding is explained in terms of three pairs of trans L–M–L hypervalent 4e–3c bonds, formed from sd^5 hybrids. This leaves three d orbitals that are set aside for back bonding to CO as the d_π set, as in ligand field theory.

Bent's rule, which helps assign geometries for main-group compounds, relies on sp^3 hybridization and therefore has to be modified for application to the d block. For example, in Me_2CCl_2, the Cl–C–Cl angle (108.3°) is less than the C–C–C angle (113.0°), since the more electronegative Cl atoms elicit a higher contribution from the less stable orbital, in this case, the carbon p orbital. The C–Cl bonds having high p orbital character also have a smaller bond angle, since p orbitals are 90° apart. In Me_2TiCl_2, in contrast, the Cl–Ti–Cl angle (116.7°) is larger than the C–Ti–C angle (106.2°) because the hybridization is now sd^3 and the d orbitals are the more stable members of the sd^3 set. The less electronegative Me substituents now elicit greater Ti d character and have the smaller bond angle.[19]

The fate of this model depends on whether it finds favor in the scientific community, and we will not use it extensively in what follows. Textbooks can give the impression that everything is settled and agreed upon, but that agreement is only achieved after much argument, leading to an evolution of the community's understanding. Ideas that come to dominate often start out as a minority view. The sd^n model may therefore either fade, flower, or be modified in future.

1.9 BACK BONDING

Ligands such as NH_3 are good σ donors but insignificant π acceptors. CO, in contrast, is a good π acceptor and relatively poor σ donor. Such π-*acid* ligands are of very great importance in organometallic chemistry. They tend to be very high field ligands and form strong M–L bonds. All have empty orbitals of the right symmetry to overlap with a filled d_π orbital of the metal: for CO, this acceptor is the empty CO π*. Figure 1.8 shows how overlap takes place to form the M–C π bond. It may

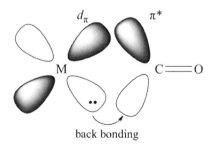

back bonding

FIGURE 1.8 Overlap between a filled metal d_π orbital and an empty CO π* orbital to give the π component of the M–CO bond. The shading refers to symmetry of the orbitals. The M–CO σ bond is formed by the donation of a lone pair on C into an empty d_σ orbital on the metal (not shown).

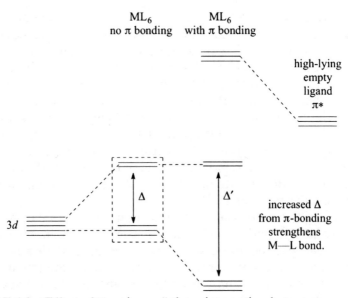

FIGURE 1.9 Effect of "turning on" the π interaction between a π-acceptor ligand and the metal. The unoccupied, and relatively unstable, π^* orbitals of the ligand are shown on the right. Their effect is to stabilize the filled d_π orbitals of the complex and so increase Δ. In $W(CO)_6$, the lowest three orbitals are filled.

seem paradoxical that an antibonding orbital such as the $\pi^*(CO)$ can form a bond, but this orbital is only antibonding with respect to C and O and can still be bonding with respect to M and C. A second CO π^*, oriented out of the image plane, can accept back bonding from a second d_π orbital that is similarly oriented.

The ligand field diagram of Fig. 1.6 has to be modified when the ligands are π acceptors, such as CO, because we now need to include the CO π^* levels (Fig. 1.9). The Md_π set now interact strongly with the empty CO π^* levels to form M-C π bonds. For d^6 complexes, such as $W(CO)_6$, where the Md_π set is filled, the Md_π electrons now spend some of their time on the ligands by back bonding.

Back bonding can occur for a wide variety of M–L bonds as long as L contains a suitable empty orbital. In one type, where the donor atom participates in one or more multiple bonds, the empty orbital is a ligand π^*, as is the case for CO or C_2H_4. As we see in detail in Sections 3.4 and 4.2, other types of ligand have suitable empty σ^* orbitals, as is the case for PF_3 or H_2. On the metal side, back bonding can only happen in d^1 or higher configurations; a d^0 ion such as Ti^{4+} cannot back bond and very seldom forms stable complexes with strong π acceptor ligands, such as CO.

Being antibonding, the CO π^* levels are high in energy, but they are able to stabilize the d_π set by back bonding as shown in Fig. 1.9. This has two important consequences: (1) The ligand field splitting parameter Δ rises, explaining why π-bonding ligands have such a strong ligand field and make such strong M–L bonds; and (2) back bonding allows electron density on a low oxidation state metal to make its way back to the π-acid ligands. This applies when low-valent or zero-valent metals form CO complexes. Such metals have a high electron density in the free state and are thus reluctant to accept further electrons from pure σ donors; this is why $W(NH_3)_6$ is not known. By back bonding, the metal can get rid of some of this excess electron density and delocalize it over the π-acid ligands. In $W(CO)_6$, back bonding is so effective that the compound is air stable and relatively unreactive; the CO groups have so stabilized the metal electrons that they have no tendency to be abstracted by an oxidant such as air. In $W(PMe_3)_6$, in contrast, back bonding is weak and the complex is reactive and air-unstable.

Their structures show that π back donation is a big contributor to the M=C bond in metal carbonyls, making the M=C bond much shorter than an M–C single bond. For example, in $CpMo(CO)_3Me$, M–CH_3 is 2.38 Å but M=CO is 1.99 Å. A true M–CO single bond would be shorter than 2.38 Å by about 0.07 Å, to allow for the higher s character of sp CO versus sp^3 CH_3, leaving a substantial shortening of 0.32 Å that can be ascribed to back bonding.

IR spectroscopy identifies the CO π^* orbital as the acceptor in back bonding. A CO bound only by its carbon lone pair — nonbonding with respect to CO — would have a $\nu(CO)$ frequency close to that in free CO. BH_3, a predominant σ acceptor, shows a slight shift of $\nu(CO)$ to *higher* energy in H_3B-CO: free CO, 2143 cm^{-1}; H_3B–CO, 2178 cm^{-1} so the shift is +35 cm^{-1}. Metal carbonyls, in contrast, show $\nu(CO)$ coordination shifts of hundreds of wavenumbers to *lower* energy, consistent with the weakening of the C–O bond as the CO π^* is partially filled by back donation; for $Cr(CO)_6$, $\nu(CO)$ is 2000 cm^{-1}, so the shift is -143 cm^{-1}. Not only is there a coordination shift, but the shift is larger in cases where we would expect stronger back donation (Table 2.10) and $\nu(CO)$ is considered a good indicator of metal basicity. In Section 4.2, we see how the $\nu(CO)$ of $LNi(CO)_3$ helps us rank different ligands L in terms of their comparative donor power to M; good donor L ligands make the Ni back donate more strongly into the CO groups.

Formation of the M–CO bond weakens the C≡O bond of free CO. This can still lead to a stable complex as long as the energy gained from the new M–C bond exceeds the loss in C≡O. Bond weakening in L on binding to M is very common in M–L complexes where back bonding is significant.

Series of compounds such as $[V(CO)_6]^-$, $Cr(CO)_6$, and $[Mn(CO)_6]^+$ are *isoelectronic* because, V(–I), Cr(0), and Mn(I) all being d^6, they have the same number of electrons similarly distributed. Isoelectronic ligands include CO and NO^+ and CN^-, for example. CO and CS are not strictly isoelectronic, but as the difference between O and S only lies in the number of core levels, while the valence shell is the same, the term is often extended to such pairs. A comparison of isoelectronic complexes or ligands can be very useful in looking for similarities and differences.[20]

Frontier Orbitals

A similar picture holds for a whole series of soft, π acceptor ligands, such as alkenes, alkynes, arenes, carbenes, carbynes, NO, N_2, and PF_3. Each has a filled orbital that acts as a σ donor and an empty orbital that acts as a π acceptor. These orbitals are almost always the highest occupied (*HOMO*) and lowest unoccupied molecular orbitals (*LUMO*) of L, respectively. The HOMO of L is normally a donor to the d_σ LUMO of the metal. The ligand LUMO thus accepts back donation from the metal HOMO, a filled metal d_π orbital. The HOMO and LUMO of each fragment, the so-called *frontier orbitals*, often dominate the bonding between the fragments. Strong interactions between orbitals require not only good overlap but also that the energy separation between them be small. The HOMO of each fragment, M and L, is usually closer in energy to the LUMO of the partner fragment than to any other vacant orbital of the partner. Strong bonding is thus expected if the HOMO–LUMO gap of both partners is small. Indeed a small HOMO–LUMO gap for any molecule gives rise to high reactivity. A small HOMO–LUMO gap also makes a ligand soft because it becomes a good π acceptor, and for d^6, makes the metal soft because it becomes a good π donor.

π-Donor Ligands

Ligands such as OR^- and F^- are π donors as a result of the lone pairs that are left after one lone pair has formed the M–L σ bond. Instead of stabilizing the d_π electrons of an octahedral d^6 ion as does a π acceptor, these d_π electrons are now destabilized by what is effectively a repulsion between two filled orbitals. This lowers Δ, as shown in Fig. 1.10, and leads to a weaker M–L bond than in the π-acceptor case, as in high-spin d^6 $[CoF_6]^{3-}$. Lone pairs on electronegative atoms such as F^- and RO^- are much more stable than the $M(d_\pi)$ level, and this is why they are lower in Fig. 1.10 than are the π^* orbitals in Fig. 1.9. Having more diffuse lone pairs, larger donor atoms pose fewer problems, and Cl^- and R_2P^- are much better tolerated by d^6 metals.

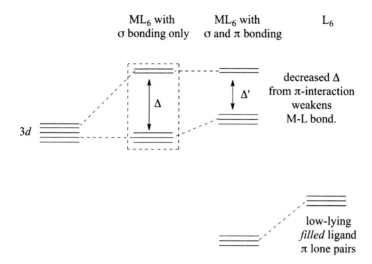

$$
\begin{array}{ccc}
\text{ML}_6\text{ with} & \text{ML}_6\text{ with} & \text{L}_6 \\
\sigma\text{ bonding only} & \sigma\text{ and }\pi\text{ bonding} &
\end{array}
$$

decreased Δ
from π-interaction
weakens
M–L bond.

3d

low-lying
filled ligand
π lone pairs

FIGURE 1.10   Effect of "turning on" the π interaction between a π-donor ligand and the metal. The occupied, and relatively stable, lone-pair (π) orbitals of the ligand are shown on the right. Their effect is to destabilize the filled d_π orbitals of the complex and so decrease Δ. In d^6, this is effectively a repulsion between two lone pairs, one on the metal and the other on the ligand, and thus unfavorable for M–L bonding. In d^0, this repulsion is no longer present, and the stabilization of the π lone pairs of L becomes a favorable factor for M–L bonding.

In sharp contrast, if the metal has empty d_π orbitals, as in the d^0 ion Ti^{4+}, π donation from the π-donor ligand to the metal d_π orbitals now leads to stronger metal–ligand bonding; d^0 metals therefore form particularly strong bonds with such ligands, as in $W(OMe)_6$ or $[TiF_6]^{2-}$, both also examples of favorable hard metal–hard ligand combinations.

1.10   ELECTRONEUTRALITY

Linus Pauling (1901–1994), a giant of twentieth-century chemistry, proposed the *electroneutrality principle* in which electrons distribute themselves in polar covalent molecules so that each atomic charge is nearly neutral. In practice, these charges fall in a range from about +1 to −1. The nonmetals tend to be negatively charged with N, O, or F being closer to −1 and Na or Al, being closer to +1. This implies that elements that have complementary preferred charges should bond best so each can satisfy the other, as in LiF or TiO_2; in contrast, elements with intermediate electronegativity prefer each other, as in H_2, HgS, and Au–Ag alloy. An isolated Co^{3+} ion is far from electroneutral so it prefers good electron donors as ligands, such as O^{2-} in Co_2O_3, or NH_3 in the Werner complexes. On the other hand, an isolated $W(0)$ atom is already neutral and is thus too electron rich for its electronegativity, so it prefers net electron-attracting ligands, such as CO that can accept electron density by π back donation so that the metal can attain a positive charge.

Oxidation State Trends

The d orbitals of transition metals are only fully available for back donation in low oxidation states. Although d^6 Co(III), for example, does have a filled d_π level, it is unavailable for back bonding—Co(III) therefore cannot bind CO. The high positive charge of Co(III) contracts all the orbitals with the result that the d_π orbital is low in energy and therefore only weakly basic. Likewise, repulsive effects of π donors such as F^- and RO^- are mild.

Periodic Trends

The orbital energies fall as we go from left to right in the transition series. For each step to the right, a proton is added to the nucleus, thus providing an extra positive charge that stabilizes all the orbitals. The earlier metals are more electropositive because it is easier to remove electrons from their less-stable orbitals. The sensitivity of the orbitals to this change is $d \sim s > p$ because the s orbital, having a maximum electron density at the nucleus, is more stabilized by the added protons than are the p orbitals, with a planar node at the nucleus. The d orbitals are also stabilized because of their lower principal quantum number, as is the case for $3d$ versus $4s$ and $4p$ in the valence shell of Fe. The special property of the transition metals is that all three types of orbital are in the valence shell with similar energies so all contribute significantly to the bonding, only omitting the $4p$ if the sd^n model is adopted. Metal carbonyls, for example, are most stable for groups 4–8 because CO requires back bonding to bind strongly and in the later groups, the needed d_π orbitals become too stable to be effective. Organometallic compounds of the electropositive early metals have a higher polar covalent character than in the later metals and thus tend to be more air-sensitive, because they are more subject both to oxidation by O_2 and hydrolysis by H_2O.

There is a sharp difference between d^0 and d^2 as in Ti(IV) versus Ti(II): d^0 Ti(IV) cannot back bond at all, while d^2 Ti(II) is a very strong back-bonder because early in the transition series, where d^2 states are most common, the d orbitals are relatively unstable for the reasons mentioned earlier. The d^0 Ti(IV) species $(C_5H_5)_2TiCl_2$ therefore does not react with CO at all, while the corresponding d^2 Ti(II) fragment, $(C_5H_5)_2$Ti, forms a very stable monocarbonyl, $(C_5H_5)_2$Ti(CO), with a low ν(CO) IR frequency, indicating very strong back bonding.

Finally, as we go down a given group in the d block from the first to the second row, the outer valence electrons become more *shielded* from the nucleus by the extra shell of electrons added. They are therefore more easily lost, making the heavier d block element more basic and more capable of attaining high oxidation states. This trend also extends to the third row, but as the f electrons that were added to build up the lanthanide elements are not as effective as s, p, or even d electrons in shielding the valence electrons from the nucleus, there is a smaller change on going from the second to the third row than from the first to the second. Compare, for example, the power-fully oxidizing Cr(VI) in Na_2CrO_4 and Mn(VII) in $KMnO_4$, with their stable second- and third-row analogs, Na_2MoO_4, Na_2WO_4, and $KReO_4$; the very weakly oxidizing character of the latter indicates an increased stability for the higher oxidation state. For the same reason, the increase in covalent radii is larger on going from the first to the second row than from the second to the third. This anomaly in atomic radius for the third row is termed the *lanthanide contraction*.

Mononuclear ionic complexes with excessively high positive or nega-tive net ionic charges are not normally seen. The majority of isolable compounds are neutral; net charges of ± 1 are not uncommon, but higher net ionic charges are rare.

1.11   TYPES OF LIGAND

Most ligands are Lewis bases and thus typically neutral or anionic, rarely cationic. Anionic ligands, often represented as X, form polar covalent M–X bonds. In addition to the σ bond, there can also be a π interaction which may be favorable or unfavorable as discussed in Section 1.9.

Among neutral ligands, often denoted L, we find lone-pair donors, such as :CO or :NH_3, π donors such as C_2H_4, and σ donors such as H_2. The first group—the only type known to Werner—bind via a lone pair. In contrast, π donors bind via donation of a ligand π-bonding electron pair, and σ donors bind via donation of a ligand σ-bonding electron pair to the metal. The relatively weakly basic σ- and π-bonding electrons of σ and π donors would form only very weak M–L bonds if acting alone. Both σ and π donors therefore normally require some back bonding to produce a stable M–L bond. Even so, the strength of binding tends to

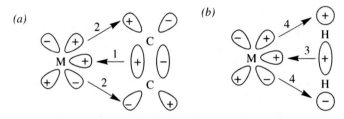

FIGURE 1.11 (a) Bonding of a π-bond donor, ethylene, to a metal. Arrow 1 represents electron donation from the filled C=C π bond to the empty d_σ orbital on the metal; arrow 2 represents the back donation from the filled M(d_π) orbital to the empty C=C π^*. (b) Bonding of a σ-bond donor, hydrogen, to a metal. Arrow 3 represents electron donation from the filled H–H σ bond to the empty d_σ orbital on the metal, and arrow 4 represents the back donation from the filled M(d_π) orbital to the empty H–H σ^*. Only one of the four lobes of the d_σ orbital is shown.

decrease as we move from lone pair to π bond to σ bond donors, other factors being equal.

For the π donor, ethylene, Fig. 1.11a illustrates how L to M donation from the C=C π orbital to M d_σ (arrow 1) is accompanied by back donation from M d_π into the C=C π^* orbital (arrow 2). For the σ donor, H_2, Fig. 1.11b shows how L to M donation from the H–H σ orbital to M d_σ (arrow 3) is accompanied by back donation from M d_π into the H–H σ^* orbital (arrow 4). As always, back bonding requires a d^2 or higher electron configuration and relatively basic M d_π electrons, usually found in low oxidation states.

Side-on binding of σ and π donors results in short bonding distances to two adjacent ligand atoms. This type of binding is represented as η^2-C_2H_4 or η^2-H_2, where the letter η (often pronounced eeta) denotes the ligand *hapticity*, the number of adjacent ligand atoms directly bound to the metal. For σ donors such as H_2,[21] forming the M–L σ bond partially depletes the H–H σ bond because electrons that were fully engaged in keeping the two H atoms together in free H_2 are now also delocalized over the metal, hence the name *two-electron, three-center* (2e,3c) bond for this interaction. Back bonding into the H–H σ^* causes additional weakening or even breaking of the H–H σ bond because the σ^* is antibonding with respect to H–H. Free H_2 has an H–H distance of 0.74 Å, but the H–H distances in H_2 complexes go all the way from 0.82 to 1.5 Å. Eventually, the H–H bond breaks and a dihydride is formed (Eq. 1.5). This is the *oxidative addition reaction* (see Chapter 6). Formation of a σ complex can be thought of as an incomplete oxidative addition, where only the addition part has occurred. Table 1.2 classifies common ligands by the nature of the M–L σ and π bonds. Both σ and

TABLE 1.2 Types of Ligand[a]

Ligand	Strong π Acceptor	Weak π Bonding	Strong π Donor
Lone-pair donor	CO, PF_3, CR_2[b]	H^-,[c]PPh_3,Me^-,Cl^-	F^-, OR^-, NR_2^-
π-Bonding electron pair donor	C_2F_4, O_2	C_2H_4, $RCH{=}O$[d]	–
σ-Bonding electron pair donor	Oxidative Addition[e]	$R_3Si{-}H$, $H{-}H$, $R_3C{-}H$	–
σ- and π-acceptor[g]	BF_3	BH_3, CO_2[f]	–

[a]Ligands are listed in approximate order of π-donor/acceptor power, with acceptors mentioned first.
[b]Fischer carbene (Chapter 11).
[c]Ligands like this are considered here as anions rather than radicals.
[d]Can also bind via an oxygen lone pair (Eq. 1.6).
[e]Oxidative addition occurs when σ-bond donors bind very strongly (Eq. 1.5).
[g]Rare.
[f]When bound η^1 via C.

π bonds bind side-on to metals when they act as ligands. Alkane C–H bonds behave similarly.[22]

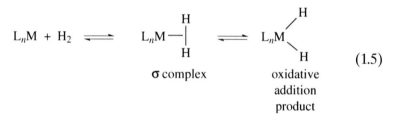

$$L_nM + H_2 \rightleftharpoons L_nM{-}\underset{H}{\overset{H}{|}} \rightleftharpoons L_nM\overset{H}{\underset{H}{\diagdown}} \qquad (1.5)$$

$$\underset{\sigma\ \text{complex}}{} \qquad \underset{\substack{\text{oxidative} \\ \text{addition} \\ \text{product}}}{}$$

Lewis acids such as BF_3 can be ligands by accepting a basic electron pair from the metal ($L_nM: \rightarrow BF_3$), in which case the ligand contributes nothing to the metal electron count: BF_3 is also a strong π-acceptor for back bonding from M d_π orbitals via the σ* orbitals, as discussed for PF_3 in Section 4.2.

Ambidentate Ligands

Alternate types of electron pair are sometimes available for bonding. For example, aldehydes have both a $C{=}O$ π bond and oxygen lone pairs. As π-bond donors, aldehydes bind side-on (Eq. 1.6, **1.21a**) like ethylene, but as lone-pair donors, they can alternatively bind end-on (**1.21b**). Thiocyanate, SCN^-, can bind via N in a linear fashion (Eq. 1.7, **1.22a**), or via S, in which case the ligand is bent (**1.22b**); in some cases, both forms are isolable.[23]

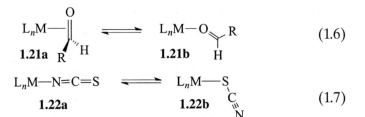

$$L_nM \overset{O}{\underset{R}{\overset{\|}{-}}} H \quad \rightleftharpoons \quad L_nM-O \overset{R}{\underset{H}{\diagdown}} \qquad (1.6)$$

1.21a   **1.21b**

$$L_nM-N=C=S \quad \rightleftharpoons \quad L_nM-S \overset{C}{\underset{\underset{N}{\|\|}}{\diagdown}} \qquad (1.7)$$

1.22a   **1.22b**

The $\{(NH_3)_5Os^{II}\}^{2+}$ fragment in Eq. 1.8 is a very strong π donor because Os(II) is soft and NH_3 is not a π-acceptor; the π-basic Os thus prefers to bind to the π acceptor aromatic C=C bond of aniline, not to the nitrogen. Oxidation to Os^{III} causes a sharp falloff in π-donor power because the extra positive charge stabilizes the *d* orbitals, and the Os(III) complex slowly rearranges to the *N*-bound aniline form.[24] This illustrates how the electronic character of a metal can be altered by changing the ligand set and oxidation state; soft Os(II) binds to the soft C=C bond and hard Os(III) binds to the hard $ArNH_2$ group.

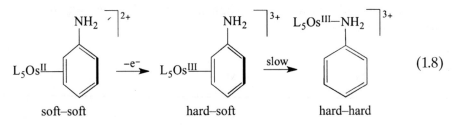

soft–soft   hard–soft   hard–hard

$$(1.8)$$

Figure 1.12 shows the typical ligands found for different oxidation states of Re, an element with a very wide range of accessible states. Low OS complexes are stabilized by multiple π-acceptor CO ligands, intermediate OSs by less π-acceptor phosphines, high OS by σ-donor anionic ligands such as Me, and very high OS by O or F, ligands that are both σ donor and π donor.

The dipyridyl phosphine ligand of Eq. 1.9 shows two distinct binding modes, depending on the conditions and anion present.[25]

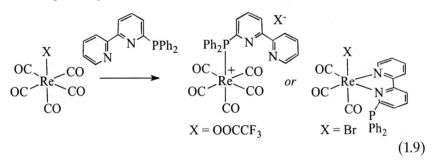

$$X = OOCCF_3 \qquad X = Br \quad Ph_2$$

$$(1.9)$$

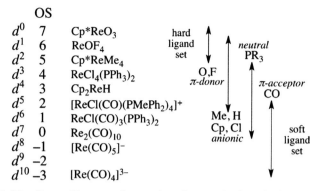

FIGURE 1.12 Some Re complexes showing typical variation of ligand type with oxidation state (OS): hard ligands with high OS and soft ligands with low OS.

Actor and Spectator Ligands

Actor ligands associate, dissociate or react in some way. They are particularly important in catalytic reactions, when they bind to the metal and engage in reactions that lead to release of a product molecule. In hydrogenation, for example, H_2 and ethylene can associate to give $[L_nMH_2(C_2H_4)]$ intermediates that go through a cycle of reactions (Section 9.3) that leads to release of the hydrogenation product, C_2H_6.

Spectator ligands remain unchanged during chemical transformations but still play an important role by tuning the properties of the metal to enhance desired characteristics. For example, in the extensive chemistry of $[CpFe(CO)_2X]$ and $[CpFe(CO)_2L]^+$ (Cp = cyclopentadienyl; X = anion; L = neutral ligand), the {CpFe(CO)$_2$} fragment remains intact. The spectators impart solubility, stabilize Fe(II), and influence the electronic and steric properties of the complex. It is an art to pick suitable spectator ligand sets to elicit desired properties. Apparently small changes in ligand can entirely change the chemistry. For example, PPh$_3$ is an exceptionally useful ligand, while the apparently similar NPh$_3$, BiPh$_3$, and P(C$_6$F$_5$)$_3$ are of very little use. The hard N-donor, NPh$_3$, is very different from PPh$_3$; the Bi-Ph bond is too easily cleaved for BiPh$_3$ to be a reliable spectator; and the electron-withdrawing C$_6$F$_5$ substituents of P(C$_6$F$_5$)$_3$ completely deactivate the P lone pair.

Steric size sets the maximum number of ligands, n, that can fit around a given metal in a d block ML$_n$ complex. Typical n values depend on the size of the ligand: H, 9; CO, 7; PMe$_3$, 6; PPh$_3$, 4; P(C$_6$H$_{11}$)$_3$, 2, and only in a trans arrangement; a few ligands are so big that $n = 1$, for

Chelate, bidentate, *cis*

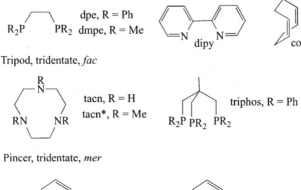

Tripod, tridentate, *fac*

Pincer, tridentate, *mer*

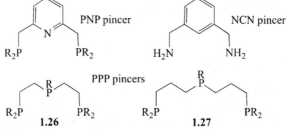

FIGURE 1.13 A selection of common ligands with different binding preferences.

example, X-Phos (**4.11**). If a big spectator ligand can occupy no more than *n* sites when the metal has *m* sites available, then *m − n* sites are kept open for smaller actor ligands. Multidentate spectator ligands can have the *n* donor atoms arranged in specific patterns and geometries, making the *m − n* available sites take up a complementary geometry. A small sample of such ligands is shown in Fig. 1.13. The tridentate ligands can bind to an octahedron either in a *mer* (meridonal) fashion **1.23** or *fac* (facial) **1.24**, or in some cases, in both ways. Ligands that normally bind in terdentate *mer* fashion are pincers. Not only do these benefit from the chelate effect, but they also allow us to control the binding at three sites of an octahedron, leaving three *mer* sites accessible to reagents.

Tetradentate ligands, such as **1.25** can also prove useful, in this case by stabilizing the unusual Pd(III) oxidation state.[26] The choice of ligand is an art because subtle stereoelectronic effects, still not fully understood, can play an important role. Ligands **1.26** and **1.27** (Fig. 1.13) impart substantially different properties to their complexes in spite of their apparent similarity, probably as a result of the greater flexibility of the three-carbon linker in **1.27**.

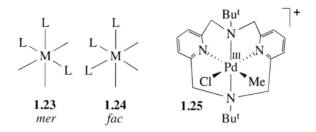

1.23   **1.24**   **1.25**
mer   *fac*

Actor ligands may allow isolation of a stable material as a precursor to a reactive species only formed after departure of the actor, that species either being too reactive to isolate or not otherwise easily accessible. A classic example is chelating 1,5-cyclooctadiene (cod) that binds to Rh(I) or Ir(I) in the $[(cod)M(PR_3)_2]^+$ hydrogenation catalysts (**1.28**). Under H_2, the cod is hydrogenated to free cyclooctane, liberating $\{M(PR_3)_2\}^+$ as the active catalyst. Cp* is a reliable spectator, except under strongly oxidative conditions, when it can degrade and become an actor. For example, the Cp* in Cp*Ir(dipy)Cl is oxidatively removed with Ce(IV) or IO_4^- to give a homogeneous coordination catalyst capable of oxidizing water or C–H bonds.[27] Similarly, the Cp* in $[Cp*Ir(OH_2)_3]SO_4$ is oxidatively degraded under electrochemical oxidation to yield a heterogeneous water oxidation catalyst that deposits on the electrode.[28]

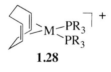

1.28

Multifunctional Ligands[29]

These more sophisticated ligands are increasingly being seen. Beyond the simple metal-binding function, numerous additional functionalities can also be incorporated. Some ligands reversibly bind protons, altering their donor properties; others have hydrogen bonding functionality for molecular recognition. Sometimes, a complex can be oxidized or reduced, but the resulting radical is ligand centered so that the metal oxidation state is unchanged.

Organometallic versus Coordination Compounds

Originally, the presence of any M–C bonds made a metal complex organometallic—their absence made it a coordination compound. Electronegativity differences (Δ_{EN}) between M and the donor atom in L were invoked. Organometallic M–L bonds, such as M–CH$_3$, typically have a lower Δ_{EN} and are thus more covalent than bonds with greater Δ_{EN} and more ionic character, such as the M–N or M–O bonds typical of coordination complexes. Mixed ligand sets are now much more common, making sharp distinctions less helpful. Ligands such as H, SiR$_3$, or PR$_3$ are now regarded as organometallic because Δ_{EN} is low and covalency predominates. In the key subfield of catalysis, coordination compounds have proved as useful as organometallics. For Wilkinson's catalyst, [RhCl(PPh$_3$)$_3$], one of the most important compounds in the history of the field (see Chapter 9), M–C bonds are only present in the intermediates formed during the catalytic cycle. Likewise, in CH activation (Section 12.4), many of the catalysts involved are again coordination compounds that operate via organometallic intermediates (e.g., [ReH$_7$(PPh$_3$)$_2$] or K$_2$[PtCl$_4$]). In an increasing number of cases, such as the metal oxo mechanism for CH activation (Sections 12.4 and 14.7), no M–C bonds are ever present, even in reaction intermediates. Today, the organometallic/coordination distinction is therefore losing importance. While still emphasizing traditional organometallics, we therefore do not hesitate to cross into coordination chemistry territory on occasion, particularly in Chapters 14–16.

- High trans effect ligands such as H or CO labilize ligands that are trans to themselves.
- In CFT (Section 1.6), the d-orbital splitting, Δ, and e$^-$ occupation determine the properties of the complex.
- Hard ligands, such as NH$_3$, have first-row donor atoms and no multiple bonds; soft ligands, such as PR$_3$ or CO, have second-row donors or multiple bonds.
- Ligands donate electrons from their HOMO and accept them into their LUMO (p. 26). LFT (Section 1.7) identifies the d$_\sigma$ orbitals as M–L antibonding.
- M–L π bonding strongly affects Δ and thus the strength of M–L bonding (Fig. 1.8, Fig. 1.9, and Fig. 1.10).
- Ligands can bind via lone pairs, π bonding e$^-$ pairs or σ bonding e$^-$ pairs (Table 1.2).
- Octahedral d^3 and d^6 are coordination inert and slow to dissociate a ligand.

REFERENCES

1. R. A. Sheldon, *Chem. Comm.*, 2008, 3352; P. Anastas, N. Eghbali, *Chem. Soc. Rev.*, **39**, 301, 2010.

2. C. Bolm, *Org. Lett.*, **14**, 2926, 2012.

3. H. -F. Chen, C. Wu, M. -C. Kuo, M. E. Thompson, and K. -T. Wong, *J. Mater. Chem.*, **22**, 9556, 2012.

4. P. G. Bomben, K. C. D. Robson, B. D. Koivisto, and C. P. Berlinguette, *Coord. Chem. Rev.*, **256**, 1438, 2012.

5. G. Jaouen and N. Metzler-Nolte (eds.), *Medical Organometallic Chemistry*, Springer, New York, 2010.

6. G. M. Whitesides and J. Deutch, *Nature*, **469**, 21, 2011.

7. S. A. Marcott, J. D. Shakun, P. U. Clark, and A. C. Mix, *Science*, **339**, 1198, 2013.

8. R. H. Crabtree, *Organometallics*, **30**, 17, 2011.

9. M. L. Helm, M. P. Stewart, R. M. Bullock, M. Rakowski DuBois, D. L. DuBois, *Science*, **333**, 863, 2011; J. D. Blakemore, N. D. Schley, G. Olack, C. D. Incarvito, G. W. Brudvig, and R. H. Crabtree, *Chem. Sci.*, **2**, 94, 2011.

10. P. J. Chirik and K. Wieghardt, *Science*, **327**, 794, 2010.

11. K. -H. Ernst, F. R. W. P. Wild, O. Blacque, and H. Berke, *Angew Chem-Int. Ed.*, **50**, 10780, 2011; E. C. Constable and C. E. Housecroft, *Chem. Soc. Rev.*, **42**, 1429, 2013.

12. N. Selander and K. J. Szabo, *Chem. Rev.*, **111**, 2048, 2011.

13. W. A. Nugent, *Angew. Chem. Int. Ed.*, **51**, 2, 2012.

14. J. Goodman, V. V. Grushin, R. B. Larichev, S. A. Macgregor, W. J. Marshall, and D. C. Roe, *J. Am. Chem. Soc.*, **131**, 4236, 2009.

15. L. C. Liang, *Coord. Chem. Rev.*, **250**, 1152, 2006.

16. H. Bolvin *Eur. J. Inorg. Chem.*, **2010**, 2221.

17. F. Weinhold and C. R. Landis, *Valency and Bonding*, Cambridge University Press, New York 2005; G. Frenking and N. Fröhlich, *Chem. Rev.*, **100**, 717, 2000.

18. B. E. Bursten, J. C. Green, N. Kaltsoyannis, M. A. MacDonald, K. H. Sze, and J. S. Tse, *Inorg. Chem.*, **33**, 5086, 1994.

19. V. Jonas, C. Boehme, and G. Frenking, *Inorg. Chem.*, **35**, 2097, 1996.

20. X. Solans-Monfort, C. Coperet, and O. Eisenstein, *Organometallics*, **31**, 6812, 2012.

21. G. J. Kubas, *Metal Dihydrogen and σ-Bond Complexes*, Kluwer/Plenum, New York, 2001.

22. A. J. Cowan and M. W. George, *Coord. Chem. Rev.*, **252**, 2504, 2008.

23. T. Brewster, W. Ding, N. D. Schley, N. Hazari, V. Batista, and R. H. Crabtree, *Inorg. Chem.*, **50**, 11938, 2011.

24. H. Taube, *Pure Appl. Chem.*, **63**, 651, 1991.

25. N. M. West, J. A. Labinger, and J. E. Bercaw, *Organometallics*, **30**, 2690, 2011.

26. J. R. Khusnutdinova, N. P. Rath, and L. M. Mirica, *J. Am. Chem. Soc.*, **132**, 7303, 2010.

27. U. Hintermair, S. W. Sheehan, A. R. Parent, D. H. Ess, D. T. Richens, P. H. Vaccaro, G. W. Brudvig, and R. H. Crabtree, *J. Am. Chem. Soc.*, **135**, 10837, 2013.

28. J. D. Blakemore, N. D. Schley, G. Olack, C. D. Incarvito, G. W. Brudvig, and R. H. Crabtree, *Chem. Sci*, **2**, 94, 2011.

29. R. H. Crabtree, *New J. Chem.*, **35**, 18, 2011.

PROBLEMS

1.1. How many isomers would you expect for a complex with an *empirical* formula corresponding to $Pt(NH_3)_2Cl_2$?

1.2. What d^n configurations should be assigned to the following and what magnetic properties—dia- or paramagnetic—are to be expected from the hexaqua complexes of Zn(II), Cu(II), Cr(II), Cr(III), Mn(II), and Co(II).

1.3. Why is $R_2PCH_2CH_2PR_2$ so much better as a chelating ligand than $R_2PCH_2PR_2$? Why is H_2O a lower-field ligand for Co^{3+} than NH_3?

1.4. How would you design a synthesis of the complex *trans-*$[PtCl_2(NH_3)(tu)]$, (the trans descriptor refers to the fact a pair of identical ligands, Cl in this case, is mutually trans), given that the trans effect order is $tu > Cl > NH_3$ ($tu = (H_2N)_2CS$, a ligand that binds via S)?

1.5. Consider the two complexes $MeTiCl_3$ and $(CO)_5W(thf)$. Predict the order of their reactivity in each case toward the following sets of ligands: NMe_3, PMe_3, and CO.

1.6. How could you distinguish between a square planar and a tetrahedral structure in a nickel(II) complex of which you have a pure sample, without using crystallography?

1.7. You have a set of different ligands of the PR_3 type and a large supply of $(CO)_5W(thf)$ with which to make a series of complexes $(CO)_5W(PR_3)$. How could you estimate the relative ordering of the electron-donor power of the different PR_3 ligands?

1.8. The stability of metal carbonyl complexes falls off markedly as we go to the right of group 10 in the periodic table. For example,

Cu complexes only bind CO weakly. Why is this? What oxidation state, of the ones commonly available to copper, would you expect to bind CO most strongly?

1.9. Low-oxidation-state complexes are often air sensitive (i.e., they react with the oxygen in the air) but are rarely water sensitive. Why do you think this is so?

1.10. MnCp$_2$ is high spin, while Mn(Cp*)$_2$ (Cp* = η^5-C$_5$Me$_5$) is low spin. How many unpaired electrons does the metal have in each case, and which ligand has the stronger ligand field?

1.11. Why does ligand **1.18** bind as a clamshell with the Me and Cl sites mutually cis, and not in a coplanar arrangement with Me and Cl trans?

1.12. Make up a problem on the subject matter of this chapter and provide an answer. This is a good thing for you to do for subsequent chapters as well. It gives you an idea of topics and issues on which to base questions and will therefore guide you in studying for tests.

2

MAKING SENSE OF ORGANOMETALLIC COMPLEXES

We now look at the 18-electron rule[1] and at the alternative ionic and covalent bonding models on which this metal valence electron counting procedure is based. We then examine the ways in which binding to the metal can perturb the chemical character of a ligand, an effect that lies at the heart of organometallic chemistry.

2.1 THE 18-ELECTRON RULE

Just as organic compounds follow the octet or eight valence electron rule, typical organometallic compounds tend to follow the 18e rule. This is also known as the noble-gas or effective atomic number (EAN) rule because the metals in an 18e complex achieve the noble-gas configuration—for example, in the Werner complexes, the cobalt has the same EAN as Kr, meaning it has the same number of electrons as the rare gas. We first discuss the covalent model that is the most appropriate one for counting compounds with predominant covalency, such as most organometallics.

Covalent Electron Counting Model

To show how to count valence electrons by forming a compound from the neutral atomic components, we first apply the method to CH_4, where the simpler octet rule applies (Eq. 2.1).

The Organometallic Chemistry of the Transition Metals, Sixth Edition.
Robert H. Crabtree.
© 2014 John Wiley & Sons, Inc. Published 2014 by John Wiley & Sons, Inc.

$$C + 4H = CH_4$$
$$4e \quad 4e \quad 8e$$

$$(2.1)$$

An octet is appropriate for carbon, where one $2s$ and three $2p$ orbitals make up the valence shell; 8e fill all four orbitals.

The 18e rule, followed by many transition metal compounds, is justified on the ligand field model by the presence of nine orbitals: five d orbitals, three p orbitals along with a single s orbital. A simple 18e case is shown in Eq. 2.2.

$$Re + 9H + 2e^- = [ReH_9]^{2-}$$
$$7e \quad 9e \quad 2e \quad 18e$$

$$(2.2)$$

The net ionic charge of 2− needs to be considered along with the nine ligands. The two electrons added for the 2− charge came from forming the counterions (e.g., $2Na \Rightarrow 2Na^+ + 2e^-$). Other anionic X ligands that also provide one electron to the metal on forming a covalent bond include CH_3^-, Cl^-, and $C_6H_5^-$.

A neutral L ligand, such as NH_3, contributes its two lone pair electrons to the metal on binding (Eq. 2.3).

$$Co + 6NH_3 - 3e^- = [Co(NH_3)_6]^{3+}$$
$$9e \quad 12e \quad -3e \quad 18e$$

$$(2.3)$$

The net 3+ ionic charge requires subtracting 3e from the count; these electrons are transferred to the anions (e.g., $1.5Cl_2 + 3e \Rightarrow 3Cl^-$).

Table 2.1 shows how most first-row carbonyls follow the 18e rule. Each metal contributes the same number of electrons as its group number, and each CO contributes 2e from its lone pair; π back bonding (Section 1.9) makes no difference to the electron count for the metal. The free atom already had the pairs of d_π electrons destined for back

TABLE 2.1 First-Row Carbonyls

$V(CO)_6$	17e paramagnetic
$[V(CO)_6]^-$	18e
$Cr(CO)_6$	Octahedral. 18e
$(CO)_5Mn-Mn(CO)_5$	M–M bond contributes 1e to each metal; all the CO groups are terminal. 18e
$Fe(CO)_5$	Trigonal bipyramidal. 18e
$(CO)_3Co(\mu\text{-}CO)_2Co(CO)_3$	μ-CO contributes 1e to each metal, as does the M–M bond. 18e
$Ni(CO)_4$	Tetrahedral. 18e

bonding; in the complex, it still has them, now delocalized over metal and ligands.

Where the metal starts with an odd number of electrons, we can never reach 18 just by adding 2e ligands, such as CO. Each carbonyl complex resolves this problem in a different way. $V(CO)_6$ is stable in spite of being 17e, but it is easily reduced to the 18e $[V(CO)_6]^-$ anion. The 17e reactive transient $Mn(CO)_5$ is not isolable but instead dimerizes to the stable 18e dimer—as a five-coordinate monomer, there is more space available to make the M–M bond than in $V(CO)_6$. This dimerization completes the noble-gas configuration for each metal because the unpaired electron in each fragment is shared with the other in forming the M–M bond, much as the 7e methyl radical dimerizes to give the 8e compound, ethane. In the 17e reactive fragment $Co(CO)_4$, dimerization also takes place to form a metal–metal bond, but a pair of CO ligands also bridge. The electron count is unchanged whether the COs are terminal or bridging because CO is a 2e ligand to the cluster in either case. On the conventional model, a ketone-like μ-CO gives 1e to each metal, so an M–M bond is still required to attain 18e. The even-electron metals are able to achieve 18e without M–M bond formation, and in each case, they do so by binding the appropriate number of CO ligands.

Ionic Electron Counting Model

An older counting convention based on the ionic model was developed early in the twentieth-century for classical Werner coordination compounds because of their more ionic bonding. The final count, d^n configuration and oxidation state is always the same for any given complex on either model—only the counting method differs. Authors invoke one or other model without identification, so we have to be able to deduce their choice from the context. Neutral L ligands pose no problem because they are always 2e donors on either model, but M–X bonds are treated differently. In the ionic model, each M–X is considered as arising from M^+ and X^- ions. To return to our organic example, whether we count octet CCl_4 by the covalent model from the atoms (Eq. 2.4) or the ionic model from the ions (Eq. 2.5), we get the same result.

$$C + 4Cl = CCl_4$$
$$4e \quad 4e \quad \quad 8e$$

(2.4)

$$C^{4+} + 4Cl^- = CCl_4$$
$$0e \quad \; 8e \quad \quad 8e$$

(2.5)

TABLE 2.2   Common Ligands and Their Electron Counts

Ligand	Type	Covalent Model	Ionic Model
Me, Cl, Ph, H, η^1-allyl, NO (bent)[a]	X	1e	2e
Lone-pair donors: CO, NH_3, PPh_3	L	2e	2e
π-Bond donors: C_2H_4	L	2e	2e
σ-Bond donors: H_2	L	2e	2e
M–Cl (bridging)	L	2e	2e
η^3-Allyl, κ^2-acetate	LX	3e	4e
NO (linear)[a]		3e	2e[a]
η^4-Butadiene	L_2[b]	4e	4e
=O (oxo)	X_2[c]	2e	4e
η^5-Cp	L_2X	5e	6e
η^6-Benzene	L_3	6e	6e

[a]Linear NO is considered as NO^+ and bent as NO^- on the ionic model; see Section 4.1.
[b]The alternative LX_2 structure sometimes adopted gives the same electron count.
[c]In some cases, a lone pair on the oxo also bonds to M, making it an LX_2 ligand (=4e covalent; 6e ionic).

Applying the ionic model to the case of Eq. 2.2 gives the result shown in Eq. 2.6. We use the covalent model in this book but we need to be familiar with both.

$$Re^{7+} + 9H^- = [ReH_9]^{2-}$$
$$0e \quad\quad 18e \quad\quad 18e$$

(2.6)

Electron Counts for Common Ligands and Hapticity

Table 2.2 shows common ligands and their electron counts on both models. Neutral ligands, L, are always 2e ligands on either model, whether they are lone-pair donors, such as CO or NH_3, π-bond donors, such as C_2H_4, or σ-bond donors such as H_2. Anionic ligands, X, such as H, Cl, or Me, are 1e X atoms or groups on the covalent model but 2e X^- ions on the ionic model. On the covalent model, a 1e X· radical bonds to a neutral metal atom; on the ionic model, a 2e X^- anion bonds to an M^+ cation. Parkin[1] and Green[2] have developed a useful extension of this nomenclature by which more complicated ligands can be classified. For example, benzene (**2.1**) can be considered as a combination of three C=C ligands, and therefore as L_3. Likewise, the η^3-allyl group, $CH_2 = CH–CH_2^-$, is an LX combination of a C=C group and an alkyl RCH_2^-. Allyl can be represented as **2.2** and **2.3** or else in a delocalized form as **2.4**. In such a case, the hapticity of the ligand, the number of contiguous ligand atoms bound to the metal, is three, and so **2.5** is bis-η^3-allyl nickel, or [Ni(η^3-C_3H_5)$_2$]. The electron count of an η^3-allyl, sometimes simply called a π-allyl, is 3e on the covalent model and 4e on the ionic model, as suggested by the LX label. The advantage of using LX is that those who follow the covalent model translate it as meaning a 3e neutral ligand, while the devotees of the ionic model translate it as meaning a 4e anionic ligand. The Greek letter kappa, κ, is normally used instead of η when describing ligands that bind via noncontiguous atoms, such as a chelating κ^2-acetate or for identifying the donor atom, as in SCN-κ-*N* versus SCN–κ-*S*. Full details of nomenclature conventions are available.[3]

The allyl group can also bind (**2.6**) via one carbon in the η^1-allyl, or σ-allyl, form. It then behaves as an X-type ligand, like a methyl group, and is therefore a 1e ligand on the covalent model and a 2e ligand on the ionic model.

Some examples of electron counting are shown in Table 2.3. Note the dissection of **2.7, 2.8, 2.9, 2.10, 2.11**, and **2.12** into atoms and radicals in the covalent model and into ions in the ionic model.

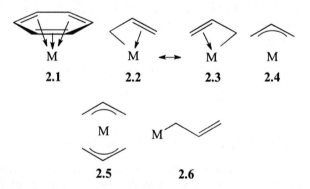

M	M ⟷ M		M
2.1	**2.2**	**2.3**	**2.4**

M	M
2.5	**2.6**

For complex ions, we adjust the count for the net ionic charge. For example, in [CoCp$_2$]$^+$ (**2.12**, Table 2.3), the Group 9 Co atom starts with 9e. On the covalent model, the two neutral Cp groups add 10e (Table 2.2) and the net ionic charge is 1+, one electron having been removed to make the cation. The electron count is therefore $9 + 10 - 1 = 18$e. Electron counting can be summarized by Eq. 2.7, for a generalized complex [MX$_a$L$_b$]$^{c+}$, where N is the group number of the metal (and therefore the number of electrons in the neutral M atom), a and b are the numbers of X and L ligands, N is the group number and c is the net positive ionic charge (if negative, then the sign of c is reversed):

$$\text{e count (covalent model)} = N + a + 2b - c \qquad (2.7)$$

TABLE 2.3   Electron Counting on Ionic and Covalent Models

Ionic Model	Compound	Covalent Model
$2C_5H_5^- + Fe^{2+}$	Cp_2Fe	$2C_5H_5\bullet + Fe(0)$
$12e + 6e = 18e$	**2.7**	$10e + 8e = 18e$
$4H^- + 4PR_3 + Mo^{4+}$	$MoH_4(PR_3)_4$	$4H\bullet + 4\ PR_3 + Mo(0)$
$8e + 8e + 2e = 18e$	**2.8**	$4e + 8e + 6e = 18e$
$2C_3H_5^- + Ni^{2+}$	$Ni(\eta^3\text{-allyl})_2$	$2C_3H_5\bullet + Ni(0)$
$8e + 8e = 16e$	**2.9**	$6e + 10e = 16e$
$2C_6H_6 + Mo$	$Mo(\eta^6\text{-}C_6H_6)_2$	$2C_6H_6 + Mo(0)$
$12e + 6e = 18e$	**2.10**	$12e + 6e = 18e$
$2C_5H_5^- + 2Cl^- + Ti^{4+}$	Cp_2TiCl_2	$2C_5H_5\bullet + 2Cl\bullet + Ti(0)$
$12e + 4e + 0e = 16e$	**2.11**	$10e + 2e + 4e = 16e$
$2C_5H_5^- + Co^{3+}$	$[Cp_2Co]^+$	$2C_5H_5\bullet + (+)^a + Co(0)$
$12e + 6e = 18e$	**2.12**	$10e - 1e + 9e = 18e$

^{a}Subtracting 1e is needed here to account for the loss of 1e to the anion on forming the organometallic cation. Note how the net ionic charge is treated on each model, explicitly in the covalent model and as the residual metal ion charge in the ionic model.

In the ionic counting model, we first calculate the oxidation state of the metal (Section 2.4). This is the ionic charge left on the metal after removal of the ligands, taking care to assign the electron pairs in the M–L bonds to the more electronegative atom in each case. (If two atoms have the same electronegativity, one electron is assigned to each; see also Section 2.4.) For $[CoCp_2]^+$, we must remove two Cp^- ions (C is more electronegative than Co); this leaves d^6 Co^{3+}. We now add back the two 6e Cp^- ions so that $[CoCp_2]^+$ has $6 + (2 \times 6) = 18$ electrons, the same count as before. For the general case of $[MX_aL_b]^{c+}$, this procedure leaves the metal as $M^{(c+a)+}$, and therefore the metal is in the oxidation state $(c + a)$, and has $N - c - a$ electrons. This number is identical to n in the d^n configuration of the ion. We now have to add 2e for each X^-, and 2e for each L in putting the complex back together—this leads to Eq. 2.8, but this simplifies to Eq. 2.7, and so the two methods of electron counting give exactly the same final result.

$$\text{e count (ionic model)} = N - a - c + 2a + 2b = N + a + 2b - c \quad (2.8)$$

Bridging Ligands

Ligands that bridge are indicated by the prefix μ. Two types can be distinguished, as discussed below on the covalent model.

Bridges with Two Independent Bridge Bonds   For monoanions μ-Cl, μ-PR$_2$ μ-SR or μ-OR and related cases, two separate two-electron, two center (2e,2c) M–(L or X) bonds are formed, each involving a distinct pair of electrons. For these, we can write a structure such as **2.13** in which the electron count for each metal is treated separately. For M$'$, we consider only the X ligand, a 1e donor on the covalent model, and for M$''$, we consider an X lone pair as equivalent to a neutral 2e L-type donor to M$''$. In **2.14**, for example, each 14e Cp*Ir reaches 18e by counting the μ-Cl in this way and assigning the 1+ ion charge to the right hand Ir. For a ligand such as Cl$^-$ with four lone pairs, bridging can also occur between three (μ^3-Cl) or even four metals (μ^4-Cl) in clusters (Chapter 13). Dianionic μ-O and μ-S can act as X ligands to each metal.

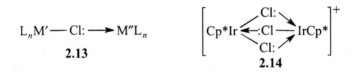

$$L_nM' \longrightarrow Cl: \longrightarrow M''L_n$$

2.13

$$\left[Cp^*Ir \begin{array}{c} Cl: \\ :Cl \\ Cl: \end{array} IrCp^* \right]^+$$

2.14

Bridges with a Single Delocalized Bridge Bond   For a small class of ligands, such as μ-H, no lone pairs are present, and the bonding situation is different. M$'$(μ-H)M$''$ is best seen as involving a single two-electron, three center (2e,3c) bond that links all three centers, M$'$, M$''$, and H, with just 2e. The classic case, B$_2$H$_6$, although coming from the main group, embodies the same bonding pattern for each B–H–B bridge. B$_2$H$_6$ can alternatively be considered as the double protonation product of the hypothetical ethylene analog, [H$_2$B=BH$_2$]$^{2-}$, where the protons add to the two lobes of the B=B π bond (Eq. 2.9). This dianion is isoelectronic with ethylene and hence an octet molecule. Protonation does not alter the e count, so B$_2$H$_6$ must also be an octet molecule. A deprotonation strategy holds for transition metals — we can count any bridged hydride by removing each μ-H as a proton, thus converting each M–H–M to an M–M bond, and counting the resulting hypothetical structure. Counting **2.15** of Eq. 2.10 on the deprotonation model requires removing 3 × H$^+$ and replacing them with three M–M bonds. 14e Cp*Ir now reaches 18e by counting 3e for the M≡M bond and assigning 1- of the 2-anionic charge to each Ir (14 + 3 + 1 = 18).

$$\begin{array}{c} H \\ H \end{array} B \begin{array}{c} H \\ \\ H \end{array} B \begin{array}{c} H \\ H \end{array} \xrightleftharpoons[+2H^+]{-2H^+} \left[\begin{array}{c} H \\ H \end{array} B{=}B \begin{array}{c} H \\ H \end{array} \right]^{2-} \tag{2.9}$$

Hypothetical

$$\left[Cp^*Ir \overset{H}{\underset{H}{\overset{|}{\longrightarrow}}} IrCp^* \right]^+ \underset{+3H^+}{\overset{-3H^+}{\rightleftharpoons}} \left[Cp^*Ir \equiv IrCp^* \right]^{2-} \qquad (2.10)$$

2.15   Hypothetical

Even though the traditional pictorial representation of B_2H_6 makes no distinction between the terminal and bridging bonds, the bridging B–H–B bonds, although shown as two separate bonds, are not the normal (2e,2c) type, but are instead half-order (2e,3c) bonds; the same holds true for M–H–M bonds in **2.15**. Ligands such as μ-CH$_3$, μ-CO, and μ-PR$_3$ can be considered in a similar way, as discussed in Chapters 3, 4, and 13.

Zero-Electron Ligands

Zero-electron neutral ligands are a growing class. For example, BR_3, having a 6e boron, completes its octet by accepting lone pairs, as in $H_3N \rightarrow BR_3$ to become an 8e boron. If the lone pair comes from a metal, we have an $L_nM \rightarrow BR_3$ bond in which BR_3 provides 0e to the metal and thus leaves the metal electron count unaltered. The $L_nM \rightarrow BR_3$ bond can alternatively be written with formal charges as $L_nM^+ - B^-R_3$ (Eq. 2.11).

Two alternative ways of assigning the oxidation state of the complex may be chosen. When the $L_nM - BR_3$ bond is formally broken, BR_3 can either be considered to dissociate as the free ligand, 6e, B(III), BR_3 to leave M(0) L_nM, or as the B(I) oxidation state octet fragment $(BR_3)^{2-}$ to leave M(II) L_nM^{2+}. The former implies the OS is unchanged on going from L_nM to $L_nM - BR_3$, while the latter assigns a +2 OS change to this step. This is therefore an example of OS ambiguity (see p. 54). The +2 OS assignment seems preferable in some cases, for example, Ir(III)B(I) explains the octahedral geometries of the complexes in Eq. 2.11,[4] expected only for d^6 Ir(III), as well as the easy loss of CO, more consistent with the weaker CO binding often seen in Ir(III) rather than Ir(I).

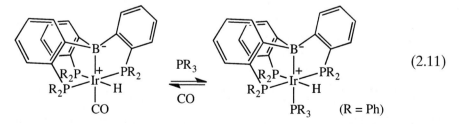

$$(2.11)$$

$(R = Ph)$

The same trisphosphine borane ligand but with R = iPr (BP$_3$) has been incorporated into an Fe complex, $[Fe(BP_3)]^+$. Both Fe(III)B(I) and

Fe(I)B(III) oxidation state assignments were considered, but computational and structural evidence made Fe(I)B(III) seem most appropriate here, in contrast to the Ir case.[5]

2.2   LIMITATIONS OF THE 18-ELECTRON RULE

Many stable complexes have an electron count other than 18; otherwise, most non-18e structures have <18e, such as $MeTiCl_3$, 8e; Me_2NbCl_3, 10e; WMe_6, 12e; $Pt(PCy_3)_2$, 14e; $[M(H_2O)_6]^{2+}$ (M = V, 15e; Cr, 16e; Mn, 17e; Fe, 18e). Much rarer are d block examples with >18e: $CoCp_2$, 19e; and $NiCp_2$, 20e are prominent cases. For the 18e rule to be useful, we need to know when it will be obeyed and when not.

The rule works best for small, high-field, monodentate ligands, such as H and CO. Such small ligands find no difficulty in binding as many times as needed to reach 18e. As high-field ligands, Δ is large, so the d_σ orbitals that would be filled if the metal had >18e are high in energy and therefore poor acceptors. On the other hand, the d_π electrons that would have to be lost if the molecule had <18e are stabilized either by π bonding (CO) or strong σ bonding (H) with the metal. The EAN rule even extends to small organometallic clusters, such as $Os_3(CO)_{12}$ (Chapter 13).

An important class of late metal complexes prefers 16e to 18e, because one of the nine orbitals is very high lying and usually empty. This can happen for the d^8 metals of groups 8–11 (Table 2.4). Group 8 shows the least and group 11 the highest tendency to become 16e. When these metals are 16e, they normally become square planar, as in $RhClL_3$, $IrCl(CO)L_2$, $PdCl_2L_2$, $[PtCl_4]^{2-}$, and $[AuMe_4]^-$ (L = PR_3).

The rule works least well for paramagnetic and high-valent metals with weak-field ligands. In the hexaaqua ions $[M(H_2O)_6]^{2+}$ (M = V, Cr, Mn, Fe, Co, Ni), the structure is much the same whatever the electron count of the metal and so must be dictated by other factors. H_2O has two lone pairs, but only one is needed to form the M–L σ bond. The remaining one acts as a π donor to the metal and so lowers Δ (Fig. 1.10); H_2O is therefore a weak-field ligand. When Δ is small, the tendency to adopt the 18e configuration is also small, because it is easy to add electrons to the low-lying d_σ or to remove them from the high-lying d_π. Early metals of groups 3–5 are often found with <18e, no single count being particularly preferred.

No rule is useful for main group elements: for example, $SiMe_4$ is 8e; PF_5, 10e; SF_6, 12e; $HgMe_2$, 14e; $MeHg(bipy)^+$, 16e; $[I(py)_2]^+$, 20e; $[SbF_6]^-$, 22e; and IF_7, 24e. Although early metal d^0 complexes can have electron counts well below 18e (e.g., 8e $TiMe_4$ and 12e WMe_6), an ambiguity

TABLE 2.4 d^8 Metals that Can Adopt a 16e Square Planar Geometry

Group Number[a]			
8	9	10	11
Fe(0)	Co(I)	Ni(II)	Cu(III)[b]
Ru(0)	Rh(I)	Pd(II)	–
Os(0)	Ir(I)	Pt(II)	Au(III)

[a]Group 8 metals prefer 18e to 16e. In group 9, tbe 16e configuration is more often seen, but 18e complexes are still common. In groups 10–11, tbe 16e configuration is much more often seen.
[b]A rare oxidation state.

often arises when the ligands have additional π-type lone pairs that can — at least in principle — be donated into empty metal d_π orbitals as shown in Fig. 1.8. For example, $W(OMe)_6$ is apparently a 12e species, but each oxygen has two π-type lone pairs for a total of 24 additional electrons that could be donated to the metal. Almost any even electron count could therefore be assigned, and for this reason, electron counting is much less useful in discussions of early metal and d^0 organometallic chemistry.

Paramagnetic complexes (e.g., $V(CO)_6$, 17e; Cp_2Fe^+, 17e; Cp_2Ni, 20e) generally do not obey the 18e rule, but many of these have reactions in which they attain an 18e configuration, for example, the 19e $CpFe(\eta^6\text{-}C_6H_6)$ is a powerful 1e reductant giving 18e $[CpFe(\eta^6\text{-}C_6H_6)]^+$ as product.

The f block metals have seven f orbitals to fill before they even start on the d orbitals, and so they are essentially never able to bind a sufficient number of ligands to raise the electron count to the full $s^2p^6d^{10}f^{14}$ count of 32e; some examples are $U(cot)_2$, 22e, and Cp_2LuMe, 28e. The stoichiometry of an f block complex instead tends to be decided by steric saturation of the space around the metal. Although coordination numbers of 8 and 9 are most common, a *CN* as high as 15 has been reported for a thorium aminodiboranate, $[Th(H_3BNMe_2BH_3)_4]$.[6]

Steric Stabilization of Reactive Species

Steric stabilization of otherwise reactive species is a standard strategy in organometallic chemistry. Steric bulk can permit formation of low electron count, low coordination number complexes, as in the isolable 14e bis-π-allyl complex **2.16**. In this case, the bulky $SiMe_3$ groups enforce a *syn,anti* conformation that minimizes the steric clash between $SiMe_3$ groups but blocks approach of additional ligands.[7] The 12e paramagnetic, high-spin Cr(II) analog has the same structure. 14e $[MePd(PtBu_3)_2]^+$ has a T-shaped geometry;[8] bulk also favors distortions from electronically preferred geometries. For example, although $[CuBr_4]^{2-}$ and $[PtI_4]^{2-}$ electronically prefer a square planar geometry, the bulky halides cause a distortion toward the less hindered tetrahedral geometry.

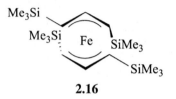

2.16

2.3   ELECTRON COUNTING IN REACTIONS

In counting a metal complex, we are free to apply covalent or ionic models, but in reactions, we no longer have a free choice because the initial reagents or intermediates are separate species with specific electron counts. For example, H^+, $H\bullet$, and H^- are all forms of the same element, but they are very different in their reactivity and in the number of electrons they bring. As a zero electron reagent, H^+ can in principle attack at any site on an 18e complex without any 18e rule limitation. H^-, in contrast, brings $2e^-$ so it cannot simply attack an 18e metal unless a ligand dissociates or else a very rare 20e complex would result. As a reagent, H^- is typically part of some hydride donor, such as $LiAlH_4$ and $NaBH_4$, but from reagent electron counting considerations, they all act as 2e H^- donors and the position of attack by H^- is restricted by the 18e rule. As we see in Chapter 7, H^- can attack a *ligand* without infringing the 18e rule. Attack on organometallics by $H\bullet$ and radicals in general is not yet well understood.

We therefore need to be very careful to specify whether a process involves proton, hydrogen atom or hydride transfer, because each has a completely different reactivity. The same holds for many other reactants, for example, different Br-containing reagents can act as Br^+, $Br\bullet$, or Br^- transfer agents; methyl transfer is typically either of Me^+ or Me^-.

Other 2e donor reagents, such as L or X^- species (e.g., PPh_3 and Cl^-), likewise cannot easily attack the metal in an 18e complex. A 2e ligand usually has to be lost first, thus giving *substitution* of one ligand by another (Chapter 4). Note that in reagent counting, L and X^- fall into the same 2e category.

Table 2.5 lists reagent electron counts and also tells us about possible isoelectronic replacements of one ligand by another. For example, an X^- group can replace an L ligand without any change in the electron

TABLE 2.5   Some Reagent Electron Counts

0e	1e	2e	3e	4e
H^+	$H\bullet^a$	H^- (LiAlH$_4$)^b	NO	$C_3H_5^-$ (C_3H_5MgBr)
Me^+ (MeI)	$Me\bullet^a$	Me^- (LiMe)		Butadiene
Br^+ (Br_2)^c		PPh_3, NO^+		NO^-
		Cl^-, CO, H_2		

^{a}These species are unstable and so they are invoked as reactive intermediates in mechanistic schemes, rather than used as reagents in the usual way.
^{b}The reagents in parentheses are the ones most commonly used as a source of the species in question.
^{c}Br$_2$, can also be a source of Br•, a 1e reagent, as well as of Br$^+$, depending on conditions.

count (Eq. 2.12) but making the ion charge, c, one unit more negative or less positive.

$$W(CO)_5(thf) + Cl^- = [W(CO)_5(Cl)]^- \qquad (2.12)$$

The reaction of Eq. 2.13 turns a 1e alkyl group into a 2e alkene group. To retain the 18e configuration, the complex must become positively charged, which implies that the H must be lost as H^- and that an electrophilic reagent such as Ph_3C^+ must be the reaction partner. In this way, the 18e rule helps us pick the right reagent type.

$$Cp(CO)_2 Fe-CH(CH_3)_2 + Ph_3C^+$$
$$= Cp(CO)_2 Fe\{\eta^2\text{-}CH_2\text{=}CH(CH_3)\}]^+ + PhCH \qquad (2.13)$$

By looking at the equations in the pages to come, you will become more familiar with electron counting of stable complexes and with counting the ligands that are gained or lost in reactions. In proposing new structures, be sure that the rules discussed in this chapter are obeyed.

2.4   OXIDATION STATE

The oxidation state (OS) of a metal in a complex is simply the integer charge that the metal would have on the ionic model. For a neutral complex, this is the number of X ligands. For example, Cp_2Fe has two L_2X ligands and so can be represented as MX_2L_4 for which the OS is $2+$, so Cp_2Fe is said to contain Fe(II). For a complex ion, we need also to take account of the net charge as shown for $[MX_aL_b]^{c+}$ in Eq. 2.14. For example, Cp_2Fe^+ is Fe(III), and $[W(CO)_5]^{2-}$ is W(−II). M–M bonding adds 1e to the electron count of each metal but does not affect the formal oxidation state, so 18e $(OC)_5Mn-Mn(CO)_5$ is d^7 Mn(0).

The d^n configuration follows from the oxidation state and is the number of valence electrons that would be present in the free metal ion that corresponds to the OS. In Cp_2Fe^+, for example, the OS is 3, Fe is in group 8, Fe(0) has 8e, and so Fe^{3+} has $8 - 3 = 5e$. Cp_2Fe^+ is therefore a d^5 Fe(III) complex. Recall that any valence electrons are always assigned to the d orbitals only, not to d and s as in the configuration for the free atom. The bare Fe atom has configuration [Ar] $3d^6$ $4s^2$ but becomes d^8 in any Fe(0) compound, when the d levels become stabilized by M–L bonding (Section 1.6).

Equation 2.15 gives the value of d^n for $[MX_aL_b]^{c+}$ and tells us to use n electrons to fill up the crystal field diagrams of Section 1.6. For example, the d^5 value for Cp_2Fe^+ implies paramagnetism because in a mononuclear complex, we cannot pair five electrons whatever the d-orbital splitting.

$$OS = c + a \qquad (2.14)$$

$$d^n = d^{\{N-(c+a)\}} = d^{(N-c-a)} \qquad (2.15)$$

Table 2.6 gives some leading characteristics of specific d configurations and shows how oxidation state and d^n configuration are linked. New oxidation states are occasionally found, for example, the d^1 Ir(VIII) species, IrO_4, was recently identified in a low temperature matrix.[9]

Most organometallic compounds occur in low or intermediate oxidation states, but high OSs are now gaining more attention (see Chapters 11 and 15). Back donation is severely reduced in high OSs because (i) there are fewer (or no) nonbonding d electrons available and (ii) the increased partial positive charge on the metal in a high OS complex strongly stabilizes the d levels so that any electrons they contain become less available. Among the most stable high OS species come from the third-row metals. The extra shielding from the f electrons added for La–Lu after the second and before the third row d block metals makes the outer electrons in the third row less tightly bound and therefore more available either for back bonding or for loss upon oxidation. High OSs are most easily accessible if the ligands are small and non-π-bonding, as is the case for H or Me in the d^0 species WMe_6 and $ReH_7(dpe)_2$.

Even d^n configurations are much more common than odd ones, particularly for the second and third row. Diamagnetic complexes are easier to study and so are more often reported, and the high Δ value for the second- and third-row metals favors electron pairing in the d_π levels. An exception exists for M–M bonded compounds, where odd

TABLE 2.6 Relationships between Oxidation States and d^n Configurations

	Ti	V	Cr	Mn	Fe	Co	Ni	Cu	
	Zr	Nb	Mo	Tc	Ru	Rh	Pd	Ag	
	Hf	Ta	W	Re	Os	Ir	Pt	Au	
d^0	4	5	6	7	8				No back donation; max. OS
d^1	3	4	5	6	7	8			Paramagnetic[a]
d^2	2	3	4	5	6				Strong back donation
d^3	1	2	3	4	5				Paramagnetic[a,b]
d^4	0	1	2	3	4	5			Commonest in groups 6–8
d^5	−1	0	1	2	3	4			Paramagnetic[a]
d^6	−2	−1	0	1	2	3	4		Commonest configuration[b]
d^7	−3	−2	−1	0	1	2	3		Paramagnetic[a]
d^8	−4	−3	−2	−1	0	1	2	3	Common in groups 8–10
d^9		−4	−3	−2	−1	0	1	2	Paramagnetic[a]
d^{10}			−4	−3	−2	−1	0	1	Common in groups 10–11; min. OS

[a]In mononuclear complexes. Odd number configurations are uncommon for organometallic complexes.
[b]Coordination inert in octahedral complexes.

electrons on each metal can pair up in the M–M bond, as shown in the d^7 Pd(III) dipalladium complex, **2.17** (X = Cl, Br);[10] only very recently have organometallic Pd(III) complexes been reported.[11]

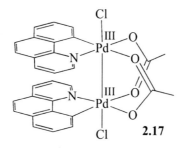

2.17

Oxidation State versus Real Charge

It is often useful to refer to the oxidation state and d^n configuration of a metal, but these only represent a formal classification and do not indicate either the real partial charge on the metal or even the trends in that charge as the ligands are changed. It is therefore important not to read too much into them. That is why ferrocene is considered as an Fe(II) and not as an Fe^{2+} compound; Fe^{2+} would only be appropriate for a predominantly ionic compound, such as $[Fe(OH_2)_6]^{2+}$. Similarly, WH_6L_3, in spite of being W(VI), is likely to be closer to $W(CO)_6$ in terms of the real charge on W than to WO_3. In real terms, W(VI) WH_6L_3 is probably more reduced and more electron rich than $W(CO)_6$, formally W(0). CO groups are excellent π acceptors, so the metal in $W(CO)_6$ has a much lower electron density than a free W(0) atom; on the other hand, the W–H bond in WH_6L_3 is only weakly polar, and so the polyhydride has a much higher electron density than the W^{6+} suggested by its W(VI) OS (a result that assumes a dissection: W^+ H^-). For this reason, the OS obtained from Eq. 2.11 is termed the *formal oxidation state*.

Even with computational assistance, real charge is not an entirely reliable value for an organometallic complex because it depends on what criterion we use to define the boundary between one covalently bound atom and another. Computational efforts have given some insight into the problem by adopting assumptions that allow assignment of atomic charges in a molecule. The early Mulliken approximation has now been largely supplanted by the natural bond orbital or NBO approach,[12] and a number of recent papers use NBO methods to calculate natural charges for atoms and groups in organometallic molecules.[13]

Ambiguous Oxidation States and Noninnocent Ligands

More problematic are cases in which even the formal oxidation state is ambiguous and cannot be specified. This problem affects any ligand that has several resonance forms that contribute to a comparable extent to the real structure but give different OS assignments; this behavior makes the ligand *noninnocent*. For example, butadiene resonance form **2.18a** is L_2, but **2.18b** is LX_2. The binding of butadiene as **2.18a** leaves the oxidation state of the metal unchanged, but as **2.18b**, it becomes more positive by two units. On the covalent model, L_2 and LX_2 each give exactly the same 4e count, so the 18e rule is unaffected. On the ionic model, L_2 is 4e and LX_2 is 6e, but the 2e change in the ligand is compensated by a formal 2e oxidation of the metal. We do not see two distinct forms of the same complex, one like **2.18a** and the other like **2.18b**.

Instead, any one given complex has a structure in a range between **2.18a** or **2.18b** as extremes. The oxidation state ambiguity can become severe—in the case of W(butadiene)$_3$, we can attribute any even oxidation between W(0) and W(VI) to by counting the butadienes as LX$_2$ or L$_2$. Fortunately, the electron count is unambiguous because it always remains the same for all resonance forms. To avoid misunderstanding, it is therefore necessary to specify the resonance form to which a formal oxidation state applies. For neutral ligands, such as butadiene, convention normally calls for the neutral form, L$_2$ in this case. Yet structural studies show that the ligand often more closely resembles **2.18b** than **2.18a**. Clearly, we can place no reliance on the formal oxidation state to tell us about the real charge on the metal in W(butadiene)$_3$. In spite of its ambiguities, the oxidation state convention is almost universally used in classifying organometallic complexes.

2.18a   **2.18b**

In **2.19a**, the S-donor behaves as a dithione L$_2$ ligand, making the metal d^6 W(0).[14] Accordingly, the central C–C bond of the dithione is long (1.49Å), the C=S short (1.69 Å), and the metal is octahedral. A small change in the substituents on the dithione leads to **2.19b**, where the metal has reduced the ligand by two electrons to give an enedithiolate, an X$_2$ ligand, making the metal d^4 W(II). The ligand C–C bond now short (1.35 Å), the C–S long (1.74 Å) and the metal geometry has converted to trigonal prismatic. Rather than the range of intermediate structures common for butadiene (**2.18**), we now see two sharply defined ligand types in different compounds, distinguishable from X-ray structural data.

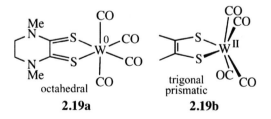

2.19a   **2.19b**

First row metals typically have stable oxidation states one unit apart and undergo 1e redox changes as a result, for example, Fe(II,III), Co(I,II,III), Ni(I,II,III), Cu(I,II). The second and third row typically have stable oxidation states two units apart and prefer 2e redox changes, e.g., Pd(0,II,IV), Ru(0,II,IV), Ir(I,III,V), W(0,II,IV,VI). When organometallics have to bring about multielectron reactions, such as we cover in Chapters 9, 12, and 14, 2e redox steps are desirable because they avoid high energy, odd-electron intermediates. Although second- and third-row metals have been preferred up to now, the price rises of the precious metals and the green chemistry aspiration of avoiding rare elements means that first row substitutes will increasingly be sought. Redox-active, noninnocent ligands may help in this respect by providing an alternative source or sink of electrons in addition to the metal, thus potentially allowing multielectron chemistry for the common metals.[15] Pincers and porphyrins are specially favored for redox activity because of the greater degree of multiring delocalization possible in these more extended ligand systems. Another noninnocent ligand class (**2.20**) can gain or lose protons easily, in this case from the NH groups. Deprotonation alters the ligand class from L_3 to LX_2 and has a big influence on the properties of the complex.[16]

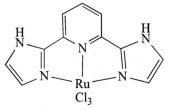

Maximum and Minimum Oxidation States

The maximum permitted oxidation state of a complex can never exceed the group number, N, of the metal. Ti can have no higher OS than Ti(IV), corresponding to minimum allowed d^n configuration of d^0. Neutral TiMe$_6$, for example, would be d^{-2} and thus implausible because Ti has only four valence electrons, not six.

Likewise, there is also a minimum OS corresponding to maximum d^n of d^{10}. $[Pt(PPh_3)_3]^{2-}$ obeys the 18e rule so it might be thought possible; however, it would be d^{12} Pt($-$II) and forbidden. These limitations need to be borne in mind when proposing intermediates in reaction mechanisms.

Net Ionic Charge

An increase in the positive ionic charge of a complex, $c+$ in $[MX_aL_b]^{c+}$ decreases any backbonding to the ligands (Section 2.7 and Table 2.10), all else being equal. It also makes the complex harder to oxidize but easier to reduce and changes the reactivity toward nucleophiles and electrophiles. For example, $[Mn(CO)_6]^+$, $Cr(CO)_6$, and $[V(CO)_6]^-$ are isoelectronic, but only the anion reacts readily with electrophilic H^+, only the cation reacts readily with the nucleophilic H_2O, while $Cr(CO)_6$ reacts readily with neither reagent.

$$V(CO)_6^- + H^+ \rightarrow HV(CO)_6 \rightarrow \text{further reaction} \qquad (2.16)$$

$$Mn(CO)_6^+ + H_2O \rightarrow HMn(CO)_5 + H^+ + CO_2 \qquad (2.17)$$

2.5   COORDINATION NUMBER AND GEOMETRY

The coordination number (CN) of a complex having only monodentate ligands is simply the number of ligands present (e.g., $[PtCl_4]^{2-}$, $CN = 4$, $W(CO)_6$, $CN = 6$). The coordination number cannot exceed 9 for the d block because the metal only has 9 valence orbitals, and each ligand needs its own orbital. If the CN is less than 9, the "unused" orbitals will then either be metal lone pairs or engaged in back bonding.

Many complexes can be discussed in terms of ideal geometries. Ignoring small distortions, each CN has one or more such associated geometries (Table 2.7). To reach the maximum CN of 9, we need relatively small ligands and a d^0 metal (e.g., $[ReH_9]^{2-}$). Coordination numbers lower than 4 are found with bulky ligands that cannot bind in greater number without prohibitive steric interference; for example, only $Pt(PCy_3)_2$ exists, not $Pt(PCy_3)_3$ or $Pt(PCy_3)_4$. The f block knows no such electronic limitations, only steric ones, and CN values up to 15 are known.

TABLE 2.7   Common Geometries with Their Typical d^n Configurations

Coordination Number, CN	Geometry	d^n Configuration[a]	Example
2	Linear	d^{10}	$[Ag(NHC)_2]^+$
3	T-shaped	d^8	$[Rh(PPh_3)_3]^+$
4	Tetrahedral	d^0, d^5 (hs), d^{10}	$Pd(PPh_3)_4$
4	Square planar	d^8	$[RhCl(PPh_3)_3]$
5	Trigonal bipyramidal	d^8, d^6{distorted}[b]	$[Fe(CO)_5]$
6	Octahedral	d^0, d^3, d^5 (ls), d^6	$[Mn(CO)_6]^+$
7	Pentagonal bipyramid	d^4	$[IrH_5(PR_3)_2]$
8	Dodecahedral	d^2	$WH_4(PMePh_2)_4$
9	TTP[c]	d^0	$[ReH_9]^{2-}$

[a]The n value in d^n cannot exceed $(9 - CN)$.
[b]A distorted version of this geometry occurs (see Section 4.4).
[c]Tricapped trigonal prism.
hs = high spin; ls = low spin.

The definition of coordination number and geometry is not clear-cut for multidentate organometallic ligands, as in Cp_2Fe. Is this molecule 2-coordinate, 6-coordinate, or 10-coordinate? Indeed, there are two ligands, but six electron pairs are involved in M–L bonding, and 10 C atoms are all within bonding distance of the metal. The definition most often seen involves counting the number of lone pairs provided by the ligands on the ionic model, making a CN of six for Cp_2Fe, and we use it in what follows (Eq. 2.15).

The maximum attainable CN is also affected by the d^n configuration. A CN of 9 requires that the ligands L or X^- (we need to use the ionic model here because the d^n concept is rooted in that model) have all nine empty s, p, and d orbitals to occupy, so d^0 is needed (Table 2.7). Eight-coordination requires d^2 or lower and similar arguments apply to the other coordination numbers.

The box below summarizes the different counting rules as applied to our generalized d^n complex $[MX_aL_b]^{c+}$, where N is the group number, CN the coordination number and OS the oxidation state. The e count is usually ≤ 18, and d^n must not stray beyond the limits of d^0 to d^{10}.

$$CN = a + b \leq 9$$

$$e \text{ count} = N + a + 2b - c$$

$$OS = a + c \leq N$$

$$d^n = d^{(N-OS)} = d^{(N-a-c)}$$

d^n Configuration and Geometry

The d^n configuration is a good guide to the expected geometry (Table 2.7 and Fig.2.1), because this is governed by ligand field effects specific to each configuration. The d^0, d^5 (high spin), and d^{10} configurations have

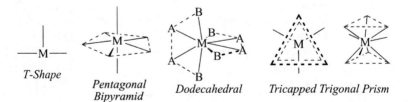

FIGURE 2.1   The T-shape geometry is typically found in d^8, e.g., $[Rh(PPh_3)_3]^+$; the pentagonal bipyramid in d^4, e.g., $[IrH_5(PCy_3)_2]^+$; the dodecahedron in d^2, e.g., $[WH_4(PR_3)_4]$; and the tricapped trigonal prism in d^0, e.g., $[ReH_9]^{2-}$.

the same number of electrons (zero, one, or two, respectively) in each d orbital. This symmetric electron distribution means there are no ligand field effects and the ligand positions are sterically determined. For example, in d^{10} PtL$_4$, we minimize repulsions by arranging the four ligands in a tetrahedral geometry. The high Δ for organometallics makes even d^n configurations strongly preferred. d^8 Ni(II) has a notorious tendency to adopt a variety of coordination geometries even with the same ligand set. For example, Ni(Ph$_2$PCH$_2$CH$_2$SR)$_2$X$_2$ can adopt square planar, tetrahedral, square pyramidal, and octahedral geometries.[17] Higher coordination numbers are associated with lower d^n configurations because on the ionic model, if CN ligand lone pairs are to find empty orbitals to fill, then the value of d^n cannot exceed $(9 - CN)$. For example, only a d^0 ion can accommodate nine ligand lone pairs in ML$_9$ because all nine d orbitals must be empty to accept them.

Fluxionality and Geometry

As we saw in Werner's work, octahedral complexes tend to be very geometrically stable—cis ligands stay cis and trans stay trans. The same holds for square planar complexes. Other common geometries tend to be more fluxional, with the ligands permuting their positions within the coordination sphere. For trigonal bipyramidal cases, for example, a number of isomers can coexist in rapid equilibrium. [RhH(CO)$_2$(PR$_3$)$_2$] has two such forms, for example.[18] Dodecahedral and TTP complexes also show fast fluxionality.

$$
\begin{array}{ccc}
\text{H} & & \text{H} \\
| \quad {}_{\diagup}\text{PR}_3 & & | \quad {}_{\diagup}\text{CO} \\
\text{OC-Rh} & \rightleftharpoons & \text{OC-Rh} \\
| \quad {}^{\diagdown}\text{PR}_3 & & | \quad {}^{\diagdown}\text{PR}_3 \\
\text{CO} & & \text{PR}_3
\end{array}
\qquad (2.18)
$$

Generalizing the 18e Rule

We can now generalize the 18e rule for complexes of any coordination number, n. Figure 2.2 shows the ligand field model for a complex ML$_n$ for $n = 4$–9, where there are n M–L σ-bonding orbitals and $(9 - n)$ nonbonding d orbitals. The value of n appropriate for this situation is the CN defined in Eq. 2.17, so for [Ni(η^3-allyl)$_2$] (**2.9**), $n = 4$; for Cp$_2$Fe (**2.7**) or [Mo(η^6-C$_6$H$_6$)$_2$] (**2.10**), $n = 6$; and for Cp$_2$TiCl$_2$ (**2.11**) or MoH$_4$(PR$_3$)$_4$ (**2.8**), $n = 8$. Filling the bonding and nonbonding levels—a total of nine orbitals—requires 18 electrons, the antibonding orbitals normally being empty. In Fig 2.2, each group of orbitals—bonding, nonbonding, and antibonding—is represented by a thick horizontal

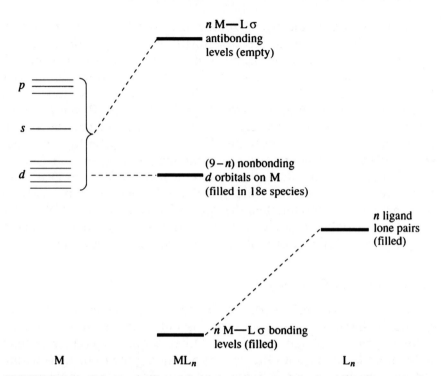

FIGURE 2.2   Schematic ligand field description of the bonding in a complex ML_n ($n = 4$–9), showing how the MOs can be divided into bonding (always filled), nonbonding (filled-in 18e complexes), and antibonding (almost always empty). Each thick horizontal line refers to one of these groups of orbitals.

line, although in reality each group is spread out in a pattern that depends on the exact geometry and ligand set. Figure 1.5 shows the nonbonding orbital pattern for tetrahedral and square planar geometries, for example.

2.6   EFFECTS OF COMPLEXATION

The chemical character of many ligands is profoundly modified on binding to the metal. For the full range of metal fragments L_nM, there is a smooth gradation of metal properties from strongly σ acceptor to strongly π basic. A typical unsaturated ligand Q is depleted of charge and made more electrophilic by a σ-acceptor L_nM fragment in a complex L_nM–Q, but made to accept electrons and therefore become more nucleophilic for a π-basic L_nM fragment. As an example, free benzene is very resistant to attack by nucleophiles, but reacts with electrophiles.

In the complex $(C_6H_6)Cr(CO)_3$, in contrast, the $Cr(CO)_3$ fragment, as a good acceptor by virtue of its three CO ligands, depletes the electron density on the aromatic ring. This makes the bound C_6H_6 susceptible to nucleophilic attack but resistant to electrophilic attack. Both Cp groups and phosphines are strong donors, and so the acetyl **2.21** in Eq. 2.19 behaves as the carbene (see Chapter 11) form **2.21b**. It is subjected to electrophilic attack to give **2.22**. Inversion of the typical reactivity pattern on binding, of which these are examples, is termed *umpolung*.

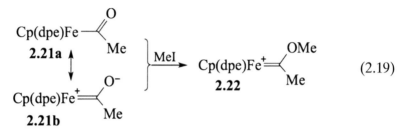

$$(2.19)$$

Ligand Polarization

If L_nM is in the middle range of electronic properties and is both a σ acceptor and a π donor, it might seem that Q in $L_nM–Q$ would differ little from free Q in chemical character. In fact, the ligand can still be strongly activated by polarization. Sigma donation from the ligand to the metal usually depletes the electron density on the ligand donor atom, but π back donation from the metal can raise the electron density on remote atoms. A good example is molecular nitrogen, N_2, where the free ligand is nonpolar and notoriously unreactive. In the N_2 complex $L_nM–N'{\equiv}N''$, σ donation to the metal comes from a lone pair on N'. The back bonding from the metal goes into a π^* orbital belonging to N' and N''. This means that N' tends to become positively charged and N'' negatively charged on binding, a polarization that enhances reactivity, facilitating protonation at N'' and nucleophilic attack at N'.

$$\overset{\partial+}{M}\!—\!\overset{}{N}\!\overset{\partial-}{\equiv\!N}$$

The general situation is summarized in Table 2.8. If a ligand is normally reactive toward, say, nucleophiles, we can deactivate it by binding to a nucleophilic metal. The metal can then act as a protecting group. A ligand that is inert toward nucleophilic attack can be activated by binding to an electrophilic metal. Protection requires a stoichiometric amount of metal to be effective, so has fallen out of favor, while activation needs only a catalytic amount.

TABLE 2.8 How the Electronic Character of a Metal Fragment Changes Reactivity

Character of Free Ligand	Character of ML$_n$ Fragment^a		
	σ Acid	Polarizing	π Base
Susceptible to electrophilic attack	Suppresses susceptibility	May enhance susceptibility	Enhances
Susceptible to nucleophilic attack	Enhances susceptibility	May enhance susceptibility	Suppresses
Unreactive	May allow nu. attack	May allow both nu. and el. attack	May allow el. attack

nu. = nucleophilic; el. = electrophilic.

Paradoxically, stronger M–L binding does not always lead to stronger ligand activation. For example, H_2 (Section 1.11) is most highly acidified on *weak* binding. The pK_a of H_2 is near 35 when free but often lies in the range 0–20 for bound H_2 with the weakly bound ligands being most acidified. This is because the strength of M–H_2 binding largely depends on the degree of π back donation, but stronger back donation reduces the positive charge on H_2 that comes from the σ donation from H_2 to M.

Free ⇄ Bound

Modification of the properties of a ligand, Q, on binding to give L$_n$M–Q is quite general. A knowledge of the behavior of free organic carbenes, dienes, or other species can be misleading in trying to understand their complexes. For example, dienes react with dienophiles in the Diels–Alder reaction, but diene complexes do not give this reaction. In a sense, the complex is already a Diels–Alder adduct, with the metal as the dienophile.

The properties of both the metal ions and the ligands are profoundly altered on complex formation. For example, Co(III) is very strongly oxidizing in simple salts, such as the acetate, which can even oxidize hydrocarbons. Werner's work showed that most of this oxidizing power can be quenched by binding six ammonias to the Co(III) ion. The presence of six strong σ-donor ligands in the resulting $[Co(NH_3)_6]^{3+}$ ion stabilizes the Co(III) state. Conversely, elemental Mo or Fe are strongly reducing, yet $Mo(CO)_6$ and $Fe(CO)_5$, also M(0), are air-stable with only modest reducing properties because CO removes electron density from the metal by back donation, thus strongly stabilizing the M(0) state.

Finally, donor and acceptor are relative terms. In a complex $L_nM–H$, where the hydride ligand bears no strong positive or negative charge, we can consider it as arising from $L_nM^+ + H^-$, $L_nM\bullet + H\bullet$, or $L_nM^- + H^+$. We would have to regard H^- as a strong donor to L_nM^+, H^+ as a strong acceptor from L_nM^-, and $H\bullet$ as being neither one nor the other with respect to $L_nM\bullet$. Normally, the ionic model is assumed, and the first type of dissection is implied, but the assumptions made are often unstated in publications.

Symbiotic and Antisymbiotic Effects

In the symbiotic effect, a hard ligand tends to form ionic M–L bonds in which L retains more negative charge than in a soft ligand case, letting the metal ion keep more of its positive charge and hence attract additional hard ligands, as in the $[M(OH_2)_6]^{2+}$ ions of the first row d block metals. In contrast, binding soft ligands makes the metal softer and hence able to bind other soft ligands, as in $[Fe(CN)_6]^{4-}$.

The antisymbiotic effect, also called transphobia,[19] applies to pairs of high trans effect, soft ligands on a soft metal. Where a choice exists, there is a strong tendency for such ligands to avoid being mutually trans by becoming cis and preferring to have low trans effect, hard ligands trans to themselves. **2.23** illustrates this point: the soft hydrides prefer to be mutually cis and to have hard aqua ligands in trans sites. The sd^n model (Section 1.8) calls for trans sets of soft ligands to be competing for covalent bonding, thus having water in trans sites allows each H to monopolize covalent bonding on its own axis.

2.23

2.7 DIFFERENCES BETWEEN METALS

A change in the metal greatly affects the properties of the resulting complexes. As we move from left to right, the electronegativity of the elements increases substantially. The orbitals in which the electrons are located start out relatively high in energy and fall steadily as we go to the right. This trend is reflected in Table 2.9, which shows the Pauling electronegativities of the d block. The early transition metals are electropositive and so readily lose all their valence electrons. These elements

TABLE 2.9   Pauling Electronegativities of the Transition Elements[a]

Sc	Ti	V	Cr	Mn	Fe	Co	Ni	Cu
1.3	1.5	1.6	1.6	1.6	1.8	1.9	1.9	1.9
Y	Zr	Nb	Mo	Tc	Ru	Rh	Pd	Ag
1.2	1.3	1.6	2.1	1.9	2.2	2.3	2.2	1.9
La	Hf	Ta	W	Re	Os	Ir	Pt	Au
1.1	1.3	1.5	2.3	1.9	2.2	2.2	2.3	2.5

[a]Lanthanides and actinides: 1.1–1.3. The electronegativities of important ligand atoms are H, 2.2; C, 2.5: N, 3.0; O, 3.4; F, 4; Si, 1.9; P, 2.2; S, 2.6; Cl, 3.1; Br, 2.9; I, 2.6. Effective electronegativities of all elements are altered by their substituents, for example, the electronegativities estimated for an alkyl C, a vinyl C, and a propynyl C are 2.5, 2.75, and 3.3, respectively.

are therefore often found in the highest permissible oxidation state, such as d^0 Zr(IV) and Ta(V). Lower oxidation states, such as d^2 Zr(II) and Ta(III), are very easily oxidized because the two d electrons, being in an orbital of relatively high energy, are easily lost either to an oxidizing agent or to the π^* orbitals of an unsaturated ligand via back donation. This often makes d^2 early metal ions air sensitive and very π basic. Ligands, such as CO, C_6H_6, and C_2H_4, that require back bonding for stability, bind only weakly, if at all, to d^0 metals, but strongly to d^2 and higher metals.

Late metals, in contrast, are relatively electronegative and tend to retain their valence electrons. The low oxidation states, such as d^8 Pd(II), tend to be stable, and the higher ones, such as d^6 Pd(IV), are often less so and tend to find ways to return to Pd(II); that is, they are oxidizing. Back donation is not so marked as with the early d^2 metals, and so any unsaturated ligand attached to the weak π-donor Pd(II) tends to accumulate a positive charge. As we see in Section 8.3, this makes the ligand subject to attack by nucleophiles and is the basis for many important applications in organic synthesis.

Real Charge

Trends in real charge at the metal can be estimated for metal carbonyls from the $\nu(CO)$ in the infrared spectrum. A metal with high negative charge is expected to back donate into bound CO and cause a decrease in $\nu(CO)$. Table 2.10 shows how $\nu(CO)$ values vary. The largest lowering (ca. 115 cm^{-1}) is caused by a change of net ionic charge by one unit to more positive values, cations being less π basic. Next comes replacement of non-π-acceptor amine ligands by COs (ca. 45 cm^{-1} lowering per replacement). Having fewer donor ligands causes a significant

TABLE 2.10 Effect of Changing Metal, Net Ionic Charge, and Ligand Set on ν(CO) in the Infrared Spectrum of Metal Carbonyls

Changing Metal across the Periodic Table					
$V(CO)_6$ 1976	$Cr(CO)_6$ 2000	$Mn_2(CO)_{10}$ 2013(av)[a]	$Fe(CO)_5$ 2023(av)[a]	$Co_2(CO)_8$ 2044(av)[b]	$Ni(CO)_4$ 2057
	$Cr(CO)_4$ 1938[c]		$Fe(CO)_4$ 1995[c]		$Ni(CO)_4$ 2057

Changing Metal Down the Periodic Table		
$Cr(CO)_6$ 2000	$Mo(CO)_6$ 2004	$W(CO)_6$ 1998

Changing Ionic Charge in an Isoelectronic Series			
$[Ti(CO)_6]^{2-}$ 1747[d]	$[V(CO)_6]^-$ 1860[d]	$Cr(CO)_6$ 2000	$[Mn(CO)_6]^+$ 2090

Replacing π-Acceptor CO by Non-π-Acceptor Amines			
$[Mn(CO)_6]^+$ 2090	$[(MeH_2N)$ $Mn(CO)_5]^+$ 2043(av)	$[(en)Mn(CO)_4]^+$ 2000(av)	$[(dien)Mn(CO)_3]^+$ 1960

Note: All values in cm^{-1}.
[a]Average of several bands.
[b]Isomer without CO bridges.
[c]Unstable species seen only in low temperature matrix studies.
[d]Band positions probably lowered by counterion binding to CO oxygen.
[e]en = $H_2NCH_2CH_2NH_2$; dien = $HN(CH_2CH_2NH_2)_2$.

reduction of back donation. Change of metal is less significant. Using the tetracarbonyl series for better comparability, this amounts to 30 cm^{-1} rise per unit increase in group number. The later elements being less effective π donors because the increase in nuclear charge stabilizes the d electrons and lowers their basicity. Surprisingly, moving down a group causes little change, perhaps as a result of cancellation of the opposing effects of the more basic but more diffuse d_π orbitals of the heavier metals.

First-row metals have lower M–L bond strengths and crystal field splittings compared with their second- and third-row analogs. They are also more likely to undergo 1e redox changes rather than the 2e changes often associated with the second and third rows. Finally, the first-row metals do not attain high oxidation states so easily as the second and specially the third row. Mn(V), (VI), and (VII) (e.g., MnO_4^-) are rare and usually highly oxidizing; Re(V) and (VII) are not unusual and the complexes are not strongly oxidizing (e.g., ReO_4^-).

- Complexes are classified by d^n configuration (Eq. 2.15), e-count (Eq. 2.7), and coordination number (Section 2.5); d^6, 18e octahedral complexes are most common.
- Specific d^n configurations are associated with specific geometries (Table 2.7).
- Complexation profoundly alters ligand properties and can even invert normal reactivity patterns seen in the free organic ligands (Section 2.6).
- Steric stabilization of reactive species is a standard strategy in organometallic chemistry.

REFERENCES

1. G. Parkin, Classification of organotransition metal compounds, in *Comprehensive Organometallic Chemistry III*, eds. D. M. P. Mingos and R. H. Crabtree, Elsevier, Amsterdam, 2007, Vol. 1.
2. M. L. H. Green and G. Parkin, *J. Chem. Educ.*, 2014, in press.
3. *Nomenclature of Inorganic Chemistry IUPAC Recommendations* 2005, eds. N. G. Connelly et al., RSC Publishing, Cambridge, UK.
4. H. Kameo, Y. Hashimoto, and H. Nakazawa, *Organometallics*, **31**, 4251, 2012.
5. J. S. Anderson, M. -E. Moret, and J. C. Peters, *J. Am. Chem. Soc.*, **135**, 534, 2013.
6. S. R. Daly, P.M.B. Piccoli, A. J. Schultz, T. K. Todorova, L. Gagliardi, and G. S. Girolami, *Angew. Chem. Int. Ed.*, **49**, 3379, 2010.
7. J. D. Smith, T. P. Hanusa, V. G. Young Jr., *J. Amer. Chem. Soc.*, **123**, 6455, 2001.
8. M. D. Walter, P. S. White, and M. Brookhart, *New J. Chem.*, **37**, 1128, 2013.
9. Y. Gong, M. Zhou, M. Kaupp, and S. Riedel, *Angew. Chem. Int. Ed.*, **48**, 7879, 2009.
10. D. C. Powers and T. Ritter, *Nature Chem.*, **1**, 302, 2009.
11. L. M. Mirica and J. R. Khusnutdinova, *Coord. Chem. Rev.*, **257**, 299, 2013.
12. F. Weinhold and C. Landis, *Valency and Bonding*, Cambridge University Press, New York, 2005.
13. A. G. Algarra, V. V. Grushin, and S. A. Macgregor, *Organometallics*, **31**, 1467, 2012.
14. Y. Yan, P. Chandrasekaran, J. T. Mague, S. DeBeer, S. Sproules, and J. P. Donahue, *Inorg. Chem.*, **51**, 346, 2012.
15. O. R. Luca and R. H. Crabtree, *Chem. Soc. Rev.*, **42**, 1440, 2013; W. Kaim, *Eur. J. Inorg. Chem.*, **2012**, 343.

16. B. G. Hashiguchi, K. J. H. Young, M. Yousufuddin, W. A. Goddard III, and R. A. Periana, *J. Am. Chem. Soc.*, **132**, 12542, 2010.

17. C. W. Machan, A. M. Spokoyny, M. R. Jones, A. A. Sarjeant, C. L. Stern, and C. A. Mirkin, *J. Am. Chem. Soc.*, **133**, 3023, 2011.

18. P. Cheliatsidou, D. F. S. White, B. de Bruin, J. N. H. Reek, and D. J. Cole-Hamilton, *Organometallics*, **26**, 3265, 2007.

19. A.-J. Martínez-Martínez, J. Vicente, and M.-T. Chicote, *Organometallics*, **31**, 2697, 2012.

PROBLEMS

A Note on Answering Problems

It is important that any intermediate you suggest in an organometallic reaction be reasonable. Does it have an appropriate electron count, coordination number, and oxidation state? If it is the only known Rh(V) carbonyl, it may be open to criticism. Check that the organic fragment is also reasonable. Sometimes, students write diagrams without stopping to consider that their structure contains five-valent carbon. Indicate the hapticity of each ligand. See also p. 473 for further points.

2.1. Give the electron counts, formal oxidation states, and d^n configurations of the following: $[Pt(NH_3)_4]^{2+}$, $PtCl_2(NH_3)_2$, $[PtCl_4]^{2-}$, $(\eta^5\text{-}C_5H_5)_2Ni$, $[(R_3P)_3Ru(\mu\text{-}Cl)_3Ru(PR_3)_3]^+$, $[ReH_9]^{2-}$, $CpIrMe_4$, $TaMe_5$, $(\eta^5\text{-}C_5H_5)_2TiCl_2$, and $MeRe(O)_3$.

2.2. A complex is found to correspond to the empirical formula $(CO)_3ReCl$. How could it attain the 18e configuration without requiring any additional ligands?

2.3. How could a complex of empirical formula $Cr(CO)_3(C_6H_5)_2$ attain the 18e configuration?

2.4. A complex $Ti(\eta^2\text{-}MeN=CH-CH=NMe)_2$ is found to be chelated via nitrogen. What oxidation state should we assign to Ti? Is any alternative assignment possible?

2.5. Count the valence electrons in the complexes shown in Problem 2.1, but using a different model (ionic or covalent) from the one you used originally.

2.6. Given the existence of $(CO)_5Mn-Mn(CO)_5$, deduce the electron counting rule that applies to M–M bonds. Verify that the same holds for $Os_3(CO)_{12}$, which contains three Os–Os bonds and only terminal CO groups. What structure do you think is most likely for $Rh_4(CO)_{12}$?

2.7. Show how the valence electron count for the carbon atom in $[CH_3NH_3]^+$ can be evaluated considering the molecule as an ammonia complex. Can the methylene carbon in $CH_2=C=O$ be treated in a similar way?

2.8. Water has two lone pairs. Decide whether both or only one of these should normally be counted, given that the following typical complexes exist: $[IrH_2(PPh_3)_2(OH_2)_2]^+$, $[Os(\eta^6\text{-}C_6H_6)(OH_2)_3]^{2+}$.

2.9. Acetone can bind in an η^2 (via C and O) and an η^1 fashion (via O). Would you expect the electron count to be the same or different in the two forms? What kind of metal fragments would you expect would be most likely to bind acetone as (a) an η^1 and (b) an η^2 ligand? Would either binding mode be expected to enhance the tendency of the carbonyl carbon to undergo nucleophilic attack?

2.10. Predict the hapticity of each Cp ring in $Cp_2W(CO)_2$, and of each "triphos" in $[Pd\{(PPh_2CH_2CH_2)_3CPh\}_2]^{2+}$.

2.11. Assign the oxidation states, d^n configurations, and electron counts for the two species shown below, which are in equilibrium in solution. Use both the covalent and ionic models.

$$W(\eta^2\text{-}H_2)(CO)_3(PR_3)_2 \leftrightarrows W(H)_2(CO)_3(PR_3)_2$$

3

ALKYLS AND HYDRIDES

Alkyls and hydrides are among the simplest organometallic species, yet transition metal alkyls remained very rare until the principles governing their stability were understood in the 1960s and 1970s. These principles make a useful starting point for our study of alkyls because they introduce some of the most important organometallic reactions that we go on to study in detail in later chapters. After alkyls, we move to hydrides and then to dihydrogen complexes, all areas with important implications for later discussions.

3.1 ALKYLS AND ARYLS

The story begins with the main-group elements when, in 1757, Louis Cadet (1731–1799) made the appallingly evil-smelling cacodyl oxide (Greek: *kakos* = bad), later shown by Robert Bunsen (1811–1899) to be $Me_2As–O–AsMe_2$. Because arsenic is a semimetal, true metal alkyls only came to light in later work by Edward Frankland (1825–1899), now considered a founder of organometallic chemistry. In an 1848 attempt to prepare free ethyl radicals by reaction of ethyl iodide with metallic zinc, he instead made a colorless liquid that proved to be diethylzinc. When Frankland added water, a greenish-blue flame several feet long

The Organometallic Chemistry of the Transition Metals, Sixth Edition.
Robert H. Crabtree.
© 2014 John Wiley & Sons, Inc. Published 2014 by John Wiley & Sons, Inc.

shot out of the sample tube.[1] It was only with the 1900 discovery by Victor Grignard (1871–1935) of the alkylmagnesium halide reagents, RMgX, that organometallic chemistry began to make a major impact through its application to organic synthesis. He later won the Nobel Prize for this, his doctoral research. The development of organolithium reagents from 1914 is associated with Wilhelm Schlenk (1879–1943) and from 1930 with Karl Ziegler (1898–1973). Ziegler also played a key role in showing the broad utility of organoaluminum reagents; today, these see service in many commercial processes but are not common laboratory reagents.

Metal Alkyls as Stabilized Carbanions

Grignard reagents, RMgX, provided the first general source of nucleophilic alkyl groups, $R^{\delta-}$, to complement the electrophilic alkyl groups, $R^{\delta+}$, long available from the alkyl halides.[2] On the ionic model, metal alkyls result from combining an alkyl anion with a metal cation. In doing so, the alkyl anion is stabilized to a different extent depending on the electronegativity of the metal concerned. Alkyls of the electropositive elements of groups 1–2, as well as Al and Zn, are sometimes called *polar organometallics*, because the alkyl anion is only weakly stabilized and retains much of the strongly nucleophilic and basic character of the free anion. Polar alkyls all react with traces of humidity to hydrolyze the M–C bond to form M–OH and release RH. Air oxidation also occurs very readily, and so polar organometallics must be protected from both air and water. Alkyls of the early transition metals, such as Ti or Zr, can also be very air and water sensitive, but as we move to less electropositive metals (see Table 2.9) by moving "southeast" in the periodic table, the compounds become much less reactive, until we reach Hg, where the Hg–C bond is so stable that $[Me–Hg]^+$ cation is indefinitely stable even in hot sulfuric acid. As we go from the essentially ionic and purely basic $NaCH_3$ via the highly polar covalent Li and Mg alkyls to the covalent late metal alkyls, the nucleophilic reactivity falls steadily along the series, showing the powerful effect of changing metal (Fig. 3.1).

The stability of the R fragment plays a role, too—as an sp^3 ion, CH_3^- is intrinsically the most reactive. Moving to sp^2 $C_6H_5^-$ and particularly to sp $RC\equiv C^-$, the carbon lone pair becomes progressively more stabilized from its increasing s character and the intrinsic reactivity falls off. The same trend governs the increase in acidity as we go from CH_4 ($pK_a = \sim 50$) to C_6H_6 ($pK_a = \sim 43$) and to $RC\equiv CH$ ($pK_a = \sim 25$), making $RC\equiv C^-$ the most stable and the least reactive anion.

Following the successful syntheses of main-group alkyls, many attempts were made to prepare transition metal alkyls. Pope and Peachey's Me_3PtI, dating from 1909, was an early but isolated example

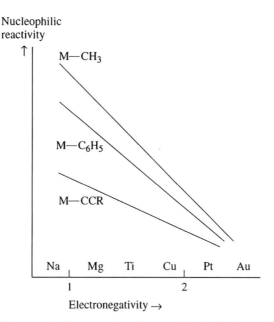

FIGURE 3.1   Schematic diagram showing qualitatively how the nucleophilic reactivity of main-group and transition metal alkyls to protons or air oxidation depends on the alkyl itself and the electronegativity of the metal. Adapted from Reference 2.

of a stable *d*-block metal alkyl. Attempts during the 1920s through 1940s to make further examples of such alkyls all failed. This was specially puzzling because by then almost every nontransition element had been shown to form stable alkyls. These early failures discouraged further work and led to the view that transition metal–carbon bonds must be abnormally weak. In fact, we now know that such M–C bonds are strong—bond strengths of 30–65 cal/mol are typical. It is the existence of several easy decomposition pathways that makes many transition metal alkyls kinetically unstable. Kinetics, not thermodynamics, was thus to blame for the synthetic failures. This is fortunate because it is much easier to manipulate the L_nM–R system to block decomposition pathways than it is to increase the bond strength. In order to be able to design stable alkyls, we must look at some of these pathways to see how they can be inhibited.

We always have to bear in mind that some of our present ideas may also be wrong. As a corrective to the textbook tendency only to teach those concepts that have survived prolonged scrutiny and omit discussion of historical developments, two authors have collected examples of once firmly held ideas in science that later proved to be wrong.[3]

β Elimination

The major decomposition pathway for alkyls, *β elimination* (Eq. 3.1), converts a metal alkyl into a hydridometal alkene complex, a step often followed by loss of the alkene. β Elimination, to be studied in detail in Section 7.5, can occur when all of the following conditions apply:

1. The β carbon of the alkyl must have a hydrogen as a substituent.
2. The M–C–C–H unit must be able to take up a roughly *syn*-coplanar conformation, which brings the β hydrogen close to the metal.
3. A vacant site on the metal cis to the alkyl, symbolized here as □, must be easily accessible, such as by loss of a labile ligand from an 18e complex.

These requirements apply because it is the β hydrogen of the alkyl that is transferred as H⁻ to a cis vacancy on the metal to give the M–H bond of the product and to form a cis-M(H)(alkene) intermediate or product. The geometry of the transfer requires a *syn*-coplanar M–C–C–H arrangement. The elimination is believed to be concerted with simultaneous C-H bond breaking and M–C and M–H bond making. The reaction is much more rapid for d^2 and higher metals than for d^0 and main-group alkyls, probably because back donation, only possible in d^2 and higher metals, promotes formation of an intermediate with a sigma bonded C–H (Section 1.11) that leads to C–H bond weakening by "back donation" into the C–H σ* orbital, as well as stabilizing the incipient alkene in the transition state.

$$\underset{\square}{\overset{H_2C-CH_2}{L_nM}}\underset{H}{\overset{}{}} \xrightarrow{\beta \text{ elimination}} \underset{H}{\overset{H_2C}{L_nM}}\overset{}{CH_2} \longrightarrow L_nM-H \ + \ H_2C{=}CH_2$$

$$\square = \text{vacancy}$$

$$(3.1)$$

The requirement for a "vacant site" is not steric but rather reflects the need for an empty orbital to accept the pair of electrons of the migrating hydride from the β-C–H bond. The electron count of $L_nM(H)$(alkene) is 2e greater than that of the starting L_nM-R. An 18e alkyl is much more reluctant to β-eliminate via a 20e intermediate than is a 16e alkyl, which can go to an 18e alkene hydride. Even if the alkene subsequently dissociates, as is common, we still have to stabilize the transition state leading to the alkene hydride intermediate for the reaction to be fast. An 18e alkyl is *coordinatively saturated* and an empty orbital is not available. Some 18e alkyls do β-eliminate, but a 2e ligand often dissociates first. Main-group alkyls can also β-eliminate (e.g., Eq. 3.2), but much more slowly.

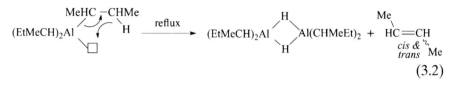

$$(3.2)$$

Stable Alkyls

The 1973 Wilkinson synthesis of WMe_6 was a big surprise. From it, we learned that an alkyl with no β hydrogens can be kinetically stable even in a complex with only 12 valence electrons. The average W–C bond strength, derived from the heat of hydrolysis, is a substantial 39 kcal/mol, disposing of the "weak M–C bond" hypothesis.[4]

To have a kinetically stable alkyl, we must therefore block the β-elimination decomposition pathway. This can happen for:

1. Alkyls that lack a β hydrogen:

WMe_6, $Ti(CH_2Ph)_4$, $W(CH_2SiMe_3)_6$,

$TaCl_2(CH_2CMe_3)_3$, $C_2F_5Mn(CO)_5$, $LAuCF_2CF_2CH_3$,

$Pt(C \equiv CCF_3)_2L_2$, $Pt(CH_2COMe)Cl(NH_3)_2$, $(CO)_4(PR_3)ReCH_2OMe$

2. Alkyls in which the β hydrogen is not easily available:
   (a) Because the M–C–C–H unit cannot become *syn*-coplanar

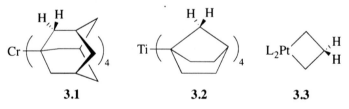

| 3.1 | 3.2 | 3.3 |

   (b) Because excessive steric bulk blocks approach of the β-H to the metal

$$Cr(CMe_3)_4, Cr(CHMe_2)_4$$

   (c) Because the β-elimination product would be unstable;

$$LPt(C \equiv CH)_2L_2, PdPh_2L_2$$

   (d) Because the 120° sp^2 angles at carbon result in a long M···β-H distance

$$CpL_3MoCH = CHCMe_3$$

3. 18e alkyls with firmly bound ligands (no suitable vacant site):

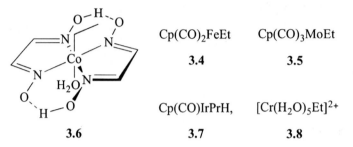

Cp(CO)$_2$FeEt	Cp(CO)$_3$MoEt
3.4	**3.5**

3.6

Cp(CO)IrPrH,	[Cr(H$_2$O)$_5$Et]$^{2+}$
3.7	**3.8**

4. Some d^0 alkyls:

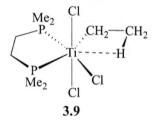

3.9

If they were to β-eliminate, **3.1** and **3.2** would give "forbidden" anti-Bredt bridgehead olefins because the C=C bond would then be twisted. The ring in **3.3** would have to fold strongly to bring the β C-H to the metal, so although not strictly forbidden, β elimination would be hard. When the aqua ligand dissociates in **3.6**, the vacancy is trans and not cis.

Further study of WMe$_6$ led to another big surprise. As predicted by Albright and Eisenstein for all d^0 MX$_6$ species (but only for X ≠ π donor), WMe$_6$ proved to have the rare 12e trigonal prismatic structure **3.10**, not the octahedral structure usually found for six-coordinate complexes. The trigonal prism also corresponds to the sd^5 hybridization that is expected on the sd^n model of Section 1.8.[5]

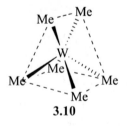

3.10

Agostic Alkyls

In some complexes, most of the criteria are favorable, but β elimination still does not occur. In **3.7**, the β-C–H bond is approaching the transition state for β elimination, but the reaction has been arrested along the way because in this d^0 system, there are no d electrons available for donation into the C–H σ* orbital, a process that would lead to completion of the C–H bond cleavage. These *agostic* alkyls can be detected by X-ray or neutron crystal structural work and by the high-field shift of the agostic H in the proton NMR. The lowering of the $J(C,H)$ and $\nu(CH)$ in the NMR and IR spectra on binding is symptomatic of the reduced C–H bond order in the agostic system.[6] Agostic C–H bonds are often seen in coordinatively unsaturated species from structural work and in transition states by theory.

We earlier saw the need for a 2e vacant site (an empty d orbital) for β elimination. Now we see that we also need an available electron pair (a filled d orbital) for breaking the C–H bond by back donation into the CH σ*. There is a very close analogy between these requirements and those for binding a soft ligand, such as CO. Both processes require a metal that is both σ acidic and π basic. In the case of CO, such π back bonding leads to a reduction in the CO bond order. In the case of the β-C–H bond of an alkyl group, this π back bonding can reduce the C-H bond order to zero, by cleavage to give the alkene hydride complex. Alternatively, if the metal is a good σ acid but a poor π base, an agostic system may be the result, and the C–H bond is only weakened, not broken. Many of the characteristic reactions of organometallic chemistry require both σ-acid and π-base bifunctional character. This is one reason why transition metals, with their *partly* filled d orbitals, give these reactions.

Theoretical work[7] has shown that σ complexation drives the β-agostic interaction (**3.11**), via electron donation of C–H bonding electrons to the metal. In α-agostic structures (**3.12**), however, this contribution is minor, particularly for early metals where α-agostic structures are the most common. Instead, the alkyl group rotates to maximize its interaction with the whole L_nM fragment. We can therefore define an agostic alkyl without implying concomitant σ complexation as an alkyl that shows a distortion that brings a C–H bond closer to the metal than normally expected.

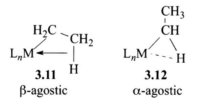

3.11	**3.12**
β-agostic	α-agostic

β Elimination of Other Groups

Only the lanthanides, actinides, and other early transition metals have very high M–F bond strengths that favor β fluoride elimination in fluoroalkyls. The late transition metals with their weak M–F bonds (Fig. 1.10) form stable fluoroalkyls; not only do these ligands lack β-H groups, but their M–C bond strengths are high because their α C–F σ^* orbitals can act as a π acceptors from late metals (Section 4.2 shows how the same applies for M–PF$_3$). The C$_6$F$_5$ group forms very strong M–C bonds, again with the late transition metals, the o-F substituents being dominant.[8a] (porphyrin)RhCH$_2$CH$_2$OH eliminates C$_2$H$_4$ to give the highly reactive (porphyrin)RhOH.[8b]

Reductive Elimination

A second very common decomposition pathway for metal alkyls is *reductive elimination* ("RE" in Eq. 3.3: X = H, Ph . . .), a reaction that we study in detail in Chapter 6.[9] It is the decrease by two units in the formal oxidation state makes it reductive. In principle, it is available to all complexes, even if they are d^0 or 18e, provided a stable oxidation state exists two units more reduced than the oxidation state in the starting alkyl. In many instances, reductive elimination is not seen. For example, if X in **3.13** is a halogen, **3.13** is usually so stable thermodynamically that the equilibrium of Eq. 3.3 lies well over toward **3.13**.

$$
\begin{array}{ccc}
 & \text{RE} & \\
\text{L}_n\text{M(Me)X} & \longrightarrow & \text{L}_n\text{M} \;+\; \text{MeX} \\
\text{18e} & & \text{16e} \\
\textbf{3.13} & & \textbf{3.14}
\end{array}
\qquad (3.3)
$$

On the other hand, when X = H, the reaction is usually both kinetically and thermodynamically favorable, so cis alkyl hydride complexes usually decompose by RE. Where X = CH$_3$, the thermodynamics still favor elimination, but the reaction is generally much slower kinetically. Reactions involving H are often much faster than those involving any other group; H has no repulsive lone pairs or substituents and its $1s$ orbital can make or break bonds in any direction in the transition state. The sp^3 orbital of the CH$_3$ fragment is directed toward the metal, and so there can often be poorer orbital overlap in the transition state.

Kinetic Stability from Bulky Substituents

Introduction of bulky ligands is a general way to stabilize organometallic complexes. They slow associative decomposition pathways, including reaction with the solvent or with another molecule of the complex, that are specially important for 16e metals. For example, square planar Ni(II) alkyls are vulnerable to attack along the z direction perpendicular to the square plane. The di-o-tolyl complex **3.15**, in which this approach is blocked, is more kinetically stable than the analogous diphenyl, **3.16**. Bulky ligands, such as pentamethylcyclopentadienyl (η^5-C_5Me_5), neopentyl (CH_2CMe_3), or trimethylsilylmethyl (CH_2SiMe_3), therefore find many uses.

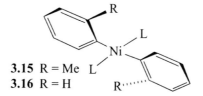

3.15 R = Me
3.16 R = H

Where β elimination cannot occur, α elimination sometimes takes over. This leads to the formation of metal *carbene* complexes with M=C double bonds. For example, the first step in the thermal decomposition of $Ti(CH_2CMe_3)_4$ is known to be α elimination to $Ti(=CHCMe_3)$-$(CH_2CMe_3)_2$. Similarly, attempts to prepare $Ta(CH_2CMe_3)_5$ led to formation of the carbene complex, t-$BuCH=Ta(CH_2CMe_3)_3$. Carbenes and α elimination are discussed in Sections 11.1 and 7.5.

With an N or O heteroatom to activate adjacent C–H bonds, double C–H bond cleavage can occur at the same carbon.[10] In Eq. 3.4, the first cleavage, an oxidative addition, and the second, an α elimination, can be observed stepwise, and even though there is a choice between α elimination and β elimination in the second step, the product still comes exclusively from α elimination.

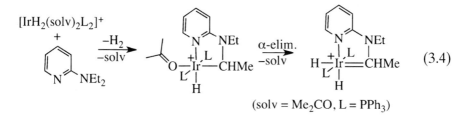

$$(3.4)$$

$$(solv = Me_2CO, L = PPh_3)$$

Preparation of Metal Alkyls

The main types of syntheses of alkyls and aryls are shown in Eq. 3.6 – 3.16.

1. From an R^- reagent (nucleophilic attack on the metal):

$$WCl_6 \xrightarrow{\ LiMe\ } WMe_6 + LiCl \qquad (3.5)$$

$$NbCl_5 \xrightarrow{\ ZnMe_2\ } NbMe_2Cl_3 + ZnCl_2 \qquad (3.6)$$

2. From an R^+ reagent (electrophilic attack on the metal):

$$Mn(CO)_5^- \xrightarrow{MeI} MeMn(CO)_5 + I^- \tag{3.7}$$

$$CpFe(CO)_2^- \xrightarrow{Ph_2I^+} CpFe(CO)_2Ph + PhI \tag{3.8}$$

$$Mn(CO)_5^- \xrightarrow{CF_3COCl} CF_3COMn(CO)_5 \xrightarrow{-CO} CF_3Mn(CO)_5 \tag{3.9}$$

3. By oxidative addition ($L = PPh_3$):

$$Ir(CO)ClL_2 \xrightarrow{MeI} MeIrICl(CO)L_2 \tag{3.10}$$

$$PtL_4 \xrightarrow{MeI} MePtIL_2 \quad (L = PPh_3) \tag{3.11}$$

$$Cr(OH_2)_6^{2+} \xrightarrow{MeI} MeCr(OH_2)_5^{2+} + ICr(OH_2)_5^{2+} \tag{3.12}$$

4. By insertion:

$$PtHClL_2 \xrightarrow{C_2H_4} PtEtClL_2 \quad (L = PEt_3) \tag{3.13}$$

$$Cp(CO)_3MoH \xrightarrow{CH_2N_2} Cp(CO)_3MoCH_3 \tag{3.14}$$

5. By cyclometalation:

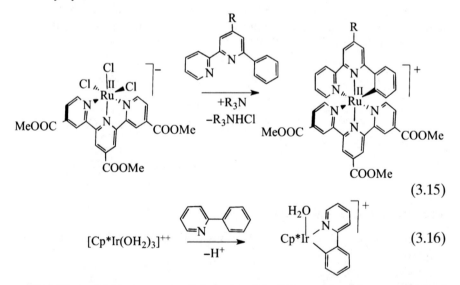

$$\tag{3.15}$$

$$\tag{3.16}$$

RMgX or RLi can react with a metal halide to give the metal alkyl via nucleophilic attack. Eq 3.5-6 show *transmetalation*, the transfer of alkyl groups between metals. Alternatively, a sufficiently nucleophilic metal can undergo electrophilic attack (Eq, 3.7–Eq. 3.9). Eq. 3.9 shows how acyl complexes can often lose CO (Section 7.2). This is particularly convenient in this case because CF_3^+ reagents are not readily available; CF_3I, for example, has a δ^- CF_3 group and a δ^+ I.

Cyclometalation

Cyclometalation (Eq. 3.15 and Eq. 3.16) relies on chelation to form metal aryls.[11] In Eq. 3.15, a phenyl-substituted dipyridyl ligand replaces three chlorides of the Ru terpyridine complex. The phenyl group of the dipyridyl binds to the metal via an agostic C–H . . . Ru bridge, the proton of which is then removed by the amine to give the NNC pincer ligand in the product. An X_3 ligand set is thus replaced by an L_2X ligand, leading to a change in the ionic charge of the complex from 1– to 1+. The product is a good sensitizer for a Grätzel solar cell, where its role is absorption of solar photons resulting in electron injection into a semiconductor electrode.[12]

Rollover cyclometalation can cause ambiguity in the mode of ligand binding. For example, 2,2′-dipyridyls, expected to be N,N′-donors, can be N,C′ donors, as in **3.17**, where the resulting NH is hard to distinguish crystallographically from the expected CH of the usual N,N′-form.[13]

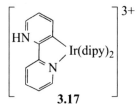

3.17

Oxidative Addition

In *oxidative addition* (OA), an important general method of making alkyls and aryls, the insertion of a metal fragment L_nM into a single R–X bond that is broken to form $L_nM(R)(X)$. X can be any one of a large number of groups, including those shown in Eq. 3.17. OA is simply the reverse of the reductive elimination we saw in Section 3.1.

$$
\begin{array}{ccc}
\begin{array}{c} X \\ | \\ Y \end{array} \quad + \quad ML_n & \longrightarrow & \begin{array}{c} X \\ \diagdown \\ Y \diagup ML_n \end{array} \\
\begin{array}{c} \text{O.S.} = 0 \\ 16e \\ \text{C.N.} = n \end{array} & & \begin{array}{c} \text{O.S.} = 2 \\ 18e \\ \text{C.N.} = n + 2 \end{array}
\end{array}
\tag{3.17}
$$

(XY = H_2, R_3C —H, Cl —H, RCO —Cl, Cl —Cl, Me —I, R_3Si —H)

The oxidation state, the coordination number, and the electron count all rise by two units in OA. This means that a metal species L_nM of oxidation state x can normally give OA only if it also has a stable OS of $(x + 2)$, can tolerate an increase in its coordination number by 2, and can accept two more valence electrons. To fulfill these requirements, the complex L_nM must be d^2 or higher and have 16 or fewer electrons. An 18e $L_nM–L'$ is still viable, provided L' is lost first. The binuclear OA of Eq. 3.12 is a variant appropriate for first row metals that prefer to change their OA, coordination number, and electron count by one unit rather than two. Equation 3.18 shows an organic process analogous to OA given by a divalent carbon species, methylene.

$$\begin{array}{ccc}
\begin{array}{c} X \\ | \\ Y \end{array} \; + \; :CH_2 & \longrightarrow & \begin{array}{c} X \\ \diagdown \\ Y \diagup \end{array}\!\!CH_2 \\
\\
6e & & 8e \\
C.N. = 2 & & C.N. = 4
\end{array} \qquad (3.18)$$

$$(XY = R_3C-H, \; R_3Si-H, \; RCO_2-H, \; RO-H)$$

Insertion

Insertion, to be studied in detail in Chapter 7, is particularly important because it allows us to form a metal alkyl from an alkene and a metal hydride. We see in Chapter 9 how this sequence occurs in an extensive series of catalytic transformations of alkenes, such as hydrogenation with H_2 to give alkanes, hydroformylation with H_2 and CO to give aldehydes, and hydrocyanation with HCN to give nitriles. Such catalytic reactions are among the most important applications of organometallic chemistry.

Olefin insertion is the reverse of the β-elimination reaction of Section 3.1. Since we insisted earlier on the kinetic instability of alkyls having β-H substituents, it might seem inconsistent that we can make alkyls of this type in this way. In practice, it is not unusual to find that only a small equilibrium concentration of the alkyl may be formed in such an insertion. This is enough to enable a catalytic reaction to proceed if the alkyl is rapidly trapped in a subsequent step. For example, in catalytic hydrogenation, the alkyl is trapped by reductive elimination with a second hydride to give the product alkane (e.g., Fig. 9.3). On the other hand, the fluoroalkyl formed from a fluoroalkene is very stable thermodynamically, accounting for the reversibility of C_2H_4 insertion versus the irreversible insertion of C_2F_4 in Eq. 3.19, driven by the high M–C bond strength in fluoroalkyls discussed in Section 3.1.

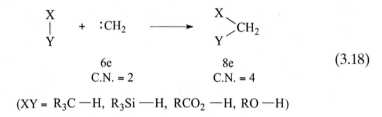

$$(3.19)$$

A vacant site opens up on the metal in an insertion because a 3e ligand set becomes a 1e alkyl, so another way to trap the alkylmetal complex is to fill this vacancy with an external ligand such as a phosphine (Eq. 3.20) or CO.

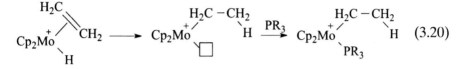

Another route to alkyls is the attack of a nucleophile on a metal alkene complex, discussed in Section 8.3. This route is more common for the synthesis of metal vinyls from alkyne complexes; vinyls are also formed from alkyne insertion into M-H bonds:

$$L_nM-H \xrightarrow{\ RC \equiv CR\ } L_nM-C\begin{matrix}R\\ \diagdown\\ CR\\ | \\ H \end{matrix} \qquad (3.21)$$

$$L_nM = P(CH_2CH_2PPh_2)_2(H_2)$$

Bridging Alkyls

The methyl group, normally terminal (i.e., M–Me), can sometimes bridge two metals in one of three ways. (i) In main-group and d^0 cases, such as Al_2Me_6 (**3.18a**), each sp^3 carbon lone pair of the CH_3^- group is shared between the two metals in a 2e,3-center bond (**3.18b**) resembling the situation for M(μ-H)M (Section 2.1). (ii) In transition metals capable of back bonding (d^2–d^{10}), the methyl group is usually a terminal ligand to one metal with a CH bond of the methyl acting as a σ complex to the second metal (**3.19**). (iii) In rare cases, a planar methyl group bridges two d^0 metals (**3.20**).

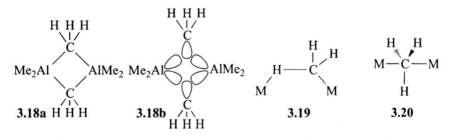

Carbenes such as CH_2 can either form single M-C bonds to each of two metals (**3.21**) or act as a terminal ligand, $M=CH_2$. The latter type has a particularly important chemistry, being the key actors in alkene metathesis, a catalytic reaction of growing importance (Chapter 12). Similarly, CH can bridge three (**3.22**) or two metals (**3.23**) or form a terminal carbyne (**3.24**). A carbon atom can bridge four different metals (**3.25**) but is more often an *interstitial* atom in a cluster (e.g., **3.26**; see Chapter 13), bound to as many as six metals.

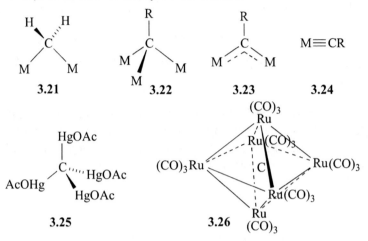

3.21   **3.22**   **3.23**   **3.24**

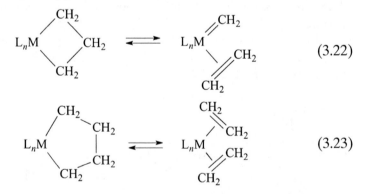

3.25   **3.26**

Metalacycles

Metallacycles, cyclic alkyls $L_nM(CH_2)_n$, are associated with two reversible reactions. In Eq. 3.22, an $n = 3$ *metallacyclobutane* rearranges to an alkene and a carbene, a key step in the important alkene metathesis reaction, and in Eq. 3.23, an $n = 4$ *metallacyclopentane* rearranges to give two alkenes. Metallacycle applications are discussed in Sections 6.8 and 12.1.

$$ \text{(3.22)} $$

$$ \text{(3.23)} $$

In Eq. 3.22 and Eq. 23, a 2e ligand converts to a 4e ligand set, so starting with a 16e metal fragment is needed to permit the rearrangement. In each case, all the β-C–H bonds are held away from the metal and thus protected from β elimination, but the β-C–C bonds are more available, being syn coplanar. Both reactions show some resemblance to a β elimination, but involving a C–C rather than a C–H bond.

In Eq. 3.24, the OA of a C–C bond is driven by strain[14] and the product is stable even though only 14e because the C–C cleavage of Eq. 3.24 would produce unstable benzynes and ligand association is disfavored by the high trans effect C-donors.

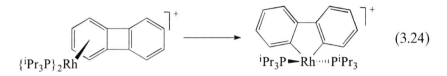

$$\{^iPr_3P\}_2Rh \longrightarrow {}^iPr_3P\!\!-\!\!Rh\!\cdots\!P^iPr_3 \tag{3.24}$$

Lacking any distinctive spectroscopy, the structure would be hard to determine without X-ray crystallography (Fig. 3.2).

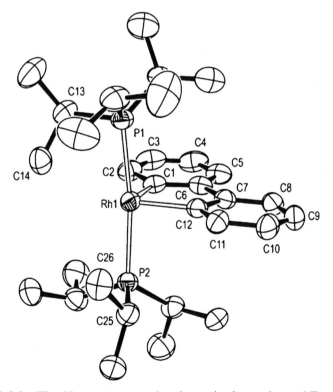

FIGURE 3.2 The X-ray structure for the cationic product of Eq. 3.24 with 50% probability ellipsoids for each atom. Hydrogens, poorly located by X-rays, are omitted. The fluoroaryl counterion, BAr_4^F (not shown), strongly resists coordination to the 14e Rh. Source: From Chaplin et al., 2010 [14]. Reproduced with permission of the American Chemical Society.

η^1 and η^2-Acyl and Vinyls

Acyls and vinyls are commonly monodentate le ligands, $L_nM- (CR=O)$ or $L_nM-(CR=CH_2)$. In a 16e complex, their $C=X$ π bonds may also bind to give 3e η^2-acyls (**3.27**, Eq. 3.25) or vinyls (**3.28**, Eq. 3.26). Eq. 3.26 also shows how cis and trans η^1-vinyls may interconvert via an η^2-form.[15]

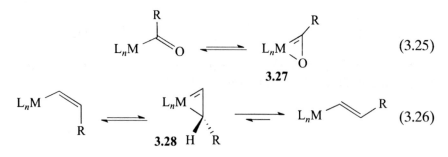

$$\text{(3.25)}$$

$$\text{(3.26)}$$

3.2 OTHER σ-BONDED LIGANDS

Group 14 Elements

Metal silyls, L_nM-SiR_3, typically have R = alkyl, aryl, halide, or OH. In $L_nM-SiMe_3$, β elimination is inhibited by the instability of $Si=C$ bonds and steric congestion is relieved by the long M–Si bond associated with the large Si atom. Being both a strong σ donor and π acceptor (Section 4.2), the $SnCl_3$ group has such a high M–X bond-strength that it can even persuade Pt(II), normally only seen in 16e, square planar complexes, to become 18e in trigonal bipyramidal $[Pt(SnCl_3)_3(cod)]^-$.

Groups 15–17

Examples involving the hard ligands, $-NMe_2$, $-OR$, and F, are $[Mo(NMe_2)_4]$, $[W(NMe_2)_6]$, $[(PhO)_3Mo\equiv Mo(OPh)_3]$, $Zr(OtBu)_4$, and Cp_2TiF_2. In an 18e, late metal d^6 or higher d^n complex without empty d_π orbitals, the heteroatom lone pairs only weaken the M–X bond by repulsion (**3.29**; see also Fig. 1.10). In an early d^0 to d^4 metal with <18e, the empty d_π orbitals can accept electrons from the lone pairs of X and so strengthen the M–X bond. These early metals, especially in their highest d^0 oxidation state, are therefore said to be *oxophilic*. The more electronegative late metals, commonly d^6 to d^{10}, tend to prefer lower oxidation states and soft π-acceptor ligands, such as CO, and to shun F, OR, and NR_2.

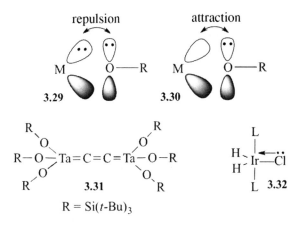

R = Si(t-Bu)$_3$

In early metal alkoxides, the M–O–R angles tend to be $>109°$. The oxygen rehybridizes from sp^3 ($109°$) to sp^2 ($120°$) or even sp (**3.31**, $180°$) so the lone pairs are now in higher energy p orbitals and thus more available for overlap with empty metal d_π orbitals. By donating both O lone pairs, a bulky linear alkoxide can be considered as a 5e (ionic model: 6e) L$_2$X donor, so (RO)$_2$NbX$_3$ having linear M–O–R groups when R is the bulky tripticyl group, somewhat resembles Cp$_2$NbX$_3$ in that Cp is also L$_2$X.

The N in L$_n$M–NR$_2$ is often planar for the early metals; this puts the N lone pair in a basic p orbital from which it can be donated to an empty metal d_π orbital just as the planar NR$_2$ group in organic RCONR$_2$ donates into the the RCO π*.

Like alkyls, L$_n$M–NR$_2$ and L$_n$M–OR can β eliminate (Eq. 3.27) to form a ketone, aldehyde, or imine. Alcohols can therefore act as reducing agents for metal complexes, especially in the presence of a base that converts the coordinated alcohol to the alkoxide, which can then β-eliminate. Alkoxides such as MOt-Bu are stable, however, because they lack β hydrogens. Amide, alkoxo, and fluoro complexes of the late metals are best known, either in high oxidation states, such as in K$_2$IrF$_6$, where the d$_\pi$ electrons are contracted, or in cases such as **3.32**, where the 16e metal has one empty d$_\pi$ orbital to accept heteroatom lone-pair electrons.

$$L_nM \overset{\displaystyle O-CH \overset{R}{\underset{H}{}}}{} \longrightarrow L_nM-H + O=CH \overset{R}{} \qquad (3.27)$$

The heavier group 15–17 elements do give terminal M–X complexes, but –PR$_2$, –SR, and –Cl have a much higher tendency to bridge than

–NR$_2$ and –OR perhaps because their longer M–X bonds decrease any steric bias against bridging.

Oxophilicity and Functional Group Tolerance

Compared with late metals, oxophilic d^0 metals are less tolerant of functional groups in their reaction with organic compounds. These groups, present in any complex organic compound, either block open sites at the metal needed for catalysis or produce undesired side reactions. This helps explain why d^8 Pd(II) is so frequently encountered in the catalysis of organic reactions (Chapters 9 and 14) where most substrate functionality must remain untouched. Early metals can still be good catalysts for hydrocarbons lacking heteroatom functionality, as in alkene polymerization (Chapter 12), typically catalyzed by d^0 Ti and Zr.

3.3   METAL HYDRIDES

Metal hydrides are important because the M–H bond enters into so many key reactions, such as undergoing insertion with a wide variety of C=X bonds to give either stable species or reaction intermediates with M–C bonds, often as part of a catalytic cycle.

Hieber's 1931 claim that H$_2$Fe(CO)$_4$ contains two Fe–H bonds long remained controversial; as late as 1950, Sidgwick still preferred the incorrect (CO)$_2$Fe(COH)$_2$ structure. Only with the discovery of Cp$_2$ReH, PtHCl(PR$_3$)$_2$, and the striking polyhydride K$_2$[ReH$_9$] in the period 1955–1964, did the reality of the M–H bond as a normal covalency become widely accepted. The landmark discovery of molecular hydrogen complexes, L$_n$M–(H$_2$), emphasizes the remarkably rich chemistry of the simplest atom, H.

Characterization

Hydrides can be detected by ^{1}H NMR spectroscopy because they resonate to high field of SiMe$_4$ in an otherwise unoccupied spectral range (0 to -60δ). They couple with a suitable metal and with cis ($J = 15$–$30\,\mathrm{Hz}$) and trans ($J = 90$–$150\,\mathrm{Hz}$) phosphines; this cis/trans difference is often useful for determining the stereochemistry of the complex. Inequivalent hydrides also couple with each other ($J = 1$–$10\,\mathrm{Hz}$). IR M–H stretching frequencies range from 1500 to 2200 cm^{-1}, but the intensities are often weak.

X-rays are scattered by electron density, not by the atomic nuclei, so crystallographic detection of hydride ligands is hard since H has no core electrons. Even when the H is seen, the M–H internuclear distance is underestimated by ~0.1 Å because the M–H bonding electrons that are detected lie between the two nuclei. Better results are obtained at low temperatures — to minimize thermal motion — and at low angles where hydride scattering is maximal. The best method to get accurate internuclear M–H distances is neutron diffraction because the H nucleus itself is now directly detected; much larger crystals (1 vs. 0.01 mm^3) are needed, however.

Synthesis

Synthetic routes are shown in Eq. 3.28–Eq. 3.34.

1. By protonation:

$$[Fe(CO)_4]^{2-} \xrightarrow{\ H^+\ } [HFe(CO)_4]^- \xrightarrow{\ H^+\ } [H_2Fe(CO)_4] \qquad (3.28)$$

$$Cp_2WH_2 \xrightarrow{\ H^+\ } [Cp_2WH_3]^+ \qquad (3.29)$$

2. From hydride donors:

$$WCl_6 + PR_3 \xrightarrow{\ Li[HBEt_3]\ } WH_6(PR_3)_3 \qquad (3.30)$$

3. From H$_2$:

$$IrCl(CO)(PPh_3)_2 \xrightarrow{\ H_2\ } IrH_2Cl(CO)(PPh_3)_2 \qquad (3.31)$$

$$WMe_6 + PMe_2Ph \xrightarrow[-CH_4]{\ H_2\ } WH_6(PMe_2Ph)_3 \qquad (3.32)$$

4. From β elimination:

$$RuCl_2L_3 + L \xrightarrow[-Me_2CO,-KCl]{\ KOCHMe_2\ } RuH_2L_4 \qquad (3.33)$$

$$Cr(CO)_6 \xrightarrow{\ OH^-\ } [Cr(CO)_5(COOH)]^- \xrightarrow[-CO_2]{} [HCr(CO)_5]^-$$

$$\xrightarrow[-CO]{\ Cr(CO)_6\ } [(CO)_5Cr\!-\!H\!-\!Cr(CO)_5]^- \qquad (3.34)$$

Protonation requires a basic metal complex, but the action of a maingroup hydride on a metal halide is very general. The third route, oxidative addition, is of particular importance in catalysis. Finally, hydrides are formed by the β elimination of a variety of groups.

Reactions

Hydrides are very reactive, giving a wide variety of transformations, as shown in Eq. 3.35–Eq. 3.38. Hydride transfer and insertion are closely related; the former implies that a hydridic hydride is attacking an electrophilic substrate.

1. Deprotonation:

$$WH_6(PMe_3)_3 \xrightarrow[-H_2]{NaH} Na[WH_5(PMe_3)_3] \tag{3.35}$$

2. Hydride transfer and insertion:

$$Cp^*_2ZrH_2 \xrightarrow{CH_2O} Cp^*_2Zr(OCH_3)_2 \tag{3.36}$$

$$Cp_2ZrHCl \xrightarrow{R} Cp_2ClZr \diagdown R \tag{3.37}$$

3. H atom transfer to form a stabilized carbon-centered radical:

$$
\begin{array}{ccc}
[HCo(CN)_5]^{3-} & & [Co(CN)_5]^{3-} \\
+ & \longrightarrow & + \\
Ph \diagup\!\!\!\diagup COOH & & Ph \diagdown_{\bullet}\!\diagup COOH
\end{array} \tag{3.38}
$$

For the electropositive early metals, the H tends to carry a significant negative charge, promoting H^- transfer to electrophiles such as aldehyde or ketone (Eq. 3.35). In contrast, The later metals impart much less negative charge to the hydride, and $HCo(CO)_4$ is even strongly acidic ($pK_a = 8.5$) because the CO groups stabilize the anionic charge of $[Co(CO)_4]^-$. Protonation of a hydride with loss of H_2 can open up a coordination site; for example, $IrH_5(PCy_3)_2$ reacts with HBF_4 in MeCN to give $[IrH_2(MeCN)_2(PCy_3)_2]^+$.

Hydricity

Hydricity refers to the tendency of $L_nM–H$ to transfer H^- to an electrophile and varies with the nature of L_nM and the solvent. Defined as ΔG° for $L_nM–H \rightarrow L_nM^+ + H^-$, it is typically determined from an electrochemical Hess's law cycle (Section 3.5) or from theoretical calculations.[16] The reactivity of a hydride also strongly depends on the nature of the reaction partner. For example, $CpW(CO)_3H$ has been shown to be an H^+ donor toward simple bases, an H donor toward styrene, and an H^- donor to a carbonium ion.

Bridging Hydrides

Hydrides have a high tendency to bridge two or more metals via a 2e,3c bond. The deprotonation counting convention was discussed in Section 2.1, but M(μ-H)M′ can also be thought of as a σ-complex (Sections 1.11 and 3.4) in which the M–H bond is a 2e donor to M′ as acceptor. The same idea can be applied synthetically, for example, a variety of L_nM-H can react with 16e M′L_n' or an 18e labile [(solv)M′L_n'] to give the bridged species L_nM–H–M′L_n'. Subsequent rearrangement can give rise to multiply bridged complexes, such as $[L_2HIr(\mu-H)_3IrHL_2]^+$ or $[H_2L_2Re(\mu-H)_4ReH_2L_2]$ (L = PPh_3).

3.4 SIGMA COMPLEXES

Sigma complexes[6,17] (**3.33**, Section 1.11) contain X–H ligands that donate the X–H σ-bonding electrons to the metal in a 2e,3c bond (**3.34**), augmented by back donation from M(d_π) into X–H σ^*. They are neutral 2e, L ligands and form for H_2 (**3.35**) and X–H, where X = H, C, Si, Sn, B, or P. The small H atom in the X–H ligand, having no lone pairs, allows X–H to approach M so that the filled M d_π orbital can back-bond into X–H σ^*, as shown in **3.34**. Even such a weak ligand as methane can bind in this way.[18] Rare examples of agostic bonds with X–Y (X and Y $\neq$ H) include agostic C–C and Si–Si complexes.[19]

Back donation into the X-H σ^* is essential for binding because pure Lewis acids such as $AlMe_3$ or BF_3 do not form isolable H_2 or HX σ complexes. On the other hand, very strong back donation breaks the X–H bond in an oxidative addition to give a classical dihydride (**3.36**).

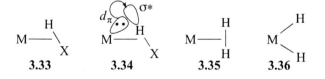

Dihydrogen Complexes[20]

Dihydrogen binds as a 2e donor σ complex to a wide variety of metal fragments (**3.35**, M = $FeH_2(PR_3)_3$, $Cr(CO)_5$, $CpRu(CNR)(PPh_3)^+$), leading to an elongation of the H–H distance from 0.74 Å in free H_2 to 0.8 to 1.1 Å. Although less common, stretched H_2 complexes (sometimes called compressed dihydrides) can have a d_{HH} up to 1.6 Å. Free H_2 is an extremely weak acid (pK_a = 35), but binding as a σ complex makes it a much better one: $[CpRe(NO)(CO)(H_2)]^+$ has a pK_a of -2.5. This great enhancement of acidification is very remarkable considering that ligands that bind via a lone pair are only mildly acidified (by 2–4 pK_a units for H_2O). The reason is that OH^-, the conjugate base from OH_2, is bound to M only a little better than is OH_2, but H^- is bound very much more strongly than H_2, providing a much bigger driving force for H^+ loss. For an acid, AH, any stabilization of A^- relative to AH translates into acidification of AH, so acidification is greatest when H_2 binds to a cationic L_nM^+ (better able to stabilize H^-) with weak back-bonding (so H_2 binds weakly). Strong acidification of H_2 paradoxically comes from weak binding to M. This easy deprotonation of coordinated H_2 provides a good route for the heterolytic activation of hydrogen: H^+ is abstracted by a base, and H^- is retained by the metal or can be transferred to a substrate in a subsequent step.[21]

On the bonding model of Fig. 1.11b, back bonding is predominant in elongating and ultimately breaking the H–H bond in a full blooded OA (**3.36**) as back bonding populates the H–H σ^* orbital.[22] Less π-basic metals tend to prefer the dihydrogen complex, **3.35**, while increasing the electron density at M favors the dihydride **3.36**. In the $MH_4(PR_3)_3$ series, the Fe and Ru species give $M(H_2)H_2(PR_3)_3$, while Os gives $OsH_4(PR_3)_3$ because π basicity rises on descending the periodic table. The role of a positive charge in reducing the basicity of a metal center is illustrated in Eq. 3.39 in which a classical pentahydride is protonated to give a bis(dihydrogen) dihydride cation. The high trans effect classical hydrides prefer to be mutually cis and to be trans to the low trans effect H_2 ligands that should be free to rotate about the M–(H_2) bond.

$$IrH_5(PCy_3)_2 \xrightarrow{\ H^+\ } \begin{bmatrix} \\ H \diagdown \underset{H}{\overset{PCy_3\ H}{\underset{|}{\overset{|\ \diagup}{Ir}}}} \diagup \underset{H}{\overset{H}{\diagdown}}\ H \\ H \diagup \quad | \quad \diagdown \\ \quad PCy_3 \end{bmatrix}^+ \qquad (3.39)$$

3.37

Coordinated H_2 can deprotonate with base, even so mild a base as NEt_3, for **3.37**. Several H_2 complexes can both exchange with free H_2 or D_2 and exchange with solvent protons and thus can catalyze isotope exchange between gas-phase D_2 and solvent protons.

Cp*FeH(dppe) shows faster protonation at the Fe–H bond, so that the nonclassical $[Cp*Fe^{II}(H_2)(dppe)]^+$ is obtained at $-80°C$; on warming above $-40°C$, the complex irreversibly converts to the classical form $[Cp*Fe^{IV}(H)_2(dppe)]^+$. The Fe–H is the better kinetic base (faster protonation), but the Fe itself is the better thermodynamic base (dihydride more stable).

Characterization

Dihydrogen complexes have been characterized by X-ray, or, much better, by neutron diffraction. An IR absorption at 2300–2900 cm^{-1}, although not always seen, is assigned to the H-H stretch. The H_2 resonance appears in the range 0 to -10δ in the ^{1}H NMR spectrum and is often broad. Partial deuteration is useful because the H–D analog shows a J_{HD} of 15–34 Hz in the ^{1}H NMR. This compares with 43 Hz for free HD and $\sim$1 Hz for classical $L_nM(H)(D)$. The empirical Morris equation reliably relates J_{HD} to the H . . . H(D) distance, d_{HD}.[23]

$$d_{HD} = 1.42 - 0.0167 J_{HD}$$
$$d_{HD} = \text{distance in Å}; J_{HD} = \text{coupling const. in Hz.} \tag{3.40}$$

Stretched H_2 complexes with H–H distances >1 Å are rare and difficult to distinguish from classical hydrides other than by neutron diffraction or J_{HD}. For example, d_{HH} in $[Re(H_2)H_5\{P(o\text{-tolyl})_3\}_2]$ is 1.36Å by neutron diffraction.

Reactions

Labile H_2 complexes are likely intermediates when protonation of a metal hydride liberates H_2. Sigma complexes can also be involved in *sigma bond metathesis* (Eq. 3.41). For example, the reaction of hydrogen with the 12e alkyl WMe_6 (Eq. 3.32) cannot go by OA because, as a d^0 species, W is already at the maximum permitted oxidation state, yet H_2 reacts readily to liberate MeH. Weak H_2 binding as a σ complex (without back donation or metal oxidation) allows protonation of the methyl groups without needing OA (Eq. 3.41, X = Me; Y = H).

$$L_nM - X + H - Y = L_nM - Y + H - X \tag{3.41}$$

Agostic Species

Sigma complexation of alkane C–H bonds is not as strong as for H_2 and examples of alkane complexes, such as $Cp(CO)_2Mn(alkane)$, are still rare.[24] If a ligand is already bound, say by an M–P bond, one of its C–H bonds can now much more easily bind to the metal in an agostic interaction that benefits from chelation, as long as the metal has $\leq$16e and can accept the additional 2e of the C–H.

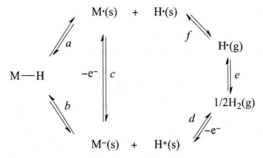

FIGURE 3.3 Thermodynamic cycle involved in one method of determining the M–H bond strength (s = solution, g = gas).

3.5   BOND STRENGTHS

Bond strengths or bond dissociation energies (BDEs) are defined as the energy required to break the M–X bond homolytically (Eq. 3.42).

$$L_nM - X = L_nM\bullet + X\bullet \qquad (3.42)$$

Bond strengths permit prediction of the thermodynamic feasibility of proposed reaction steps. For example, oxidative addition of a C–H bond to a metal would require that the inevitable loss of the large C–H bond energy of $\sim$90–100 kcal/mol be compensated by the formation of sufficiently strong M–C and M–H bonds. Several methods can be used, but Fig. 3.3 illustrates a Hess's law cycle for finding BDEs of metal hydrides. If we measure ΔGs for all but one step in the cycle, the remaining one, the M–H BDE, can be deduced. By measuring the acid dissociation constant of the hydride and the potential required for oxidizing the conjugate base, the metal anion, the ΔG values corresponding to steps b and c can be estimated from Eq. 3.43 and Eq. 3.44.

$$\Delta G = -RT \ln K \qquad (3.43)$$

$$\Delta G = \frac{RT}{F} \ln E_0 \qquad (3.44)$$

The H^+/H_2 potential gives ΔG for step d, leaving the well-known bond strength of H_2 and solvation energy of H. The M–H BDE follows from the requirement that $\Delta G = 0$ for the full cycle. For M–C BDEs, kinetic methods are discussed by Jones.[25]

Typical data for M–C BDEs of various types are shown in Fig. 3.4, in which the relative M–C BDE is plotted against the known H–X BDE. A good correlation exists, but the slope of the line varies depending on

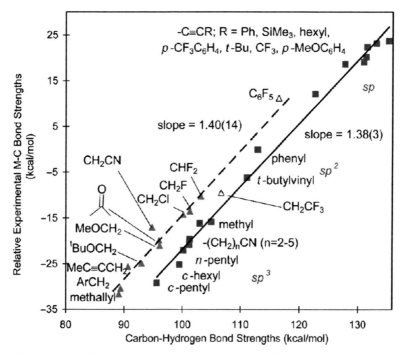

FIGURE 3.4 Relative experimental M–R BDEs versus R–H BDEs for alkyls (solid line and squares), a substituted alkyls (dashed and triangles), and outliers (open triangles) that were not included in either fit. Alkyne and nitrile BDEs were calculated (DFT). Source: From Jiao et al., 2013 [25c]. Reproduced with permission of the American Chemical Society.

the system studied.[26] Alkyls follow one correlation but resonance stabilized radicals follow another (Fig. 3.4). In an exception to the linear trend, $L_nM–H$ is normally stronger than $L_nM–CH_3$ by 15–25 cal/mol even though Me–H and H–H have almost the same BDEs.

Unlike organic chemistry, where the same bond often has a similar bond strength in a variety of compounds, organometallics behave differently. Organic compounds, with four-coordinate carbon and many C–H bonds, do not have the strong intersubstituent repulsions. In organometallics, often much more crowded, relief of intraligand repulsions promotes ligand loss and weakens the M–L BDEs. For example, in $Cp^*Ru(PMe_3)_2Cl$ (**3.38**), eight atoms are directly bound to the metal. PMe_3 loss in **3.38** is also promoted by Cl being a π-donor; on loss of PMe_3, the chloride helps stabilize the 16e fragment **3.39** by π-donation. Relative to the non-π-bonding $Cp^*Ru(PMe_3)_2Me$, the M–PMe_3 bond energy is thus also lowered by the π-donor Cl. No one set of BDE values is therefore generally applicable. Indeed, reported M–CO bond energies go all the way from 22 to 84 cal/mol in metal carbonyls.

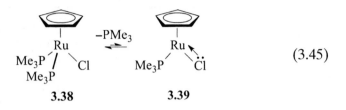

$$(3.45)$$

3.38   **3.39**

Supramolecular Interactions

These are much weaker. In hydrogen bonding, critically important in biology, a weak acid such as an N–H or O–H bond binds via a 5–10 cal/mol interaction with a weak base, typically an N or O lone pair to give linear structures such as O–H . . . O or N–H . . . N, where the dotted bond shows the weak hydrogen bond. Hydrogen bonds (HBs) are typical where good proton donors and acceptors coexist in solution or in a crystal. In trying to grow a crystal for structure determination, account must be taken of the need for HBs to be formed. Solvents with complementary HB preferences can be useful by becoming incorporated into the crystal to satisfy the HB requirements of the complex. Aromatic stacking between aryl groups is another stabilizing interaction often seen in crystal structures.

The hydrogen bond, A–H . . . B, forms between a weak acid AH and a weak base, B. The M–H bond is a weak base, so it can form hydrogen bonds with N–H . . . H–M or O–H . . . H–M structures, commonly known as dihydrogen bonds (DHBs), and involving proton–hydride attractions. The bond strengths do not differ much from conventional HBs, and the H . . . H distance is typically 1.8 Å, much shorter than the sum of their van der Waals radii (2.4 Å). The N–H or O–H acid approaches side-on to the M–H base because the proton has to get close to the pair of electrons in the M–H bond that constitute the weak base.[27]

- Alkyls typically decompose by β elimination or reductive elimination (Section 3.1).
- Stable d-block metal alkyls usually either lack β-H groups or lack cis vacant sites.
- Metal hydrides and $M(H_2)$ complexes are key reactive intermediates.
- Sigma complexation serves to activate H_2, C–H, and related bonds.

REFERENCES

1. E. Frankland, *Experimental Researches in Pure, Applied and Physical Chemistry*, John Van Voorst, London, 1877; M. Sutton, *Chemistry World*, June 2012, 56.

2. M. Schlosser, *Organometallics in Synthesis*, 3rd ed., Wiley, New York, 2012.

3. W. A. Nugent, *Angew Chem Int Ed*, **51**, 8936, 2012; M. Livia, *Brilliant Blunders*, Simon & Schuster, New York, 2013.

4. F. A. Adedeji, J. A. Connor, H. A. Skinner, L. Galyer, and G. Wilkinson, *Chem. Commun.*, **159**, 1976.

5. F. Weinhold and C. Landis, *Valence and Bonding*, Cambridge University Press, New York, 2005.

6. M. Brookhart, M. L. H. Green, and G. Parkin, *Proc. Nat. Acad. Sci. USA*, **104**, 6908, 2007.

7. D. A. Pantazis, J. E. McGrady, M. Besora, F. Maseras, and H. Etienne, *Organometallics*, **27**, 1128, 2008; W. Scherer, V. Herz, A. Bruck, C. Hauf, F. Reiner, S. Altmannshofer, D. Leusser, and D. Stalke, *Angew. Chem. Int. Ed.*, **50**, 2845, 2011.

8. (a) E. Clot, C. Mégret, O. Eisenstein, and R. N. Perutz, *J. Am. Chem. Soc.*, **131**, 7817, 2009. (b) C. S. Chan, S. Y. Lee, and K. S. Chan, *Organometallics*, **32**, 121, 2013.

9. P. S. Hanley, S. L. Marquard, T. R. Cundari, and J. F. Hartwig, *J. Amer. Chem. Soc.*, **134**, 15281, 2012 and refs cited.

10. E. Clot, J. Y. Chen, D. H. Lee, S. Y. Sung, L. N. Appelhans, J. W. Faller, R. H. Crabtree, and O. Eisenstein, *J. Am. Chem. Soc.*, **126**, 8795, 2004.

11. M. Albrecht, *Chem. Rev.*, **110**, 576, 2010; I. Omae, *J. Organometal. Chem.*, **696**, 1128, 2011.

12. K. C. D. Robson, P. G. Bomben, and C. P. Berlinguette, *Dalton Trans.*, **41**, 7814, 2012; P. G. Bomben, K. C. D. Robson, B. D. Koivisto, and C. P. Berlinguette, *Coord. Chem. Rev.*, **256**, 1438, 2012.

13. B. Butschke and H. Schwarz, *Chem. Sci.*, **3**, 308, 2012.

14. A. B. Chaplin, R. Tonner, and A. S. Weller, *Organometallics*, **29**, 2710, 2010.

15. R. H. Crabtree, *New J. Chem.*, **27**, 771, 2003.

16. Y. Matsubara, E. Fujita, M. D. Doherty, J. T. Muckerman, and C. Creutz, *J. Am. Chem. Soc.*, **134**, 15743, 2012.

17. J. C. Green, M. L. H. Green, and G. Parkin, *Chem. Commun.*, **48**, 11481, 2012.

18. W. H. Bernskoetter, C. K. Schauer, K. I. Goldberg, and M. Brookhart, *Science*, **326**, 553, 2009.

19. A. B. Chaplin and A. S. Weller, *J. Organometal. Chem.*, **730**, 90, 2013. P. Gualco, A. Amgoune, K. Miqueu, S. Ladeira, and D. Bourissou, *J. Am. Chem. Soc.*, **133**, 4257, 2011; N. Takagi and S. Sakaki, *J. Am. Chem. Soc.*, **134**, 11749, 2012.

20. G. J. Kubas, *Metal–Dihydrogen and σ-Bond Complexes: Structure, Theory, and Reactivity*, Kluwer Academic, New York, 2001.

21. G. Dobereiner, A. Nova, N. D. Schley, N. Hazari, S. Miller, O. Eisenstein, and R.H. Crabtree, *J. Am. Chem. Soc.*, **133**, 7547, 2011.

22. D. Devarajan and D. H. Ess, *Inorg. Chem.*, **51**, 6367, 2012.

23. R. H. Morris, *Coord. Chem. Rev.*, **252**, 2381, 2008.

24. J. A. Calladine, S. B. Duckett, M. W. George, S. L. Matthews, R. N. Perutz, O. Torre, and K. Q. Vuong, *J. Am. Chem. Soc.*, **133**, 2303, 2011; S. D. Pike, A. L. Thompson, A. G. Algarra, D. C. Apperley, S. A. Macgregor, and A. S. Weller, *Science*, **337**, 1648, 2012.

25. (a) M. E. Evans, T. Li, A. J. Vetter, R. D. Rieth, and W. D. Jones, *J. Org. Chem.*, **74**, 6907, 2009; (b) G. Choi, J. Morris, W. W. Brennessel, and W. D. Jones, *J. Am. Chem. Soc.*, **134**, 9276, 2012; (c) Y. Jiao, M. E. Evans, J. Morris, W. W. Brennessel, and W. D. Jones, *J. Am. Chem. Soc.*, **135**, 6994, 2013.

26. E. Clot, O. Eisenstein, N. Jasim, S. A. Macgregor, J. E. McGrady, and R. N. Perutz, *Acct. Chem. Res.*, **44**, 333, 2011.

27. N. V. Belkova, L. M. Epstein, and E. S. Shubina, *Eur. J. Inorg. Chem.*, **3555**, 2010.

PROBLEMS

3.1. [Pt(Ph$_3$P)$_2$(RC≡CR)] reacts with HCl to give **3.37**. Propose a mechanism for this process to account for the fact that the H in the product vinyl is endo with respect to the metal, as shown in **3.37**.

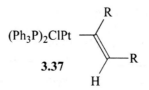

3.2. In which direction would you expect a late transition metal hydride to undergo insertion with CH$_2$=CF$_2$ to give the most stable alkyl product?

3.3. Suggest an efficient method for preparing IrMe$_3$L$_3$ from IrCIL$_3$. LiMe, and MeCl.

3.4. Propose three alkoxides, which should be as different in structure as possible, that you would examine in trying to make a series of stable metal derivatives, say, of the type Mo(OR)$_6$. Would you expect CpFe(CO)$_2$(OR) to be linear or bent at O? Explain.

3.5. What is the metal electron count for $H_2Fe(CO)_4$ and ReH_9^{2-}? Would the electron count be changed if any of these species had a nonclassical structure?

3.6. Ligands of type X–Y only give 2e three-center bonds to transition metals if X=H and Y lack lone pairs. Why do you think this is so? (*Hint:* Consider possible alternative structures if X and Y are nonhydrogen groups.)

3.7. Reductive eliminations can sometimes be encouraged to take place by oxidizing the metal. Why do you think this is so?

3.8. Give the electron counts, oxidation states, and d^n configurations in the following: $L_3Ru(\mu\text{-}CH_2)_3RuL_3$, $[(CO)_5Cr(\mu\text{-}H)Cr(CO)_5]^-$, and WMe_6.

3.9. $Me_2CHMgBr$ reacts with $IrClL_3$ to give $IrHL_3$. How can this be explained, and what is the organic product formed?

3.10. Certain 16e metal hydrides catalytically convert free 1-butene to free 2-butene. Propose a plausible mechanism, using the symbol [M]–H to represent the catalyst. Would an 18e metal hydride be able to carry out this reaction?

3.11. Why does hydricity (p. 88) depend strongly on the solvent, but BDE (Eq. 3.42) show much less solvent dependence.

4

CARBONYLS, PHOSPHINES, AND SUBSTITUTION

We now move to the key 2e ligand types, CO, phosphines, and N-heterocyclic carbenes. We see how one ligand replaces another in a *substitution* reaction, Eq. 4.1, specifically for the classic case of the substitution of CO groups in metal carbonyls by phosphines, PR_3. The principles involved will appear again later, for example, in catalysis.

$$L_nM-CO + PR_3 = L_nM-PR_3 + CO \qquad (4.1)$$

4.1 METAL CARBONYLS

A chance 1884 observation by Ludwig Mond (1839–1909)[1] led to an important advance of both practical and theoretical interest. On finding that some hot nickel valves in his chemical works had been eaten away by CO, he deliberately heated Ni powder in a CO stream to form $Ni(CO)_4$, the first metal carbonyl. In the Mond nickel refining process, the volatile carbonyl easily separates and is then decomposed by strong heating to give pure Ni. Kelvin was so impressed by this result, that he remarked that Mond "gave wings to nickel."

Whenever the donor atom of a ligand engages in a multiple bond, as in C≡O, we have an unsaturated ligand. Along with PR_3, these are

The Organometallic Chemistry of the Transition Metals, Sixth Edition.
Robert H. Crabtree.
© 2014 John Wiley & Sons, Inc. Published 2014 by John Wiley & Sons, Inc.

soft π acceptors because they can accept metal d_π electrons by back bonding (Section 1.9). In contrast, hard ligands have electronegative donor atoms with no donor atom unsaturation and are often π donors, too (e.g., H_2O and –OR).

The frontier orbitals, d_σ and d_π for M and C(lp) and CO(π^*) for CO, dominate the M–CO bonding. As shown in Fig. 4.1a and b, both C and O are sp hybridized in free CO. The singly occupied sp and p_z orbitals on each atom form a σ and a π bond, respectively. This leaves carbon p_y empty, and oxygen p_y doubly occupied, and so the second π bond is dative, formed by transfer of the O(p_y) lone pair to the empty C(p_y) orbital. This transfer leads to a $C^-–O^+$ polarization of the molecule, which is almost exactly balanced by a partial $C^+–O^-$ polarization of all three bonding orbitals because of the higher electronegativity of oxygen. The free CO molecule therefore has a triple bond and a net dipole moment very close to zero.

O being much more electronegative than C, the energy of O(p_z) is much lower than C(p_z) in Figure 4.1c. The resulting C–O π bond has more O(p_z) than C(p_z) character, CO(π) being closer in energy to O(p_z) than to C(p_z), thus polarizing the π bond toward O. In general, any bonding orbital is oppositely polarized to its corresponding antibonding orbital, so the π^* antibonding orbital, CO(π^*), is polarized toward C. The resulting CO molecule has a structure shown in VB terms in Fig. 4.1d (upper).

Figure 4.1e shows the M–CO bonding in the complex. The C(sp) lone pair donates 2e to the empty M(d_σ) orbital, raising the electron count on the metal by 2e but not much affecting the CO bond. The filled M(d_π) orbital back bonds into the CO π^*, a process that raises the M–C and lowers the C–O bond order, because any filling of a π^* orbital weakens the corresponding π bond. If the back bonding is strong, the M–C can be raised from single to double, and the CO bond can be correspondingly weakened from triple to double, resulting in the VB picture of Fig. 4.1d (lower).

The metal binds to C, not O, because the ligand HOMO is the C lone pair; O being more electronegative, its orbitals have lower energy and the O lone pair is less basic. Because the CO(π^*) LUMO is polarized toward C, M–CO π overlap is also optimal at C. While CO to M_σ donation removes electron density from C, back donation increases electron density at both C and O because CO(π^*) has both C and O characters. This results in the C becoming $\partial+$ on coordination, while O becomes $\partial-$, translating into a polarization of CO on binding. The infrared spectrum shows a big increase in the intensity of the CO stretching band on binding because the intensity depends on the bond dipole, as discussed in Section 10.8.

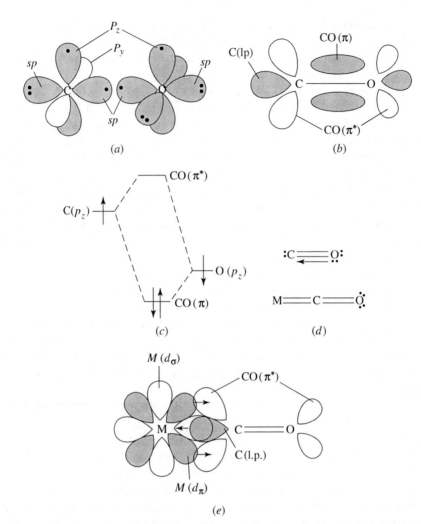

FIGURE 4.1 Electronic structure of CO and carbonyl complexes. Shading represents occupied orbitals. (*a*) and (*b*) Building up CO from C and O, each atom having two *p* orbitals and two *sp* hybrids. In (*a*), the dots represent the electrons occupying each orbital in the C and O atoms. In (*b*), only one of the two mutually perpendicular sets of π orbitals is shown. (*c*) An MO diagram showing a π bond of CO. (*d*) Valence bond representations of CO and the MCO fragment. (*e*) An MO picture of the MCO fragment. Again, only one of the two mutually perpendicular sets of π orbitals is shown.

This metal-induced polarization also activates the bound CO for chemical reactions, making the carbon subject to nucleophilic and the oxygen subject to electrophilic attack. The other ligands, L_n, modulate the polarization, as does the net charge on the complex. In $L_nM(CO)$, the CO carbon becomes particularly ∂^+ with strongly π acidic L, as in $Mo(CO)_6$, or in a cation such as $[Mn(CO)_6]^+$, because the CO-to-metal σ-donor transfer is enhanced over metal to CO back donation. If the L groups are good donors or the complex is anionic, as for $Cp_2W(CO)$ or $[W(CO)_5]^{2-}$, enhanced back donation depletes the ∂^+ charge on C but the O now becomes more ∂^-. The extreme in which CO acts as a pure σ donor can be represented in valence bond terms as **4.1**,* the mid-range as **4.2**, while **4.3** represents the extreme in which both the π_x^* and π_y^* are fully engaged in back bonding. Neither extreme, **4.1** nor **4.3**, is reached in practice, but each can be considered to contribute to a variable extent to the real structure. On the covalent model, the electron count of CO in **4.1**, **4.2**, and **4.3** is always 2e — indeed, resonance forms of a complex always have the same electron count. Polarization effects such as these also determine the reactivity of other unsaturated ligands, with nuances depending on the particular ligand.

$$M \longleftarrow C \equiv O^+ \qquad M = C = O \qquad M^+ \equiv C - O^-$$

$$\textbf{4.1} \qquad\qquad \textbf{4.2} \qquad\qquad \textbf{4.3}$$

We can tell from the IR spectrum where any particular CO lies on the continuum between **4.1** and **4.3**. Because **4.3** has a lower C=O bond order than **4.1**, the greater the contribution of **4.3** to the real structure, the lower the observed CO stretching frequency, $v(CO)$, the normal range being 1820–2150 cm^{-1}. The MO picture leads to a similar conclusion: CO π^* back bonding populates an orbital that is C=O π antibonding and so lengthens and weakens the CO bond. The position of the $v(CO)$ band is thus a measure of the the π basicity of the metal. From the number and pattern of the bands, we can tell the number and stereochemistry of the COs present (Section 10.8).

Carbonyls bound to very weakly π-donor metals, where **4.1** is predominant, have very high $v(CO)$ bands. Some, termed "nonclassical carbonyls," even appear to high energy of the 2143 cm^{-1} band of free CO.[2] Even d^0 species can bind CO, for example, the formally d^0 Ti(IV) carbonyl, $[Cp_2Ti(CO)_2]^{2+}$, has $v(CO)$ bands at 2099 and 2119 cm^{-1}. One of the most extreme weak π-donor examples is $[Ir(CO)_6]^{3+}$ with $v(CO)$ bands at 2254, 2276, and 2295 cm^{-1}. The X-ray structure of the related complex $[IrCl(CO)_5]^{2+}$ shows the long M–C [2.02(2) Å] and short C–O [1.08(2) Å] distances expected from structure **4.1**. The highest oxidation state carbonyl known is $trans$-$[OsO_2(CO)_4]^{2+}$, with $v(CO) = 2253$ cm^{-1}. Conversely, carbonyls with exceptionally low $v(CO)$ frequencies, found

---

*The + and − in **4.1**, **4.2**, and **4.3** are *formal charges* and do not necessarily reflect the real charge, which is shown here by ∂^+ or ∂^- signs.

in negative oxidation states (e.g., $[Ti(CO)_6]^{2-}$ ($v(CO) = 1747$ cm^{-1}) or where a single CO is accompanied by non-π-acceptor ligands (e.g., $[ReCl(CO)(PMe_3)_4]$ ($v(CO) = 1820$ cm^{-1}), show relatively short M–C and long C–O bonds. With such a wide range of behavior, there is considerable looseness in the way carbonyls are commonly represented. We may see M–CO or M–C=O, but whatever picture is chosen, the bonding picture discussed above still applies.

Preparation of Carbonyls

Typical examples are shown in Eq. 4.2 – Eq. 4.7:

1. From CO and a low-valent metal species:

$$Fe \xrightarrow[\substack{200° \\ 200 \text{ atm}}]{CO} Fe(CO)_5 \tag{4.2}$$

$$IrCl(cod)L_2 \xrightarrow[\substack{25° \\ 1 \text{ atm}}]{CO} IrCl(CO)L_2 \underset{}{\overset{CO}{\rightleftharpoons}} IrCl(CO)_2L_2 \tag{4.3}$$
$$(L = PMe_3)$$

2. From CO and a reducing agent (reductive carbonylation):

$$NiSO_4 \xrightarrow[{[S_2O_4]^{2-}}]{CO} Ni(CO)_4 \tag{4.4}$$

$$Re_2O_7 \xrightarrow[-CO_2]{CO} (OC)_5Re-Re(CO)_5 \tag{4.5}$$

$$[Cr(CO)_4(tmeda)] \xrightarrow{Na} Na_4[Cr(CO)_4] \tag{4.6}$$
$$(tmeda = Me_2NCH_2CH_2NMe_2)$$

3. From a reactive organic carbonyl compound and a low-valent metal species:

$$RhClL_3 \xrightarrow[OA]{RCHO} L_3ClRh\overset{COR}{\underset{H}{<}}$$

$$\downarrow -L \; \; \begin{matrix} \text{Alpha} \\ \text{Elimination} \end{matrix} \tag{4.7}$$

$$RhCl(CO)L_2 \xleftarrow[RE]{RH} L_2ClRh\overset{CO}{\underset{H}{-}}R$$

The first method (Eq. 4.2 and Eq. 4.3) needs a low oxidation state metal because only π-basic metals can bind CO well. A high-oxidation-state complex can be the starting material, if we reduce it first (Eq. 4.4). On occasion, CO itself can even be the reductant, as shown in Equation 4.5 for Re(VII). Eq. 4.6 shows how strong π acceptor COs can stabilize polyanionic species by delocalizing the negative charge over the oxygens. $Na_4[Cr(CO)_4]$ has the extraordinarily low $v(CO)$ of 1462 cm^{-1}, the extremely high anionic charge on the complex, and the ion pairing of Na^+ with the carbonyl oxygen contribute by favoring the M≡C–ONa resonance form of type **4.3**.

Equation 4.7 illustrates abstraction of CO from an organic compound, an aldehyde in this case. There are three steps; (i) an oxidative addition of the C-H bond, (ii) an α elimination (or reverse migratory insertion), and iii) a reductive elimination. The success of the reaction relies in part on the thermodynamic stability of the final metal carbonyl, which provides a driving force for the CO abstraction.

Reactions of Carbonyls

CO can act as an unreactive spectator or a reactive actor ligand. The reactions of Eq. 4.8–Eq. 4.12 all depend on the polarization of the CO on binding and thus also on the coligands and net charge change. For example, types 1 and 3 are promoted by the electrophilicity of the CO carbon and type 2 by nucleophilicity at CO oxygen.

1. Nucleophilic attack at carbon:

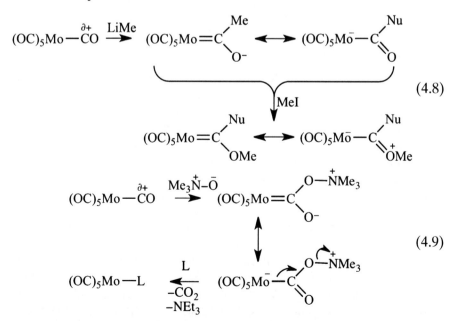

$$[Cp(NO)(PPh_3)Re-CO]^+ \xrightarrow{\text{LiBHEt}_3} Cp(NO)(PPh_3)Re-C{\overset{\displaystyle O}{\underset{\displaystyle H}{\Big\backslash}}} \qquad (4.10)$$

2. Electrophilic attack at oxygen:

$$Cl(PR_3)Re-CO \xrightarrow{\text{AlMe}_3} Cl(PR_3)Re-CO{:}\longrightarrow AlMe_3 \qquad (4.11)$$

3. Migratory insertion:

$$MeMn(CO)_5 \xrightarrow{\text{PMe}_3} MeCOMn(PMe_3)(CO)_5 \qquad (4.12)$$

Equations 4.8 and 4.9 give carbene complexes (Section 11.1) or carbenelike intermediates with M=C multiple bonds. Eq. 4.10 shows one of the rare ways in which tightly bound CO can be removed to generate an open site at the metal. In this case, CO can be replaced by a weak ligand L that would otherwise not be able to displace CO. As a 2e reagent, H$^-$ cannot attack the 18e Re in Eq. 4.10 but instead attacks the CO carbon to give a formyl ligand, stable in this case because the 18e complex has no empty site to allow rearrangement to a hydrido-carbonyl complex. In Eq. 4.11, the bulky acid and 0e reagent, AlMe$_3$, prefers to bind at CO oxygen, rather than attack the metal, as does H$^+$. Equation 4.12 shows a migratory insertion reaction (Section 7.2). When the initial 3e (Me)(CO) ligand set becomes a 1e (COMe) ligand in the course of this reaction, a 2e vacancy is created at the metal; binding of a 2e PMe$_3$ ligand at this vacant site then traps the product.

Bridging Carbonyls

CO often bridges, but the electron count is usually unchanged on going from terminal to bridging (e.g., **4.4** and **4.5**). On the traditional bonding model, the 15e CpFe(CO)$_2$ fragment is completed in **4.4** by an M–M bond, counting 1e for each metal, and a terminal CO counting 2e. In **4.5**, each of the two bridging ketone-like μ^2-CO groups adds 1e to each metal and 1e comes from an M–M bond, a feature very often accompanying a μ^2-CO on this model.

In an alternative μ^2-CO bonding model that is more consistent with theoretical work, the CO can either be ketone-like with no M–M bond or be considered as bridging via a 2e,3c bond analogous to the case of M(μ-H)M discussed in Section 2.1. In this type of μ-CO, the C lone pair is thus a 2e ligand to *each* metal and there is again no M–M bond.[3] Structures **4.6a–b** show how this model applies to compound **4.4**, which has one CO of each type. While this model is more realistic, it is too new to have gained general assent, and since the literature of the area still uses the traditional model, we do so in this book.

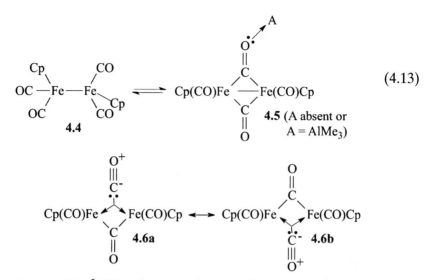

(4.13)

Consistent with μ^2-CO being more ketone-like, the IR $\nu(CO)$ stretching frequency falls to 1720–1850 cm^{-1}, and μ^2-CO is more basic at O than terminal CO. For example, a Lewis acid binds more strongly to the μ^2-CO oxygen and so displaces the equilibrium of Eq. 4.13 toward **4.5**. Triply and even quadruply bridging CO groups with $\nu(CO)$ in the range 1600–1730 cm^{-1} are also known in metal cluster compounds, for example, $(Cp^*Co)_3(\mu^3\text{-}CO)_2$ (**4.7**).

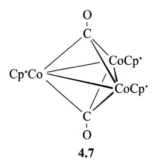

4.7

Isonitriles

Replacement of the CO oxygen with RN gives isonitrile, RNC. As a better electron donor than CO, it is more common than CO in cationic, high oxidation-state and high coordination number complexes such as $[Pt(CNPh)_4]^{2+}$ and $[Mo(CNR)_7]^{2+}$, where no CO analogs are known. It tends to bridge less readily than does CO, is more sensitive to nucleophilic attack at carbon to give aminocarbene complexes (Eq. 11.6), and has a higher tendency for migratory insertion. Unlike $\nu(CO)$ in carbonyls, the $\nu(CN)$ in isonitrile complexes is often at higher energy than in the free ligand. Unlike the C,O nonbonding C lone pair of CO, the C lone pair of CNR is C,N antibonding. Donation to the metal therefore has little effect on $v(CO)$, but raises $v(CN)$. Back bonding lowers both $v(CO)$ and $v(CN)$. Depending on the balance of σ versus π bonding, $v(CN)$ is raised for weak π-donor metals, such as Pt(II), and lowered for strong π-donor metals, such as Ni(0). Isonitrile normally remains linear on binding (**4.8**), but in strong donor cases, such as d^2 NbCl(CO)(CNR)-(dmpe)$_2$, RNC is bent at N (**4.9**, C–N–C angle: 129–144°). The M–C bond is also unusually short in **4.9** (2.05 Å vs. 2.32 Å for an Nb–C single bond), and the $v(CN)$ is low (1750 cm^{-1} for **4.9** but ~2100 cm^{-1} for **4.8**). The appalling stench of volatile isonitriles may also be a result of complexation—their binding to a Cu ion receptor in the nose.

$$M-C\equiv N-R \qquad M=C=\ddot{N}\!\!\!\diagdown_{R}$$

$$\textbf{4.8} \qquad\qquad\qquad \textbf{4.9}$$

Thiocarbonyls

Free CS is unstable above −160°C, and although a number of complexes are known, such as RhCl(CS)(PPh$_3$) (Eq. 4.14) and Cp(CO)-Ru(μ^2-CS)^{2}RuCp(CO), so far none are "pure" or *homoleptic* examples, M(CS)$_n$. They are usually made from CS$_2$ or by conversion of a CO to a CS group. Perhaps because of the lower tendency of the second-row elements such as S to form double bonds, the M$^+\equiv$C–S$^-$ form analogous to **4.3** is more important for M(CS) than for M(CO). The MC bond therefore tends to be short and CS is a better π acceptor than CO and CO is more substitutionally labile than CS.

$$RhCl(PPh_3)_3 \xrightarrow{\;CS_2\;} RhCl(CS)(PPh_3)_2 + SPPh_3 \qquad (4.14)$$

Typical $v(CS)$ ranges for CS complexes are 1273 cm^{-1} for free CS, 1040–1080 cm^{-1} for M$_3$(μ_3–CS), 1100–1160 cm^{-1} for M$_2$(μ_2–CS), and 1160–1410 cm^{-1} for M–CS.

Nitrosyls

Like CO, NO$^+$ is a 2e ligand, and nitrosyls are often made from the salt [NO]BF$_4$ (Eq. 4.15).[4] Being isoelectronic with CO, NO$^+$ binds in a linear fashion. Its positive charge and electronegative N atom makes it an even more strongly π-acceptor ligand than CO. In some cases, NO can also bind in a bent structure, in which case, it is considered as an anionic ligand NO$^-$.

$$\text{CpMo(CO)}_2(\eta^3\text{-allyl}) \xrightarrow{\text{NOBF}_4} [\text{CpMo(CO)(NO)}(\eta^3\text{-allyl})]\text{BF}_4 \qquad (4.15)$$

NO can be considered as a 3e reagent that lowers the metal oxidation state by one unit in forming a linear NO (*lin*-NO) complex because NO donates an electron to the metal in becoming bound NO^+. For example, if we remove the *lin*-NO ligands as NO^+ from W(*lin*-NO)$_4$, we have d^{10} W(-IV); counting $2 \times 4e$ for the two NO^+ ligands, we get 18e. Alternatively, we could start with a 6e W atom and add $4 \times 3e$ for the four neutral NOs, also making 18e.

NO is a redox-active ligand: Eq. 4.16 shows how 2e from the Co(I) can be transferred to *lin*-NO in an internal redox reaction to give a bent NO complex. Bent NO, considered as an X ligand, raises the oxidation state by two units. The product of Eq. 4.16 is thus Co(III) and 16e, because the NO^+ has been transformed into an X-type NO^-. Just as the lone pair of a halide can help stabilize a 16e metal (Section 3.5), the nominally uncoordinated N lone pair of a bent NO may do the same here because bent NO complexes are most often 16e (becoming 18e if the π lone pair is also counted).

The deliberately ambiguous Feltham–Enemark notation is useful because it does not matter whether the NO is linear or bent. We consider just the M(NO)$_y$ part of the molecule and sum the number of electrons in the metal d_π and NO π^* orbitals. For example, in [(tacn*)-Fe(NO)(N$_3$)$_2$] (**4.10**), we remove the non-NO ligands as L$_3$-type tacn* and two X-type N_3^- to obtain {Fe(NO)}$^{2+}$. On the covalent model, neutral Fe is d^8, and neutral NO has one π^* electron, making 9 in all; now adjusting for the 2+ ion charge of the {Fe(NO)}$^{2+}$ fragment, we arrive at 7 valence electrons, making the complex {Fe(NO)}^{7} on this notation.

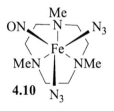

4.10

For Eq. 4.16, the linear complex has v(NO) at 1750 cm^{-1} and the bent form at 1650 cm^{-1}; unfortunately, the typical v(NO) ranges for the two types overlap. Equation 4.16 also shows that it is not always possible to decide whether NO is linear or bent by finding which structure leads to an 18e configuration. Only if a *lin*-NO complex would be 20e, as in 18e, Co(III) [CoCl(*bent*-NO)(diars)$_2$]$^+$ (diars = Me$_2$AsCH$_2$CH$_2$AsMe$_2$), can we safely assign a bent structure. Equation 4.17 shows a synthesis from NO; unlike most ligands, NO can replace all the COs in a metal carbonyl to give a homoleptic nitrosyl.

In one case, $[Fe(lin\text{-}NO)_2(bent\text{-}NO)_2]^-$, a homoleptic nitrosyl contains both linear and bent NO ligands.[5] NS and NSe have also been recently identified as new ligands.[6]

$$CoCl_2L_2(lin\text{-}NO) \rightleftharpoons CoCl_2L_2(bent\text{-}NO) \qquad (4.16)$$

18e, Co(I)   16e, Co(III)

$$Cr(CO)_6 \xrightarrow{\text{NO, } h\nu} Cr(lin\text{-}NO)_4 \qquad (4.17)$$

18e, Cr(0)   18e, Cr(–IV)

NO and CO are both of great biological importance, particularly NO, which aids in the maintenance of vascular tone, in neurotransmitter function and in mediation of cellular defence.[7] NO binding to iron also occurs in nitrophorins, ferric heme proteins produced by blood-sucking insects that transport and release NO with the aim of facilitating blood flow from victim to insect.

Cyanide

The CO analog, CN^-, has recently gained attention as a ligand for the active-site iron in many hydrogenases (Section 16.4), but its synthetic metal complexes date back to early times.[8] In 1706, Diesbach, a Berlin draper, boiled beef blood in a basic medium to obtain the pigment, Prussian blue, $K[Fe_2(CN)_6]$, still in common use. Later shown to be a coordination polymer containing $Fe^{II}-C\equiv N-Fe^{III}$ units in which the softer Fe(II) binds the softer C end of cyanide, Prussian blue can be considered as both the first organometallic and the first coordination compound. The boronyl ligand $(B\equiv O^-)$, recently discovered in $[\{(C_6H_{11})_3P\}_2Pt(BO)Br]$, is the latest cyanide analog to be identified.[9]

In gold and silver mining, the metals are extracted from their ores with an aqueous KCN solution in which the elemental metals dissolve as linear $[M(CN)_2]^-$ complexes on air oxidation (Eq. 4.18, M = Ag or Au). The toxicity of soft CO and CN^- is associated with irreversible binding to key soft Fe(II) active sites of metalloproteins such as hemoglobin and cytochrome c oxidase.[10]

$$M \xrightarrow[\text{–OH}^-]{CN^-,\ O_2,\ H^+} [M(CN)_2]^- \qquad (4.18)$$

Other CO Analogs

Dinitrogen (N_2), the key ligand in biological nitrogen fixation—conversion of N_2 to NH_3—is discussed in Section 16.3.[11] N_2 binds to metals much less strongly than CO because it is both a weaker σ donor and a weaker π acceptor. BF groups, also isoelectronic with CO, can formally replace three COs in $Fe_3(CO)_{12}$ to give $Fe_3(BF)_3(CO)_8$, where it strongly prefers μ^2 and μ^3 bridging positions.[12]

4.2 PHOSPHINES

Phosphines, PR_3, form one of the few series of ligands in which electronic and steric properties can be altered in a systematic and predictable way over a wide range, in this case by varying R. As spectator ligands, PR_3 also stabilize a wide variety of other M–L units, as their phosphine complexes $(R_3P)_n$M–L. Phosphines that are air sensitive,[13] typically trialkylphosphines, need to be handled under N_2 or Ar.

As ligands of intermediate hardness and π-acceptor power (Fig. 1.12), phosphines are able to stabilize a broad range of oxidation states and promote important catalytic reactions where redox cycling of the metal occurs in the reaction. Only cyclopentadienyl and N-heterocyclic carbenes rival phosphines in promoting organometallic catalysis.

Structure and Bonding

Phosphines can donate their P lone pair to a metal to give the monodentate terminal M–PR_3 group. They are also mild π acids, to an extent that depends on the nature of the R groups in the PR_3 ligand. For alkyl phosphines, the π acidity is weak; aryl, dialkylamino, and alkoxy groups are successively more effective in promoting π acidity, and in the extreme case, PF_3 is more π acid than CO.

We saw for M–CO that CO π^* orbitals accept back bonding from the metal. In M–PR_3, the P–R σ^* orbitals play the same role.[14] As the R group becomes more electronegative, the atomic orbital (a.o.) it uses to form the P–R bond becomes more stable and thus lower in energy (Fig. 4.2), in turn stabilizing the P–R σ^* orbital. The larger the energy gap between them, the more the stabler a.o. contributes to σ, and the least stable to σ^*. As the P contribution to P–R σ^* increases, the size of the σ^* lobe that points toward M also increases. Both energy and overlap factors thus make the empty σ^* more accessible for back donation. The final order of increasing π-acid character is:

$$PMe_3 \sim P(NR_2)_3 < PAr_3 < P(OR)_3 < PCl_3 < CO \sim PF_3$$

$P(NR_2)_3$ is a better donor than it should be based on Fig. 4.2, probably because the basic N lone pairs compete with $M(d_\pi)$ in donating to PR σ^*.

Occupation of the P–R σ^* by back bonding from M should make the P–R bonds lengthen slightly on binding. In practice, this is masked by

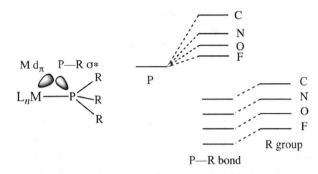

FIGURE 4.2 The empty P–R σ^* orbital plays the role of acceptor in metal complexes of PR_3. As the atom attached to P becomes more electronegative, the empty P–X σ^* orbital becomes more stable and so moves to lower energy and becomes a better acceptor from the filled metal d_π.

a shortening of the P–R bond due to donation of the P lone pair to the metal, reducing P(lone pair)–R(bonding pair) repulsions. To eliminate this complication, Orpen has compared the structures of pairs of complexes, such as $[(\eta^3\text{-}C_8H_{13})Fe\{P(OMe)_3\}_3]^{n+}$, where $n = 0$ or 1. The M–P σ bonds are similar in both cases, but the cationic iron in the oxidized complex is less π basic and so back-donates less to the phosphite; this leads to a longer Fe–P ($\Delta r = +0.0151 \pm 0.003 Å$), and a shorter P–O ($\Delta r = -0.021 \pm 0.003 Å$). As for CO, the M–L π bond is made at the expense of a bond in the ligand, but this time, it is the P–R σ, not the C=O π bond.

Tolman Electronic Parameter and Cone Angle

The reactivity of a complex can be varied by tuning the electron-donor power and steric effect of PR_3.[15] In his ligand map (Fig. 4.3), Tolman quantified both effects. The electronic effect of L comes from comparing $v(CO)$ for an $LNi(CO)_3$ series having different PR_3 ligands. Stronger donor PR_3 increase the electron density on Ni, increasing back donation to CO and lowering $v(CO)$. Computational $v(CO)$ values for $LNi(CO)_4$ avoid the need to work with toxic $Ni(CO)_4$.[16] For chelates, $v(CO)$ data for $(L–L)Mo(CO)_4$ are useful. The Lever[17] parameter, based on electrochemical data, is preferred for coordination compounds, but all these scales correlate well together.[16]

The steric bulk of PR_3 is also adjusted by changing R. Bulky PR_3 ligands favor low coordination, making room for small but weakly binding ligands that would be excluded by competition with small

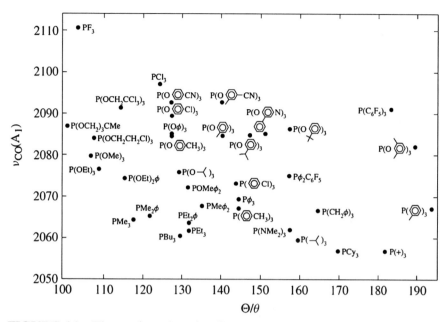

FIGURE 4.3 Electronic and steric effects of common P-donor ligands plotted on a map according to Tolman (v in cm^{-1}, θ in degrees). Source: From Tolman, 1977 [15a]. Reproduced with permission of the American Chemical Society.

ligands such as PMe$_3$. The usual maximum number of phosphines that can bind to a single metal is 1 for X-Phos (**4.11**), 2 for PCy$_3$ or P(i-Pr)$_3$, 3–4 for PPh$_3$, 4 for PMe$_2$Ph, and 5–6 for PMe$_3$. Examples include coordinatively unsaturated species stabilized by bulky phosphines, Pt(PCy$_3$)$_2$ and [Rh(PPh$_3$)$_3$]$^+$, and high CN species only possible with a small ligand, [Ir(PMe$_3$)$_5$]$^+$ and W(PMe$_3$)$_6$.[18]

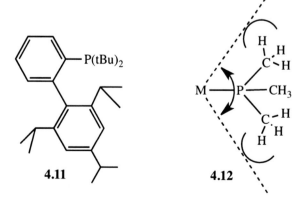

Tolman also quantified the steric effects of phosphines with his *cone angle*. This was first obtained by taking a space-filling model of the $M(PR_3)$ group, folding back the R substituents, and measuring the angle of the cone that will just contain all of the ligand, when the apex of the cone is at the metal (**4.12**). Computational methods are again available[19] and cone angles have been very successful in rationalizing the behavior of a wide variety of complexes.[20]

Variation of ligand steric and electronic properties is central to optimizing any reactivity property of interest in the complex as a whole. We can relatively easily change electronic effects without changing steric effects—for example, by moving from PBu_3 to $P(O^iPr)_3$—or change steric effects without changing electronic effects—for example, by moving from PMe_3 to $P(o\text{-tolyl})_3$. Increasing the ligand electron donor strength, for example, can favor a higher OS and thus perturb an oxidative addition/reductive elimination equilibrium in favor of the oxidative addition product. We can therefore expect the chemistry of a phosphine-containing complex to vary with the position of the phosphine in the Tolman map. Heteroatom-substituted P donors are much less often employed, however, perhaps because they can hydrolyse, for example with loss of ROH from phosphites, $P(OR)_3$.

Bite Angle

Bite angle preferences in chelate ligands (i.e., the P–M–P angle) can strongly influence reactivity.[20] These are calculated from molecular mechanics with a dummy metal atom that has no angular preferences—for example, $Ph_2P(CH_2)_nPPh_2$ has natural bite angles of 73, 86, 91, and 94° for $n = 1, 2, 3$, and 4 but there is some flexibility, at least for higher n. In contrast, very rigid diphosphines, such as the phenoxathiin **4.13**, enforce a specific bite angle, $\sim107°$ in this case. A trans-spanning ligand with bite angle $\sim180°$, shown as its Rh complex, **4.14**, is also unusual in having a phosphabenzene donor.[21]

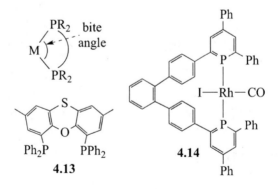

4.3   N-HETEROCYCLIC CARBENES (NHCs)

Carbenes in general are discussed in Chapter 11, but NHCs (**4.15**) are covered here because they have attained equal importance with phosphines as spectator ligands, particularly in catalysis.[22] Compared with phosphines, the range of accessible steric parameters is equally wide, but most NHCs are much stronger donors, best quantified in DFT calculations.[23] NHCs also seem to be modest π acceptors.[24]

As early as 1968, Öfele and Wanzlick found the first NHC complexes (**4.17**), but there was a long lag time before the area became active. The trigger was Arduengo's synthesis of an isolable example of **4.15** stabilized by steric protection with a bulky R group, 1-adamantyl. Useful catalytic properties were seen by Herrmann, but it was the finding that replacement of a phosphine by an NHC greatly improved the properties of Grubbs' alkene metathesis catalyst (Chapter 11) that ignited a major wave of NHC research. NHCs have now been incorporated into a impressive number of ligand architectures, including chelates pincers and tripods.[25]

NHCs are most commonly derived from *N,N'*-disubstituted imidazolium salts (**4.16**) by deprotonation at C2 to give the free NHC, **4.15**. This acts as a very powerful 2e donor, binding to a variety of ML_n fragments to give NHC complexes (**4.17**). These are often seen represented in one of two ways depending on whether we want to emphasize the carbene character of the product (**4.17a**) or else the alternative picture of a metal substituted arenium ring (**4.17b**).

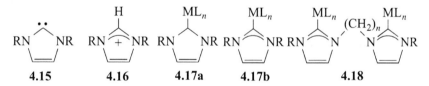

Because the M–NHC bond is so strong, the NHC does not normally dissociate from the metal. This causes problems in the case of potentially chelating NHCs, where 2 : 1 complexes like **4.18** can easily be formed as kinetic products in attempts to make chelates. In a similar diphosphine case, reversible dissociation/association would soon convert this 2 : 1 complex into the thermodynamically favored chelate, but NHCs do not rearrange if they initially form the "wrong" complex.[26]

NHC complexes can be synthesized in a wide variety of ways:[27] from a metal precursor and the imidazolium salt (Eq. 4.19 and Eq. 4.20) a free NHC (Eq. 4.21); or by transmetallation from the silver NHC complex, often conveniently available under mild conditions from the imidazolium salt and Ag_2O (Eq. 4.22).[28] Various other methods are available (Eq. 4.23–Eq. 4.26).

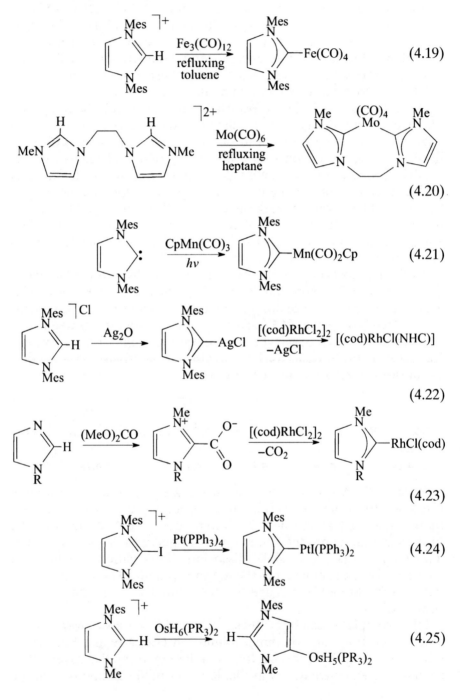

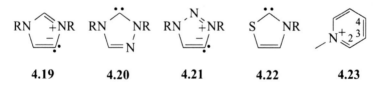

$$(4.26)$$

An attractive feature of NHCs is the very wide range of stuctures that can be accessed.[29] One class of specially strong donors, *abnormal* or *mesoionic* NHCs (Eq. 4.25),[30] derives from carbenes that are mesoionic in the free form, meaning that no structure with all-neutral formal charges can be written. For example, **4.15** is all neutral as written, but **4.19** is not. The extra nitrogen in the 1,2,4-triazole ring reduces the donor power of the NHC (**4.20**), but 1,2,3-triazole gives an abnormal NHC (**4.21**) that is a stronger donor than **4.20**. Complexes of **4.19** are thermodynamically less stable than the normal **4.15** complex, but the strong M–aNHC bond prevents any such rearrangement. The thiazole-based NHC, **4.22**, lacks one R substituent next to the carbene center, as is also the case for **4.19** and **4.21**. **4.23** can give carbenes at positions 2, 4 (normal), and 3 (abnormal) (e.g., Eq. 4.26). Many other related structures are also known.[31]

Steric bulk is easily achievable with NHCs because the R groups point toward the metal, not away as in M–PR$_3$. For example, in IMes, the mesityl groups play this role, leading to easy access to low coordinate complexes, such as [PtMe(IMes)$_2$]$^+$.[32] The *I* of this naming convention means that an imidazole ring is involved and the *Mes* refers to the mesityl substituents at N.

| 4.19 | 4.20 | 4.21 | 4.22 | 4.23 |

4.4 DISSOCIATIVE SUBSTITUTION

The mechanisms of CO substitution by PR$_3$ in metal carbonyls are the basis for the understanding of organometallic substitution in general. Two extreme mechanisms are invoked, one dissociative, D, and the other associative, A. In the D mechanism, a CO first dissociates, leaving a vacant site at which PR$_3$ subsequently binds. This is typical of 18e complexes because the intermediate is then 16e after CO loss. In the A mechanism of Section 4.5, PR$_3$ binds first and only subsequently does the CO depart. This is typical of 16e complexes because the intermediate is then 18e after association of PR_3. The D and A paths are analogous to the S_N1 and S_N2 paths in organic substitution.

Kinetics

In the D mechanism, initial CO loss to generate a vacant site at M is usually slow, followed by fast trapping by the incoming ligand L'. The rate-determining step is thus independent of the concentration and identity of L'. This leads to the simple rate equation of Eq. 4.27.

$$\text{Rate} = k_1[L_nM(CO)] \tag{4.27}$$

$$L_nM\!-\!CO \underset{+CO,\, k_{-1}}{\overset{-CO,\, k_1}{\rightleftharpoons}} L_nM\!-\!\square \xrightarrow{+L',\, k_2} L_nM\!-\!L' \tag{4.28}$$

$$\text{Rate} = \frac{k_1 k_2 [L'][L_nM(CO)]}{k_{-1}[CO] + k_2[L']} \tag{4.29}$$

In some cases, the back reaction, k_{-1} of Eq. 4.28, becomes important and the intermediate, $L_nM\text{-}\square$ ($\square$ = vacancy), now partitions between the forward and back reactions. Increasing the concentration of L' does now have an effect on the rate because k_2 now competes with k_{-1}. The rate equation derived for Eq. 4.28, shown in Eq. 4.29, also appears in a wide variety of reactions beyond substitution. It reduces to Eq. 4.27, if the concentration of CO, and therefore the rate of the back reaction, is negligible.

If k_{-1} is smaller than k_1, the overall rate in Eq. 4.27 is entirely controlled by the rate at which the leaving ligand dissociates. Ligands that bind less well to the metal dissociate faster than does CO. For example, $Cr(CO)_5L$ shows faster rates of substitution of L in the order $L = CO < Ph_3As < py$. For a series of $M\text{–}PR_3$ complexes, the larger the cone angle, the faster the PR_3 dissociates. In a series of similar D reactions, we expect the rates to go up as the $M\text{–}L^d$ bond to the departing ligand, L^d, becomes weaker.

For an 18e complex, the alternative A process would generate a disfavored 20e species. While a 20e transition state is not forbidden — after all, $NiCp_2$ is stable with 20e — the 16e intermediate of Eq. 4.28 provides a lower-energy path in most cases. The activation enthalpy for the reaction is close to the M–CO bond strength because this bond is broken in going to the transition state. $\Delta S^{\ddagger}$ is usually positive and in the range 10–15 eu, as expected for a dissociative process with less order in the transition state.

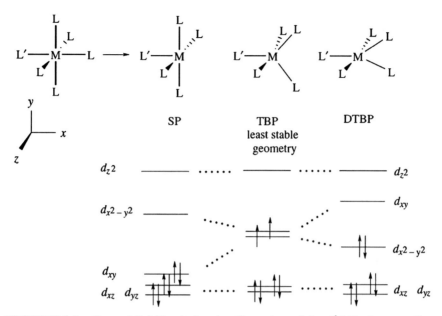

FIGURE 4.4 Crystal field basis for the distortion of the d^6 ML$_5$ intermediate formed after initial dissociation of L from a d^6 ML$_6$ complex in dissociative substitution. Pure TBP (LML = 120°) is the least stable geometry, and distortion occurs to DTBP (LML = 75°) if L′ is a π donor or to SP (LML = 180°) if L′ is a high trans-effect ligand.

Stereochemistry and Trans Effect

A dissociative substitution of a d^6 ML$_6$ complex may go with retention, inversion, or loss of the starting stereochemistry depending on the behavior of the d^6 ML$_5$ intermediate formed after initial dissociation of L. Unlike the d^8 ML$_5$ situation, where a trigonal bipyramid (TBP) is preferred, a d^6 ML$_5$ species is unstable in a TBP geometry and tends to undergo a distortion. Figure 4.4 shows why this is so. The pure TBP geometry requires that two electrons occupy the two highest filled orbitals. Hund's rule predicts a triplet paramagnetic ground state for such a situation. A distortion from TBP may take place in one of two ways, either to the square pyramidal (SP) geometry or to the distorted TBP (DTBP) geometry. In either case, the system is stabilized because the two electrons can pair up and occupy the lower-lying orbital. In the SP and DTBP structures, the equatorial ligands form the letters T and Y, respectively, hence the names T and Y for the geometries.

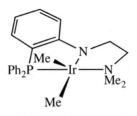

FIGURE 4.5   A d^6, Ir(III), 16e DTBP (Y) complex of an anionic PNN pincer, showing the close approach of two methyl groups (Me-Ir-Me = 77.6°).[33]

An SP or T geometry is favored when L' is a high trans-effect ligand such as H and a DTBP, or Y geometry when L' is a π donor such as Cl. If the SP geometry of Fig. 4.4 is preferred for the intermediate in Eq. 4.30, the incoming ligand can simply replace the leaving group, and we expect retention of stereochemistry. Thus, a high trans-effect ligand is one that favors the SP geometry. On the other hand, if the DTBP geometry is favored, inversion of the stereochemistry is expected. Complications can occur because SP and DTBP can both be fluxional, in which case different products can be obtained. Crystal structures of the rare stable examples of d^6 ML$_5$ species show SP, DTBP, or even intermediate geometries (e.g., Fig 3.2), but never pure TBP. The structure of an isolable DTBP complex is shown in Fig. 4.5.[33] Both T and Y geometries also occur in 14e d^8 complexes, such as [Rh(PPh$_3$)$_3$]$^+$ (T) and [(NHC)Pt(SiMe$_2$Ph)$_2$] (Y), where the NHC is located at the foot of the Y and the Si–Pt–Si angle is 80°.[34]

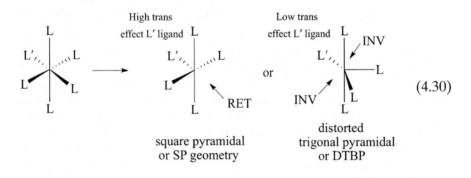

$$(4.30)$$

Electronic and Steric Factors

The dissociative mechanism is favored in d^8 TBP > d^{10} tetrahedral > d^6 octahedral. For example, d^8 Co$_2$(CO)$_8$ has a half-life for CO dissociation of a few tens of minutes at $0°$, but for d^6 $Mn_2(CO)_{10}$ at room temperature, the half-life is about 10 years! Substitution rates follow the order third row < second row > first row. For example, at $50°$, the rate constants for CO dissociation in $M(CO)_5$ are Fe 6×10^{-11}, Ru 3×10^{-3}, and Os 5×10^{-8} s^{-1}. The rate for Fe is exceptionally slow, perhaps because $Fe(CO)_4$, but not the Ru or Os analog, has a high-spin ground state with low thermodynamic stability, leading to a higher activation energy for CO loss.

While 18e complexes are usually diamagnetic, non-18e intermediates may have more than one accessible spin state.[35] Sixteen electron $M(CO)_4$ (M = Fe, Ru, and Os), for example, has singlet ($\downarrow\uparrow$) and triplet ($\uparrow\uparrow$) states, each state having a different structure and reactivity. Transitions between spin states are generally thought to be very fast, but data are sparse. This is an aspect of transition metal chemistry that is still far from well understood (Section 15.1).

To form the triplet, an electron has to be promoted from HOMO to LUMO, hard to do in an 18e organometallic where Δ is large. With 16 or fewer electrons, at least one low-lying d orbital is available for this promotion, and for $Fe(CO)_4$, the triplet is more stable than the singlet.

Phosphines do not replace all the carbonyls in a complex, even with a small phosphine; $Mo(CO)_6$ rarely proceeds beyond the fac-$Mo(CO)_3L_3$ stage. This is in part because the phosphines are much more electron donating than the carbonyls they replace. The remaining COs therefore benefit from increased back donation and are more tightly held. The fac stereochemistry (**1.24**) is preferred electronically to the mer arrangement (**1.23**), because CO has a higher trans effect than PR_3, and substitution trans to CO continues until no trans OC–M–CO groups are left. The mer arrangement is less hindered, however, and is seen for bulky PR_3.

In the 18e NiL_4 catalyst series, L dissociation liberates the open site needed for catalytic activity. Since dissociation is promoted by steric bulk, it is not surprising that the very bulky phosphite $P(O$-o-$tolyl)_3$ gives one of the very best catalysts. Triphenylphosphine is very useful in a wide variety of catalysts for the same reason.

Dissociation can be encouraged in various ways. For example, Cl^- can often be removed from M–Cl by Ag^+ via AgCl precipitation. Protonation can remove RH or H_2 from M–R or M–H. Weakly bound solvents are also readily displaced. As a π donor, thf is a poor ligand for W(0), and $W(CO)_5(thf)$, obtainable from photolysis of $W(CO)_6$ in thf, readily reacts with a wide range of ligands L to give $W(CO)_5L$. Substitution of halide by alkyl or hydride is often carried out with RMgX or $LiAlH_4$.

In some cases, dissociation is hard: the chelate effect prevents polydentate ligands from dissociating easily, for example. Carbon-donor L_nX ligands such as η^5-Cp (L_2X) or Me (X), tend to dissociate less easily than otherwise analogous L_n ligands, such as η^6-C_6H_6 (L_3) or CO (L). L_n ligands are often stable in the free state, but L_nX ligands would have to dissociate as less stable radicals or ions. M–Hal only spontaneously dissociates in a polar solvent where Hal$^-$ is stabilized. The electronic configuration of the metal is also important: substitution-inert d^6 octahedral complexes are much less likely to dissociate than are substitution-labile d^8 TBP metals (Section 1.4). Redox catalysis of substitution is possible if an 18e complex is oxidized or reduced (Section 4.5).

4.5   ASSOCIATIVE SUBSTITUTION

Associative substitution differs from dissociative in that the incoming ligand binds to the complex before the departing ligand leaves. This is typical of 16e complexes because the intermediate is then 18e and is analogous to the associative S_N2 organic reaction.

Kinetics

The slow step in associative substitution is the attack of the incoming ligand L^i on the complex to form an intermediate that only subsequently expels one of the original ligands L. The rate of the overall process is now controlled by the rate at which the incoming ligand can attack the metal in the slow step, and so $[L^i]$ appears in the rate equation (Eq. 4.31) and the rate also depends on the nature of L^i.

$$\text{Rate} = k_1[L^i][L_nM] \tag{4.31}$$

$$L_nM \xrightarrow{+L^i, k_1} L_nM\text{---}L^i \xrightarrow[\text{fast}]{-L, k_2} L_{(n-1)}M\text{---}L^i \tag{4.32}$$

For 16e complexes, the 18e intermediate of an A mechanism usually provides a lower energy route than the 14e intermediate of a D substitution. The entropy of activation is negative ($\Delta S^\ddagger = -10$ to -15 eu), as expected for a more ordered transition state.

Trans Effect

Classic examples of the A mechanism are seen for 16e, d^8 square planar complexes of Pt(II), Pd(II), and Rh(I). The 18e intermediate is a standard trigonal bipyramid with L^i in the equatorial plane (**4.24**). By microscopic reversibility, if L^i enters an equatorial site, the departing ligand, L^d, must leave from an equatorial site. Only loss of an equatorial ligand can give a stable d^8 square planar species — loss of an axial ligand would leave a tetrahedral fragment, much less favorable in d^8. This affects the stereochemistry of the product and explains how the trans effect works (Section 1.4). L^t is the highest trans-effect ligand because it also has the highest tendency to occupy the equatorial sites in the intermediate. This ensures that the ligand L^d, trans to L^t, will also be in an equatorial site. Either L^t or L^d may in principle now be lost but since L^t, as a good π-bonding ligand, is likely to be firmly bound, L^d, as the most labile equatorial ligand, in fact leaves to give the final product; L^d is thus labilized by L^t. Good π-acid ligands are high in the trans effect series because they prefer the more π-basic equatorial sites as a result of the metal being a better π donor to equatorial ligands in the TBP intermediate.

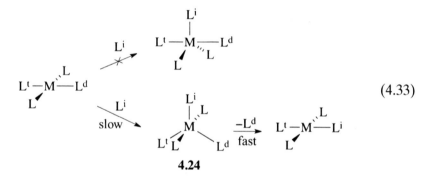

(4.33)

4.24

Solvent Participation

The solvent, present in high molarity, can act as an incoming ligand and expel L^d to give a solvated four-coordinate intermediate. A subsequent associative substitution with L^i then gives the final product. Substitutions of one halide by another on Pd(II) and Pt(II) can follow this route (Eq. 4.34).

$$L_2MCl_2 \xrightarrow[\text{slow}]{+\text{solv}} [L_2MCl(\text{solv})]^+ \xrightarrow[\text{fast}]{+Br^-} L_2MClBr$$

$$(M = \text{Pd or Pt}) \quad (4.34)$$

$$\text{Rate} = k_s[L_2MCl_2] + k_a[L_2MCl_2][Br^-] \quad (4.35)$$

Because it is cationic, the intermediate is much more susceptible to Br^- attack than the starting complex. Since the solvent concentration cannot be varied without introducing rate changes due to solvent effects, the [solv] term does not appear in the experimental rate equation, Eq. 4.35, where the first term refers to the solvent-assisted associative route, and the second to the direct associative reaction; k_s becomes less important for less strongly ligating solvents. When $k_a[L^i] << k_s$, the reaction can wrongly appear to be dissociative because the rate equation is now indistinguishable from Eq. 4.27. These types of complication have led to the dictum that mechanisms can never be said to be unambiguously proved, only that they have not yet been disproved.

Ligand Rearrangements

Eighteen-electron complexes can undergo associative substitution without forming an unfavorable 20e intermediate if a ligand can rearrange to leave a 2e vacancy to allow L^i to bind. Nitrosyls, with their 3e linear to 1e bent rearrangements, can do this. For example, $Mn(CO)_4(lin\text{-}NO)$ shows a second-order rate law (Eq. 4.37) and a negative entropy of activation, $\Delta S^{\ddagger}$, as expected for Eq. 4.36.

$$(CO)_4Mn(lin\text{-}NO) \xrightarrow[\text{slow}]{+L^i} (CO)_4Mn(bent\text{-}NO)L^i \xrightarrow[\text{fast}]{-CO} (CO)_3Mn(lin\text{-}NO)L^i$$
$$\text{18e} \qquad\qquad\qquad \text{18e} \qquad\qquad\qquad \text{18e}$$

$$\text{(4.36)}$$

$$\text{Rate} = k_a[\text{complex}][L^i] \qquad\qquad \text{(4.37)}$$

Likewise, η^5-indenyl complexes undergo associative substitution much faster than their Cp analogs. This results from the indenyl easily slipping from an η^5 to an η^3 structure (Eq. 4.38), favorable because the C_6 ring regains its full aromatic stabilization as the 8 and 9 carbons dissociate and participate fully in the C_6 ring aromaticity. Several other 3e/1e rearrangements are known, such as η^3/η^1-allyl, M(R)(CO)/ M(COR) and κ^2/κ^1-OAc.

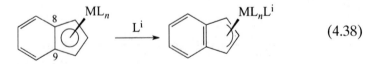

$$\text{(4.38)}$$

4.6   REDOX EFFECTS AND INTERCHANGE SUBSTITUTION

Two ways to make a coordination inert 18e complex give fast substitution are oxidation or reduction to a coordination labile 17e[36] or 19e intermediate. These are often too reactive to be isolable, but some are known, such as 19e Cp_2Co and 17e $[Re(CO)_3\{P(C_6H_{11})_3\}_2]$, where dissociation (19e) or association (17e) is inhibited.

17e and 19e Species

With an electron in an M–L σ^* orbital, 19e species[37] are much more dissociatively labile than their 18e counterparts. For example, $Fe(CO)_5$ can be substituted with electrochemical catalysis by a D mechanism, where 19e $[Fe(CO)_5]\cdot^-$ is the chain carrier in the catalytic cycle of Eq. 4.39. The initial product radical reduces the starting $Fe(CO)_5$ so the cycle can continue.

$$Fe(CO)_5 \xrightarrow{+e^-} [Fe(CO)_5]\cdot^- \xrightarrow[\text{fast}]{-CO} [Fe(CO)_4]\cdot^-$$

$$\text{18e} \qquad\qquad \text{19e} \qquad\qquad \text{17e}$$

$$[LFe(CO)_4] \xleftarrow{\quad Fe(CO)_5 \quad} [LFe(CO)_4]\cdot^-$$

$$\text{18e} \qquad\qquad\qquad\qquad \text{19e}$$

$$\downarrow L$$

(4.39)

The D substitution of $[(\eta^6\text{-ArH})Mn(CO)_3]^+$ by PPh_3 to give $[(\eta^6\text{-ArH})Mn(CO)_2(PPh_3)]^+$ goes in the same way.

Similarly, oxidation of a d^6 18e complex gives a coordination labile d^5 17e species that can give associative substitution. Very large rate accelerations can be seen: 17e $V(CO)_6$, for example, undergoes second-order, associative ligand exchange at 25°, while 18e $[V(CO)_6]^-$ does not do so even in molten PPh_3 (m.pt. 80°C). Substitution in an 18e species can often be catalyzed by oxidation and even a trace of air is sometimes enough, leading to irreproducibility problems. Electrochemical oxidation of $CpMn(CO)_2(MeCN)$ causes A substitution of MeCN by PPh_3 in a chain reaction with up to 250 molecules substituted for each electron abstracted. In Eq. 4.40, the initial product radical reoxidizes the starting material so the cycle can continue.

$$CpMn(CO)_2(MeCN) \xrightarrow{-e^-} [CpMn(CO)_2(MeCN)]\cdot^+ \xrightarrow[\text{fast}]{L} [CpMn(CO)_2(L)(MeCN)]\cdot^+$$

$$\text{18e} \qquad\qquad \text{17e} \qquad\qquad\qquad \text{19e}$$

$$CpMn(CO)_2(L) \xleftarrow{\quad CpMn(CO)_2(MeCN) \quad} [CpMn(CO)_2(L)]\cdot^+$$

$$\text{18e} \qquad\qquad\qquad\qquad\qquad \text{17e}$$

$$\downarrow -MeCN$$

(4.40)

Thermal M–M bond cleavage of L_nM–ML_n or abstraction of $X\bullet$ from L_nM–X by a radical initiator, $Q\bullet$, also provide ways of accessing substitutionally labile 17e ML_n intermediates.

Most 19e species are reactive transients but some are isolable, such as Tyler's $(\eta^5$-$Ph_4C_5H)Mo(CO)_2L_2$ [L_2 = 2,3-bis(diphenylphosphino)-maleic anhydride] and Astruc's[37] $CpFe(\eta^6$-arene) are stable 19e species. Mössbauer and EPR (electron paramagnetic resonance) data for the Fe(I) case suggested that the 19th electron is largely located on the metal; the X-ray crystal structure shows that all 11 carbons of both rings are coordinated, but the Fe–C(Cp) distances are 0.1Å longer than in analogous 18e species. Sometimes, the nineteenth electron is largely ligand based, as in $CoCp_2$. Likewise, some 17e species are isolable, such as $V(CO)_6$ and $[Cp*Fe(C_6Me_6)]^{2+}$, but most are only transients or else seen only at very low temperatures, such as $[Mn(CO)_5]$ or $[Co(CO)_4]$.

The Interchange Mechanism

Certain soft nucleophiles show a second-order, A component for substitution of $Mo(CO)_6$, a molecule that cannot rearrange to avoid 20e on L^i binding. Although 20e intermediates are not favored, a 20e transition state seems possible. An intermediate has to survive for many molecular vibrations, while a transition state need only survive for one ($\sim10^{-13}$ s). Although both L^i and L^d bind simultaneously to the metal in a 20e t.s., they do so weakly. This is the *interchange mechanisms of substitution*, designated I, with subcategories I_a and I_d, according to whether the transition state is closer to the A or D extreme. I_a and I_d are also invoked where the independent existence of the true A or D intermediate is doubtful; it is hard to detect a very short-lived intermediate.

4.7 PHOTOCHEMICAL SUBSTITUTION

Photochemical reactions can occur when light is absorbed by a compound. The ground-state electronic configuration is changed to that of one of the excited states by the resulting promotion of an electron. Promotion from the singlet ground state, S_0, initially gives the excited singlet, S_1. This can undergo *intersystem crossing*, with formation of the triplet state, T_1, which is now slow to return to the S_0 state because a spin flip would be needed. T_1 is therefore longer lived than S_1, but even so it only lives for 10^{-6}–10^{-9} s, and so if any photochemistry is to occur, T_1 must react very quickly and bimolecular steps involving external reagents are usually too slow to contribute; ligand dissociation is thus most often seen. If a molecule of product is formed for every photon absorbed, the quantum yield, Φ, is said to be unity or 100%. Otherwise the electron falls back to the ground state and the compound either emits light (luminescence) or produces heat; in this case, no chemistry occurs and Φ becomes <1.

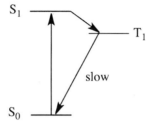

Carbonyls

In photochemical substitution of $W(CO)_6$, UV irradiation in thf gives $W(CO)_5(thf)$. This useful synthetic intermediate subsequently reacts in the dark with a variety of ligands L to give $W(CO)_5L$ cleanly, rather than the mixture of $W(CO)_{(6-n)}L_n$, obtained thermally. Light-induced promotion of a d_π electron to a d_σ M–L σ-antibonding level weakens the M–L bonds, allowing rapid dissociative substitution in the excited state. Knowing the UV–visible spectrum of the starting material is essential in planning the experiment. The complex must absorb light at the wavelength to be used, but if the product also absorbs at that wavelength, then subsequent photochemistry may occur. The buildup of highly absorbing decomposition products can also stop the photoreaction by absorbing all the light.

The photolysis of $W(CO)_5L$ can lead either to loss of L or of a CO group cis to L, depending on the wavelength. This can be understood in terms of a crystal field diagram (Fig. 4.6). Since the symmetry is lower than octahedral because of the presence of L, both the d_σ and the d_π levels split up in a characteristic pattern. The L ligand, conventionally placed on the z axis, is usually a lower-field ligand than CO, and so the d_{z^2} orbital is stabilized with respect to the $d_{x^2-y^2}$. As we saw in Section 1.7, these are really M–L σ^* orbitals, $d_{x^2-y^2}$ (σ^*_{xy}) playing this role for ligands in the xy plane, and d_{z^2} (σ^*_z) for the ligands along the z axis. This means that irradiation at v_1 tends to populate the σ^*_z, which labilizes L because it lies on the z axis. Irradiation at v_2 populates σ^*_{xy}, so a cis CO is labilized because it lies in the xy plane, cis to L. Where L is pyridine, the appropriate wavelengths are ~400 nm (v_1) and <250 nm (v_2), respectively. The

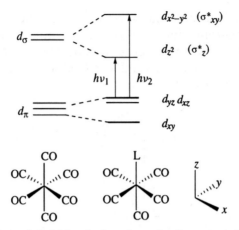

FIGURE 4.6 Crystal field basis for the selective photolysis of $M(CO)_5L$ complexes. Irradiation at a frequency ν_1 raises an electron from the filled $d\pi$ level to the empty $\sigma^*(z)$, where it helps to labilize ligands along the z axis of the molecule. Irradiation at ν_2 labilizes ligands in the xy plane.

method has often been used to synthesize *cis*-$M(CO)_4L_2$ complexes (M = W, Mo).

$W(CO)_4(phen)$ has near-UV and visible absorptions at 366 and 546 nm. The first corresponds to promotion of a d_π electron to the d_σ level and is referred to as a ligand field (LF) band. The 546-nm band is a *metal-to-ligand charge transfer* (or MLCT) band and corresponds to promoting a metal d_π electron to a π^* level of the dipy ligand; the long-lived MLCT excited state therefore contains a 17e substitution-labile metal and a reduced ligand $W^{\cdot+}(CO)_4(phen^-)$. Irradiation in either band leads to substitution by PPh_3, for example, to give $W(CO)_3(PPh_3)(phen)$.

Increased pressure accelerates an associative process because the volume of the transition state $L_nM\cdots L'$ is smaller than that of the separated L_nM and L'; the reverse is true for a dissociative process because $L_{n-1}M\cdots L$ has a larger volume than L_nM. Several hundred atmospheres are required to see substantial effects, however. Van Eldik has shown that pressure accelerates the MLCT photosubstitution of $W(CO)_4(phen)$ but decelerates the LF photosubstitution. As the MLCT excited state is effectively a 17e species, an A mechanism is reasonable for this process; the LF process is evidently a D mechanism, probably as a result of populating the M–L σ^* levels.

Thermal substitution in $(\eta^6\text{-}C_7H_8)Cr(CO)_3$ goes by loss of C_7H_8 because the triene binds much more weakly than CO. In contrast, photochemical substitution (366 nm) gives $(\eta^6\text{-}C_7H_8)Cr(CO)_2L$ because monodentate ligands are more affected by occupation of "their" σ^* orbital than the triene that binds simultaneously along all three axes of the molecule. Chelate ligands thus tend to be much more photostable than monodentate ones. The arene is lost in photosubstitution of $[CpFe(\eta^6\text{-}PhCH_3)]PF_6$, however, because the Cp is also polydentate and more strongly bound.

Other Photochemical Processes

In the photochemical homolysis of M–M bonds in L_nM–ML_n, the resulting $L_nM\bullet$ fragments are usually 17e and substitutionally labile. For example, photosubstitution of CO in Mn_2CO_{10} by PPh_3 to give $Mn_2(PPh_3)CO_9$ goes via 17e $\bullet Mn(CO)_5$. In substitution by NH_3, the replacement of three COs by the non-π-acceptor NH_3 leads to a buildup of electron density on the metal. This is relieved by an electron transfer from a 19e $Mn(CO)_3(NH_3)_3$ intermediate to 17e $\bullet Mn(CO)_5$ to give the disproportionation product **4.24** in a chain mechanism (Eq. 4.41). Soft PPh_3 is fully compatible with both Mn(0) and Mn(–I), but the hard NH_3 drives the conversion of Mn(0) to Mn(I).

$$\underset{Mn(0)}{Mn_2(CO)_{10}} \xrightarrow{h\nu,\ NH_3} \underset{Mn(I)}{[Mn(CO)_3(NH_3)_3]^+} \underset{Mn(-I)}{[Mn(CO)_5]^-} \tag{4.41}$$

$$\textbf{4.24}$$

Photolytic reductive elimination of H_2 can be followed by oxidative addition of a solvent C–H bond (Eq. 4.42).

$$Cp_2WH_2 \xrightarrow[-H_2]{h\nu,\ C_6H_6} Cp_2W(Ph)(H) \tag{4.42}$$

4.8   COUNTERIONS AND SOLVENTS IN SUBSTITUTION

Solvents and counterions can be coordinating and must be chosen so as not to interfere with substitution. Common solvents that are most likely to bind are MeCN, pyridine, Me_2SO (dimethylsulfoxide, DMSO), and Me_2NCHO (dimethylformamide, DMF). Several complexes dissolve only in such solvents, as a result of the solvent binding to the metal. DMF, Me_2NCHO, bonds via the O lone pair because the N lone pair is tied up by resonance ($Me_2N^+=CH-O^-$).

DMSO can bind either via the S or the O depending on both steric and hard and soft effects. Unhindered, soft Ru(II) gives S-bound

$[Ru(DMSO)_2(bipy)_2]^{2+}$ that converts to the O-bound form on photolysis and then reverts to the S-bound form in the dark. CS_2 finds restricted use in organometallic chemistry because it reacts with most complexes; liquid SO_2 can be useful as a low-temperature NMR solvent.

Tetrahydrofuran (THF), acetone, water, and ethanol are less strongly ligating and often used for late metals. Early d^0 complexes can react with protic solvents, however. Ketones usually bind in the η^1 mode via O, but the C=O bond can also bind in the η^2 mode, where back donation is strong and steric hindrance low. For example, the $[Os^{III}(NH_3)_5]^{3+}$ fragment prefers η^1-acetone binding, but reduction to the very strong π-donor $[Os^{II}(NH_3)_5]^{2+}$ leads to rearrangement to the η^2 form.

Halocarbon solvents tend to be oxidizing and can destroy sensitive compounds. $PhCF_3$ is a useful, less oxidizing alternative to CH_2Cl_2. Halocarbons can form stable complexes, some of which have been crystallographically characterized, such as $[IrH_2(IMe)_2(PPh_3)_2]^+$.

Arenes can in principle bind to metals, but the reaction is usually either sufficiently slow or thermodynamically unfavorable to permit the satisfactory use of arenes as solvents without significant interference. Alkanes are normally reliably noncoordinating (but see Section 12.4). Many complexes do not have sufficient solubility in the usual alkanes, but solvents such as ethylcyclohexane are significantly better because the solvent molecules pack poorly, allowing easier formation of pores in the liquid structure that provide homes for solute molecules. IR spectra are best recorded in alkanes because the weak solvent–solute interactions give minimal interference with the solute and thus yield the sharpest absorption peaks.

The rise of green chemistry has led to development of lists of solvents[38] ranked according to hazard and sustainability criteria that are now followed for process development work by the main pharmaceutical industries. This means that future organic method development research also needs to take this factor into account when designing procedures.

"Noncoordinating" Anions

In complex salts, counterion choice is important to prevent undesired reactions. BF_4^-, although useful, can form a B–F–M bridge or undergo F^- abstraction to give an M–F complex, particularly with d^0 metals. PF_6^- is less reactive but can still give problems.[39] BPh_4^- can form η^6-PhBPh$_3$ complexes. The "barf" anion (**4.25**), one of our very best noncoordinating anions,[40] permits isolation of such electrophilic, low-coordinate cations as 14e $[IrH_2(PR_3)_2]^+$. Even so, undesired aryl transfer to M is sometimes seen.[41] Among noncoordinating cations, $[PPh_4]^+$ and $[Ph_3P=N=PPh_3]^+$ are useful. In all cases, the counterions of choice are large, so as to stabilize the ionic lattice of the organometallic counterion, also large. In the low dielectric medium of an organic solvent like CH_2Cl_2, ion pairs readily form, affecting reactivity.[42]

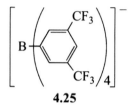

4.25

- Increased back bonding to CO lowers $v(CO)$, weakens the CO bond, and decreases the electrophilicity of carbon (Section 4.1).
- PR_3 ligands can be predictably tuned sterically and electronically (Fig. 4.3).
- First-order dissociative substitution (Section 4.4) is typical for 18e complexes, otherwise second-order associative substitution is often seen (Section 4.5).

REFERENCES

1. L. Mond, C. Langer, and F. Quincke, *J. Chem. Soc.*, **57**, 749, 1889.

2. A. J. Lupinetti, S. H. Strauss, and G. Frenking, *Progr. Inorg. Chem.*, **49**, 1, 2001.

3. J. C. Green, M. L. H. Green, and G. Parkin, *Chem. Commun.*, **48**, 11481, 2012.

4. T. W. Hayton, P. Legzdins, and W. B. Sharp, *Chem. Rev.*, **102**, 935, 2002.

5. Z.-S. Lin, T.-W. Chiou, K.-Y. Liu, C.-C. Hsieh, J. S. K. Yu, and W.-F. Liaw, *Inorg. Chem.*, **51**, 10092, 2012.

6. M. G. Scheibel, I. Klopsch, H. Wolf, P. Stollberg, D. Stalke, and S. Schneider, *Eur. J. Inorg. Chem.*, **2013**, 3454.

7. S. Moncada and E. A. Higgs, *Brit. J. Pharmacol.*, **147**, S193, 2006.

8. M. Shatruk, C. Avendano, and K. R. Dunbar, *Progr. Inorg. Chem.*, **56**, 155, 2009.

9. H. Braunschweig, K., Radacki, and A. Schneider, *Science*, **328**, 345, 2010; X. Gong, Q. Li, Y. Xie, R. B. King, and H. F. Schaefer III, *Inorg. Chem.*, **49**, 10820, 2010.

10. D. Blum, *The Poisoner's Handbook*, Penguin, New York, 2010; Y. Alarie, *Crit. Rev. Toxicol.*, **32**, 259, 2002.

11. M. M. Rodriguez, E. Bill, W. W. Brennessel, and P.L. Holland, PL, *Science*, **334**, 780, 2011.

12. L. Xu, Q. Li, Y. Xie, R.B. King, and H.B. Schaefer III, *New J. Chem.*, **34**, 2813, 2010.

13. B. Stewart, A. Harriman, and L. J. Higham, *Organometallics*, **30**, 5338, 2011.

14. M. Yamanaka and K. Mikami, *Organometallics*, **24**, 4579, 2005.

15. (a) C. A. Tolman, *Chem. Rev.*, **77**, 313, 1977; (b) K. A. Bunten, L. Z. Chen, A. L. Fernandez, and A. J. Poe, *Coord. Chem. Rev.*, **233**, 41, 2002.

16. L. Perrin, E. Clot, O. Eisenstein, J. Loch, and R. H. Crabtree, *Inorg. Chem.*, **40**, 5806, 2001.

17. A. B. P. Lever, *Inorg. Chem.*, **30**, 1991, 1980.

18. J. H. Rivers and R. A. Jones, *Chem. Commun.*, **46**, 4300, 2010.

19. K. D. Cooney, T. R. Cundari, N. W. Hoffman, K. A. Pittard, M. D. Temple, and Y. Zhao, *J. Am. Chem. Soc.*, **125**, 4318, 2003.

20. M. -N. Birkholz, F. Zoraida, and P. W. N. M. van Leeuwen, *Chem. Soc. Rev.*, **38**, 1099, 2009.

21. C. Müller, Z. Freixa, M. Lutz, A. L. Spek, D. Vogt, and P. W. N. M. van Leeuwen, *Organometallics*, **27**, 834, 2008.

22. G. C. Fortman and S. P. Nolan, *Chem. Soc. Rev.*, **40**, 5151, 2011; T. Droege, F. Glorius, *Angew. Chem. Int. Ed.*, **49**, 6940, 2010.

23. D. G. Gusev, *Organometallics*, **28**, 6458, 2009.

24. M. Alcarazo, T. Stork, A. Anoop, W. Thiel, and A. Fürstner, *Angew. Chem. Int. Ed.*, **49**, 2542, 2010.

25. J. A. Mata, M. Poyatos, and E. Peris, *Coord. Chem. Rev.*, **251**, 841, 2007.

26. J. A. Mata, A. R. Chianese, J. R. Miecznikowski, M. Poyatos, E. Peris, J. W. Faller, and R. H. Crabtree, *Organometallics*, **23**, 1253, 2004.

27. E. Peris, *Top. Organomet. Chem.*, **21**, 83, 2007.

28. I. J. B. Lin and C. S. Vasam, *Coord. Chem. Rev.*, **251**, 642, 2007.

29. Y. Han and H. V. Huynh, *Dalton Trans.*, **40**, 2141, 2011.

30. J. Müller, K. Öfele, G. Krebs, *J. Organometal. Chem.*, **82**, 383, 1974; R. H. Crabtree, *Coord. Chem. Rev.*, **257**, 755, 2013.

31. M. Melaimi, M. Soleilhavoup, and G. Bertrand, *Angew. Chem. -Int. Ed.*, **49**, 8810, 2010; O. Schuster, L. R. Yang, H. G. Raubenheimer, and M. Albrecht, *Chem. Rev.*, **109**, 3445, 2009.

32. O. Rivada-Wheelaghan, M. A. Ortuno, J. Diez, A. Lledos, and S. Conejero, *Angew. Chem. -Int. Ed.*, **51**, 3936, 2012.

33. N. G. Leonard, P. G. Williard, and W. H. Bernskoetter, *Dalton Trans.*, **40**, 4300, 2011.

34. G. Berthon-Gelloz, B. de Bruin, B. Tinant, and I. E. Markó, *Angew. Chem., Int. Ed.*, **48**, 3161, 2009; N. Takagi and S. Sakaki, *J. Am. Chem. Soc.*, **134**, 11749, 2012.

35. M. Besora, J. -L. Carreón-Macedo, A. J. Cowan, M. W. George, J. N. Harvey, P. Portius, K. L. Ronayne, X. -Z. Sun, and M. Towrie, *J. Am. Chem. Soc.*, **131**, 3583, 2009; S. Shaik, H. Hirao, and D. Kumar, *Acc. Chem. Res.*, **40**, 532, 2007.

36. F. Zobi, O. Blacque, R. A. Jacobs, M. C. Schaub, and A. Y. Bogdanova, *Dalton Trans.*, **41**, 370, 2012.

37. D. Astruc, *Acc. Chem. Res.*, **33**, 287, 2000.

38. R. K. Henderson, C. Jiménez-González, D. J. C. Constable, S. R. Alston, G. G. A. Inglis, G. Fisher, J. Sherwood, S. P. Binksa, and A. D. Curzons, *Green Chem.*, **13**, 854, 2011.

39. P. de Fremont, N. Marion, and S. P. Nolan, *J. Organometal. Chem.*, **694**, 551, 2009.

40. W. E. Geiger, F. Barriere, *Acc. Chem. Res.*, **43**, 1030, 2010.

41. H. Salem, L. J. W. Shimon, G. Leitus, L. Weiner, and D. Milstein, *Organometallics*, **27**, 2293, 2008.

42. E. Clot, *Eur. J. Inorg. Chem.*, **2319**, 2009.

PROBLEMS

4.1. (a) Would you expect 18e metal carbonyl halides $M(CO)_nX$, X = halide, to dissociate into halide anions and the metal carbonyl cation as easily as the hydrides, X = H, dissociate into H^+ and the metal carbonyl anion? (b) Given that we have a case where both of the above processes occur, contrast the role of the thf solvent in the two cases.

4.2. $Ni(CO)_4$ and $Co(lin\text{-}NO)(CO)_3$ are both tetrahedral. Why does the Ni compound undergo dissociative substitution and the Co compound undergo associative substitution?

4.3. List the following in the order of decreasing reactivity you would predict for the attack of trimethylamine oxide on their CO groups: $Mo(CO)_6$, $Mn(CO)_6^+$, $Mo(CO)_2(dpe)_2$, $Mo(CO)_5^{2-}$, $Mo(CO)_4(dpe)$, and $Mo(CO)_3(NO)_2$.

4.4. What single piece of physical data would you choose to measure as an aid to establishing the reactivity order of the carbonyl complexes of Problem 4.3?

4.5. What are the oxidation states and d^n configurations numbers of the metals in all the species depicted in Eq. 4.39 and Eq. 4.40?

4.6. Amines, NR_3, are usually only weakly coordinating toward low-valent metals. Why is this so? Do you think that NF_3 would be a better ligand for these metals? Discuss the factors involved.

4.7. Ligand dissociation from NiL_4 is only very slight for $L = P(OMe_3)$, yet for $L = PMe_3$, it is almost complete. Given that the two ligands have essentially the same cone angle, discuss the factors that might be responsible.

4.8. Determine whether associative or dissociative substitution is more likely for the following species (not all of which are stable): $CpFe(CO)_2L^+$, $Mn(CO)_5$, $Pt(PPh_3)_4$, $ReH_7(PPh_3)_2$, $PtCl_2(PPh_3)_2$, and $IrCl(CO)(PPh_3)_2$.

4.9. Propose plausible structures for complexes with the following empirical formulas: $Rh(cod)(BPh_4)$, $(indenyl)_2W(CO)_2$, $PtMe_3I$, $(cot)(PtCl_2)_2$, and $(CO)_2RhCl$.

4.10. Given a complex $M(CO)_6$ undergoing substitution with an entering ligand L', what isomer(s) would you expect to find in the products if L' were (a) monodentate and a higher-trans-effect ligand than CO, or (b) L' were bidentate and had a lower trans effect than CO?

4.11. NO^+ is isoelectronic with CO and often replaces CO in a substitution reaction, so it might seem that Eq. 4.43 should be a favorable reaction. Comment on whether the process shown is likely.

$$Mo(CO)_6 + NOBF_4 \rightarrow Mo(NO)_6(BF_4)_6 + 6CO \qquad (4.43)$$

4.12. $Fe(CO)_5$ loses CO very slowly, but in the presence of an acid, substitution is greatly accelerated. Suggest possible explanations. For dissociative CO substitutions, the rate tends to be higher as the $v(CO)$ stretching frequency of the carbonyl increases. Suggest a reason.

4.13. Use the data of Table 2.10 to predict the position of the highest frequency $v(CO)$ band in $[Co(CO)_6]^{3+}$ and comment on the result in connection with deciding whether this hypothetical species would be worth trying to synthesize.

4.14. Tertiary amines, such as NEt_3, tend to form many fewer complexes with low-valent metals (e.g., $W(0)$) than PEt_3. What factors make two cases so different? In spite of this trend, $(Et_3N)W(CO)_5$ is isolable. What factors are at work to make this species stable?

4.15. Given a suitable L_nM fragment, would you expect X-phos (**4.11**) to be able to cyclometallate at the aryl C–H bond? What factors are relevant?

4.16. Draw all the resonance forms for the free carbenes **4.19**, **4.20**, **4.21**, **4.22**, and **4.23** to justify their classification as normal or abnormal.

5

Pi-COMPLEXES

Continuing our survey of the different types of ligand, we now turn to π complexes in which the metal interacts with the π bonding electrons of a variety of unsaturated organic ligands.

5.1 ALKENE AND ALKYNE COMPLEXES

In 1827, the Danish chemist William Zeise (1789–1847) obtained a new compound from the reaction of K_2PtCl_4 and EtOH that he took to be the solvated double salt, $KCl \cdot PtCl_2 \cdot EtOH$. Only in the 1950s was it established that Zeise's salt is really a π complex of ethylene, $K[PtCl_3(\eta^2\text{-}C_2H_4)] \cdot H_2O$, the ethylene being formed by dehydration of the ethanol. In Zeise's anion, **5.1**, the metal is located out of the C_2H_4 plane so that it can interact with the alkene π bond. The $M–(C_2H_4)$ σ bond involves donation of the C=C π electrons to an empty $M(d_\sigma)$ orbital, so this electron pair is now delocalized over three centers, M, C, and C′. The $M–(C_2H_4)$ back bond involves donation from $M(d_\pi)$ to the C=C π* orbital (**5.2**). As we saw for CO, a σ bond is insufficient for significant M–L binding, and so only d^2–d^{10} metals, capable of back donation, bind alkenes well.

The Organometallic Chemistry of the Transition Metals, Sixth Edition.
Robert H. Crabtree.
© 2014 John Wiley & Sons, Inc. Published 2014 by John Wiley & Sons, Inc.

The applicable bonding model depends on the strength of the back donation. The *Dewar–Chatt* (D-C, **5.3**) model holds for weak back bonding and the *metalacyclopropane* (MCP, **5.4**) model for strong back bonding. Experimental structures can fall anywhere between the two extremes. For Zeise's salt and other intermediate oxidation state late metals, the D-C model fits best, while for Pt(0), the MCP model applies.[1] Both cases are considered η^2 structures.

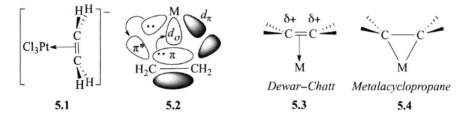

		Dewar–Chatt	Metalacyclopropane
5.1	**5.2**	**5.3**	**5.4**

The alkene C=C bond length, d_{CC}, increases on binding for two reasons. The M−alkene σ bond depletes the C=C π bond by donation to M and so slightly weakens and lengthens d_{CC}. The major factor in raising d_{CC}, however, is back donation from the metal that lowers the alkene C–C bond order by filling C=C π^*. For weakly π-basic Pt(II) (**5.1**), the D-C model means this reduction is slight, d_{CC}, being 1.375 Å, closely resembling free C_2H_4 ($d_{CC} = 1.337$ Å). In contrast, for strongly π basic Pt(0), as in $[Pt(PPh_3)_2C_2H_4]$, the MCP model applies, d_{CC} lengthens to 1.43 Å, and the C–H bonds fold back strongly. An MCP C_2H_4 resembles the $[C_2H_4]^{2-}$ dianion with the carbons rehybridized from sp^2 (D-C) toward sp^3 (MCP). The MCP extreme resembles **5.4**, with L_nM replacing one CH_2 in cyclopropane, hence the name of the model. Electron-withdrawing substituents on carbon encourage back donation and strengthen the M-(alkene) bond; for example, $Pt(PPh_3)_2(C_2CN_4)$ has a d_{CC} of 1.49 Å, approaching the C–C single bond d_{CC} of 1.54 Å. The bonds to the four substituents of the alkene, H atoms in the case of ethylene, are bent away from the metal to a small extent in the D-C case but to a much bigger extent in an MCP complex.

In the D-C extreme, the ligand predominantly acts as a simple L donor like PPh_3, but in the MCP extreme, we have a cyclic X_2 dialkyl, as if an oxidative addition of the C=C π bond had taken place. In both cases, we have a 2e ligand on the covalent model, but while the D-C formulation (L), **5.3**, leaves the oxidation state unchanged, the MCP picture (X_2), **5.4**, adds two units to the formal oxidation state. By convention, the D-C model is always adopted for the assignment of the formal oxidation state to avoid ambiguity, because there is no sharp boundary between the D-C and MCP extremes.

TABLE 5.1 Dewar–Chatt versus Metalacyclopropane Bonding Models

Property	Dewar–Chatt (D-C)	Metalacyclopropane (MCP)
Back bonding	Weak	Strong
C=C bond order	1.5–2	1–1.5
Charge on vinyl carbon	∂^+	∂^-
Vinyl C–H bonds	Near coplanar with C=C	Strongly folded back
Hybridization of carbon	Near sp^2	Near sp^3
Typical metal	Late metal, intermediate OS	Early metal or low OS

Note: OS = oxidation state.

The d_{CC} helps determine where any given alkene complex lies on the D-C/MCP continuum. The coordination-induced shift of any vinyl protons, or of the vinyl carbons in the ^{1}H and ^{13}C NMR spectra, also shows a correlation with the structure. For example, at the MCP extreme, the vinyl protons can resonate 5 ppm, and the vinyl carbons 100 ppm to high field of their position in the free ligand, owing to change of hybridization from sp^2 to $\sim sp^3$ at carbon. Coordination shifts are much lower for the D-C extreme.

Greater MCP character is favored by strong donor coligands, a net negative charge on a complex ion, and a low metal oxidation state. This means that Pd(II), Hg(II), Ag(I), and Cu(I) alkene complexes tend to be D-C, while those of Ni(0), Pd(0), and Pt(0), tend to be MCP.

Dewar–Chatt alkenes have a ∂^+ charge on carbon because the ligand-to-metal σ donation that depletes charge on the C=C ligand is not compensated by back donation. The vinyl carbons are therefore subject to nucleophilic attack but are resistant to electrophilic attack, Pd(II), being the classic case in which this applies. Since free alkenes are subject to electrophilic but not nucleophilic attack, binding therefore inverts the chemical character of the alkene, a phenomenon known as *umpolung*. The metal can either promote nucleophilic attack or inhibit electrophilic attack at the ethylene carbons, and so can either act as an activating group or a protecting group, depending on the substituents, metal, and coligands.

Strained alkenes, such as cyclopropene or norbornene (**5.5**), bind more strongly than unstrained ones. When the C–C=C angles are constrained to be much smaller than the sp^2 ideal of 120° (e.g., 107° in **5.5**), relief of strain on complexation strengthens metal binding because the ideal angles at the metal-bound vinylic carbons drop from the sp^2 ideal of 120° much closer to the sp^3 ideal of 109°, reducing the C–C=C angle strain.

5.5

Synthesis

Alkene complexes are usually synthesized by the methods shown in Eq. 5.1–Eq. 5.7:

1. Substitution in a low-valent metal:

$$[Ag(OH_2)_2]O_3SCF_3 + C_2H_4 \rightleftharpoons [Ag(C_2H_4)(OH_2)]O_3SCF_3 + H_2O \quad (5.1)$$

$$PtCl_4^{2-} + C_2H_4 \rightarrow [PtC_3(C_2H_4)]^- + Cl^- \quad (5.2)$$

$$Cp(CO)_2Fe \overset{+}{-} \|\quad \xrightarrow{nBu} \quad Cp(CO)_2\overset{+}{Fe} - \| \quad (5.3)$$
$$_{nBu}$$

2. Reduction of a higher-valent metal in the presence of an alkene:

$$\text{Pt} \overset{Cl}{\underset{Cl}{<}} \quad \xrightarrow{[C_8H_8]^{2-}} \quad \text{Pt} \quad (5.4)$$

$$RhCl_3 + CH_3CH_2OH \longrightarrow [\quad Rh(\mu\text{-}Cl)]_2 + CH_3CHO \quad (5.5)$$

3. From β-elimination of alkyls and related species:

$$(CO)_5Mn \diagdown \diagup H^+ \longrightarrow (CO)_5\overset{+}{Mn} - \| \diagup_H \quad (5.6)$$

$$Cp_2TaCl_3 \xrightarrow{BuMgX} Cp_2TaBu_3 \diagdown Cp_2TaHBu_2$$
$$\text{unstable}$$
$$\qquad \qquad \qquad \qquad \qquad \qquad \qquad \qquad (5.7)$$
$$\overset{H}{\underset{Cp_2Ta}{|}} - \| \diagup \quad \longleftarrow \quad Cp_2TaBu$$

Reversible binding of alkenes to Ag^+ (Eq. 5.1) leads to alkene separation on Ag^+-doped gas chromatography columns. Eq. 5.3 shows how
less hindered alkenes usually bind more strongly. The reducing agent
in Eq. 5.4 is the $[C_8H_8]^{2-}$ anion, which the authors may have intended
to act as a ligand. On reduction, the square planar d^8 Pt(II) converts
to tetrahedral d^{10} Pt(0). Ethanol is the reductant in Eq. 5.5 by the
β-elimination mechanism of Eq. 3.27. Protonation at the terminal methylene in the η^1-allyl manganese complex of Eq. 5.6 creates a carbonium
ion having a metal at the β position. Since the carbonium ion is a zero-
electron ligand like a proton, it can coordinate to the 18e metal to give
the alkene complex. Equation 5.7 (Bu = *n*-butyl) shows β-elimination,
a common result of trying to make a metal alkyl with a β hydrogen.

Reactions

Alkene insertions into M–X bonds to give alkyls (Eq. 3.20 and Eq. 3.21)
go very readily for X=H; insertion into other M–X bonds is harder.
Strained alkenes, fluoroalkenes, and alkynes insert most readily—relief
of strain is again responsible.

$$PtHCl(PEt_3)_2 + C_2H_4 \rightleftharpoons PtEtCl(PEt_3)_2 \qquad (5.8)$$

$$AuMe(PPh_3) + C_2F_4 \rightleftharpoons Au(CF_2CF_2Me)(PPh_3) \qquad (5.9)$$

With a weakly basic metal, the D-C model **(5.3)** applies, the vinylic
carbons become δ^+ and often undergo nucleophilic attack (e.g., Eq.
5.10). This is an example of a more general reaction type—nucleophilic
attack on polyenes or polyenyls (Section 8.3).

$$(NHMe_2)Cl_2Pt \overset{}{-}\!\!\parallel\!\!\overset{}{-}:NHMe_2 \longrightarrow (NHMe_2)Cl_2\bar{Pt} \diagdown\!\!\diagup \overset{+}{N}HMe_2 \qquad (5.10)$$

Alkenes with allylic hydrogens can undergo C–H oxidative addition to
give an allyl hydride complex. In the example of Eq. 5.11, a base is also
present to remove HCl from the metal.

$$[Cl_3Pt\!-\!Cl]^{2-} \longrightarrow Cl_3\bar{Pt}\!-\!\parallel \underset{-Cl^-}{\overset{base}{\underset{-HCl}{\longrightarrow}}} \quad Pt \overset{Cl}{\underset{Cl}{<}} Pt \qquad (5.11)$$

Other X=Y ligands can bind in the same way, for example, O_2 usually
gives MCP adducts, such as $[(\eta^2\text{-}O_2)IrCl(CO)L_2]$ with an O–O single
bond, but it can also form D-C adducts where the ligand is best considered a singlet O=O group, as in $[(\eta^2\text{-}O_2)RhCl(NHC)_2]$.[2]

Alkyne Complexes

The MCP model (**5.6**) is the most appropriate description when alkynes act as 2e donors. Having more electronegative sp carbons, they get more back donation and bind more strongly than alkenes. The substituents fold back from the metal by 30°–40° in the complex, and the M–C distances are slightly shorter than for alkene complexes. A few homoleptic examples exist, such as $[M(cyclooctyne)_n]^+$ ($M = Au$, $n = 2$; $M = Cu$, $n = 3$). More interestingly, alkynes can form what appear to be coordinatively unsaturated complexes. For example, **5.7** is 16e if we count the alkyne as a 2e donor. In such cases, the alkyne can be a 4e donor by involving its second C=C π-bonding e pair, which lies at right angles to the first.[3] **5.7** can now be formulated as an 18e complex. An extreme valence bond formulation of the 4e donor form is the bis-carbene (**5.8**). Four electron alkyne complexes are rare for d^6 metals because of a 4e repulsion between the filled metal d_π and the second alkyne C=C π-bonding pair.

Cyclohexyne and benzyne, highly unstable in the free state, bind very strongly to metals, as in $[(Ph_3P)_2Pt(\eta^2\text{-cyclohexyne})]$ or the product in Eq. 5.12; strain is again partially relieved on binding. Cyclobutyne, inaccessible in the free state, has been trapped as its triosmium cluster complex.

$$Cp*Me_2Ta\diagup^{Me} \xrightarrow[-MeH]{heat} Cp*Me_2Ta \qquad (5.12)$$

Alkynes readily bridge an M–M bond, in which case they are 2e donors to each metal (**5.9**). The alternative tetrahedrane form (**5.10**) is the equivalent of the MCP picture for such a system. 1-Alkynes, RCCH, can easily rearrange by an intramolecular proton transfer process to vinylidenes, $RHC=C=M$.[4]

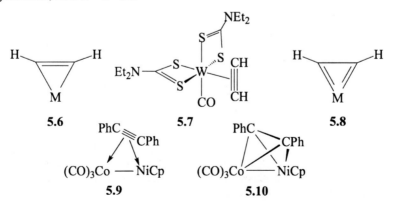

5.2 ALLYLS

The allyl group is commonly a reactive actor ligand in catalysis by undergoing nucleophilic attack.[5] It either binds in the monohapto form as a 1e X ligand (**5.11**) or in the trihapto form (**5.12**) as a 3e LX enyl ligand with resonance forms **5.13a** and **5.13b**.

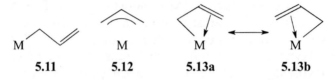

Figure 5.1*a* shows that the allyl ψ_1 can interact with a suitable metal d_σ orbital and ψ_2 with an $M(d_\pi)$ orbital, the filling of the MOs of the allyl radical being shown in Fig. 5.1*b*. Two structural peculiarities of η^3-allyl

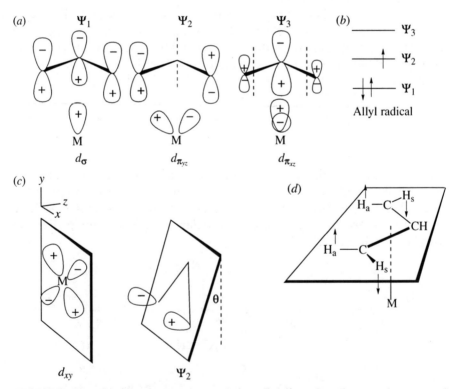

FIGURE 5.1 Electronic structure of the allyl ligand and some features of metal–allyl bonding. Nodes are shown as dotted lines in (*a*). Electron occupation in the allyl radical is shown in (*b*). The canting of the allyl is seen in (*c*), and the twisting of the CH₂ groups in (*d*).

complexes can be understood on this picture. First, the plane of the allyl is canted with respect to the xy plane at an angle θ—usually 5–10°—thus improving the interaction between ψ_2 and the d_{xy} orbital on the metal, as seen in Fig. 5.1c. Second, the terminal CH_2 group of the allyl rotates in the direction shown by the arrows in Fig. 5.1d. This allows the p orbital on this carbon to point more directly toward the metal, thus further improving the overlap.

The η^3-allyl group often shows exchange of the syn and anti substituents. Note the nomenclature of these substituents, which are syn or anti with respect to the central C–H. A common mechanism goes through an η^1-allyl intermediate, as shown in Eq. 5.13. This kind of exchange can affect the appearance of the ^1H NMR spectrum (Section 10.2), and also means that in an allyl complex of a given stereochemistry, R_{syn} may rearrange to R_{anti}.

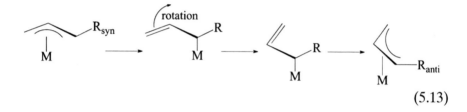

$$(5.13)$$

Synthesis

Typical routes to allyl complexes follow.

1. From an alkene (see also Eq. 5.11):

$$(5.14)$$

2. From attack of an allyl nucleophile on the metal:

$$(5.15)$$

3. From attack of an allyl electrophile on the metal:

$$(5.16)$$

4. From a conjugated diene:

$$\left[\begin{array}{c}\text{—Fe(CO)}_3\end{array}\right] \xrightarrow{\text{H}^+} \left[\begin{array}{c}\text{Fe(CO)}_3\end{array}\right]^+ \tag{5.17}$$

$$\text{Cp}_2\text{TiCl} \xrightarrow{i\text{PrMgBr}} \text{Cp}_2\text{Ti}(i\text{Pr}) \xrightarrow{\hspace{1cm}} \text{Cp}_2\text{TiH} \xrightarrow{\hspace{1cm}} \text{Cp}_2\text{Ti}\!-\!\!\left\langle\!\!\left\langle \tag{5.18}\right.\right.$$

$$(\text{R}_3\text{P})_2\overset{+}{\text{Pt}}\overset{\text{H}}{\underset{\text{O}=\!\!\!<}{\diagup}} \xrightarrow{\text{H}_2\text{C}=\text{C}=\text{CH}_2} (\text{R}_3\text{P})_2\overset{+}{\text{Pt}}\!-\!\!\left\rangle\!\!\left\rangle \tag{5.19}$$

The first route we saw in Section 5.1; the second and third resemble the synthetic reactions most commonly used for alkyl complexes. In Eq. 5.15 and Eq. 5.16, the metal reacts with the sterically slim terminal CH_2 group, and Eq. 5.17 shows an electrophilic attack on a diene complex. Equation 5.18 shows that when a C=C group of a diene undergoes insertion into an M–H bond, the hydrogen tends to add to the terminal carbon (Markovnikov's rule). The resulting methylallyl can become η^3 if a vacant site is available. In Eq. 5.19, when an allene inserts into an M–H bond, the hydride adds to the central carbon to give an allyl.

Reactions

The key reactions of allyls follow (Eq. 5.20–Eq. 5.23):

1. With nucleophiles:

$$\text{Cp(NO)COMo}\!-\!\!\left\rangle\!\!\left\rangle^+ \xrightarrow{\text{Nu}^-} \text{Cp(NO)COMo}\underset{\hspace{0.5cm}}{\overset{\text{Nu}}{\diagdown}} \tag{5.20}$$

2. With electrophiles:

$$\text{Cp(CO)}_2\text{Fe}\!-\!\!\diagup\!\!=\!\! \xrightarrow{\text{E}^+} \text{Cp(CO)}_2\text{Fe}\!-\!\!\left[\!\!\right.^{\text{E}} \tag{5.21}$$

3. By insertion:

$$\left\langle\!\!\left\langle-\text{Ni}-\!\!\right\rangle\!\!\right\rangle \xrightarrow{\text{CO}_2} \left\langle\!\!\left\langle-\text{Ni}\overset{\text{O}}{\underset{\text{O}}{\diagdown}}\!\!\diagdown\!\!\diagup\!\!=\!\! \right. \tag{5.22}$$

4. With reductive elimination (Eq. 5.23):

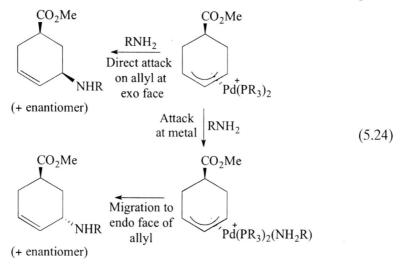

$$(5.23)$$

Nucleophilic attack at an allyl normally takes place from the exo face — the one opposite to the metal. A nucleophile that first attacks the metal, however, can transfer to the endo face of the allyl but this can only happen if a 2e vacancy is made available at the metal; both routes occur in Eq. 5.24.

$$(5.24)$$

Related Ligands

If a 2e vacancy is available, η^1-benzyl groups can convert to η^3, but the aromatic C=C bond is a weak ligand, so reversion to η^1 is easy. The η^3-benzyl complex of Eq. 5.25 is formed via arene ring CH oxidative addition, followed by rearrangement. Propargyl $(CH_2-C\equiv CH)^-$ can either be η^1 or convert to an η^3-allenyl $(CH_2=C=CH)^-$. The η^3-propargyl complex of Eq. 5.26 is formed by hydride abstraction from the methyl group of an η^2-2-butyne. Cyclopropenyl complexes, such as $(\eta^3\text{-}Ph_3C_3)Co(CO)_3$, are rare.

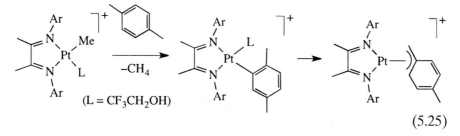

$$(5.25)$$

$$Cp^*(CO)_2Re \longrightarrow \| \quad \xrightarrow[-Ph_3CH]{Ph_3C^+} \quad Cp^*(CO)_2Re \longrightarrow \left. \right\rangle\!\!\rangle \quad \rceil^+ \qquad (5.26)$$

5.3 DIENE COMPLEXES

Nonconjugated dienes, such as 1,5-cyclooctadiene (cod), and norborna-diene (nbd), can chelate and thus bind more strongly than monoenes, but conjugated dienes behave differently. Butadiene usually acts as a 4e donor in its s-cis conformation, **5.14**. Weak back donation favors the D-C L_2 diene form, **5.14**, while the MCP LX_2 (enediyl) form **5.15**, results from strong back bonding. Compared with C_2H_4, butadiene has a lower π^* energy, and is thus a better π acceptor, so the diene D-C form is less important. In the typical case of $[(\eta^4\text{-butadiene})Fe(CO)_3]$, intermediate D-C /MCP character is evident from the near-equality of the C_1-C_2, C_2-C_3, and C_3-C_4 distances (~ 1.46 Å) and the longer M-C_1 and -C_4 distances versus M–C_2 and –C_3. In contrast, bound to the strongly back-donating d^2 Hf(PMe$_3$)$_2$Cl$_2$ group, 1,2-dimethylbutadiene shows a more pronounced LX_2 enediyl pattern. The C_1 and C_4 substituents twist 20–30° out of the plane of the ligand and bend back so that the C_1 and C_4 p orbitals can overlap better with Hf (**5.16**). The C_1–C_2, and C_3–C_4 distances (av. 1.46 Å) are longer than C_2–C_3 (1.40 Å), and M–C_2 and –C_3 are longer than M–C_1 and –C_4 in this case.

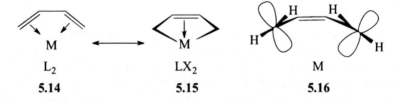

M	M	M
L_2	LX_2	
5.14	**5.15**	**5.16**

The butadiene frontier orbitals, ψ_2 (HOMO) and ψ_3 (LUMO), domi-nate bonding to the metal. The MO diagram of Fig. 5.2 shows that both the depletion of electron density in ψ_2 by σ donation to the metal and population of ψ_3 by back donation from the metal should lengthen C_1–C_2 and shorten C_2–C_3 because ψ_2 is C_1C_2 bonding and ψ_3 is C_2C_3 antibonding. Protonation can occur at C_1 (Eq. 5.17) where the HOMO, ψ_2, has its highest coefficient.

This bonding pattern is general for soft ligands: M–L binding usually depletes the ligand HOMO and back bonding partially fills the ligand

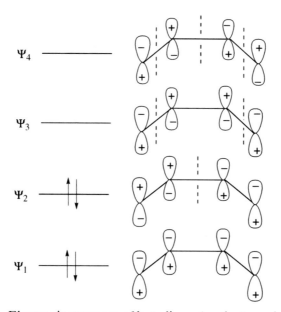

FIGURE 5.2 Electronic structure of butadiene. An electron-rich metal tends to populate Ψ_3; an electron-poor metal tends to depopulate Ψ_2.

LUMO, resulting in a profound change of the chemical character of L (Section 2.6). In another general p M–L, the structure of bound L often resembles its first excited state, L*, because to reach L* from L we promote an electron from HOMO to LUMO, thus depleting the former and filling the latter, as is also the case in M–L bonding. For example, CO_2 is linear in the free state but bent both in the first excited state and as an η^2 ligand.

Diene complexes can be synthesized from the free diene or by nucleophilic attack on a cyclohexadienyl complex (Eq. 5.27).

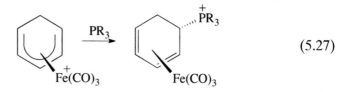

(5.27)

Butadiene occasionally binds in the s-trans conformation.[6] In $Os_3(CO)_{10}(C_4H_6)$, **5.17**, the diene is η^2-bound to two different Os, but in $Cp_2Zr(C_4H_6)$ and $Cp^*Mo(NO)(C_4H_6)$, **5.18**, the diene is η^4-bound to one metal. In the Zr case, the s-cis conformation also exists, but rearranges to a 1:1 thermodynamic mixture on standing; photolysis restores the trans form.

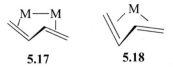

5.17   **5.18**

Cyclobutadiene Complexes

Most neutral ligands are stable in the free state, but free cyclobutadiene, with four π electrons, is antiaromatic, rectangular, and highly unstable. Bound cyclobutadiene is square and aromatic because the metal stabilizes the diene by populating the LUMO by back donation, giving it an aromatic sextet. This is another good example of the free and bound forms of the ligand being substantially different (Section 2.6). Some synthetic routes are shown in Eq. 5.28 and Eq. 5.29.

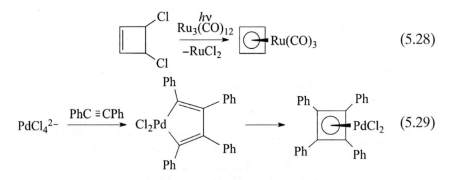

$$(5.28)$$

$$(5.29)$$

The Ru case may involve oxidative addition of the dihalide to $Ru(CO)_3$, formed by photolysis. Eq. 5.29 illustrates an important general reaction, *oxidative coupling* (Section 6.8) of alkynes to give a metalacycle, followed in this case by a reductive elimination to give the cyclobutadiene.

Trimethylenemethane

Also very unstable in the free state is ligand **5.19**, best pictured as an LX_2 enediyl (**5.20**) on binding. An umbrella distortion from the ideal planar conformation moves the central carbon away from the metal. Delocalization within the ligand favors planarity, but the distortion improves M–L overlap because the *p* orbitals of the terminal carbons can now point more directly toward the metal. Some synthetic routes are illustrated in Eq. 5.30.

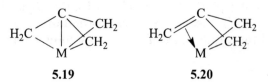

5.19   **5.20**

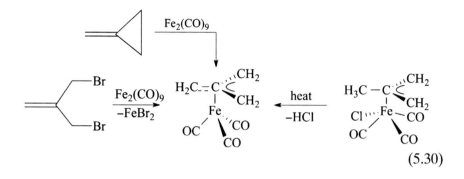

$$(5.30)$$

5.4 CYCLOPENTADIENYL COMPLEXES

The celebrated discovery[7] of the sandwich structure of ferrocene, Cp_2Fe, by Wilkinson, Woodward, and Fischer prompted a "gold rush" into organometallic transition metal π complexes. The cyclopentadienyl group (Cp) is of central importance to the field, being the most firmly bound polyenyl and the most inert to both nucleophiles and electrophiles, although not to strong oxidants (Section 12.4). This makes it a reliable spectator ligand in a vast array of Cp_2M (metallocene) and $CpML_n$ complexes (two-, three-, or four-legged piano stools where $n = 2$–4). The most important application of metallocenes today is alkene polymerization (Section 12.2).

The steric bulk of a Cp can be varied by substitution, as reflected by the following cone angles: η^5-$C_5(i$-$Pr)_5$, $\theta = 167°$; η^5-$C_5H(i$-$Pr)_4$, $\theta = 146°$; η^5-C_5Me_5, $\theta = 122°$; η^5-$C_5H_4SiMe_3$, $\theta = 104°$; η^5-C_5H_4Me, $\theta = 95°$; η^5-C_5H_5, $\theta = 88°$.[8] Substituent electronic effects in a series of $Cp_2Zr(CO)$ complexes have also been documented from $\nu(CO)$, electrochemistry and computational data.[9]

The η^1-Cp structure is also found where the coligands are sufficiently firmly bound so that the Cp cannot become η^5 (e.g., **5.21**). η^1-Cp groups show both long and short C–C distances, as appropriate for an uncomplexed diene. The aromatic η^5 form has essentially equal C=C distances, and the substituents bend very slightly toward the metal. Trihapto-Cp groups as in $(\eta^5$-Cp$)(\eta^3$-Cp$)W(CO)_2$ are rather rare; the η^3-Cp folds so the uncomplexed C=C group can bend away from the metal. The tendency of an η^5 Cp group to "slip" to η^3 or η^1 is small. Nevertheless, 18e piano stool complexes can undergo associative substitution, suggesting that the Cp can slip in the reaction (Eq. 5.31).

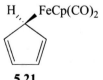

5.21

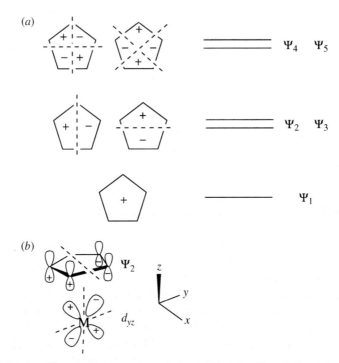

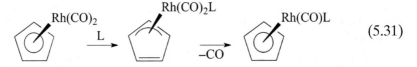

FIGURE 5.3 Electronic structure of the cyclopentadienyl ligand and one of the possible M–Cp bonding combinations.

$$Rh(CO)_2 \xrightarrow{L} Rh(CO)_2L \xrightarrow{-CO} Rh(CO)L \tag{5.31}$$

Diamagnetic η^5-Cp complexes show a ^{1}H NMR resonance at 3.5–5.5δ, a position appropriate for an arene. Woodward first showed that ferrocene, like benzene, undergoes electrophilic acylation.[8] In η^1-Cp groups, the α hydrogen appears at ~3.5δ, and the β and γ hydrogens at 5–7δ. As we see in Chapter 10, the η^1-Cp group can be fluxional, in which case the metal rapidly moves around the ring so as to make all the protons equivalent.

In the MO scheme of Fig. 5.3 for M–C$_5$H$_5$, the five carbon p orbitals lead to five MOs for the C$_5$H$_5$ group. Only the nodes are shown in Fig. 5.3a, but Fig. 5.3b shows the orbitals in full for one case. The most important overlaps are ψ_1 with the metal d_{z^2}, and ψ_2 and ψ_3 with the d_{xz} and d_{yz} orbitals, as shown explicitly in Fig. 5.3b; ψ_4 and ψ_5 do not interact very strongly with metal orbitals, and the Cp group is therefore not a

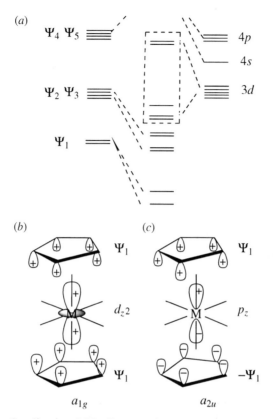

FIGURE 5.4 Qualitative MO diagram for a first-row metallocene. (a) The box shows the crystal field splitting pattern, only slightly distorted from its arrangement in an octahedral field. Because we now have two Cp groups, the sum and difference of each MO has to be considered. For example, Ψ_1 gives $\Psi_1 + \Psi_1'$, of symmetry a_{1g}, which interacts with the metal d_{z^2}, as shown in (b), and $\Psi_1 - \Psi_1'$, of symmetry a_{2u}, which interacts with p_z, as shown in (c). For clarity, only one lobe of the Cp p orbitals is shown.

very good π acceptor. This and the anionic charge makes Cp complexes basic, and this encourages back donation to the non-Cp ligands.

The MO diagram for a Cp$_2$M metallocene (Fig. 5.4) requires consideration of both Cp groups. We therefore look at the symmetry of *pairs* of Cp orbitals to see how they interact with the metal. As an example, a pair of ψ_1 orbitals, one from each ring (Fig. 5.4b), has a_{1g} symmetry and can thus interact with the metal $d - z^2$, also a_{1g}. The opposite combination of ψ_1 orbitals, now a_{2u}, (Fig. 5.4c), interacts with the metal p_z, also a_{2u}. Similarly, ψ_2 and ψ_3 combinations are strongly stabilized by interactions with the metal d_{xz}, d_{yz}, p_x, and p_y. Although the

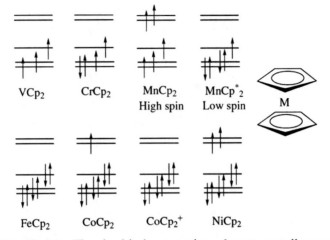

FIGURE 5.5 The *d*-orbital occupation of some metallocenes.

details are more complex for Cp_2M, the bonding scheme retains both the L→M direct donation and the M→L back donation that we saw for $M(CO)_6$, as well as a *d*-orbital splitting pattern that broadly resembles the two-above-three pattern characteristic of an octahedral crystal field and highlighted in a box in Fig. 5.4*a*. The different choice of axes in this case (Fig. 5.4*c*) make the orbital labels (d_{xy}, d_{yz}, etc.) different here from what they were before, but this is just a matter of definitions.

In the case of Cp_2Fe itself, the bonding and nonbonding orbitals are all exactly filled, leaving the antibonding orbitals empty, making the group 8 metallocenes the stablest of the series. The MCp_2 unit is so intrinsically stable that the same structure is adopted for numerous first-row transition metals even when this results in a paramagnetic, non-18e complex (Fig. 5.5). Metallocenes from groups 9 and 10 have one or two electrons in antibonding orbitals; this makes $CoCp_2$ and $NiCp_2$ paramagnetic and much more reactive than $FeCp_2$. Nineteen electron $CoCp_2$ also has an 18e cationic form, $[Cp_2Co]^+$. Chromocene and vanadocene have fewer than 18e and are also paramagnetic, as Fig. 5.5 predicts. Predominantly ionic $MnCp_2$ is very reactive because the high spin d^5 Mn ion provides no crystal field stabilization. The higher-field C_5Me_5, denoted Cp*, on the other hand, gives a much more stable, low-spin $MnCp^*_2$.

Bent Metallocenes

Metallocenes of group 4, and of the heavier elements of groups 5–7 can bind up to three additional ligands, in which case the Cp groups bend

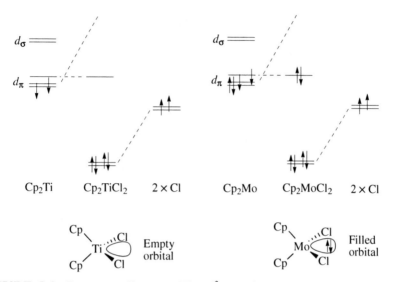

FIGURE 5.6   Bent metallocenes. The d^2 Cp$_2$Ti fragment can bind two Cl atoms to give the metallocene dichloride Cp$_2$TiCl$_2$, in which the single non-bonding orbital is empty and located as shown between the two Cl ligands; this empty orbital makes the final complex a hard 16e species. The d^4 Cp$_2$Mo fragment can also bind two Cl atoms to give the metallocene dichloride Cp$_2$MoCl$_2$, in which the single nonbonding orbital is now full and located as before; this filled orbital, capable of back donation, makes the final 18e complex soft.

back as shown in Fig. 5.6. This bending causes mixing of the d, s, and p orbitals so that the three hybrid orbitals shown in **5.22** point out of the open side of the metallocene toward the additional ligands. In ferrocene itself, these are all filled, but one may still be protonated to give bent Cp$_2$FeH$^+$. The Cp$_2$Re fragment is 17e, and so requires one 1e ligand to give a stable 18e complex, such as Cp$_2$ReCl. The Cp$_2$Mo and Cp$_2$W fragments, being 16e, can bind two 1e ligands or one 2e ligand to reach 18e, as in Cp$_2$MH$_2$ or Cp$_2$M(CO). Only two of the three available orbitals are used in Cp$_2$MH$_2$, which leaves a lone pair between the hydrides that can be protonated to give the water-soluble cations, [Cp$_2$MH$_3$]$^+$. This lone pair can alternatively provide back donation to stabilize any unsaturated ligands present, as in [Cp$_2$M(C$_2$H$_4$)Me]$^+$. Cp$_2$M fragments from the group 5 metals have 15e and can bind three X ligands (e.g., Cp$_2$NbCl$_3$).

5.22

For group 4 metals, the maximum permitted oxidation state of M(IV) means the 14e Cp_2M fragments can bind only two X ligands, making the resulting Cp_2MX_2 electron-deficient 16e species. This leaves us with an empty orbital in Cp_2TiCl_2, rather than a filled one as in 18e Cp_2MoCl_2 and accounts for striking differences in the metallocene chemistry of the two groups, 4 and 6. The group 4 metallocenes act as hard Lewis acids and tend to bind π-basic ligands such as –OR that can π-donate from O lone pairs into the empty orbital, but the group 6 metallocenes act as soft π bases and tend to bind π-acceptor ligands such as ethylene, where back donation comes from the same orbital, now filled.

The orbital pattern of Fig. 5.6 is consistent with the discussion of Fig. 2.2. Since the virtual CN $(a + b)$ of Cp_2MX_2 is 8 (Cp_2MX_2 is an MX_4L_4 system), we expect $(9 - 8)$ or one nonbonding orbital, as shown in Fig. 5.6.

Cp^*, or $\eta^5\text{-}C_5Me_5$, the most important variant of Cp, is not only higher field but also more electron releasing, bulkier, and gives more soluble derivatives. It also stabilizes a wider range of organometallic complexes than Cp. This reflects a general strategy for stabilizing unstable compounds by introducing steric hindrance. Cp^* has reactions not shared by Cp, for example, conversion to a fulvene complex by H^- abstraction from the Cp^* methyl (Eq. 5.32).[10] Other differences are discussed in Sections 11.1 and 15.4.

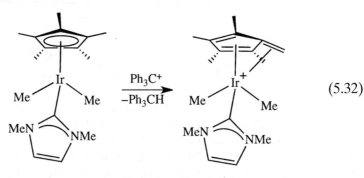

$$(5.32)$$

Synthesis

The synthesis of cyclopentadienyls follows the general pattern shown in Eq. 5.33–Eq. 5.38. TlCp, an air-stable reagent capable of making many Cp complexes from the metal halides, is often avoided in recent practice because of the toxicity of Tl.

1. From a source of Cp^-:

$$FeCl_2 \xrightarrow{\text{NaCp}} FeCp_2 \qquad (5.33)$$

$$MoCl_5 \xrightarrow[-100° \text{ to } 25°]{\text{NaCp, NaBH}_4} Cp_2MoH_2 \qquad (5.34)$$

2. From a source of Cp^+:

$$[CpFe(CO)_2]^- \xrightarrow{C_5H_5Br} [CpFe(CO)_2(\eta^1\text{-Cp}] \xrightarrow[-CO]{\text{heat}} FeCp_2 \qquad (5.35)$$

3. From the diene or a related hydrocarbon:

$$MeRe(CO)_5 \xrightarrow{C_5Me_5H} Cp^*Re(CO)_3 \qquad (5.36)$$

$$[IrH_2(acetone)_2(PPh_3)_2]^+ \xrightarrow[-2H_2]{} [CpIrH(PPh_3)_2]^+ \qquad (5.37)$$

The high reactivity of paramagnetic metallocenes, such as 20e $NiCp_2$, is illustrated in Eq. 5.38, where a Cp^- from $NiCp_2$ deprotonates the C2 proton of the imidazolium ion to give an NHC complex.

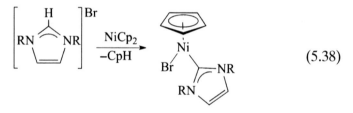

$$(5.38)$$

Cp Analogs

Two close L_2X analogs are cyclohexadienyl **5.23** and pentadienyl **5.24**. In **5.23**, the uncomplexed ring CH_2 is bent 30–40° out of the ligand plane. Pentadienyl, being acyclic, is more easily able to shuttle back and forth between the η^1, η^3, and η^5 structures.

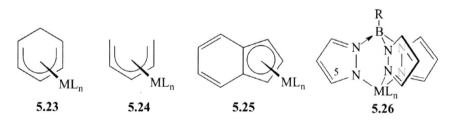

5.23   **5.24**   **5.25**   **5.26**

Indenyl (**5.25**) is a better π acceptor than Cp: for example, $[(\eta^5\text{-Ind})IrHL_2]^+$ is deprotonated by NEt_3, but the Cp analog is not deprotonated even by t-BuLi.

Tris-pyrazolyl borate (**5.26**), often denoted Tp, is a useful tridentate *fac* N-donor spectator ligand. Tp complexes have some analogy with

Cp, although this is not as close as once thought. Tp has a lower field strength, for example, and Tp_2Fe, unlike Cp_2Fe, is high spin and paramagnetic. As an L_3 ligand with a negative charge, Tp behaves as an L_2X ligand at least from the point of view of electron count. Tp ligands with substituents at the 5-position can be so bulky that they only permit a single additional ligand to bind, in which case they are considered *tetrahedral enforcers*.[11]

5.5   ARENES AND OTHER ALICYCLIC LIGANDS

$[Cr(\eta^6\text{-}C_6H_6)_2]$ holds a special place in the field because Fischer and Hafner identified its "sandwich" structure as early as 1955, just after having proposed the same type of structure for ferrocene.[12] Closely related compounds had been made by Hein from 1918, but their structures remained mysterious in an era before X-ray crystallography became routine.[13]

Arenes usually bind in the 6e, η^6-form **5.27**, but η^4 (**5.28**) and η^2 (**5.29**) structures are also seen. An η^2 or η^6 arene is planar, but the η^4 ring is strongly folded. The C–C distances are usually essentially equal, but slightly longer than in the free arene. Arenes are much more reactive than Cp groups, and they are also more easily lost from the metal so arenes are more often actor rather than spectator ligands.

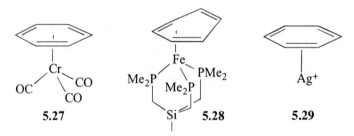

Synthesis

Typical synthetic routes resemble those used for alkene complexes:

1. From the arene and a complex of a reduced metal:

$$Cr(CO)_6 \xrightarrow[\text{heat}]{C_6H_6} (\eta^6\text{-}C_6H_6)Cr(CO)_3 \qquad (5.39)$$

$$FeCp_2 \xrightarrow{C_6H_6,\ AlCl_3} [CpFe(\eta^6\text{-}C_6H_6)][AlCl_4] \qquad (5.40)$$

2. From the arene, a metal salt and a reducing agent:

$$CrCl_3 \xrightarrow[\text{heat}]{C_6H_6,\ Al,\ AlCl_3} [Cr(\eta^6\text{-}C_6H_6)_2]^+ \xrightarrow{\text{reduction}} [Cr(\eta^6\text{-}C_6H_6)_2] \quad (5.41)$$

3. From the diene:

$$RuCl_3 \xrightarrow{} [(\eta^6\text{-}C_6H_6)RuCl(\mu\text{-}Cl)]_2 \quad (5.42)$$

Arene binding in $(C_6H_6)Cr(CO)_3$ depletes the electron density on the ring, which becomes subject to nucleophilic attack. In addition, the metal encourages deprotonation both at the ring protons, because of the increased positive charge on the ring, and α to the ring (e.g., at the benzylic protons of toluene), because the negative charge of the resulting carbanion can be delocalized on to the metal, where it is stabilized by back bonding to the CO groups.

Other Arene Ligands

For naphthalene, η^6 binding is still common, but the tendency to go η^4 is enhanced because this allows the uncomplexed ring to be fully aromatic. If one ring is differently substituted from the other, isomers called *haptomers* have the metal bound to one ring or the other, often with metal exchanging between sites.[14]

TpW(NO)(PMe$_3$) gives an η^2 complex with naphthalene, where the stabler 1,2-bound form is in equilibrium with the 2,3-form, which has the character of a quinodimethane and can give the Diels-Alder reaction of Eq. 5.43.[15]

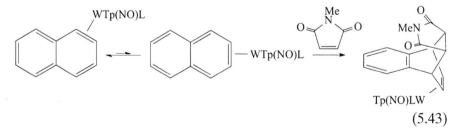

$$(5.43)$$

In the fullerene series, Fig. 5.7 shows how the ellipsoidal molecule C_{70} binds to Vaska's complex. Free C_{70} itself does not give crystallographically useful crystals, and so this result on the complex confirmed the ellipsoidal structure previously deduced from the NMR spectrum of C_{70}. The junctions between six-membered rings seem to

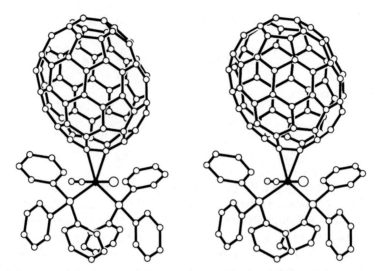

FIGURE 5.7   Stereoscopic view of $(\eta^2\text{-}C_{70})Ir(CO)Cl(PPh_3)_2$. Source: From Balch et al., 1991 [20]. Reproduced with permission of the American Chemical Society.

be the most reactive in the fullerenes, and this is where the metal binds. It is almost always the Cl and CO groups in the planar Vaska complex that bend back to become cis when an alkene or alkyne binds; here, the PPh$_3$ groups bend back, presumably from steric repulsion by the bulky C$_{70}$ group. Figure 5.7 is a stereoscopic diagram of a type commonly seen in research papers. With practice, it is possible to relax the eyes so that the two images formed by each eye are fused to give a three-dimensional representation of the molecule. Mass spectral evidence suggests that the small C$_{28}$ fullerene binds Ti^{4+}, for which the structure shown in Fig. 5.8 has been proposed by computation; the Ti is predicted to be off-center within the cage.[16]

η^7 Ligands

η^7-Cycloheptatrienyl ligands, as in CpTa(η^7-C$_7$H$_7$), have a planar ring with equal C–C distances.[8] The C–H bonds are tilted about 6° toward the metal to improve the overlap between the C p orbitals and Ta. An OS ambiguity arises since the ligand might be the aromatic [C$_7$H$_7$]$^+$ or [C$_7$H$_7$]$^{3-}$, [C$_7$H$_7$]$^-$ being excluded as antiaromatic. The L$_2$X$_3$ trianion seems most appropriate choice for CpTa(η^7-C$_7$H$_7$), making it Ta(IV). A common synthesis is abstraction of H$^-$ from an η^6 cycloheptatriene

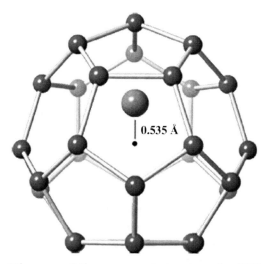

FIGURE 5.8 Diagram of the proposed structure for TiC_{28}, formed in the vapor phase, showing the displacement of the Ti from the center of the C_{28} cage toward a C_5 ring that is predicted from computational work. Source: From Dunk et al., 2012 [16]. Reproduced with permission of the American Chemical Society.

complex with Ph_3C^+ (Eq. 5.44) or Et_3O^+, the second being preferred because the by-products, Et_2O and EtH, are volatile. The stable, aromatic $[C_7H_7]BF_4$ salt is also synthetically useful (Eq. 5.45).

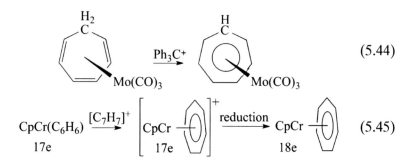

$$\text{CpCr}(C_6H_6) \xrightarrow{[C_7H_7]^+} \left[\text{CpCr} \right]^+ \xrightarrow{\text{reduction}} \text{CpCr} \qquad (5.45)$$

CpCr(C_6H_6) $[C_7H_7]^+$ $[$CpCr$]^+$ reduction   CpCr
17e   17e   18e

η^8 Ligands

The antiaromatic 8π electron, nonplanar hydrocarbon, η^8-cyclooctatetraene (cot), can form complexes as the reduced, aromatic $10\pi e$ cot^{2-} dianion (L_2X_2), the classic example being $U^{IV}(cot)_2$.

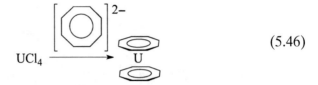

$$UCl_4 \longrightarrow \qquad\qquad (5.46)$$

Early metals that need many electrons to achieve an 18e structure can also give η^8-C_8H_8 complexes, such as $[(\eta^8$-$C_8H_8)Ti^{IV}(=N t Bu)]$.[17]

5.6   ISOLOBAL REPLACEMENT AND METALACYCLES

The chemical character of any fragment depends on the symmetry and electron occupation of the frontier orbitals. Fragments that are very dissimilar in composition can therefore have very similar frontier orbitals. Hoffmann named such fragments *isolobal* (Section 13.2), a concept that has often proved useful.[18] For example, $Fe(CO)_4$ and CH_2 are isolobal in having one empty LUMO and one filled HOMO of comparable symmetries and so form many analogous compounds, such as $(H)_2Fe(CO)_4$ and CH_4 or $Fe(CO)_5$ and $CH_2=$ $C=O$. Isolobality helps in understanding metallabenzenes (**5.30**), in which we replace one CH of benzene by a metal fragment isolobal with CH (e.g., **5.31**).[19] Metalabenzenes have a planar MC_5 ring without the alternating CC bond lengths expected for a nonaromatic metalacyclohexatriene. The extent of the aromaticity in such rings is still under discussion, but reactions characteristic of arenes are seen, such as nitration and bromination.

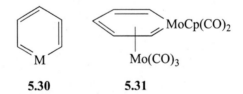

5.30   **5.31**

Related to metalabenzenes are metalloles, where the metal fragment replaces the NH of the heteroarene, pyrrole. On a strongly back-donating metal, the metallole of Eq. 5.47 has bis-carbene character. The X-ray structure shows that the complex has the bis-carbene structure, **5.32,** and not the usual metallole structure, **5.33**. The carbocycle in **5.32** is a 4e ligand, but in **5.33** is a 2e ligand, so this conversion can happen only if the metal can accept 2e. On the ionic model, both ligands are counted as 4e ligands, but the metal is counted as d^6 Os(II) in **5.32** and d^4 Os(IV) in **5.33**, on both models. **5.32** is an 18e complex and **5.33** is a 16e complex.

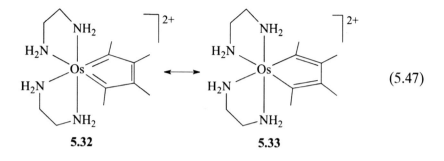

$$5.32 \qquad\qquad 5.33 \qquad\qquad (5.47)$$

5.7 STABILITY OF POLYENE AND POLYENYL COMPLEXES

Polyene complexes L_n more easily dissociate than polyenyl complexes L_nX because the free polyene is usually a stable species, but the polyenyl must dissociate as a radical or an anion, both likely to be less stable than a neutral polyene. The strongest π-back bonding and most electron-rich metal fragments generally bind polyenes and polyenyls most tightly. For example, butadiene complexes of strongly π-basic metal fragments have more LX_2 character than those of less basic fragments and so less resemble the free ligand and dissociate less easily. Electron-withdrawing substituents also encourage back donation and can greatly increase complex stability, as we have seen for C_2F_4 in Section 5.1. Conversely, d^0 metals incapable of back donation, such as Ti(IV) and Nb(V), normally bind L_nX ligands such as Cp (e.g., Cp_2NbCl_3 or $[Ti(\eta^3\text{-}C_3H_5)_4]$), but only rarely L_n ligands such as CO, C_2H_4, and C_6H_6.

For each step to the right in the d block, similar ML_n fragments gain one electron. This makes it more difficult for the larger polyenes, such as cot, to bind without exceeding 18e. Uranium, not limited by the 18e rule from having f orbitals, is able to accept two $[\eta^8\text{-cot}]^{2-}$ ligands in uranocene, $U(\eta^8\text{-}C_8H_8)_2$. Because the two $[\eta^8\text{-cot}]^{2-}$ ligands bring 20e, no d-block element could do the same. Ti is known with one $\eta^8\text{-}C_8H_8$ ring, Cr with one $\eta^6\text{-}C_8H_8$ ring, but Rh does not accept more than 4e from cot in the $\mu\text{-}\eta^4\text{-}C_8H_8$ acetylacetonate complex, **5.34**.

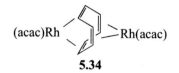

5.34

Although the problem is less severe for η^5-Cp and $(\eta^6\text{-}C_6H_6)$ complexes, these are notably less stable on the right-hand side of the periodic table, for example, for Pd and Pt. The η^4-butadiene and η^3-allyl groups do not seem to be affected until we reach group 11. Stability of polyene complexes also increases in lower oxidation states. In Eq. 5.48, Co(–I) back-donates so strongly that it gives the η^4-anthracene ligands significant LX$_2$ enediyl character.

$$\text{CoBr}_2 \xrightarrow{\text{K/anthracene}} \text{K}\left[\underset{}{\text{(anthracene)}} \overset{-1}{\text{Co}} \underset{}{\text{(anthracene)}} \right] \tag{5.48}$$

- L$_n$X ligands such as Cp tend to bind more strongly than comparable L$_n$ ligands such as benzene.
- Increased back bonding to π-bound ligands (e.g., Sections 5.1–5.3) weakens bonds within the ligand and decreases the tendency for nucleophilic attack on the ligand.

REFERENCES

1. D. M. P. Mingos, *J. Organometal. Chem.*, **635**, 1, 2001.
2. J. M. Praetorius, D. P. Allen, R. Y. Wang, J. D. Webb, F. Grein, P. Kennepohl, and C. M. Crudden, *J. Am. Chem. Soc.*, **130**, 3724, 2008.
3. H. Nuss, N. Claiser, S. Pillet, N. Lugan, E. Despagnet-Ayoub, M. Etienne, and C. Lecomte, *Dalton Trans.*, **41**, 6598, 2012.
4. C. -M. Che, C. -M. Ho, and J. -S. Huang, *Coord. Chem. Rev.*, **251**, 2145, 2007.
5. B. M. Trost, *Acct. Chem. Res.*, **35**, 695, 2002.
6. T. Tran, C. Chow, A. C. Zimmerman, M. E. Thibault, W. S. McNeil, and P. Legzdins, *Organometallics*, **30**, 738, 2011.
7. P. Laszlo and R. Hoffmann, *Angew. Chem. Int. Ed.*, **39**, 123, 2000; H. Werner, *Angew. Chem. Int. Ed.*, **51**, 6052, 2012.
8. A. Glöckner, H. Bauer, M. Maekawa, T. Bannenberg, C. G. Daniliuc, P. G. Jones, Y. Sun, H. Sitzmann, M. Tamm, and M. D. Walter, *Dalton Trans.*, **41**, 6614, 2012.
9. C. E. Zachmanoglou, A. Docrat, B. M. Bridgewater, G. Parkin, C. G. Brandow, J. E. Bercaw, C. N. Jardine, M. Lyall, J. C. Green, and J. B. Keister, *J. Am. Chem. Soc.*, **124**, 9525, 2002.
10. J. M. Meredith, K. I. Goldberg, W. Kaminsky, and D. Michael Heinekey, *Organometallics*, **31**, 8459, 2012.
11. A. Kunishita, T. L. Gianetti, and J. Arnold, *Organometallics*, **31**, 372, 2012.
12. E. O. Fischer and R. Jira, *J. Organomet. Chem.*, **637**, 7, 2001.
13. D. Seyferth, *Organometallics*, **21**, 1520 and 2800, 2002.

14. T. A. Albright, S. Oldenhof, O. A. Oloba, R. Padillab, and K. P. C. Vollhardt, *Chem. Commun.*, **47**, 9039, 2011.

15. L. Strausberg, M. Li, D. P. Harrison, W. H. Myers, M. Sabat, and W. D. Harman, *Organometallics*, 2013.

16. P. W. Dunk, N. K. Kaiser, M. Mulet-Gas, A. Rodríguez-Fortea, J. M. Poblet, H. Shinohara, C. L. Hendrickson, A. G. Marshall, and H. W. Kroto, *J. Am. Chem. Soc.*, **134**, 9380, 2012.

17. S. C. Dunn, N. Hazari, N. M. Jones, A. G. Moody, A. J. Blake, A. R. Cowley, J. C. Green, and P. Mountford, *Chem. Eur. J.*, **11**, 2111, 2005.

18. H. G. Raubenheimer and H. Schmidbaur, *Organometallics*, **31**, 2507, 2012 and reference cited.

19. G. R. Clark, P. M. Johns, W. R. Roper, T. Sohnel, and L. J. Wright, *Organometallics*, **30**, 129, 2011.

20. A. L. Balch, V. J. Catalano, J. W. Lee, M. M. Olmstead, and S. R. Parkin, *J. Am. Chem. Soc.*, **113**, 8953, 1991.

PROBLEMS

5.1. Suggest a mechanism for the following transformation and say how you would test it.

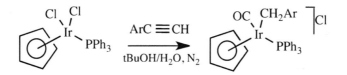

5.2. Although $L_nMCH_2CH_2ML'_n$ can be thought of as a 1,2-bridging ethylene complex in which each carbon is bound to a different metal atom, examples of this type of structure are rarely made from ethylene itself. Propose a general route that does not involve ethylene and explain how you would know that the complex had the 1,2-bridging structure without using crystallography. What might go wrong with the synthesis?

5.3. Among the products formed from PhC≡CPh and $Fe_2(CO)_9$, is 2,3,4,5,-tetraphenylcyclopentadienone. Propose a mechanism for the formation of this product. Do you think the dienone would be likely to form metal complexes? Suggest a specific example and how you might try to make such a complex.

5.4. Suggest a synthesis of $Cp_2Mo(C_2H_4)Me^+$ from Cp_2MoCl_2. What orientation would you expect for the ethylene ligand? Given that there is no free rotation of the alkene, how would you show what orientation is adopted?

5.5. What structural distortions would you expect to occur in the complex $L_nM(\eta^4\text{-butadiene})$ if the ligands L were made more electron releasing?

5.6. 1,3-Cod (= cyclooctadiene) can be converted into free 1,5-cod by treatment with $[(C_2H_4)IrCl]_2$, followed by $P(OMe)_3$. What do you think is the mechanism? Since 1,5-cod is thermodynamically unstable with respect to 1,3-cod (why is this so?), what provides the driving force for the rearrangement?

5.7. How many isomers would you expect for $[PtCl_3(\text{propene})]^-$?

5.8. [TpCoCp] is high spin (Tp is shown in structure **5.26**). Write its d-orbital occupation pattern following Fig. 5.5 and predict how many unpaired electrons it has (see *Chem. Comm.* 2052, 2001).

5.9. $[IrH_2(H_2O)_2(PPh_3)_2]^+$ reacts with indene, C_9H_8 (**5.35**), to give $[(C_9H_{10})Ir(PPh_3)_2]^+$. On heating, this species rearranges with loss of H_2 to give $[(C_9H_7)IrH(PPh_3)_2]^+$. Only the first of the two Ir species mentioned reacts with ligands such as CO to displace C_9H_7. What do you think are the structures of these complexes?

5.35

5.10. From the information in Eq. 5.26, deduce how many electrons the η^3-propargyl ligand contributes to the electron count. The C–C–C angle in the propargyl ligand is 153°. Why does this differ from the ideal 120° of the allyl ligand and from the 180° of simple propargyl compounds such as $HC\equiv C-CH_2OH$?

6

OXIDATIVE ADDITION AND REDUCTIVE ELIMINATION

Moving on from discussing ligand types, we now return to reactivity questions by looking at two reactions that play a key role in most catalytic cycles as well as in many synthetic pathways.

6.1 INTRODUCTION

In Chapter 4, we saw how neutral ligands such as C_2H_4 or CO enter the coordination sphere by substitution. We now see how pairs of anionic ligands, A and B, do this by *oxidative addition* (OA) of A–B. In OA, A–B molecules such as H–H or CH_3–I add to a low valent metal, L_nM, to produce $L_nM(A)(B)$ (Eq. 6.1). The equally important reverse reaction, *reductive elimination* (RE), leads to the release of A–B from $L_nM(A)(B)$. In the oxidative direction, the A–B bond breaks to form bonds from M to A and B. Since A and B are X-type ligands, the oxidation state, electron count, and coordination number all increase by two units during OA, the reverse taking place during RE. These changes in formal oxidation state (OS) justify the oxidative and reductive parts of the reaction names. In a catalytic cycle, a reactant often binds via OA and the product dissociates via RE.

The Organometallic Chemistry of the Transition Metals, Sixth Edition.
Robert H. Crabtree.
© 2014 John Wiley & Sons, Inc. Published 2014 by John Wiley & Sons, Inc.

$$L_nM \; + \; A—B \; \underset{\substack{\text{reductive elimination} \\ \text{(RE)}}}{\overset{\substack{\text{oxidative addition} \\ \text{(OA)}}}{\rightleftharpoons}} \; L_nM\overset{\displaystyle A}{\underset{\displaystyle B}{\big<}} \quad \begin{array}{l} \Delta\text{O.S.} = +2 \\ \Delta\text{C.N.} = +2 \end{array} \qquad (6.1)$$

16e   18e

Oxidative additions go by a variety of mechanisms, but since the metal electron count increases by two, a vacant 2e site is always needed. We may start with a 16e complex, or a 2e site may be opened up by initial ligand loss from an 18e complex. The change in oxidation state means that to undergo Eq. 6.1, a complex must have a stable OS two units more positive (and vice versa for RE).

First row metals typically prefer a one-unit change in oxidation state, electron count, and coordination number. Equation 6.2 shows how *binuclear oxidative addition* conforms to this pattern. We start with a 17e complex or an M—M bonded 18e complex that can dissociate into 17e fragments. The metal must now have a stable OS more positive by one unit for OA. Table 6.1 shows common types of OAs by d^n configuration and position in the periodic table. Whatever the mechanism, two electrons from M transfer into the A–B σ^*, while the A–B σ bonding pair donate to M. This cleaves the A–B

TABLE 6.1 Common Oxidative Additions by d^n Configuration

Change in d^n Configuration	Examples	Group	Remarks
$d^{10} \to d^8$	Au(I) $\to$ (III)	11	
	Pt, Pd(0) $\to$ (II)	10	
$d^8 \to d^6$	M(II) $\to$ (IV)	10	M = Pd, Pt
	Rh, Ir(I) $\to$ (III)	9	Very common
	M(I) $\to$ (III)	9	
	M(0) $\to$ (II)	8	
$d^7 \to d^6$	2Co(II) $\to$ (III)	8	Binuclear
	2Co(II) $\to$ (III)	8	Binuclear
$d^6 \to d^4$	Re(I) $\to$ (III)	7	
	M(0) $\to$ (II)	6	
	V(−I) $\to$ (I)	5	
$d^4 \to d^3$	2Cr(II) $\to$ (III)	6	Binuclear
	2Cr(II) $\to$ (III)	6	Binuclear
$d^4 \to d^2$	Mo, W(II) $\to$ (IV)	6	
$d^2 \to d^0$	M(III) $\to$ (V)	5	
	M(II) $\to$ (IV)	4	

Note: Common reductive eliminations follow the reverse paths.

bond and makes the new bonds to A and B. OA is favored in low oxidation states and is rare for M(III) and higher, except with powerful oxidants, such as Cl_2. OA is also favored where the A–B bond is weak relative to M–A and M–B. The opposite holds for RE, where a high OS metal and a product with a strong A–B bond are favorable for the reaction.

$$L_nM\text{-}ML_n \; \rightleftharpoons \; 2L_nM \; \xrightarrow{\;A-B\;} \; L_nM\text{-}A \; + \; L_nM\text{-}B \qquad \begin{array}{l} \Delta O.S. = +1 \\ \Delta C.N. = +1 \end{array} \qquad (6.2)$$
$$\quad\; \text{18e} \qquad\quad \text{17e} \qquad\qquad\quad \text{18e} \qquad\; \text{18e}$$

OA is favored in the second and third rows, where M–A and M–B bonds are strongest or for alkyl halides, with their relatively weak C–Hal bonds. Conversely, the strong C–H bonds of alkanes encourages RE in $L_nM(R)(H)$. An OA/RE equilibrium is sometimes seen (Eq. 6.3).

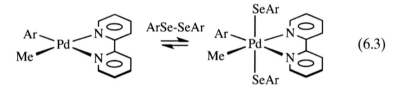

$$(6.3)$$

OA is also favored by strongly donor coligands, L_n, because these stabilize the oxidized $L_nM(A)(B)$ state. While the formal ΔOS for Eq. 6.1 is always $+2$, the real change in metal charge is less than this because A and B do not have full -1 charges in $L_nM(A)(B)$. The change in real charge depends mostly on the electronegativity of A and B, so that $H_2 < HCl < Cl_2$ are increasingly oxidizing. This order comes from measuring the IR spectral change in $\nu(CO)$ on going from $IrCl(CO)L_2$ to $Ir(A)(B)Cl(CO)L_2$ (Table 6.2), where a high $\Delta\nu(CO)$ during OA corresponds to a greater degree of oxidation by raising the positive charge on M and so reducing M–CO back bonding.

These reactions are not limited to transition metals—perhaps the most familiar oxidative addition is the formation of Grignard reagents (Eq. 6.4). Indeed, OA can usually occur whenever an element has two accessible oxidation states two units apart.

$$Me-Br + Mg \rightarrow Me-Mg-Br \qquad (6.4)$$

A wide range of A–B reagents can give OA, including such relatively unreactive ones as silanes, H_2, and even alkanes. Oxidative additions are also very diverse mechanistically, so we need to consider the main types separately.

TABLE 6.2   Carbonyl Stretching Frequencies in Oxidative Addition with Vaska's Complex

Reagent	$\nu(CO)$ (cm^{-1})	$\Delta\nu(CO)$ (cm^{-1})
None	1967	0
O_2	2015	48
$D_2{}^a$	2034	67
HCl	2046	79
MeI	2047	80
C_2F_4	2052	85
I_2	2067	100
Cl_2	2075	108

^{a}The D isotope is used because the Ir–H stretching vibrations have a similar frequency to $\nu(CO)$ and so couple with CO stretching and cause $\nu(CO)$ to shift for reasons that have nothing to do with the change in the electronic character of the metal (see Chapter 10).

6.2   CONCERTED ADDITIONS

Concerted, three-center OA starts out as an associative substitution in which an incoming ligand, A–B, binds as a σ complex but then undergoes A–B bond breaking if back donation from the metal into the A–B σ* orbital is strong enough. This mechanism applies to nonpolar reagents, such as H_2, R_3C–H or R_3Si–H (**6.1**; A = H; B = H, C, or Si). The associative step *a* of Eq. 6.5 forms the σ complex; if this is stable, the reaction stops here. Otherwise, metal electrons are transferred to the A–B σ* in step *b*, the oxidative part of the reaction. The classic examples, from the Estonian-American chemist, Lauri Vaska (1925–), involve OA to the 16e square planar d^8 species, IrCl(CO)(PPh$_3$)$_2$, known as *Vaska's complex*. The 18e d^6 octahedral dihydride of Eq. 6.6, has mutually cis hydrides; conversely, in an RE such as the loss of H_2 from a dihydride, the two H ligands need to become mutually cis.

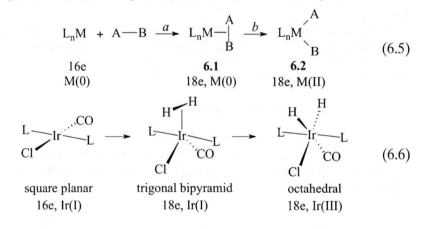

$$L_nM + A—B \xrightarrow{a} L_nM—\overset{A}{\underset{B}{|}} \xrightarrow{b} L_nM\overset{A}{\underset{B}{\diagup}} \qquad (6.5)$$

16e	6.1	6.2
M(0)	18e, M(0)	18e, M(II)

$$(6.6)$$

square planar	trigonal bipyramid	octahedral
16e, Ir(I)	18e, Ir(I)	18e, Ir(III)

In OA of H_2 to Vaska's complex, the initially *trans*-Cl–Ir–(CO) set of ligands folds back to become cis both in a proposed transient H_2 complex and in the final product (Eq. 6.6). As a powerful π acceptor, the CO prefers to be in the equatorial plane of the resulting TBP transient, following the same pattern we saw in A substitution. This tendency for a trans pair of very strong π-acceptor ligands on a strongly π donor metal to fold back can be so great that a d^8 ML_4, normally expected to be square planar, distorts toward TBP even in the absence of an fifth ligand, as in **6.3** (L = P(t-Bu)$_2$Me).[1] Bending enhances the π donor power of the metal by raising the energy of the relevant d orbitals, as well as avoiding the CO ligands being mutually trans (see transphobia in Section 2.6).

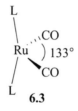

6.3

In 18e complexes, a ligand may be lost to give the 2e site needed for OA, as occurs in initial CO loss from $[Ir(CO)_3L_2]^+$ in the OA of H_2 to give $[Ir(H)_2(CO)_2L_2]^+$. Equation 6.7 shows how η^6 to η^4 arene slip also allows OA of H_2:

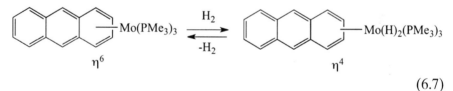

$$\text{(6.7)}$$

The reactions are usually second order with negative entropies of activation ($\Delta S^{\ddagger} \sim -20$ eu) consistent with an ordered transition state resembling **6.1**. They are little affected by the polarity of the solvent but are accelerated by electron-releasing ligands.

Agostic complexes, σ complexes of C–H bonds, can be thought of as lying along the OA pathway C–H + M → C–M–H, but arrested at different points. The C–H bond is thought to approach with the H atom pointing toward the metal. The C–H bond then pivots around the hydrogen to bring the carbon closer to the metal in an increasingly side-on arrangement, followed by C–H bond cleavage. The addition goes with retention of stereochemistry at carbon, as expected on this mechanism. Even though H–H and hydrocarbon C–H bonds are very strong, they readily oxidatively add to metals because of the high M–H and M–C bond strengths in the product.

Carbon–carbon OA is rare, but Eq. 6.8 is driven by ring strain and the high trans effect of the biphenyl leads to an unusual 16e Ir(III) product. For R–CN, OA of the C–C bond is favored by formation of a strong M–CN bond (Eq. 6.9).[2]

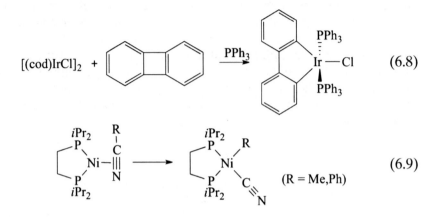

$$[(cod)IrCl]_2 \quad + \qquad\qquad \xrightarrow{\text{PPh}_3} \qquad\qquad \tag{6.8}$$

$$(R = Me, Ph) \tag{6.9}$$

Aryl halides can also react via a concerted mechanism. For example, $[Pd(P\{Ar\}_3)_2]$ reacts with Ar'Br in this way (Ar = o-tolyl; Ar' = t-BuC$_6$H$_4$). Prior loss of PAr$_3$ is required to give the very reactive 1-coordinate intermediate, Pd(PAr$_3$), that goes on to give $[(PAr_3)(Ar')Pd(\mu\text{-}Br)]_2$ as final product.[3] The more reactive ArI compounds do not need prior dissociation of L and can give OA with L$_2$Pd(0).[4] In one case, an apparent OA of the C-F of CH$_3$F in fact goes by initial C-H OA, followed by rearrangement via a transient methylene complex.[5]

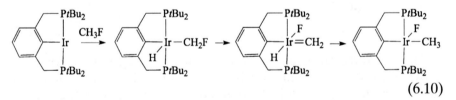

$$\tag{6.10}$$

6.3   S$_N$2 PATHWAYS

In OA, a pair of electrons from a nucleophilic metal transfers to the A–B σ* orbital to break that bond and oxidize the metal. In the S$_N$2 pathway (Eq. 6.11), adopted by polar substrates such as alkyl halides, the metal electron pair of L$_n$M^N (N = oxidation state) directly attacks the R–X σ* at carbon, the least electronegative atom, because σ* is predominant there, to give $[L_nM^{(N+2)}(A)(B)]$.

The S_N2 mechanism is often found in the addition of methyl, allyl, acyl, and benzyl halides, as is the case for Vaska's complex. Like the concerted type, they are second-order reactions, but they are accelerated in polar solvents that stabilize the polar transition state, and they show negative entropies of activation ($\Delta S^{\ddagger} = -40$ to -50 eu). This is consistent with an ordered, polar transition state, as in organic S_N2 reactions. Inversion at carbon has been found in suitably substituted halides. Equation 6.11 shows how the stereochemistry at the carbon of the oxidative addition product was determined by carbonylation to give the metal acyl followed by methanolysis to give the ester. Both of these reactions are known to leave the configuration at carbon unchanged; the configuration of the final ester can be determined from the optical rotation. R and X may end up mutually cis or trans, as expected for the recombination of the ion pair formed in the first step. The product is trans in Eq. 6.12 because the high-trans-effect Me group prefers to remain trans to the vacancy in the 16e square pyramidal intermediate, reminiscent of dissociative substitution trans to a high trans effect ligand (Section 4.4).

$$(6.11)$$

$$(6.12)$$

The first of the two steps in Eq. 6.12 involves oxidation by two units but no change in the electron count, Me^+ being a 0e reagent; since I^- is a 2e reagent, the second involves an increase by 2e in the electron count, but no change in the OS. Only the two steps together constitute the full OA. When an 18e complex is involved, the first step can therefore proceed without the necessity of initial ligand loss; only the second step requires a vacant 2e site. In some cases, the product of the first step is stable and does not lose a ligand to admit the halide anion, for example, Eq. 6.13. This is sometimes loosely called an oxidative addition, but it is better considered as an electrophilic addition to the metal (Section 8.5).

$$Cp^*Ir \overset{L}{\underset{CO}{\diagup}} \xrightarrow{\text{MeI}} \left[Cp^*Ir \overset{L}{\underset{CO}{-}Me} \right] I \qquad (6.13)$$

The more nucleophilic the metal, the greater its reactivity in S_N2 additions, as illustrated by the following reactivity order for a series of Ni(0) complexes (R = alkyl; Ar = aryl).

$$Ni(PR_3)_4 > Ni(PAr_3)_4 > Ni(PR_3)_2(alkene)$$
$$> Ni(PAr_3)_2(alkene) > Ni(cod)_2$$

Steric hindrance at carbon slows the reaction, giving the following reactivity order:

$$MeI > EtI > i\,PrI$$

A better leaving group, X at carbon, accelerates the reaction for this mechanism, which gives rise to the reactivity order:

$$ROSO_2(p-tolyl) > RI > RBr > RCl.$$

Halide ions can increase the nucleophilicity of the metal and hence exert a powerful acceleration on S_N2 OA, as happens for iodide ions in the OA of MeI to $RhI(CO)(PPh_3)_2$ to give $Rh(Me)I_2(CO)(PPh_3)_2$. Iodide ion initially replaces PPh_3 at the metal to give an intermediate $[RhI_2(CO)(PPh_3)]^-$ that reacts very rapidly with MeI.[6]

R_3Sn-X, another reagent with a strong tendency to give S_N2 additions (X = Cl, Br, I), gives the following rapid, reversible addition/elimination equilibrium.[7]

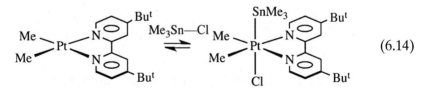

$$(6.14)$$

6.4 RADICAL MECHANISMS

Less desirable are oxidative additions involving radicals,[8] because these reactive intermediates tend to give undesired side-reactions. Minor changes in the structure of the substrate, the complex, or even the impurity level can be enough to affect the rate. The alkyl group always loses any stereochemistry at the α carbon because $RR'R''C\cdot$ is planar. In radical reactions, the solvent must not react fast with $R\cdot$ intermediates; alkane, C_6H_6 AcOH, CH_3CN, and water are usually suitable.

Two types of radical process can be distinguished: the nonchain and the chain. The nonchain variant applies to OA of alkyl halides, RX, to $Pt(PPh_3)_3$ (Eq. 6.15, RX = MeI, EtI, or $PhCH_2Br$).

$$PtL_3 \rightleftharpoons PtL_3 \xrightarrow{RX} \cdot PtXL_2 + R\cdot \xrightarrow{fast} RPtXL_2 \qquad (6.15)$$

A one-electron transfer from M to the RX σ^* forms $\cdot PtXL_2$ and R$\cdot$. This radical pair rapidly recombines to give the product. Like the S_N2 process, the radical mechanism is faster the more basic the metal, and the more readily electron transfer takes place, which gives the reactivity order shown.

$$RI > RBr > RCl \gg ROTs$$

The reaction goes faster as R$\cdot$ becomes more stable and easier to form, giving rise to increasing reactivity in the order: Me $< 1° < 2° < 3°$. In the reaction of NiL_3 with aryl halides, the Ni(I) intermediate, $NiXL_3$, formed in the first step, is sufficiently stable to survive as an observable reaction product because the Ar$\cdot$ radical abstracts an H atom from the solvent to give ArH before it can combine with the Ni.

The second kind of reaction, the *radical chain*, is seen for OA of EtBr or $PhCH_2Br$ with the Vaska's PMe_3 analog (Eq. 6.16). A radical *initiator*, Q$\cdot$ (e.g., a trace of air or peroxide in the solvent), may be required to substitute for R$\cdot$ in the first cycle to set the process going. Chain termination steps, such as recombination of two R$\cdot$ to give R_2, limits the number of possible cycles.

$$
\begin{array}{c}
Ir^ICl(CO)L_2 + R\cdot \longrightarrow RIr^{II}Cl(CO)L_2 \\
\rule{0pt}{1.5em} \\
RXIr^{III}Cl(CO)L_2
\end{array}
\qquad R\cdot \diagup RX \qquad (6.16)
$$

The well-defined relative stereochemistry at the α and β carbons in **6.4** helps us tell if the stereochemistry at the α carbon changes during OA. There is no need to resolve anything, both enantiomers of **6.4** being present. We assume that the reaction cannot affect the β carbon, so we can look at the configuration at the α position relative to the β. This is easily done by ^1H NMR spectroscopy because the favored conformation has the two bulky groups, *t*-Bu and ML_n or *t*-Bu and X, mutually anti. Depending on whether the α stereochemistry has been retained, inverted or scrambled, the α and β protons will therefore be mutually gauche or anti or both of these. The Karplus relationship between the HCCH' dihedral angle and $^3J(H, H')$ predicts a very different coupling constant in the gauche and anti cases. For example, inversion would form **6.4b**, identified by its large $^3J(H, H')$ coupling of $\sim$15 Hz.

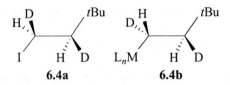

A useful test for radicals relies on the fact that some free radicals rearrange at known rapid rates and thus serve as *radical clocks* (Eq. 6.20).[9] For example, if hexenyl bromide OA gives a cyclopentylmethyl metal complex (radical cyclization rate: 2.5×10^5 s^{-1} at 20°), then the hexenyl radical intermediate must live much longer than 10^{-5} s to give it time to cyclize. Cyclopropylmethyl radicals ($C_3H_5CH_2\cdot$), rearrange by a much faster ring opening (rate: 1.5×10^8 s^{-1}) to give $CH_2{=}CHCH_2CH_2\cdot$. Other common test reactions for R· radicals are Br atom abstraction from a CCl_3Br to give RBr, and dimerization to give R–R. An NMR method, *chemically induced dynamic nuclear polarization* (CIDNP),[10] can also be useful. The method relies on the product of a radical recombination having a very unusual distributions of α and β spins that can lead to very large signal enhancements in the ^{1}H NMR spectra. The intensity of the effect is variable and difficult to predict, preventing easy quantitation of the radicals.

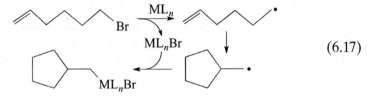

$$(6.17)$$

Because they involve 1e rather than 2e OS changes at the metals, binuclear oxidative additions often go via radicals (e.g., Eq. 6.18).

$$2[Co^{II}(CN)_5]^{3-} \xrightarrow{\ RX\ } [RCo^{III}(CN)_5]^{3-} + [XCo^{III}(CN)_5]^{3-} \quad (6.18)$$

The rate-determining step is net abstraction of a halogen atom from RX by the odd-electron d^7 Co(II) forms R· that subsequently combines with a second Co(II).

6.5   IONIC MECHANISMS

In a polar solvent, where HX (X = Cl, Br, I) can dissociate, X$^-$ and H$^+$ often give a two-step OA with L$_n$M. The metal usually protonates first, followed by X$^-$ binding to give L$_n$M(H)(X) (Eq. 6.19); rarer is X$^-$ attack, followed by protonation (Eq. 6.20). The first path is favored by basic ligands, an 18e complex and a low-OS metal, the second by electron-acceptor ligands, a 16e complex, and by a net positive ionic charge, $[L_nM]^+$.

$$PtL_4 \xrightarrow[-L]{H^+} [HPtL_3]^+ \xrightarrow[-L]{Cl^-} [HClPtL_2]^+ \qquad (6.19)$$

$$[(cod)IrL_2]^+ \xrightarrow{Cl^-} [(cod)IrClL_2] \xrightarrow{H^+} [(cod)IrHClL_2]^+ \qquad (6.20)$$

$$Rate = k[complex][Q] \qquad (6.21)$$

The rate of the first type follows Eq. 6.21 (Q = H+), when protonation is the slow step. Switching from HX to HBF_4 provides a test, because an intermediate, $[L_nMH]BF_4$, is then expected; only the first step of Eq. 6.19 is viable, BF_4^- being noncoordinating.

The rate of the second type (Eq. 6.20) usually follows Eq. 6.21 (Q = X^-), suggesting that anion addition is the slow step. If so, this step should occur with LiCl alone, but no reaction is expected with HBF_4 alone.

Table 6.3 summarizes the information in Sections 6.2–6.5.

6.6 REDUCTIVE ELIMINATION

Reductive elimination, the reverse of oxidative addition, is most often seen in higher oxidation states because the formal OS of the metal drops by two units in RE. The reaction is particularly efficient for the group 10–11 d^8 metals, Ni(II), Pd(II), and Au(III), and Group 9-10 d^6 metals, Pt(IV), Pd(IV), Ir(III), and Rh(III).[11] RE can be stimulated by oxidation or photolysis as in photoextrusion of H_2 from L_nMH_2 (Section 12.4). For chelates, wide ligand bite angles (Section 4.2) can also favor RE, as in BISBI **6.5** (bite angle 122°), because the transition state for RE often has a wide P–M–P angle to compensate for the narrow angle separating the groups trans to the P-donor chelate that are being eliminated.[12]

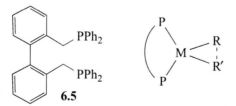

6.5

Thermodynamics dictates if OA or RE will dominate, for example, Eq. 6.22 typically goes to the right for X = alkyl or aryl and Y = H,

TABLE 6.3   Oxidative Addition: One Reaction but Many Mechanisms

Mechanistic Class	Typical Metal, M	Typical Substrate, A–B	Config.[a]	ΔOS, Δe	Remarks
Concerted	d^8, ML$_4$	H$_2$, R$_3$Si–H, R$_3$C–H	Retention	+2, +2	A and B are cis in the product
Nucleophilic[b] (S$_N$2)	d^8, ML$_4$; d^{10}, ML$_4$	R–Hal, RCOCl	Inversion[c]	+2, +2	A and B typically trans in a d^6 product; a polar solvent favors S$_N$2
Radical	First row d block	RHal	Racemized	+2, +2	Can be a chain process
Binuclear	First row d block or L$_n$M–ML$_n$	RHal, Hal$_2$	Racemized	+1,[d] +1[d]	Radical intermediates can occur
Ionic	Any M[e]	HHal		+2, +2	Polar solvent favors this OA

[a]Configurational fate of carbon center.
[b]Several mechanisms are possible within this class (S$_N$2, S$_N$2′, S$_N$Ar, etc.).
[c]As in the organic S$_N$2 mechanism.
[d]At each M starting from L$_n$M fragment.
[e]Having an OS two units more positive.
ΔOS = oxidation state change; Δe = electron count change; Hal = halide.

CHO or SiR$_3$. Kinetics are hard to predict but reactions that involve H are particularly fast. Not only can the H 1s orbital form partial bonds equally in any direction in the transition state, but a relatively stable intermediate σ complex L$_n$M(H–X) can also form.

$$L_nM(X)(Y) \xrightarrow{RE} L_nM + X{-}Y \qquad (6.22)$$

In catalysis (Chapter 9), RE is often the last step in the cycle, and the resulting L$_n$M fragment need only survive long enough to react with the substrates to reenter the catalytic cycle. The eliminations of Eq. 6.22 resemble concerted OAs in going by the same nonpolar, nonradical three-center transition state of type **6.6**, with retention of any stereochemistry at carbon.

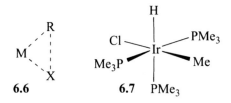

6.6   **6.7** PMe$_3$

Octahedral Complexes

Just as there are several mechanisms for OA (Table 6.3) reversibility arguments suggest that REs should show the same variety. For example, octahedral d^6 complexes of Pt(IV), Pd(IV), Ir(III), and Rh(III) readily undergo RE, often with initial ligand loss to generate a five-coordinate intermediate or else from the initial six-coordinate complex.[13] Without ligand dissociation, RE can be slow even when otherwise expected. For example, complexes with a cis M(R)(H) group are rare because RE of R–H is so thermodynamically favorable. A stable example, *mer*-[IrH(Me)Cl(PMe$_3$)$_3$], **6.7**, having H cis to Me survives heating to 100°C because PMe$_3$ does not dissociate. The Rh analog, **6.8** in Fig. 6.1, with its weaker M–PMe$_3$ bonds, gives RE even at 30°C. The PMe$_3$ trans to the high-trans-effect hydride dissociates because this site is labeled by reaction of **6.8** with P(CD$_3$)$_3$ at 30°C. The five-coordinate intermediate can more readily distort to reach the transition state for RE. If it becomes a Y-type distorted trigonal bipyramidal structure, **6.9** (Fig. 6.1 and Section 4.4), favored where one π-donor ligand, Cl in this case, is located at the basal position of the Y (**6.9**), the two groups to be eliminated, R and H, are brought very close together. The typical R–M–H angle in such cases, ~70°, facilitates achievement of the transition state (**6.10**) for RE. After RE, a T-shaped three-coordinate species is formed, a species known to be particularly active in OA, consistent

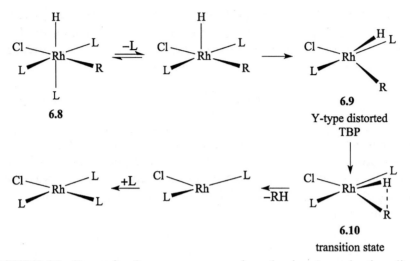

FIGURE 6.1 Example of a common general mechanism for reductive elimination in Milstein's octahedral d^6 species (L = PMe₃; R = CH₂COMe). The
reverse mechanism often holds for oxidative addition to square planar d^8
species (e.g., R = H).

with microscopic reversibility. Indeed, RhCl(PPh₃)₂, formed by loss of
a PPh₃ group from RhCl(PPh₃)₃, gives oxidative addition with hydrogen
at a rate at least 10^4 times faster than the four-coordinate complex.

Reversibility also holds for RE of alkyl halides where an S_N2 pathway
(Fig. 6.2) applies for the OA direction. Iodide attacks the coordinated
methyl trans to the open site and nucleophilically displaces the Pt(II)
complex, a good leaving group. The reactive five-coordinate intermediate, isolable in some cases, can also undergo concerted reductive elimination of ethane if the I⁻ concentration is low.[14]

Other Complexes

Square planar d^8 complexes show a variety of RE mechanisms: dissociative, nondissociative, and associative. Sometimes, a ligand dissociates
from M(R)(X)L₂, and the elimination occurs from the three-coordinate
M(R)(X)L intermediate, resulting in initial formation of a one-
coordinate ML metal fragment; this happens for PdR₂L₂ and several
Au(III) species. In some cases, the four-coordinate *trans*-M(R)(X)L₂
can reductively eliminate after initial trans to cis isomerization to bring
R and H close together. A fifth ligand can associate to form a five-
coordinate TBP intermediate that gives RE, as seen for Ni(II).[15]
Hartwig[16] has analyzed the kinetics for *trans*-[PdAr(N{tolyl}₂)(PPh₃)₂]

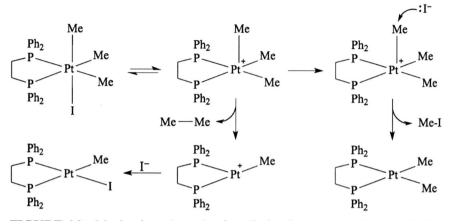

FIGURE 6.2 Mechanisms for reductive elimination to form C–C and C–Hal bonds in octahedral d^6 species in Goldberg's complex. These are the reverse of the mechanisms that apply for oxidative addition of nonpolar (C–C) and polar (Me–I) bonds to square planar d^8 species.

(**6.11**), where RE of Ar–N{tolyl}$_2$ takes place via competing dissociative and nondissociative pathways.

Mechanisms are probed via the kinetics; for example, in the dissociative RE of Me–Me from *trans*-[PdMe$_2$(PPh$_3$)$_2$] (**6.12**), added PPh$_3$ retards the reaction in an inverse first-order way (the rate is proportional to 1/[PPh$_3$]), suggesting that loss of PPh$_3$ takes place to give the three-coordinate intermediate PdMe$_2$(PPh$_3$). The retardation might alternatively have been due to stoichiometric formation of PdMe$_2$(PPh$_3$)$_3$, which would have to be less reactive than PdMe$_2$(PPh$_3$)$_2$ itself; NMR data shows that this is not the case, however.

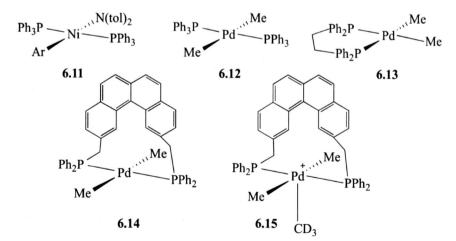

A *crossover* experiment is an important mechanistic test to distinguish between inter- and intramolecular reactions. For this particular case, a mixture of *cis*-Pd(CH$_3$)$_2$L$_2$ and *cis*-Pd(CD$_3$)$_2$L$_2$ is thermolyzed with the result that only C$_2$H$_6$ and C$_2$D$_6$ are formed, showing that the reaction is *intramolecular*—that is, R groups can couple only within the same molecule of starting complex. This experiment rules out coupling between R groups originating in different molecules of the complex (the *intermolecular* route). The crossover product, CH$_3$CD$_3$, would have been formed if alkyl groups eliminated in a binuclear way, or if free methyl radicals had been involved and lived long enough to migrate from one molecule to the next. Proper controls are needed, however; even if CH$_3$CD$_3$ is formed, the CH$_3$ and CD$_3$ groups may already have exchanged in the starting materials before RE takes place. Looking at the starting materials after partial conversion to products is needed to ensure that no significant amount of Pd(CH$_3$)-(CD$_3$)L$_2$ is present.

Dissociation of a monodentate phosphine in **6.12** is much easier than going from bidentate to monodentate ligation in the chelating diphosphine analog, **6.13**. As a result, RE is ~100 times slower in **6.13** versus **6.12**. The "transphos" complex **6.14** does not eliminate ethane at all, even under harsher conditions in which the cis **6.13** readily does so. The groups to be eliminated therefore need to be cis, but transphos locks them in a trans arrangement. Oxidation can induce RE, for example, the Pd(II) transphos complex **6.14** reacts with CD$_3$I to give CD$_3$CH$_3$, probably via the Pd(IV) intermediate, **6.15**.

Reductive elimination involving acyl groups is easier than for alkyls. For example, the cobalt dimethyl shown in Eq. 6.23 does not lose ethane but undergoes migratory insertion with added CO, no doubt via reversible loss of L to generate a 2e site for CO binding. The intermediate acyl alkyl complex subsequently gives acetone via RE. A crossover experiment with the mixed protonated d_0 and perdeuterated d_6 dialkyls showed that this reaction is intramolecular by giving no d_3 but only d_0 and d_6 acetone.

$$\underset{L}{\overset{Me}{CpCo}}\!\!-Me \xrightarrow{CO} \underset{L}{\overset{COMe}{CpCo}}\!\!-Me \xrightarrow{CO} \underset{L}{\overset{CO}{CpCo}} + \ \rangle\!\!=\!O \qquad (6.23)$$

In some cases, RE can be induced by oxidation, such as RE of C–Cl from Pd(IV), formed from Pd(II) and PhICl$_2$;[17] this is no doubt because Pd(IV), an unusually high OS for Pd, has a higher driving force for RE.

Binuclear Reductive Elimination

We saw earlier that binuclear OA is important for first row metals that prefer to change their oxidation state by one rather than two units. The same holds for RE as shown in Eq. 6.24 (L = PBu$_3$) and Eq. 6.25.

$$2MeCH = CH - CuL \rightarrow MeCH = CH - CH = CHMe \qquad (6.24)$$

$$ArCOMn(CO)_5 + HMn(CO)_5 \rightarrow ArCHO + Mn_2(CO)_{10} \qquad (6.25)$$

Reductive Elimination of C–F, –O, and –N

These reductive eliminations tend to have a higher kinetic barrier than for C–H or C–C. In RE of C–X (X = F, OR, NR$_2$),[18] the π donor X group prefers to locate at the base of the Y in the Y-shaped intermediate mentioned earlier, and thus is remote from the RE partner. Numerous cases of such REs have been reported in recent years, however,[19] notably in connection with the Buchwald-Hartwig coupling procedure to form C–X bonds (Section 9.7 and 14.1).

6.7 σ-BOND METATHESIS

Apparent OA/RE sequences can in fact go by a different route, *σ-bond metathesis* or *σ-bond complex-assisted metathesis*.[20] These are most easily identified for d^0 early metal complexes, such as Cp$_2$ZrRCl or WMe$_6$, where OA is forbidden, since the product would have to be d^{-2} (Section 2.4). When a d^0 complex reacts with H$_2$ (Eq. 3.32), path *a* of Fig. 6.3 is therefore forbidden and path *b* or *c* must take over. Path *b*

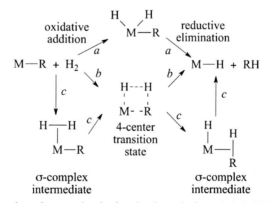

FIGURE 6.3 Sigma bond metathesis (paths *b* and *c*) and OA/RE (path *a*) are hard to distinguish for d^2–d^{10} complexes, but for d^0 cases, only sigma bond metathesis is allowed because OA would produce a forbidden oxidation state.

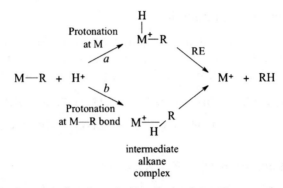

FIGURE 6.4 Protonation has similar limitations. Protonation at the metal (path *a*) and at the M–R bond (path *b*) are hard to distinguish for d^2–d^{10} complexes, but for d^0 cases, only path *a* is allowed because protonation at the metal would produce a forbidden oxidation state.

and *c* differ only in that *c* explicitly postulates an intermediate σ-complex. In d^2–d^{10} transition metals, both OA and σ-complex formation is usually permitted, but distinguishing between them is hard since both the products and the kinetics are identical.[21] In a Rh(III) alkyl, path *a* is technically allowed, but Rh(V) is an unusual oxidation state, so paths *b* or *c* would be preferred. Pathway *a* is typical when OA occurs readily.

In the same way, to avoid forbidden oxidation states, reaction of d^0 alkyls with acids cannot go via initial protonation at the metal (step *a* in Fig. 6.4) because as a d^0 system, the metal has no $M(d_\pi)$ lone pairs. Instead, protonation of the M–R bond must take place. Formation of an alkane σ-complex would then lead to loss of alkane. For d^2–d^{10} metals, where all pathways are allowed, it is again hard to tell which is followed; pathway *a* is normally assumed to operate in the absence of specific evidence to the contrary.

- Reductive elimination, the reverse of oxidative addition, decreases both the oxidation state, and the coordination number by two units.
- σ-Bond metathesis gives the same outcome as oxidative addition/ reductive elimination; the two situations are hard to tell apart.

6.8   OXIDATIVE COUPLING AND REDUCTIVE FRAGMENTATION

In oxidative coupling, Eq. 6.26, a metal couples two alkenes to give a metalacycle or two alkynes to give a *metallole*. Even CO and CN multiple bonds can sometimes participate.[22] The formal oxidation state of the metal increases by two units; hence the "oxidative" part of the name. The electron count decreases by two, but the coordination number stays the same. The reverse reaction, "reductive fragmentation," is much rarer. It cleaves a relatively unactivated C–C bond to give back the two unsaturated ligands. From the point of view of the metal, these are cyclic OA and RE reactions; the new aspect is the remote C–C bond formation or cleavage (Eq. 6.26).

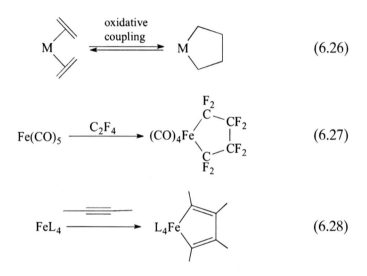

$$\text{(6.26)}$$

$$\text{(6.27)}$$

$$\text{(6.28)}$$

Alkynes undergo the reaction more easily than do alkenes unless activated by the substituents or by strain. C_2F_4 undergoes the reaction easily because F prefers the sp^3 C–F of the product where the high electronegativity of F is better satisfied by the less electronegative sp^3 carbon and repulsion between the F lone pairs and the C=C pi bond is relieved.

- Oxidative addition needs a metal that can undergo a 2e oxidation and a 2e (4e ionic model) change in electron count.
- Many mechanisms are seen (Table 6.3): concerted (Section 6.2), S_N2 (Section 6.3), radical (Section 6.4), and ionic (Section 6.5).
- Reductive elimination goes by the reverse of these mechanisms (Section 6.6).
- Sigma bond metathesis is an alternative pathway to oxidative addition followed by reductive elimination (Section 6.7).

REFERENCES

1. T. Gottschalk-Gaudig, J. C. Huffman, K. G. Caulton, H. Gerard, and O. Eisenstein, *J. Am. Chem. Soc.,* **121**, 3242, 1999.

2. M. E. Evans, T. Li, and W. D. Jones, *J. Amer. Chem. Soc.,* **132**, 16278, 2010.

3. J. F. Hartwig, *Synlett.,* **2006**, 1283.

4. F. Barrios-Landeros, B. P. Carrow, J. F. Hartwig, *J. Amer. Chem. Soc.,* **131**, 8141, 2009.

5. J. Choi, D. Y. Wang, S. Kundu, Y. Choliy, T. J. Emge, K. Krogh-Jespersen, and A. S. Goldman, *Science,* **332**, 1545, 2011.

6. C. M. Thomas and G. Süss-Fink, *Coord. Chem. Rev.,* **243**, 125, 2003.

7. C. J. Levy and R. J. Puddephatt, *J. Am. Chem. Soc.,* **119**, 10127, 1997.

8. C. D. Hoff, *Coord. Chem. Rev.,* **206**, 451, 2000.

9. M. Newcomb and P. H. Toy, *Acc. Chem. Res.,* **33**, 449, 2000.

10. J. R. Woodward, *Progr. React. Kinet.,* **27**, 165, 2002; C. T. Rodgers, *Pure Appl. Chem.,* **81**, 19, 2009.

11. A. Yahav-Levi, I. Goldberg, A. Vigalok, and A. N. Vedernikov, *Chem. Commun.,* **46**, 3324, 2010.

12. M. N. Birkholz, Z. Freixa, and P. W. N. M van Leeuwen, *Chem. Soc. Rev.,* **38**, 1099, 2009.

13. A. Ariafard, Z. Ejehi, H. Sadrara, T. Mehrabi, S. Etaati, A. Moradzadeh, M. Moshtaghi, H. Nosrati, N. J. Brookes, and B. F. Yates, *Organometallics,* **30**, 422, 2011.

14. U. Fekl and K. I. Goldberg, *Adv. Inorg. Chem.,* **54**, 259, 2003.

15. M. S. Driver and J. F. Hartwig, *J. Am. Chem. Soc.,* **119**, 8232, 1997.

16. J. F. Hartwig, *Acct. Chem. Res.* **41**, 1534, 2008; *Nature,* **455**, 314, 2008.

17. S. R. Whitfield and M. S. Sanford, *J. Am. Chem. Soc.,* **129**, 15142, 2007.

18. J. M. Racowski, J. B. Gary, and M. S. Sanford, *Angew. Chem. Int. Ed.,* **51**, 3414, 2012.

19. P. S. Hanley, S. L. Marquard, T. R. Cundari, and J. F. Hartwig, *J. Am. Chem. Soc.,* **134**, 15281, 2012.

20. R. N. Perutz and S. Sabo-Etienne, *Angew Chem. -Int. Ed. Engl.,* **46**, 2578, 2007.

21. B. Butschke, D. Schroeder, and H. Schwarz, *Organometallics,* **28**, 4340, 2009.

22. M. Takahashi and G. C. Micalizio, *J. Am. Chem. Soc.,* **129**, 7514, 2007.

PROBLEMS

6.1. An oxidative addition to a metal complex **A** is found to take place with MeOSO$_2$Me but not with *i*-PrI. A second complex, **B**, reacts with *i*-PrI but not with MeOSO$_2$Me. What mechanism(s) do you think is (are) operating in the two cases? Which of the two complexes, **A** or **B**, would be more likely to react with MeI? What further tests could you apply to confirm the mechanism(s)?

6.2. Suppose we are able to discover that the equilibrium constants for Eq. 6.1 are in the order $CH_3–H < Ph–H < H–H < Et_3Si–H$ for a given square planar Ir(I) complex. Can we say anything about the relative metal–ligand bond strengths in the adducts? Justify any assumptions that you make.

6.3. A given complex ML_n forms only a dihydrogen complex $(\eta^2\text{-}H_2)$-ML_n, not the true oxidative addition product H_2ML_n with H_2. Would the true oxidative addition product be more or less likely to form as we move to (a) more electron-releasing ligands L, (b) from a third- to a first-row metal, M, or (c) to the 1e oxidation product $H_2ML_n^+$? Would you expect the same metal fragment to form an ethylene complex, $(C_2H_4)ML_n$, with predominant Dewar–Chatt or metalacyclopropane character? Explain.

6.4. Complexes of the type $Pt(PR_3)_4$ can form $PtCl_2(PR_3)_2$ with HCl. How do you explain this result? The same product can also be formed from t-BuCl and $Pt(PR_3)_4$. What do you think is happening here? In each case, a different nonmetal-containing product is also formed; what do you think they are?

6.5. A 16e metal complex L_nM is found to react with ethylene to give 1-butene and L_nM. Provide a reasonable mechanism involving oxidative coupling.

6.6. Predict the order of reactivity of the following in oxidative addition of HCl: **A**, $IrCl(CO)(PPh_3)_2$; **B**, $IrCl(CO)(PMe_3)_2$; **C**, $IrMe(CO)(PMe_3)_2$; **D**, $IrPh(CO)(PMe_3)_2$. How do you expect the $\nu(CO)$ frequencies of **A-D** (i) to vary within the series and (ii) to change in going to the oxidative addition products? Explain and justify any assumptions you make.

6.7. The products from HCl addition to **C** and **D** in Problem 6.6 are unstable, but the addition products to **A** and **B** are stable. Explain and state how **C** and **D** will decompose.

6.8. WMe_6 reacts with H_2 and PMe_3 to give $WH_2(PMe_3)_5$. Propose a reasonable mechanism.

6.9. H_2 adds to $Ir(dppe)(CO)Br$ to give a kinetic product **A**, in which the cis H ligands are trans to P and CO, and a thermodynamic product **B**, in which the cis H ligands are trans to P and Br. Write the structures of **A** and **B**. How would you tell whether the rearrangement of **A** to **B** occurs by initial loss of H_2 or by a simple intramolecular rearrangement of **A**?

6.10. $Pt(PEt_3)_2$, generated electrochemically, reacts with the PhCN solvent to give $PhPt(CN)(PEt_3)_2$. Oxidative addition of a C–C bond is very rare. Discuss the factors that make it possible in this case.

6.11. Complex **6.16** is formed by the route of Eq. 6.29. Suggest a plausible pathway for this reaction if epoxide **6.17** gives complex **6.18** (Eq. 6.30).

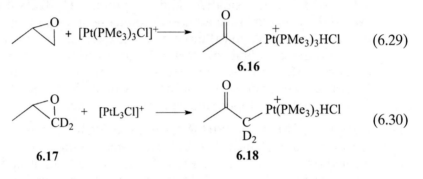

$$(6.29)$$

6.16

$$(6.30)$$

6.17   **6.18**

7

INSERTION AND ELIMINATION

Oxidative addition and substitution allow us to introduce a variety of 1e and 2e ligands into the coordination sphere of a metal. With insertion, and its reverse, elimination, we can combine and transform these ligands, ultimately to expel these transformed ligands to give useful products, often in the context of a catalytic cycle. In this way, organometallic catalysis can convert organic reagents into organic products with regeneration of the metal species for subsequent reaction cycles.

7.1 INTRODUCTION

By insertion, a π-bound 2e ligand, A=B, inserts into an M–X bond to give M–(AB)–X, where AB has formed a new bond with both M and X. There are two main types of insertion, either 1,1 (Eq. 7.1) or 1,2 (Eq. 7.2). In 1,1 insertion, M and X end up bound to the same atom of AB, but in the 1,2 type, M and X end up on adjacent atoms of AB. The type of insertion in any given case depends on the nature of A=B. For example, CO gives only 1,1 insertion where both M and X end up bound to CO carbon. On the other hand, ethylene gives only 1,2 insertion, where the product, MCH_2CH_2X, has M and X on adjacent atoms of the ligand.

The Organometallic Chemistry of the Transition Metals, Sixth Edition.
Robert H. Crabtree.
© 2014 John Wiley & Sons, Inc. Published 2014 by John Wiley & Sons, Inc.

In general, η^1 ligands give 1,1 insertion, and η^2 ligands give 1,2 insertion. SO_2 is the only common ligand that can give both types of insertion and accordingly, SO_2 can either be an η^1 (S) or η^2 (S,O) ligand.

$$
\begin{array}{cc}
\underset{\substack{|\\ M-\underset{B}{\overset{A}{\|}}\\ 18e}}{X} & \xrightarrow[\text{insertion}]{\text{1,2-migratory}} \quad \underset{16e}{M-A\overset{\overset{\displaystyle X}{\diagdown}\;B}{\diagup}}
\end{array}
\qquad (7.1)
$$

$$
\underset{18e}{\overset{\displaystyle X}{\underset{|}{L_nM-A=B}}} \xrightarrow[\text{insertion}]{\text{1,1-migratory}} \underset{16e}{L_nM-A\underset{\overset{\displaystyle}{\diagdown}\;B}{\overset{\diagup\;X}{}}} \qquad (7.2)
$$

In principle, insertion is reversible, and reversibility is indeed seen experimentally,[1] but just as we saw for OA and RE in Chapter 6, in many cases, only the thermodynamically favored direction is ever observed. For example, SO_2 commonly inserts into M–R bonds to give alkyl sulfinate complexes, but these rarely eliminate SO_2. Conversely, N_2 readily eliminates from diazoarene complexes, but the reverse is not seen.

$$M-R + SO_2 \rightarrow M-SO_2R \qquad (7.3)$$

$$M-N{=}N-R \rightarrow M-R + N_2 \qquad (7.4)$$

Both the 1e and 2e ligands normally need to coordinate to the metal before insertion. This means that a 3e set of ligands in the intermediate converts to a 1e ligand in the insertion product, so that a 2e vacant site (□) is generated (Eq. 7.2). Binding of an external 2e ligand can trap the insertion product (Eq. 7.5). Conversely, the elimination requires a 2e vacant site, so that an 18e complex cannot undergo the reaction unless a ligand first dissociates. The insertion also requires a cis arrangement of the 1e and 2e ligands, while the elimination generates a cis arrangement of these ligands. The formal oxidation state does not change during the reaction.

$$
\underset{18e}{\overset{\displaystyle R}{\underset{|}{M-C{=}O}}} \xrightleftharpoons[]{\substack{\text{1,1 migratory}\\ \text{insertion}}} \underset{16e}{\overset{\displaystyle \Box}{\underset{|}{M-C\underset{O}{\overset{R}{\diagup\diagdown}}}}} \xrightarrow{\;L\;} \underset{18e}{\overset{\displaystyle L}{\underset{|}{M-C\underset{O}{\overset{R}{\diagup\diagdown}}}}} \qquad (7.5)
$$

In one useful picture of insertion, the X ligand migrates with its M–X bonding electrons (e.g., as H^- or Me^-) to attack the π^* orbital of the A=B ligand. In this intramolecular nucleophilic attack on A=B, the migrating group, R, retains its stereochemistry. This picture also justifies the term "migratory insertion," often applied to these reactions, in that the X migrates to the A=B group. A component of M–(A=B) bonding is back donation, in which an M d_π electron pair is partially transferred to the A=B π^*; in an insertion, an M–X bonding electron pair is fully transferred to the A=B π^*.

7.2 CO INSERTION

CO shows a strong tendency to insert into metal–alkyl bonds to give metal acyls, a reaction that has been carefully studied for a number of systems. Although the details may differ, most follow the pattern set by the best-known case:

$$
\begin{array}{c}
Me \\
| \\
(CO)_4Mn-C=O
\end{array}
\quad
\underset{-CO}{\overset{CO}{\rightleftarrows}}
\quad
\begin{array}{c}
CO \\
| \quad Me \\
(CO)_4Mn-C \\
\quad\quad\quad O
\end{array}
\qquad (7.6)
$$

The usual mechanism of migratory insertion is shown in Eq. 7.7. The alkyl group in the reagent (Rgt) undergoes a migration to the CO to give an acyl intermediate (Int.) that is trapped by added ligand, L, to give the final product (Pdct).

$$
\begin{array}{ccc}
\begin{array}{c}
O \\
\| \\
C \\
| \\
Me-Mn(CO)_4 \\
\text{Rgt.}
\end{array}
&
\underset{k_{-1}}{\overset{k_1}{\rightleftarrows}}
\begin{array}{c}
Me \quad O \\
\backslash \quad // \\
C \\
| \\
\square-Mn(CO)_4 \\
\text{Int.}
\end{array}
&
\underset{\text{slow}}{\overset{L,\, k_2}{\longrightarrow}}
\begin{array}{c}
Me \quad O \\
\backslash \quad // \\
C \\
| \\
L-Mn(CO)_4 \\
\text{Pdct.}
\end{array}
\end{array}
\qquad (7.7)
$$

The kinetics are reminiscent of dissociative substitution (Section 4.4) except that the 2e site is formed at the metal in the migratory step, not by loss of a ligand. Using the usual steady-state method, the rate is given by Eq. 7.8.

$$
\text{Rate} = \frac{-d[\text{Rgt}]}{dt} = \frac{k_1 k_2 [\text{L}][\text{Rgt}]}{k_{-1} + k_2[\text{L}]}
\qquad (7.8)
$$

There are three possible regimes,[2] each of which can be found in real cases:

1. If k_{-1} is very small relative to $k_2[\text{L}]$, [L] cancels and Eq. 7.8 reduces to Eq. 7.9.

$$\text{Rate} = \frac{-d[\text{Rgt}]}{dt} = k_1[\text{Rgt}] \tag{7.9}$$

Because k_{-1} is small, L always traps the intermediate; this means the rate of the overall reaction is governed by k_1, and we have a first-order reaction.

2. If k_{-1} is very large relative to $k_2[\text{L}]$, then Eq. 7.8 reduces to Eq. 7.10.

$$\text{Rate} = \frac{-d[\text{Rgt}]}{dt} = \frac{k_1 k_2[\text{L}][\text{Rgt}]}{k_{-1}} \tag{7.10}$$

In this case, the intermediate almost always goes back to the starting reagent, and the second step, attack by L, governs the overall rate, so we have second-order kinetics.

3. If k_{-1} is comparable with $k_2[\text{L}]$, then the situation is more complicated and the equation is usually rewritten as Eq. 7.11, where a new term, k_{obs}, is defined by Eq. 7.12.

$$\text{Rate} = \frac{-d[\text{Rgt}]}{dt} = k_{obs}[\text{Rgt}] \tag{7.11}$$

$$k_{obs} = \frac{k_1 k_2[\text{L}]}{k_{-1} + k_2[\text{L}]} \tag{7.12}$$

The intermediate is now trapped by L at a rate that is comparable with the reverse migration. This is handled by plotting $1/k_{obs}$ versus $1/[\text{L}]$ to find $1/k_1$ from the intercept and $k_{-1}/(k_1 k_2)$ from the slope (Eq. 7.13). Dividing the slope by the intercept gives k_{-1}/k_2, which tells us how the intermediate partitions between the forward (k_2) and back (k_{-1}) reactions.

$$\frac{1}{k_{obs}} = \underbrace{\left(\frac{k_{-1}}{k_1 k_2}\right) \frac{1}{[\text{L}]}}_{\text{slope}} + \underbrace{\frac{1}{k_1}}_{\text{intercept}} \tag{7.13}$$

When the incoming ligand in Eq. 7.7 is ^{13}CO, the product contains only one labeled CO, cis to the newly formed acetyl. This suggests that the acetyl group is initially formed cis to a vacant site in the intermediate. The labeled CO can be located in the product by NMR and IR spectroscopy.

In an example of a useful general strategy, we can learn about any forward process by looking at the reverse reaction—here, α elimination of CO from $Me^{13}COMn(CO)_5$ (Eq. 7.14; $C^* = {}^{13}C$). We can easily label the acyl carbon with ^{13}C by reaction of $[Mn(CO)_5]^-$ with $Me^{13}COCl$ and find that after α elimination of CO, the label ends up in a CO cis to the methyl in the product.

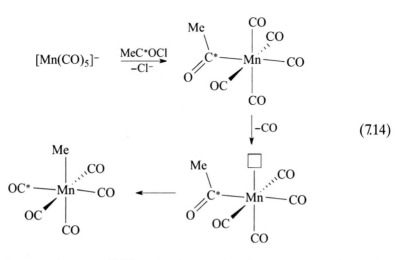

$$(7.14)$$

By microscopic reversibility, the forward and reverse reactions of a thermal process must follow the same path. In this case, if the labeled CO ends up cis to Me in the elimination direction, the CO to which a methyl group migrates in the insertion direction must also be cis to methyl. We are fortunate in seeing the kinetic products of these reactions. If a subsequent scrambling of the COs had been fast, we could have deduced nothing.

We now know that Me and CO must be mutually cis to insert, but we do not yet know if Me migrates to the CO site or vice versa. It is also possible to use reversibility arguments to show that it is Me, not CO, that moves. To do this, we look at CO elimination in *cis*-(MeCO)Mn(CO)$_4$(^{13}CO), in which the labeled CO is cis to the acetyl. If the acetyl CO migrates during the elimination, then the methyl in the product will stay where it is and so remain cis to the label. If the methyl migrates, then it will end up both cis and trans to the label, as is in fact observed (Eq. 7.15).

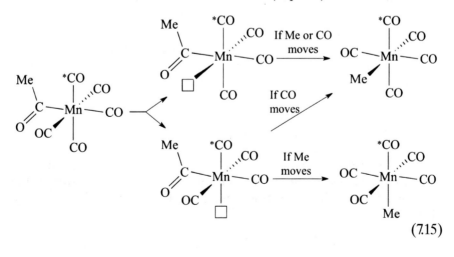

$$(7.15)$$

This implies that the methyl also migrates in the insertion direction. The *cis*-(MeCO)Mn(CO)$_4$(^{13}CO) required for this experiment can be prepared by the photolytic method discussed in Section 4.7. This migration of Me not CO is one feature of migratory insertion that does not reliably carry over to other systems, where the product acyl is occasionally found at the site originally occupied by the alkyl. Consistent with this mechanism, any stereochemistry at the alkyl carbon is retained both on insertion and on elimination.

Enhancing Insertion Rates

Steric bulk in the L_n ligand set of $L_nM(Me)(CO)$ accelerates insertion, no doubt because the acetyl in the $L_nM(COMe)$ product, occupying one coordination site, is far less bulky than the alkyl and carbonyl, occupying two sites in the starting complex, $L_nM(Me)$ (CO). Lewis acids such as AlCl$_3$ or H$^+$ can increase the rate of migratory insertion by as much as 10^8-fold, where k_2 is the slow step.[3] Metal acyls (**7.1**) are more basic at oxygen than are the corresponding carbonyls by virtue of the resonance form **7.2**. By binding to the oxygen, the Lewis acid would be expected to stabilize the transition state and speed up trapping by L and therefore speed up the reaction. Polar solvents such as acetone also significantly enhance the rate.

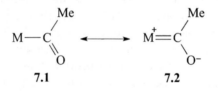

7.1   **7.2**

Another important way of promoting insertion is oxidation of the metal.[4] Cp(CO)$_2$Fe^{II}Me is normally very slow to insert, but 1e oxidation at −78°C in MeCN electrochemically or with Ce(IV), gives the 17e, Fe(III) acyl [CpFe^{III}(MeCN)(CO)(COMe)]$^+$, in which the solvent plays the role of incoming ligand. As we saw in Chapter 4, 17e complexes can be very labile, but another factor here may be the increased electrophilicity (decreased π basicity) of the oxidized metal enhancing the partial positive charge on the CO carbon. The migration of Me$^-$ to a now more electron-deficient CO carbon is expected to be faster.

Early d^0 metals are Lewis acids that prefer O-donor ligands (for the oxophilicity of d^0 metals, see Section 3.2); they can therefore act as their own Lewis acid catalysts for insertion, the product being an η^2-acyl (Eq. 7.16).

$$Cp^*_2Zr \overset{R}{\underset{R}{\big<}} \quad \xrightarrow{\ CO\ } \quad Cp^*_2Zr \overset{\overset{R}{|}}{\underset{R}{\longleftarrow}} \overset{C}{\underset{}{\big\Vert}} O \qquad (7.16)$$

$$\underset{16e,\ d^0}{} \qquad\qquad\qquad \underset{18e,\ d^0}{}$$

By altering the thermodynamics in favor of the adduct, this effect is even sufficient to promote the normally unfavorable CO insertion into an M–H bond, as shown in Eq. 7.17.

$$Cp^*_2(RO)Th\text{-}H \quad \xrightarrow{\ CO\ } \quad Cp^*_2(RO)Th \overset{\overset{H}{|}}{\underset{O}{\big\Vert}} C \qquad (7.17)$$

In each of these reactions, the formation of an intermediate carbonyl complex is proposed but d^0 Zr(IV) and Th(IV) are both poor π bases, so these intermediates must be very unstable; in compensation, the limited back bonding makes the CO much more reactive for insertion, however. In rare cases, CO inserts directly into an M–R bond without first binding to the metal, as seems to be the case for a Re(V) oxo alkyl where the high valent Re is poorly adapted to bind CO.[5]

Apparent Insertions

An insertion that appears to be migratory can in fact go by an entirely different route (Eq. 7.18). Since MeO^- is a good π donor bound to a d^6 π-donor metal, the MeO^- group easily dissociates to give an ion pair with a 2e vacancy at the metal. The free CO present then binds to this 2e site and is strongly activated toward nucleophilic attack at the CO carbon owing to the positive charge on the metal. The product is the interesting metalloester shown in Eq. 7.18.

$$L_2(CO)Ir\text{—}OMe \xrightarrow{\ CO\ } [L_2(CO)Ir\text{—}CO]^+OMe^- \longrightarrow L_2(CO)Ir\text{—}C \overset{\overset{OMe}{\diagup}}{\underset{O}{\diagdown}} \qquad (7.18)$$

Genuine migratory insertions into M–O bonds are also possible. For *trans*-[Pt(Me)(OMe)(dppe)], CO inserts into the Pt–OMe bond, while for [Ni(Me)(O-p-C_6H_4CN)(bipy)], CO inserts into Ni–Me. For nickel, the M–Me bond is significantly stronger than M–OMe, but migratory insertion with M–Me is marginally preferred owing to the weaker C–O bond of the aryloxycarbonyl. For platinum, M–Me and M–OMe bonds are equally strong, so the stronger methoxycarbonyl C–O bond favors reaction with the M–OMe bond.[6]

Double Insertion

Given that the methyl group migrates to the CO, why stop there? Why does the resulting acyl group not migrate to another CO to give an MeCOCO ligand? To see why, we can treat $[Mn(CO)_5]^-$ with MeCO-COCl to give $[MeCOCOMn(CO)_5]$, which easily and irreversibly eliminates CO to give $MeCOMn(CO)_5$. This means that the double-insertion product does not form because it is thermodynamically unstable with respect to $MeCOMn(CO)_5 + CO$. The –CHO and CF$_3$CO– groups also eliminate CO irreversibly to give M–H and M–CF$_3$ complexes, implying that these insertions cannot occur thermally. Thermodynamics drives these eliminations because the M–COMe, M–H, and M–CF$_3$ bonds are all distinctly stronger than the M–CH$_3$ bond that is formed in CO elimination from the acetyl. In contrast to CO, isonitriles can undergo repeated migratory insertion to give $R(CNR)_mM$ polymers, with m as high as 100. The instability of $R(CO)_mM$ is associated with having successive δ^+ carbonyl carbons mutually adjacent; =NR being less electronegative than =O, the problem is less severe for RNC than for CO. We look at 1,1 insertions involving carbenes in Chapter 11.

7.3 ALKENE INSERTION

The insertion of coordinated alkenes into M–H bonds leads to metal alkyls and constitutes a key step in a variety of catalytic reactions (Chapter 9). For example, the commercially important alkene polymerization reaction (Chapter 12) involves repeated alkene insertion into the growing polymer chain.

As η^2-ligands, alkenes give 1,2 insertion in the reverse of the familiar β elimination (Eq. 7.19). Some insertions give agostic (**7.3**) rather than classical alkyls, and species of type **7.3** probably lie on the pathway for insertion into M–H bonds. The position of equilibrium depends not only on whether an incoming ligand, L in Eq. 7.19, is available to trap the alkyl, but also very strongly on the alkene and the insertion thermodynamics. For simple alkenes, such as ethylene (Eq. 7.18), the equilibrium tends to lie to the left and the alkyl prefers β elimination, but for alkenes such as C_2F_4, which form strong M–R bonds, insertion is preferred and the product alkyl $L_nMCF_2CF_2H$ does not β-eliminate.

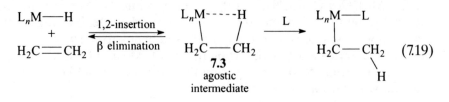

The transition state for insertion, **7.4**, resembles **7.3** in having an essentially coplanar M–C–C–H arrangement, and this implies that both insertion and elimination also require the M–C–C–H system to become coplanar. We have seen in Section 3.1 how we can stabilize alkyls against β elimination by having a noncoplanar M–C–C–H system. The same principles apply to stabilizing alkene hydride complexes. Compound **7.5** undergoes insertion at least 40 times more rapidly than **7.6**, although the alkene and M–H groups are cis in both cases, only in **7.6** is there a noncoplanar M–C–C–H arrangement.

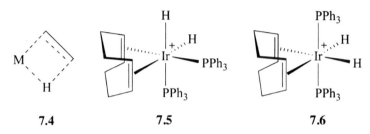

7.4	**7.5**	**7.6**

Regiochemistry of Insertion

In hydrozirconation of alkenes by Cp_2ZrHCl,[7] terminal alkenes insert in the anti-Markovnikov direction to give a stable 1° alkyl. Internal alkenes, such as 2-butene, insert to give an unstable 2° alkyl, that β-eliminates to give 1- and 2-butene. The 1-butene can now give a stable 1° alkyl that is the final product. This is particularly noteworthy because the free terminal alkene is less stable than the internal alkene. The outcome arises because the 1° alkyl is thermodynamically more stable than a 2° alkyl for steric reasons. The 1° alkyl, R, can subsequently be functionalized in a number of ways to give a variety of RX derivatives. Hydrozirconation is also effective with less reactive substrates, such as nitriles, where addition of Zr–H across the C≡N bond is possible.[8]

For $ArCH=CH_2$, the preferred L_nM–H insertion product tends to have the metal bound at the benzylic position in spite of the resulting steric disadvantage; not just Ph but electron-withdrawing groups in general prefer to locate at the α-carbon on insertion. Equilibration of the two regioisomers (Eq. 7.20)[9] also favors **7.8**, showing that this is indeed the thermodynamic product. Traditionally, this outcome of insertion has been ascribed to the new M–C bond being stronger in **7.8** than in **7.7**, but Jones[10] has called attention to the strength of the newly formed C–H′ bonds as a key factor. In **7.7**, the new C–H′ bond, being benzylic, is weak, while in **7.8**, the new C–H′ bond is no longer benzylic, so much stronger. The new benzylic M–C bond in **7.8** is typically weaker than the M–C bond in **7.7**, not stronger as once thought. Breaking the

M–C bond in **7.8** homolytically gives a stabilized C radical so is easier than breaking the M–C bond in **7.7**, where the resulting radical is not specially stabilized. **7.8** is nevertheless preferred as product, probably because its M–C bond is a little stronger than might be expected without back donation from metal d_π orbitals into the C–Ar σ^*, less favorable in **7.7**, where the M–C bond has no electronegative substituent. The same arguments probably apply to a variety of other electronegative substituents, such as –CN, –F, and –CHO. This reflects the general principle that we must consider *all* the bonds broken and formed in order to successfully interpret reactivity trends.

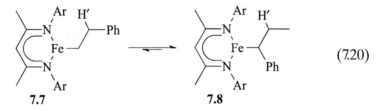

$$(7.20)$$

7.7   **7.8**

Simple α-olefins, where the two ends of the C=C bond are not well differentiated electronically, may give insertion with a mixed regiochemistry, although steric effects can bias the outcome in suitable cases.[11]

Syn versus Apparent Anti Insertion

In the usual syn insertion, the stereochemistry at both carbons is retained. This is best seen for alkynes, where the vinyl product can preserve the syn disposition of M and H. If the initially formed *cis*-vinyl complex remains 16e, it can rearrange to the sterically less hindered trans isomer, via an 18e η^2-vinyl. This can lead to an apparent anti addition of a variety of X–H groups (Eq. 7.21) to alkynes.[12]

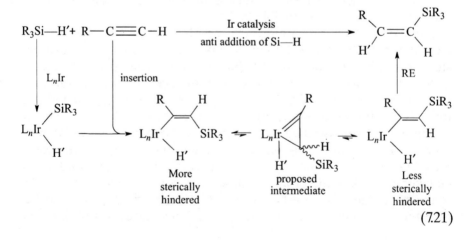

$$(7.21)$$

TABLE 7.1 Comparison of Barriers (kcal/mol) for Insertion in [Cp*{(MeO)$_3$P}MR(C$_2$H$_4$)]$^+$ for R = H and R = Et[11]

M	R = H[a]	R = Et[b]	Difference
Rh	12.2	22.4	10.2
Co	6–8 (est.)	14.3	6–8 (est.)

[a] ±0.1 kcal/mol.
[b] ±0.2 kcal/mol.

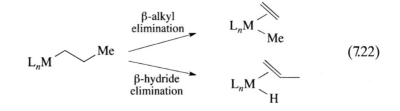

$$(7.22)$$

Insertion into M–H versus M–R

For thermodynamic reasons, CO insertion generally takes place into M–R, but not into M–H bonds. Alkene insertion, in contrast, is common for M–H, but much less common for M–R. The thermodynamics still favor the reaction with M–R, so its comparative rarity must be due to kinetic factors. Brookhart and Templeton[13] have compared the barriers for insertion of ethylene into the M–R bond in [Cp*{(MeO)$_3$P} MR(C$_2$H$_4$)]$^+$, where R is H or Et and M is Rh or Co. The reaction involving M–H has a 6- to 10-kcal/mol lower barrier (Table 7.1). This corresponds to a migratory aptitude ratio k_H/k_{Et} of 10^6–10^8. As we have seen before, reactions involving M–H are almost always kinetically more facile than reactions of M–R. This means that an alkene probably has less intrinsic kinetic facility for insertion than does CO. Looking at the reverse reaction (Eq. 7.22), elimination, we see that this implies that β-H elimination in an alkyl will be kinetically very much easier than β-alkyl elimination, and it will also give a thermodynamically more stable product, so it is not surprising that β-alkyl elimination is extremely rare. In those cases where it is observed, there is always some special factor that modifies the thermodynamics or the kinetics or both. For example, for *f*-block metals M–alkyl bonds appear to be comparable in strength, or stronger than M–H bonds, and both β-H and β-alkyl elimination is seen.

Strain, or the presence of electronegative substituents on the alkene, or moving to an alkyne are some of the other factors that can bias both the thermodynamics and the kinetics in favor of insertion, as shown in Eq. 7.23 for a strained bridgehead alkene.

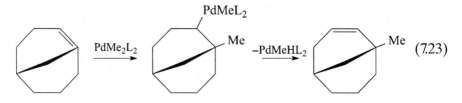

$$(7.23)$$

Radical Pathways

Styrene can insert into the M–M bond of [Rh(OEP)]$_2$ (OEP = octaethylporphyrin) via initial M–M bond homolysis to give the 15e metalloradical [Rh(OEP)]•. This adds to the alkene to give [PhCH(•)CH$_2$Rh(OEP)], stabilized by benzylic resonance, followed by the sequence of Eq. 7.24. [Rh(OEP)]$_2$ also initiates radical photopolymerization of CH$_2$=CHCOOR, where the intermediate C radicals add repetitively to acrylate rather than recombine with a metalloradical as in Eq. 7.24.

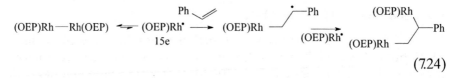

$$(7.24)$$

As we saw in Sections 5.2–5.3, butadiene and allene react with a variety of hydrides by 1,2 insertion, but butadienes also react with HMn(CO)$_5$ to give an apparent 1,4 insertion. Since this 18e hydride has no vacant site and CO dissociation is slow, an indirect mechanism is proposed: H atom transfer to give a 1,1-dimethylallyl radical that is subsequently trapped by the metal (Eq. 7.25). Only substrates such as a 1,3-diene that form particularly stable radicals, such as an allyl, can react in this way; CIDNP effects (Chapter 10) arising from the radical pathway are sometimes seen in the NMR spectra of the reacting mixture.

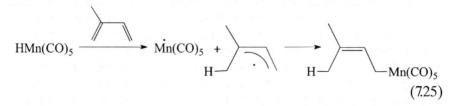

$$(7.25)$$

Insertion of O$_2$ into (dipy)PdMe$_2$ to give (dipy)PdMe(OOMe), has highly irreproducible rates because a radical chain is initiated by trace impurities, but addition of the radical initiator AIBN gives reproducible rates.[14]

Alternating CO/Alkene Insertion

$[(phen)PdMe(CO)]^+$ can copolymerize CO and ethylene to give a strictly alternating copolymer, $(CH_2CH_2CO)_n$.[15] This is of practical interest because its carbonyl functionality permits useful chemical modification. The polymerization reaction is also of mechanistic interest because of the essentially perfect alternation of alkene and CO insertions.

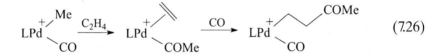

$$(7.26)$$

Of the possible erroneous insertions, double CO insertion is forbidden for the thermodynamic reasons discussed in Section 7.2, and double alkene insertion is rare because of its much slower intrinsic rate and the high affinity of the catalyst for CO, together amounting to a rate enhancement of 2000 versus CO insertion into M–R.

7.4 OUTER SPHERE INSERTIONS

In some cases, the A=B bond does not need to coordinate to the metal prior to insertion and can undergo the reaction with an 18e complex. The weakly binding ligand, CO_2, can insert into an M–H bond in this way. The nucleophilic hydride first attacks the carbon of free CO_2 to give a 16e M^+ unit and free $HCOO^-$. The formate then binds to the metal to give the 1,2-insertion product, M–OCHO.

Sulfur dioxide is a much stronger electrophile than CO_2 and also needs no vacant site. If SO_2 electrophilically attacks the α carbon of an 18e alkyl from the side opposite the metal, an alkyl sulfinate ion is formed with inversion at carbon. Since the anion has much of its negative charge on the oxygens, it is not surprising that the kinetic product of ion recombination is the O-bound sulfinato complex. On the other hand, the thermodynamic product is usually the S-bound sulfinate, as is appropriate for a soft metal binding. This sequence constitutes a 1,2 (O bound sulfinate) or a 1,1 insertion of SO_2 (S bound).

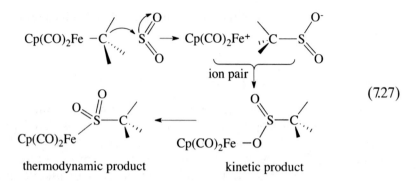

$$(7.27)$$

thermodynamic product   kinetic product

As expected for this mechanism, the reactivity falls off for bulky alkyls and electron attracting substituents. A crossover reaction of a mixture of RS and SR isomers of [CpFe*(CO)L{CH₂C*H(Me)Ph}], chiral at both Fe and the β-carbon, forms very little of the crossover products, the R,R and S,S isomers of the sulfinate complex. This shows both that the intermediate must stay ion-paired, and that the intermediate iron cation must have stereochemical stability. Ion pairing is very common in organic solvents of relatively low polarity, such as CH_2Cl_2, and ion pairs can have a well-defined solution structure, and such pairing can affect reaction outcomes.[16] O_2 can insert into M–H to give M–O–O–H; in some cases, an H atom abstraction mechanism by O_2 via M• and •O–O–H has been identified.[17] Insertions of CO_2 are discussed in Section 12.3.

7.5   α, β, γ, AND δ ELIMINATION

β Elimination

Continuing the discussion of β elimination from Section 3.1, we now look at the kinetics. An 18e complex has to lose a ligand to open up a site for elimination, but this may or may not be rate limiting. In either case, the addition of an excess of ligand can inhibit the reaction by quenching the open site. A significant kinetic isotope effect k_H/k_D in the elimination rate of $L_nMC_2H_5$ versus $L_nMC_2D_5$ suggests that the elimination itself is rate limiting since C–H(D) bond breaking must be important in the slow step.

In 16e complexes, a 2e site is usually available for β elimination. For example, 16e d^8 trans-[PdL₂Et₂] complexes (L = PR₃), can decompose by β elimination via an 18e transition state, but PR₃ dissociation is still required for elimination in trans-[PtL₂Bu₂], where the preference for 16e over 18e structures is more marked than for Pd(II).[18] The related metalacycle **7.9** β-eliminates 10⁴ times more slowly than [PtL₂Bu₂], presumably because a coplanar M–C–C–H arrangement is much harder to achieve (Eq. 7.28).

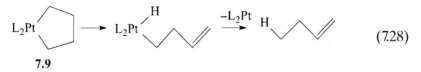

$$(7.28)$$

7.9

In a series of analogous nickel complexes in the presence or absence of excess phosphine, three different decomposition pathways are found, one for each of the different intermediates, 14e, 16e, and 18e, that can be formed (Eq. 7.29).

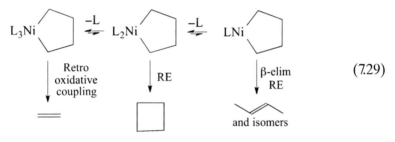

$$(7.29)$$

An alkyl and its alkene hydride elimination product can occasionally be seen in equilibrium together (Eq. 7.30).[19]

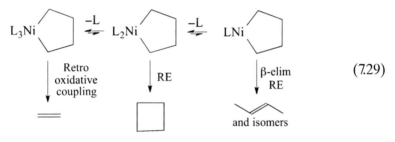

$$(7.30)$$

Alkoxide complexes β eliminate readily to give ketones or aldehydes, accounting for the ability of basic isopropanol to reduce many metal halides to hydrides with formation of acetone by the pathway of Eq. 3.27. β Elimination of amides and amines to imines also occurs but tends to be slow.[20]

α Elimination

Common for alkyls that lack β hydrogens, this is the reverse of 1,1 insertion (e.g., Eq. 7.14). β elimination being impossible, $L_nM–Me$ can only undergo an α elimination to give $L_nM(=CH_2)H$. While any β process gives an alkene, a stable species that can dissociate from the metal, an alkylidene ligand from an α elimination is unstable in the free state and cannot dissociate. Methylene hydride complexes are therefore rarely seen because they are thermodynamically unstable with respect to the corresponding methyl complex, but α elimination can still occur reversibly in a reaction sequence. For this reason, the α process is less well characterized than β elimination. Isotope exchange studies on both Mo and Ta alkyls suggest that α elimination can be up to 10^6 times faster than β elimination even in cases in which both α- and β-H substituents are present.[21] A coordinatively unsaturated methyl complex can be in equilibrium with a methylene hydride,[22] that can be trapped either by nucleophilic attack at the carbene carbon (Eq. 7.31) or by removing the hydride by reductive elimination with a second alkyl (Eq. 7.32):

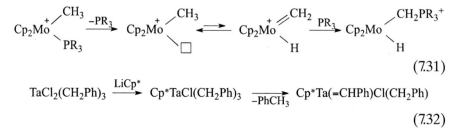

$$(7.31)$$

$$\text{TaCl}_2(\text{CH}_2\text{Ph})_3 \xrightarrow{\text{LiCp*}} \text{Cp*TaCl}(\text{CH}_2\text{Ph})_3 \xrightarrow[-\text{PhCH}_3]{} \text{Cp*Ta}(=\text{CHPh})\text{Cl}(\text{CH}_2\text{Ph})$$

$$(7.32)$$

Other Eliminations

A great variety of other ligands may lack β-Hs but possess γ- or δ-H's and can thus undergo γ or δ elimination to give cyclic products (Eq. 7.33).

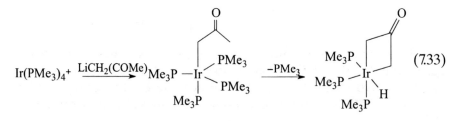

$$(7.33)$$

- 1,1-Insertion occurs for η^1 ligands such as CO and 1,2 insertion occurs for η^2 ligands such as C_2H_4. In each case an X ligand migrates from M to L (Eq. 7.1 and Eq. 7.2).
- Insertions are kinetically favored for X = H over X = R, but for CO, insertion into M–H is thermodynamically disfavored (Eq. 7.1 and Eq. 7.2).

REFERENCES

1. A. J. Pontiggia, A. B. Chaplin, and A. S. Weller, *J. Organometal. Chem.*, **696**, 2870, 2011.
2. A. Derecskei-Kovacs and D. S. Marynick, *J. Am. Chem. Soc.*, **122**, 2078, 2000.
3. K. Fukumoto and H. Nakazawa, *J. Organometal. Chem.*, **693**, 1968, 2008; M. Rubina, M. Conley, and V. Gevorgyan, *J. Am. Chem. Soc.*, **128**, 5818, 2006.
4. Z. X. Cao, S. Q. Niu, and M. B. Hall, *J. Phys. Chem.*, *A* **104**, 7324, 2000.
5. C. P. Lilly, P. D. Boyle, and E. A. Ison, *Organometallics*, **31**, 4295, 2012.
6. S. A. Macgregor and G. W. Neave, *Organometallics*, **23**, 891, 2004.
7. Y. H. Zhang, R. J. Keaton, and L. R. Sita, *J. Am. Chem. Soc.*, **125**, 8746, 2003.
8. C. Lu, Q. Xiao, and P. E. Floreancig, *Org. Lett.*, **12**, 5112, 2010.
9. J. Vela, S. Vaddadi, T. R. Cundari, J. M. Smith, E. A. Gregory, R. J. Lachicotte, C. J. Flaschenriem, and P. L. Holland, *Organometallics*, **23**, 5226, 2004.
10. M. E. Evans, T. Li, A. J. Vetter, R. D. Rieth, and W. D. Jones, *J. Org. Chem.*, **74**, 6907, 2009; G. Choi, J. Morris, W. W. Brennessel, and W. D. Jones, *J. Am. Chem. Soc.*, **134**, 9276, 2012 and personal communication, 2012.
11. E. A. Standley and T. F. Jamison, *J. Am. Chem. Soc.*, **135**, 1585, 2013.
12. R. H. Crabtree, *New J. Chem.*, **27**, 771, 2003.
13. L. Luan, P. S. White, M. Brookhart, and J. L. Templeton, *J. Am. Chem. Soc.*, **112**, 8190, 1990.
14. L. Boisvert, M. C. Denney, S. Kloek Hanson, and K. I. Goldberg, *J. Am. Chem. Soc.*, **131**, 15802, 2009.
15. C. Lu and J. C. Peters, *J. Am. Chem. Soc.*, **124**, 5272, 2002.
16. A. Macchioni, *Chem. Rev.*, **105**, 2039, 2005.
17. M. M. Konnick, N. Decharin, B. V. Popp, and S. S. Stahl, *Chem. Sci.*, **2**, 326, 2011.
18. J. Zhao, H. Hesslink, and J. F. Hartwig, *J. Am. Chem. Soc.*, **123**, 7220, 2001.
19. K. Umezawa-Vizzini and T. R. Lee, *Organometallics*, **23**, 1448, 2004.
20. J. Louie, F. Paul, and J. F. Hartwig, *Organometallics*, **15**, 2794, 1996.
21. R. R. Schrock, S. W. Seidel, N. C. Mosch-Zanetti, K. Y. Shih, M. B. O'Donoghue, W. M. Davis, and W. M. Reiff, *J. Am. Chem. Soc.*, **119**, 11876, 1997.
22. H. Hamilton and J. R. Shapley, *Organometallics*, **19**, 761, 2000.

PROBLEMS

7.1. Predict the structures of the products (if any would be expected) from the following: (a) $CpRu(CO)_2Me$ + PPh_3, (b) Cp_2Zr-HCl + butadiene, (c) $CpFe(CO)_2Me$ + SO_2, and (d) $Mn(CO)_5CF_3$ + CO.

7.2. Me$_2$NCH$_2$Ph reacts with PdCl$_2$ to give **A**; then **A** reacts with 2,2-dimethylcyclopropene and pyridine to give a mixture of **C** and **D**. Identify **A** and explain what is happening. Why is it that Me$_2$NPh does not give a product of type **A**, and that **A** does not insert ethylene?

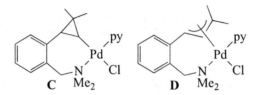

7.3. In the pyrolysis of TiMe$_4$, both ethylene and methane are observed; explain.

7.4. Suggest mechanisms for the following:

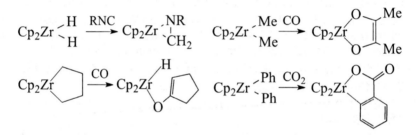

7.5. The reaction of *trans*-PdAr$_2$L$_2$ (**A**, Ar = *m*-tolyl, L = PEt$_2$Ph) with MeI gives 75% of *m*-xylene and 25% of 3,3'-bitolyl. Explain how these products might be formed and list the possible Pd-containing products of the reactions. When the reaction of **A** was carried out with CD$_3$I in the presence of d_0-PdMeIL$_2$ (**B**), both d_0- and d_3-xylene were formed. **A** also reacts with **B** give *m*-xylene and 3,3'-bitolyl. How does this second result modify your view of the mechanism?

7.6. [Cp*Co{P(OMe)$_3$}Et]$^+$ has an agostic interaction involving the β-H of the ethyl group. Draw the structure. It reacts with ethylene to form polyethylene. How might this reaction proceed? RhCl$_3$/EtOH and other late metal systems usually only dimerize ethylene to a mixture of butenes. Given that a Rh(I) hydride is the active catalyst in the dimerization, what mechanism would you propose? Try to identify and explain the key difference(s) between the two systems.

7.7. Design an alkyl ligand that will be resistant to β elimination (but not the ones mentioned in the text; try to be as original as possible). Design a second ligand, which may be an alkyl or an aryl-substituted alkyl, that you would expect to be resistant to β elimination but have a high tendency to undergo β–C–C bond cleavage. What products are expected?

7.8. Given the existence of the equilibrium shown:

$$L_nM \underset{}{\overset{Me}{\underset{CO}{\rightleftharpoons}}} \xrightarrow{\text{solvent}} L_nM \overset{solv}{\underset{\underset{O}{\overset{\parallel}{C}}}{\diagdown}} Me$$

how would you change L, M, and the solvent to favor (a) the right-hand side and (b) the left-hand side of the equation?

7.9. *trans*-PtCl(CH₂CMe₃){P(C₅H₉)₃}₂ gives 1,1-dimethylcyclopropane on heating. What mechanism is most likely, and what Pt-containing product would you expect to be formed? If the neopentyl group is replaced by –CH₂Nb (Nb = 1-norbornyl), then CH₃Nb is formed instead. What metal complex would you expect to find as the other product?

7.10. In mononuclear metal complexes, β elimination of ethyl groups is almost always observed, rather than α elimination to the ethylidene hydride $L_nM(=CHCH_3)H$. In cluster compounds, such as $HOs_3(CO)_{10}(Et)$, on the other hand, α elimination to give the bridging ethylidene $H_2Os_3(CO)_{10}(\eta^1,\mu_2\text{-}CHCH_3)$ is observed in preference to β elimination. Suggest reasons for this difference.

7.11. Consider the three potential rate-accelerating effects on CO insertion mentioned in Section 7.2: steric, Lewis acid, and oxidation. For each effect, discuss whether an acceleration of the overall reaction rate is to be expected if the reaction in question is (a) first order, (b) second order, (c) an intermediate case, and (d) an apparent insertion of the type shown in Eq. 7.18.

8

ADDITION AND ABSTRACTION

In reductive elimination or migratory insertion, ligand transformations occur within the coordination sphere of the metal. In contrast, we now look at outer sphere processes in which direct attack of an external reagent can take place on a ligand without prior binding of the reagent to the metal.

8.1 INTRODUCTION

The attacking reagent can be a nucleophile or an electrophile, but for reasons discussed here, the nucleophilic version is much more controllable and generally applicable. Nucleophilic attack on L′ is favored when the metal fragment L_nM–L′ is a poor π base but a good σ acid, for example, if the complex bears a net positive charge or has electron-withdrawing ligands. In such a case, L′ is depleted of electron density to such an extent that the nucleophile, Nu^- (e.g., LiMe or OH^-), can attack. Electrophilic attack is favored when the metal is a weak σ acid but a strong π base, for example, if the complex has a net anionic charge, a low oxidation state, and good donor ligands, L. The electron density on L′ is so much enhanced by back donation that it now becomes susceptible to attack by electrophiles, E^+ (H^+, MeI, etc.).

The Organometallic Chemistry of the Transition Metals, Sixth Edition.
Robert H. Crabtree.
© 2014 John Wiley & Sons, Inc. Published 2014 by John Wiley & Sons, Inc.

Both nucleophiles and electrophiles can give either addition or abstraction. In addition, the reagent becomes covalently attached to L′, and the newly modified ligand stays on the metal. In abstraction, the reagent detaches a part or even the whole of ligand L′ and leaves the coordination sphere of the metal. A nucleophile abstracts a cationic fragment, such as H^+ or Me^+, while an electrophile abstracts an anionic fragment, such as H^- or Cl^-. Often, reaction with an electrophile generates a positive charge on the complex and prepares it for subsequent attack by a nucleophile. We will see an example of alternating Nu^-/E^+ reactivity steps in Eq. 8.10; Eq. 8.17 shows the reverse sequence of reagents.

Equation 8.1–Equation 8.9 show some examples of these reactions and reagents. In Eq. 8.1 and Eq. 8.2, the nucleophiles reduce the hapticity of the ligands because they displace the metal from the carbon to which they add. In Eq. 8.2, we convert an η^5-L_2X into an η^4-L_2 ligand and subtract one unit from the net ionic charge, for a zero net change in the metal valence electron count. In general, an L_nX ligand is converted to an L_n ligand, and an L_n ligand is converted to an $L_{n-1}X$ ligand. Electrophilic reagents, in contrast, tend to increase the hapticity of the ligand to which they add (Eq. 8.6 and Eq. 8.7). Electrophilic attack on a ligand depletes the electron density on that ligand, often compensated by the attack of a metal lone pair on the ligand. For instance, in Eq. 8.7, an η^4-L_2 diene ligand becomes an η^5-L_2X pentadienyl. At the same time, a net positive charge is added to the complex, which leaves the overall electron count unchanged. In general, an L_nX ligand is converted to an L_{n+1} ligand and an L_n ligand is converted to an L_nX ligand. Equations 8.3 and 8.4 show that nucleophilic abstraction of H^+ is simply ligand deprotonation. Nucleophilic abstraction of a methyl cation from Pt(IV) by iodide was the key step in the reductive elimination mechanism of Fig. 6.2:

1. Nucleophilic addition:[1]

$$Cl_2(py)Pt-\!\!\!\parallel \xrightarrow{\text{py}} Cl_2(py)\bar{P}t \diagup\diagdown \overset{+}{N} \bigcirc \tag{8.1}$$

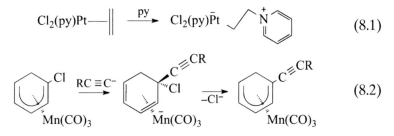

$$\tag{8.2}$$

2. Nucleophilic abstraction:[2]

$$[Cp_2TaMe_2]^+ \xrightarrow[-Me_4P^+]{Me_3PCH_2} Cp_2Ta\!\diagup^{CH_2}_{\diagdown Me} \tag{8.3}$$

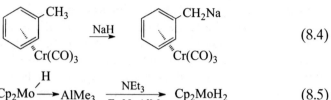

$$\text{(8.4)}$$

$$\underset{\underset{H}{|}}{\overset{\overset{H}{|}}{Cp_2Mo}}\!\!-\!\!AlMe_3 \xrightarrow[-Et_3N-AlMe_3]{NEt_3} Cp_2MoH_2 \qquad (8.5)$$

3. Electrophilic addition:[3]

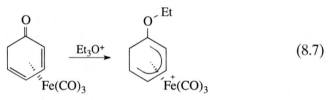

$$\text{(8.6)}$$

$$\text{(8.7)}$$

4. Electrophilic abstraction:[4]

$$Cp(CO)_3Mo \diagup\!\!\!\diagdown\!\!H \;,{}^+CPh_3 \xrightarrow[-Ph_3CH]{} Cp(CO)_3\overset{+}{Mo}\!\!-\!\!\| \qquad (8.8)$$

$$Cp(CO)_2Fe \diagdown\!\!\!\diagup\!\!\!OH \;, H^+ \xrightarrow[-H_2O]{} Cp(CO)_2\overset{+}{Fe}\!\!-\!\!\| \qquad (8.9)$$

Attack often occurs at the metal rather than at the ligand. For a nucleophile, this is simply associative substitution (Section 4.5) and can lead to the displacement of an existing ligand. If the original metal complex is 16e, nucleophilic attack may take place directly on the metal; if 18e, a ligand must usually dissociate first. In an 18e complex, a nucleophile is therefore more likely to attack a ligand, rather than the metal. The pyridine in Eq. 8.1 is a potential 2e ligand, but it does not attack the metal because the resulting 18e configuration is unfavorable for Pt(II). By attacking the ligand, the nucleophile does not raise the metal electron count.

For an electrophile, the situation is different. As a 0e reagent, an electrophile does not increase the electron count of the metal whether it attacks the metal or the ligand. Attack at the metal is thus always a possible alternative pathway even for an 18e complex except for d^0 complexes, which have no lone pairs on the metal. Of course, large electrophiles, such as Ph_3C^+, may still have steric problems that prevent attack at the metal. This lack of selectivity has made electrophilic attack less useful.

Organic free radicals are a third class of reagent that can give addition and abstraction reactions, but these reactions are less well understood and have not been widely employed. Radicals are typically reactive transients, so addition and abstraction steps tend to occur only as part of a multistep reaction scheme (e.g., Section 16.2).

8.2 NUCLEOPHILIC ADDITION TO CO

When bound to weakly π basic metals, CO becomes very sensitive to nucleophilic attack at carbon;[5] L-to-M σ donation not being compensated by M-to-L back donation, the CO carbon becomes positively charged. RLi can now convert a number of metal carbonyls to the corresponding anionic acyls. The resulting net negative charge now promotes electrophilic addition to the acyl oxygen to give the Fischer (heteroatom-stabilized) carbene complex, **8.1**. Equation 8.10 also illustrates a common pattern in synthetic pathways—alternation of nucleophilic and electrophilic attack. Addition of one prepares the system for attack by the other.

$$\text{Fe(CO)}_5 \xrightarrow{\text{LiNEt}_2} \text{(CO)}_4\text{Fe} = \text{C} \begin{smallmatrix} \nearrow \text{NEt}_2 \\ \searrow \text{OLi} \end{smallmatrix} \xrightarrow{\text{Me}_3\text{O}^+} \text{(CO)}_4\text{Fe} = \text{C} \begin{smallmatrix} \nearrow \text{NEt}_2 \\ \searrow \text{OMe} \end{smallmatrix} \quad (8.10)$$

$$\textbf{8.1}$$

The cationic charge makes $[\text{Mn(CO)}_6]^+$ much more sensitive to nucleophilic attack than $[\text{Mo(CO)}_6]$. Hydroxide, or even water, can now attack coordinated CO to give an unstable metalacarboxylic acid intermediate that decomposes to CO_2 and the metal hydride by β elimination (Eq. 8.11). This can be a useful way of removing one CO from the metal.

$$\overset{+}{\text{Mn(CO)}}_6 \xrightarrow{\text{H}_2\text{O}} \text{(CO)}_5\text{Mn} - \text{C} \begin{smallmatrix} \nearrow \overset{+}{\text{O}}\text{H}_2 \\ \searrow \text{O} \end{smallmatrix} \xrightarrow[-\text{H}^+]{} \text{(CO)}_5\text{Mn} - \text{C} \begin{smallmatrix} \nearrow \text{OH} \\ \searrow \text{O} \end{smallmatrix} \xrightarrow[-\text{CO}_2]{} \text{(CO)}_5\text{MnH}$$

$$(8.11)$$

Nucleophilic attack of methanol instead of water can give a metala-ester, $L_nM(COOR)$, stable from having no β-H.

The polar solvent favors loss of Cl^- over loss of PPh_3 ($=L$) in the first step of Eq. 8.12. The resulting $1+$ ionic charge sets the stage for a subsequent nucleophilic attack by MeOH on the activated CO. Acid can reverse the addition reaction by protonating the methoxy group, leading to loss of MeOH. This methoxide abstraction reaction is a case of a nucleophilic addition being reversed by a subsequent electrophilic abstraction and shows how the Nu/E alternation strategy can fail, perhaps from unsuitable workup conditions. For example, the product of a nucleophilic addition may revert to starting materials if excess acid is added to the reaction mixture with the object of neutralizing the excess nucleophile.

$$L_2PtCl_2 \xrightarrow[-Cl^-]{CO} [L_2PtCl(CO)]^+ \xrightarrow[-R_3NH^+]{\overset{MeOH}{R_3N}} L_2ClPt-C\overset{\displaystyle OMe}{\underset{\displaystyle O}{\diagdown\!\!\!\backslash}} \tag{8.12}$$

We saw in Chapter 4 that Et_3NO can remove coordinated CO from 18e metal complexes.[6] Its nucleophilic oxygen ($Et_3N^+-O^-$) can attack the CO carbon with subsequent breakdown to Et_3N, CO_2, and a 16e metal fragment (Eq. 8.13). The cis-disubstituted product is obtained selectively because a CO trans to another CO is activated toward a nucleophilic attack by receiving less back donation. Two problems arise: the amine also formed can sometimes coordinate if no better ligand is available and successive carbonyls become harder to remove as the back bonding to the remaining CO groups increases, and so only one CO is usually removable in this way.

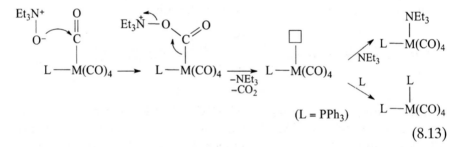

$$\text{(L = PPh}_3\text{)} \tag{8.13}$$

A complexed isonitrile is more easily attacked by nucleophiles than is CO (Eq. 8.14); isonitriles are not only intrinsically less π-acidic but also tend to bind to higher oxidation state metals where back donation is reduced.[7]

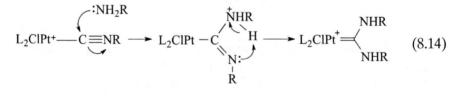

$$\tag{8.14}$$

8.3   NUCLEOPHILIC ADDITION TO POLYENES AND POLYENYLS

Free polyenes, such as benzene and butadiene, normally undergo electrophilic, not nucleophilic attack. In a complete reversal of their chemical character, called *umpolung*, complexation enhances nucleophilic, but suppresses electrophilic attack. The metal can therefore be considered either as an activating group for nucleophilic attack or a protecting group against electrophilic attack.

The nucleophile normally adds to the face of the polyene opposite to the metal. Since the metal is likely to have originally bound to the least hindered face of the free polyene, we expect to see selective attack at what was the more hindered face of the free polyene, a useful selectivity pattern.

Davies–Green–Mingos Rules

In a complex with several polyene or polyenyls, we often see selective attack at one site only. Davies, Green, and Mingos[8] systematized these reaction outcomes in terms of rules that usually correctly predict the site of addition:

Rule 1: Polyenes (even or L_n ligands) react before polyenyls (odd or L_nX ligands).

Rule 2: Open ligands with interrupted conjugation react before closed ligands with cyclic conjugation. Rule 1 takes precedence over rule 2 if they conflict.

Rule 3: Open polyenes give terminal addition. Open polyenyls usually give terminal attack, but nonterminal if L_nM is particularly electron donating.

Rule 4: A cation $[L_nM]^{c+}$ with an $c+$ net ionic charge is often subject to attack c times, but the selectivity for later steps has to be considered in light of the structure produced by the preceding addition.

Polyenes or even ligands have an even-electron count on the covalent model and include η^2-C_2H_4 and η^6-C_6H_6; odd ligands with an odd-electron count include η^3-C_3H_5 and η^5-C_5H_5. Closed ligands include Cp or η^6-C_6H_6, while open ligands include allyl or cyclohexadienyl. Some ligands and their classification are illustrated in **8.2–8.5**.

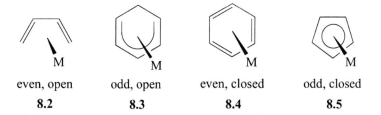

even, open	odd, open	even, closed	odd, closed
8.2	**8.3**	**8.4**	**8.5**

Diagrams **8.6**, **8.7 8.8**, **8.9**, **8.10**, **8.11**, **8.12**, and **8.13** show the rules in action with the point of attack indicated by the arrow(s) in the diagram.

In **8.6**, addition of a variety of nucleophiles takes place at the arene ring, as predicted by rule 1. By rule 4, a second nucleophile may also be added, but by rule 1, it must take place at the other ring. Rule 4 also requires double addition to dications **8.6** and **8.9**, where rules 1 and 3 specify the location of the second addition.

In **8.7**, addition takes place to the even, open butadiene, rather than to the even, closed arene (rule 2), and does so at the terminal position (rule 3). In **8.8**, we apply rule 1 before rule 2 so that the even, closed arene is attacked in preference to the odd, open allyl. Cp is usually very resistant to attack, but **8.9** shows a rare example of attack at a Cp ring, in a case where the 2+ ionic charge is strongly activating. As an odd, closed polyenyl, this only happens to Cp if there are no other π-bonded ligands.

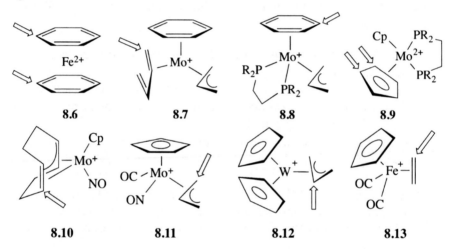

In **8.10**, the even alkene is attacked because we treat the alkene and the allyl parts of the bicyclooctadienyl as independent entities. CO is an even ligand but, not being π-bound, is among the least reactive, as shown in **8.11** and **8.13**.

Although developed empirically, MO studies show that the rules often successfully predict where the LUMO is predominantly located. Under kinetic control, we would expect addition at the point where this empty acceptor orbital is concentrated. Qualitatively, we can understand the rules as follows. Ligands having a higher X character tend to be more negatively charged and therefore tend to resist nucleophilic attack over L ligands. A coordinated allyl, as an LX ligand, has more anionic character than ethylene, an L ligand. This picture even predicts the relative reactivity of different ligands in the same class, a point not covered by the original rules. For example, pentadienyl (L_2X) reacts before allyl (LX) because the former has the least X character. Ethylene reacts before butadiene, and as we saw in Section 5.3, the LX_2 form is always a more significant contributor to butadiene complexes than the X_2 form is for ethylene.

The terminal carbons of even, open ligands are the sites of addition because the LUMO is predominant there, as can be seen from Fig. 5.2, where ψ_3 in butadiene is depicted. An odd, open polyenyl gives terminal addition only if the metal is sufficiently electron withdrawing. The MO picture for the allyl group (Fig. 5.1) shows that the LUMO of the free ligand, ψ_2, prefers the terminus, but ψ_3 is predominant at the central carbon. As we go to a less electron-withdrawing metal, we tend to fill ψ_2, and to the extent that ψ_3 becomes the new LUMO, we may no longer see terminal attack. An example of nonterminal attack in an allyl is shown by $[Cp_2W(\eta^3\text{-}C_3H_5)]^+$ (Eq. 8.15)—as a d^2 fragment, Cp_2W is strongly electron donating.

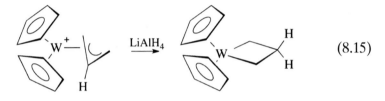

$$(8.15)$$

Although these simple rules do so well in most cases, the situation can sometimes be much more complicated. In Eq. 8.16, the methoxide attacks at every possible site, as the mixture is warmed from −80°C to room temperature. Initial addition is at the metal—with an η^7 to η^5 shift of the C_7H_7 to generate an open site—and later at the CO and C_7H_7 sites. Above 0°C, only the normal product would have been observed, and the complications would have escaped detection. This is a general point—if we halt experimentation when we have achieved the "right" result, we may miss new and worthwhile aspects of our system.

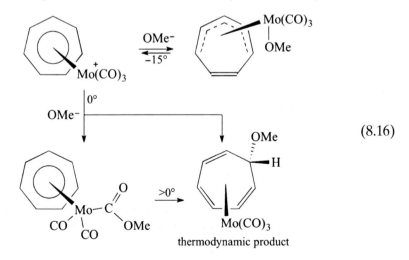

$$(8.16)$$

thermodynamic product

Cyclohexadienyl complexes react with nucleophiles to give 1,3-diene complexes, as in Eq. 8.17. The synthesis of the starting complex by electrophilic abstraction activates the ligand for nucleophilic attack. Once again, directing effects can be used to advantage: a 2-OMe substituent directs attack to the C-5 position of the cyclohexadienyl, for example.

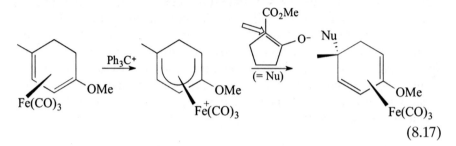

$$(8.17)$$

Dienes give allyls on nucleophilic attack. The s-cis conformation of the butadiene in Eq. 8.18 gives rise to an anti methylallyl—in the specialized nomenclature of allyl complexes, a substituent is considered as syn or anti with respect to the central CH proton. Equation 8.19 is interesting in that the amine acts in this case as a carbon, not as a nitrogen nucleophile.

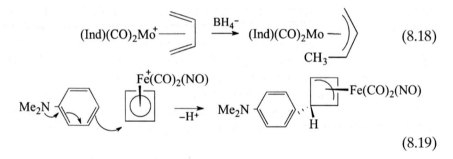

$$(8.18)$$

$$(8.19)$$

Wacker Process

The Wacker process, an important industrial procedure now used to make ~6 million tons a year of aldehydes, relies on nucleophilic attack on an alkene complex.[9] The fact that aqueous $PdCl_2$ oxidizes ethylene to acetaldehyde had been known—although not understood—since the nineteenth century; the reaction consumes the $PdCl_2$ as oxidant and deposits metallic Pd(0). It took considerable imagination to see that such a reaction could ever be useful on an industrial scale because $PdCl_2$ is obviously far too expensive to use stoichiometrically. It is often useful to find a way to convert a stoichiometric to a catalytic process. J. Smidt of Wacker Chemie realized in the late 1950s that it is possible to intercept the Pd with $CuCl_2$ before Pd(0) has a chance to precipitate. Cu(II) reoxidizes the palladium and is itself reduced to cuprous chloride, which is reoxidized back to Cu(II) with air, allowing the Pd to be recycled almost indefinitely. The resulting set of reactions (Eq. 8.20) is an elegantly simple solution to the problem and resembles the coupled reactions of biochemical catalysis.

$$\frac{1}{2}O_2 \quad 2Cu^{II} \quad Pd^{II} \quad CH_3CHO$$
$$H_2O \quad 2Cu^{I} \quad Pd^0 \quad C_2H_4$$

(8.20)

Later mechanistic work revealed the rate equation of Eq. 8.21 for the industrially relevant conditions of low $[Cl^-]$ and low $[Cu]$. The exact mechanism depends on the conditions, however, and the details are still a matter of debate.

$$\text{Rate} = \frac{k[PdCl_4^{2-}][C_2H_4]}{[Cl^-]^2[H^+]}$$

(8.21)

Equation 8.21 implies that the complex, in going from its normal state in solution, $[PdCl_4]^{2-}$, to the transition state of the slow step of the reaction, has to gain a C_2H_4 and lose two Cl^- ions and a proton. It was originally argued that the proton must be lost from a coordinated water, and so $[Pd(OH)(C_2H_4)Cl_2]^-$ was invoked as the key intermediate; it was assumed that this might undergo olefin insertion into the Pd–OH bond, or the OH might attack the coordinated ethylene as a nucleophile. The resulting hydroxyethyl palladium complex might β-eliminate to give vinyl alcohol, $CH_2=CHOH$, known to tautomerize to acetaldehyde.

In fact, this mechanism is wrong,[10] something that was only discovered 20 years later as a result of stereochemical work by Bäckvall and by Stille. According to the original intramolecular mechanism, whether the reaction goes by insertion or by nucleophilic addition from a coordinated OH, the stereochemistry at each carbon of the ethylene should remain unchanged. This can be tested if we use cis- or trans-CHD=CHD as the alkene and trap the intermediate alkyl. We have to trap the alkyl because the rearrangement to acetaldehyde destroys the stereochemical information. Equation 8.22 shows trapping with CO: once the hydroxyethyl is carbonylated, the OH group can cyclize by a nucleophilic abstraction of the acyl to give a free lactone. The lactone stereochemistry can be determined by a number of methods, including NMR and microwave spectroscopy. In fact, the stereochemistry of the two carbons in the product is not the same as that of the starting material, which rules out the older mechanism. This result is now supported by molecular dynamics computations.[11]

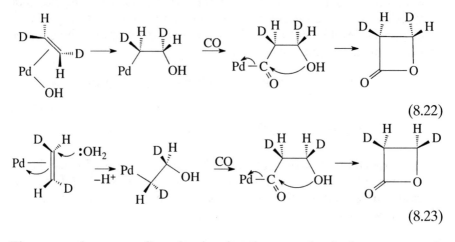

$$(8.22)$$

$$(8.23)$$

The currently accepted mechanism involves attack of a free water molecule from the solvent on the coordinated ethylene. Equation 8.23 shows how this inverts the stereochemistry at one of the carbons, as opposed to the old insertion mechanism (Eq. 8.22).

The loss of two Cl^- ions removes the anionic charge from the metal, which would otherwise inhibit nucleophilic attack. Equation 8.24 shows the sequence of events as now understood. This mechanism implies that an $[H_2O]^2$ term should be present in the rate equation, and if it could have been seen, the mechanistic problem would have been solved earlier, but one cannot normally alter the concentration of a solvent and get meaningful rate data because changing the solvent composition leads to unpredictable solvent effects on the rate.

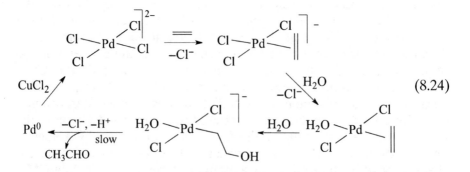

$$(8.24)$$

If vinyl alcohol were released from Pd in D_2O as solvent, we would expect deuterium incorporation into the acetaldehyde, but none is seen. This requires that the vinyl alcohol never leaves the Pd until it has had time to rearrange to CH_3CHO by multiple insertion–elimination steps:

$$(8.25)$$

Wacker chemistry normally gives branched products via Markovnikov olefin hydration, yet linear alcohols are highly desirable. Grubbs[12] has now found conditions that give the long-sought anti-Markovnikov olefin hydration, based on a multistep scheme in which a pair of catalysts cooperate.

Alkyne Hydration

Nucleophilic addition of H_2O to coordinated alkyne is the key step of catalytic alkyne hydration with traditional Hg(II) or Au(I) catalysts (Eq. 8.26) that convert terminal alkynes $RC{\equiv}CH$ to the methyl ketones RCOMe. As part of a general trend associated with the rise of green chemistry (Section 1.1), toxicity and expense concerns have led to the advent of base metal catalysts, such as a water-soluble Co(III) porphyrin.[13]

$$(8.26)$$

8.4 NUCLEOPHILIC ABSTRACTION IN HYDRIDES, ALKYLS, AND ACYLS

Hydrides

Deprotonation of a metal hydride can produce a nucleophilic metal anion. For example, ReH_7L_2 ($L_2 = $ dppe) does not lose H_2 easily as does the L = PPh_3 complex. To generate the reactive ReH_5(dppe) fragment, the anion must first be formed with BuLi and treated with MeI to give the methyl hydride, which gives ReH_5(dppe) and CH_4 by RE (Eq. 8.27), the driving force for methane loss being higher than for H_2 loss. The resulting ReH_5(dppe) was intercepted with cyclopentadiene to give $CpReH_2L_2$ as final product.

$$\text{ReH}_7\text{L}_2 \xrightarrow[-\text{BuH}]{\text{BuLi}} \text{LiReH}_6\text{L}_2 \xrightarrow[-\text{LiI}]{\text{MeI}} \text{MeReH}_6\text{L}_2 \xrightarrow{-\text{MeH}} \text{ReH}_5\text{L}_2$$

transient 16e (8.27)

$$\text{CpReH}_2\text{L}_2 + \text{C}_5\text{H}_{10} \xleftarrow{\text{CpH}} \text{intermediate}$$

Alkyls and Acyls

Alkyl groups can be exchanged between metals, typically with inversion at carbon. This transmetalation reaction provides a route for the racemization of a metal alkyl during the early stages of an oxidative addition reaction, while there is still some of the low-valent metal left in the reaction mixture. In Eq. 8.28, exchange of a $(\text{CR}_3)^+$ fragment between the metals turns the Pd(0) partner into Pd(II), and the Pd(II) into Pd(0). The stereochemical outcome of an OA can be clouded by exchange reactions such as these.

$$\text{L}_3\text{Pd:} \overset{R''}{\underset{R'}{\big\rangle}}\!\!-\!\text{PdClL}_2 \longrightarrow \text{L}_3\text{Pd}^+ \!-\!\!\overset{R''}{\underset{R'}{\big\langle}}\!\!R + \text{PdL}_2 + \text{Cl}^- \quad (8.28)$$

Acyls undergo abstraction by nucleophiles in the last steps of Eqs. 8.22 and 8.23. As in the abstraction of Eq. 8.28, the reaction goes with reduction of the metal by two units, so a Pd(II) acyl is ideal because the Pd(0) state is easily accessible.

The recurrence of Pd(II) in this chapter is no accident—it has a very high tendency to encourage nucleophilic attack at the ligands in its complexes. Being on the far right-hand side of the d block, it is very electronegative (Pauling electronegativity: 2.2), and its d orbitals are very stable. This means that polyene-to-metal electron donation is more important than metal-d_π-to-polyene-π^* back donation, and so the polyene is left with a net positive charge.

8.5 ELECTROPHILIC ADDITION AND ABSTRACTION

In common with 2e nucleophiles, 0e electrophiles, such as H^+ or Me^+, can attack a ligand. Unlike nucleophiles, however, they can also attack the M–L bond or the metal itself because, as zero electron reagents, wherever they attack, they do not alter the electron count of the complex. The resulting mechanistic complexity and unreliable selectivity makes electrophilic attack far less controllable and less useful than nucleophilic attack. Polysubstitution is also more common in the electrophilic case.[14]

Addition to the Metal

Oxidative addition by the S_N2 or ionic mechanisms involves two steps: initial electrophilic addition to the metal (Eq. 8.29 and Sections 6.3 and 6.5), followed by substitution.

$$L_nM \xrightarrow[\text{MeI}]{\substack{\text{electrophilic} \\ \text{addition}}} [L_nMMe]^+I^- \xrightarrow[-L]{\text{substitution}} [L_{(n-1)}MMeI] \qquad (8.29)$$

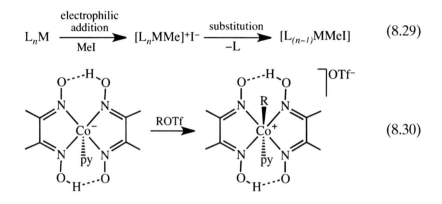

$$(8.30)$$

Without the second step, the reaction becomes a pure electrophilic addition. An example is the reaction of the highly nucleophilic Co(I) anion, $[Co(dmg)_2py]^-$, with an alkyl triflate, a reaction known to go with inversion at carbon (Eq. 8.30). Protonation of metal complexes to give metal hydrides is also very common (Eq. 3.28 and Eq. 3.29).

The addition of any zero-electron ligand to the metal is also an electrophilic addition: $AlMe_3$, BF_3, $HgCl_2$, Cu^+, and even η^1-CO_2, when it binds via carbon, all act in this way. Each of these reagents has an empty orbital by which it can accept a lone pair from the metal.

Addition to a Metal–Ligand Bond

Protonation reactions are common—for example, in Eq. 8.31, protonation of L_nM–H can give a dihydrogen complex $[L_nM–(H_2)]^+$.[15] Early metal alkyls, such as Cp_2TiMe_2, are readily cleaved by acid to liberate the alkane via a transient alkane complex.

$$[Pt(diphos)_2] \xrightarrow{H^+} [PtH(diphos)_2]^+ \xrightarrow{H^+} [Pt(H_2)(diphos)_2]^{2+} \quad (8.31)$$

Protonation of the alkene complex shown below can occur by two simultaneous paths: (1) direct protonation at the metal and (2) initial protonation at the alkene followed by β elimination. Path 2 leads to incorporation of label from DCl into the alkene ligands of the resulting pentagonal bipyramidal hydride complex.[16]

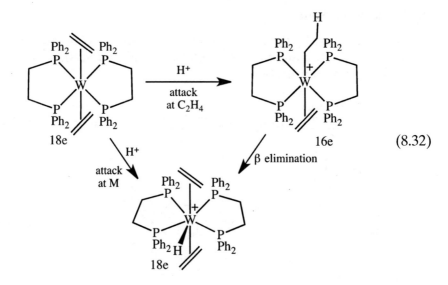

(8.32)

Addition to Ligand

Simple addition to the ligand occurs in protonation of Cp_2Ni, as shown by the exo attack and lack of scrambling of the deuterium label.

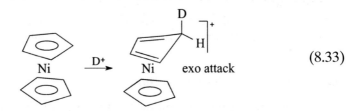

(8.33)

Unlike nucleophiles, where exo attack is the rule, an endo addition is also possible for electrophiles via attack at the metal, followed by transfer to the endo face of the ligand, particularly favored for soft electrophiles, for example, $Hg(OAc)_2$. Exo-proton abstraction by OAc^- completes the sequence (Eq. 8.34).

The hard electrophile CH_3CO^+ gives exo attack at the ligand in Eq. 8.35. The preference for exo-proton abstraction means that an endo-deuterium has to be transferred to the endo position of the other ring. This leads to loss of the resulting exo-proton, so that all five D atoms are retained by the complex.[17]

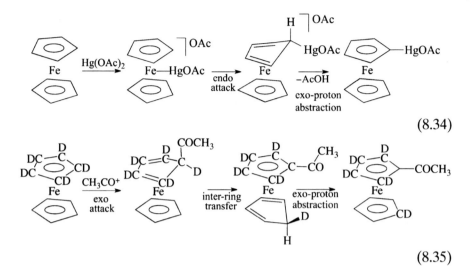

$$(8.34)$$

$$(8.35)$$

Abstraction of Alkyl Groups

Electrophilic metal ions, notably Hg^{2+}, can cleave an M–alkyl bond. Two main pathways are seen: (i) attack at the α carbon of the alkyl with inversion at carbon (Eq. 8.36) and (ii) attack at the metal or at the M–C bond with retention (Eq. 8.37). The difference has been ascribed to the greater basicity of the metal in the CpFe case. The unpredictable stereochemistry again makes the reaction less useful.

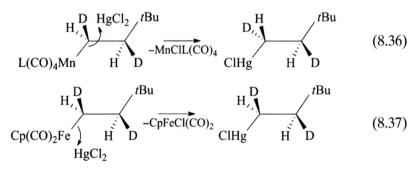

$$(8.36)$$

$$(8.37)$$

8.6 SINGLE-ELECTRON TRANSFER AND RADICAL REACTIONS

It is often difficult to differentiate between a true electrophilic abstraction or addition, a one-step process in which a pair of electrons is implicated (Eq. 8.36 and Eq. 37), from a two-step process involving single-electron transfer (SET) from the metal to E^+ going via the radical intermediate, E• (Eq. 8.38 and Eq. 8.39).[18] First row metals prefer 1e to 2e OS changes (Co(I), (II), (III) versus Ir(I), (III), (V)), and are therefore more likely to give radical pathways. For example, halogens, X_2, give electrophilic cleavage of M–R to form RX. One common mechanism involves SET oxidation of the metal, which increases the electrophilic character of the alkyl and generates halide ion, so that, paradoxically, it is *nucleophilic* abstraction of the alkyl group by halide ion that leads to the final products. Co(III) alkyls are known to behave in this way, and the intermediate Co(IV) species, formed via Ce(IV) oxidation, are stable enough to be detected by EPR at −50°C (Eq. 8.38, R = *n*-hexyl). Addition of Cl⁻ leads to the nucleophilic abstraction of the alkyl with inversion.

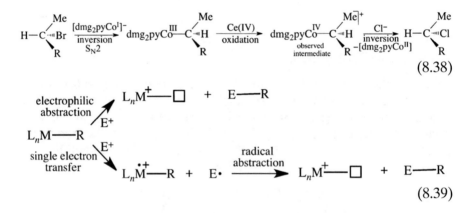

$$(8.38)$$

$$(8.39)$$

Nucleophiles can also give SET reactions, for example, [Cp*Mo-(CO)$_3$(PMe$_3$)]$^+$ reacts with LiAlH$_4$ to give paramagnetic [Cp*Mo(CO)$_3$-(PMe$_3$)], observed by EPR. Loss of CO, easy in this 19e species, leads to Cp*Mo(CO)$_2$(PMe$_3$), which abstracts H•, probably from the THF solvent, to give the final product, Cp*MoH(CO)$_2$(PMe$_3$).

Radical traps, such as galvinoxyl, TEMPO, and DPPH (Q•), are sometimes used as a test for the presence of radicals, R•, in solution; in such a case, the adduct Q–R is expected as product. Unfortunately, this procedure can be misleading in organometallic chemistry because typical Q• abstract H from some palladium hydrides at rates competitive with those of typical organometallic reactions; [PdHCl(PPh$_3$)$_2$] reacts in this way but [PdH(PEt$_3$)$_3$]BPh$_4$ is stable.[19]

Radical abstraction from a ligand is also possible. For example, in Eq. 8.40, an alkyl radical abstracts a hydrogen atom from a coordinated water.[20] In fact, the process is better seen as a concerted H$^+$ transfer from the water and an e$^-$ transfer from the Ti(III) center.

$$\text{(8.40)}$$

- Nucleophilic addition is more predictable (Section 8.3) than the other pathways considered.
- Nucleophilic attack at a ligand is favored by weak and electrophilic attack by strong back bonding to that ligand.

REFERENCES

1. B. Jacques, J. P. Tranchier, F. Rose-Munch, E. Rose, G. R. Stephenson, and C. Guyard-Duhayon, *Organometallics*, **23**, 184, 2004.
2. S. E. Gibson and H. Ibrahim, *Chem. Commun.*, **2002**, 2465.
3. M. A. Todd, M. L. Grachan, M. Sabat, W. H. Myers, and W. D. Harman, *Organometallics*, **25**, 3948, 2006.
4. Y. H. Kim, T. H. Kim, N. Y. Kim, E. S. Cho, B. Y. Lee, D. M. Shin, and Y. K. Chung, *Organometallics*, **22**, 1503, 2003.
5. J. Barluenga, K. Muniz, M. Tomas, A. Ballesteros, and S. Garcia-Granda, *Organometallics*, **22**, 1756, 2003.
6. L. C. Song, L. X. Wang, G. J. Jia, Q. L. Li, and J. B. Ming, *Organometallics*, **31**, 5081, 2012.
7. A. G. Tskhovrebov, K. V. Luzyanin, F. M. Dogushin, M. F. C. G. da Silva, A. J. L. Pombeiro, and V. Y. Kukushkin, *Organometallics*, **30**, 3362, 2011.
8. S. G. Davies, M. L. H. Green, and D. M. P. Mingos, *Tetrahedron*, **34**, 3047, 1978.
9. D. Harakat, J. Muzart, and J. Le Bras, *RSC Advances*, **2**, 3094, 2012 and references cited; V. Imandi, S. Kunnikuruvan, and N. N. Nair, *Chem. Eur. J.*, **19**, 4724, 2013.
10. J. M. Takacs and X. T. Jiang, *Curr. Org. Chem.*, **7**, 369, 2003.
11. A. Comas-Vives, A. Stirling, A. Lledos, and G. Ujaque, *Chem. Eur. J.*, **16**, 8738, 2010.
12. G. Dong, P. Teo, Z. K. Wickens, and R. H. Grubbs, *Science*, **333**, 1609, 2011.
13. T. Tachinami, T. Nishimura, R. Ushimaru, R. Noyori, and H. Naka, *J. Am. Chem. Soc.*, **135**, 50, 2013.
14. G. R. Clark, P. M. Johns, W. R. Roper, T. Sohnel, and L. J. Wright, *Organometallics*, **30**, 129, 2011.

15. A. Miedaner, J. W. Raebiger, C. J. Curtis, S. M. Miller, and D. L. DuBois, *Organometallics*, **23**, 2670, 2004.

16. R. A. Henderson, *Angew. Chem. -Int. Ed. Engl.*, **35**, 947, 1996.

17. M. J. Mayor-Lopez, J. Weber, B. Mannfors, and A. F. Cunningham, *Organometallics*, **17**, 4983, 1998.

18. K. M. Smith, *Organometallics*, **24**, 778, 2005.

19. A. C. Albeniz, P. Espinet, R. Lopez-Fernandez, and A. Sen, *J. Am. Chem. Soc.*, **124**, 11278, 2002.

20. J. Jin and M. Newcomb, *J. Org. Chem.*, **73**, 7901, 2008.

PROBLEMS

8.1. Where would a hydride ion attack each of the following?

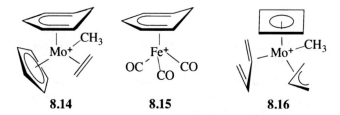

| 8.14 | 8.15 | 8.16 |

8.2. Predict the outcome of the reaction of CpFe(PPh₃)(CO)Me with each of the following: HCl, Cl₂, HgCl₂, and HBF₄/THF.

8.3. Explain the outcome of the reaction shown below:

$$
\text{\raisebox{0pt}{$\diagup\!\!\diagdown\!\!\diagup$}} + \text{PhI} + \text{R}_2\text{NH} \xrightarrow[\text{PPh}_3]{\text{Pd(OAc)}_2}
\begin{array}{c}
\text{PhCH}_2\text{CH=CHCH}_2\text{NR}_2 \\
+ \\
\text{PhCH=CHCH=CH}_2
\end{array}
\qquad (8.41)
$$

8.4. [CpCo(dppe)(CO)]²⁺ (A) reacts with 1° alcohols, ROH, to give [CpCo(dppe)(COOR)]⁺, a reaction known for very few CO complexes. The ν(CO) frequency for A is 2100 cm⁻¹, extremely high for a CO complex. Br⁻ does not usually displace CO from a carbonyl complex, but it does so with A. Why is A so reactive?

8.5. Nucleophilic addition of MeO⁻ to free PhCl is negligibly slow under conditions for which the reaction with (η⁶-C₆H₅Cl)Cr(CO)₃ is fast. What product would you expect, and why is the reaction accelerated by coordination?

8.6. Given a stereochemically defined starting material (either erythro or threo), what stereochemistry would you expect for the products of the following electrophilic abstraction reaction:

$$CpFe(CO)_2(CHDCHDCMe_3) \xrightarrow{Ph_3C^+} \begin{array}{c} CHD=CHCMe_3 \\ + \\ CHD=CDCMe_3 \end{array} \qquad (8.42)$$

Let us say that for a related 16e complex $L_nM(CHDCHDCMe_3)$ gave precisely the same products, but of opposite stereochemistries. What mechanism would you suspect for the reaction?

8.7. You are trying to make a methane complex $L_nM(\eta^1\text{-}H-CH_3)^+$ (**8.17**), by protonation of a methyl complex L_nMMe with an acid HA. Identify three things that might go wrong and suggest ways to guard against each.

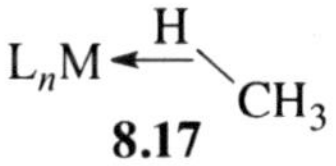

$$L_nM \xleftarrow{\;\;\overset{H}{\diagup}\;} CH_3$$

8.17

8.8. (cod)PtCl$_2$ reacts with MeOH/NaOAc to give a species [{C$_8$H$_{12}$(OMe)}PtCl]$_2$. This in turn reacts with PR$_3$ to give 1-methoxycyclooctadiene (**8.18**) and PtHCl(PR$_3$)$_2$. How do you think this might go?

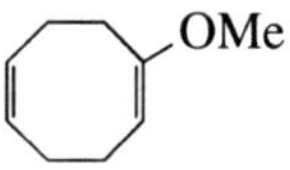

8.18

8.9. [CpFe(CO)(PPh$_3$)(MeC≡CMe)]$^+$ reacts with (i) LiMe$_2$Cu (a source of Me$^-$) and (ii) I$_2$ to give compound **8.19**; explain this reaction. What product do you think might be formed from LiEt$_2$Cu?

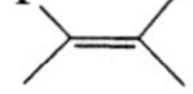

8.19

8.10. Equation 8.24 and Equation 8.26 involve substrates with different oxidation states of carbon in the substrate hydrocarbon, one starts from ethylene, the other from acetylene. Explain how both reactions can give the same product, CH$_3$CHO, when both are hydration reactions where we do not expect the oxidation state of carbon to change.

9

HOMOGENEOUS CATALYSIS

The catalysis of organic reactions[1] is one of the most important applications of organometallic chemistry and has been a significant factor in the rapid development of the field as a whole. Organometallic catalysts now have numerous applications in the pharmaceutical,[2] fine chemical, and commodity chemical industries and are beginning to contribute to the rising topics of energy and green chemistry. By bringing about a reaction at lower temperature, a catalyst can save energy input and, by improving selectivity, can minimize product separation problems and waste formation. With growing regulatory pressure to market drugs in enantiopure form, asymmetric catalysis has come to the fore as a practical way to make such products on a large scale from racemic or achiral reactants.

9.1 CATALYTIC CYCLES

A catalytic cycle consists of a set of reactions that occurs only in the presence of a catalyst and that leads to product formation from reactants, or *substrates*. The catalyst can mediate an indefinite series of

The Organometallic Chemistry of the Transition Metals, Sixth Edition.
Robert H. Crabtree.
© 2014 John Wiley & Sons, Inc. Published 2014 by John Wiley & Sons, Inc.

cycles and is thus only needed in substoichiometric amount relative to reactants or products. The catalyst loading is typically from 1 ppm to 1% relative to reactants, meaning that the number of cycles initiated by each molecule of catalyst runs from 10^6 to 10^2, respectively. Catalysis can be useful either by speeding up a reaction or modifying its selectivity, or both. The same reactants can give quite different products depending on the catalyst: ethylene oxidation with O_2, for example, can give the epoxide or acetaldehyde.

Of interest here are soluble complexes, or *homogeneous catalysts,* as opposed to insoluble materials, or *heterogeneous catalysts*, so named because the catalyst and substrates are in the same phase only in the first case. Some reactions, such as hydrogenation, are amenable to both types of catalysis, but others are currently limited to one or the other, for example, O_2 oxidation of ethylene to the epoxide over a heterogeneous Ag catalyst or Wacker air oxidation of ethylene to acetaldehyde with homogeneous Pd(II) catalysts. Homogeneous catalysis extends far beyond organometallics to cover acid or base catalysis, organocatalysis, and coordination catalysis, such as H_2O_2 decomposition by Fe^{2+}. Electrocatalysis is also a rising area.[3]

Catalytic mechanisms are easier to study in homogeneous cases, where powerful methods such as NMR can assign structures and follow reaction kinetics. Homogeneous catalysts are at a disadvantage, however, in being difficult to separate from the product. Sometimes, this requires special techniques, but in polymer synthesis, the catalyst still remains in the final product. Homogeneous catalysts are also heterogenized by covalently grafting onto solid supports to aid separation. Although now technically heterogeneous, the catalyst often retains the characteristic reactivity of the homogeneous form. We can distinguish between homogeneous or heterogenized homogeneous catalysts that have a single type of active site, or a small number of them (homotopic), from metal and metal oxide surfaces that can have a cocktail of sites (heterotopic). The first case tends to give higher selectivity than the second. Homogeneous catalysts are also amenable to tuning by change of ligand.

Homogeneity

A homogeneous precursor can give rise to a homogeneous catalyst; however, it can also decompose to give catalytically active solid material. A particularly dangerous form of decomposition gives rise to suspended nanoparticles, of typical diameter 10–1000 Å. These can mislead by masquerading as homogeneous catalysts. Many early "homogeneous" catalysts, formed by reduction of metal salts in polar solvents, may well have given active nanoparticles, and even today, ambiguities can easily arise.[4] One might think an asymmetric catalyst *has* to be homogeneous, but in one recent case, an impressive level of asymmetric induction (90% e.e.) was achieved by modification of a nanoparticle surface with an asymmetric "ligand."[5] Two catalytic reactions not normally seen for true homogeneous catalysts can be considered a "red flag": nitrobenzene reduction and arene hydrogenation. Careful work in homogeneous catalysis should include tests for heterogeneity; sometimes, both types even occur together.[6] The possibility that the true catalyst is very different from the complex originally introduced into the reaction mixture has led authors to term the original complex the *catalyst precursor* (or precatalyst).

Thermodynamics

Before trying to find a catalyst for a given reaction, we need to check that the reaction itself has favorable thermodynamics, as is the case for the alkene isomerization of Eq. 9.1, for example. If a reaction is disfavored, as in splitting H_2O to H_2 and O_2, then no catalyst, however efficient, can bring it about without energy input. To get round this problem, we might couple an unfavorable reaction to a strongly favorable process or provide energy in the form of photons, as in photosynthesis, or a voltage, as in electrolysis. In the absence of these effects, the catalyst only increases the rate but does not change the position of equilibrium, decided by the thermodynamics of substrates, S, versus products, P. In the energy diagram of Figure 9.1*a*, for example, S is slightly less stable than P, so the reaction favors P. For **9.1** → **9.2**, the additional conjugation in **9.2** is sufficient to make the reaction favorable. Normally, the substrate binds to form a substrate–catalyst complex, M.S (Fig. 9.1). Stronger M–S binding might be thought to be better, but this is not always so. If binding is too strong, M.S will be too stable, and the activation energy to get to "M.TS" may be just as large as it was in going from S to TS in the uncatalyzed reaction, so no rate acceleration would be achieved. Nor can S bind too weakly because it would then be excluded from the metal and fail to be activated. The product P, initially formed as M.P, must be the least strongly bound of all so that S can displace P to give back M.S and start a new cycle. Many of these ideas also apply to enzymes.[7]

$$\text{Ph} \diagup\!\!\!\diagup \xrightarrow{\text{catalyst}} \text{Ph} \diagup\!\!\!\diagdown\!\!\!\backsim \qquad (9.1)$$

9.1   **9.2**

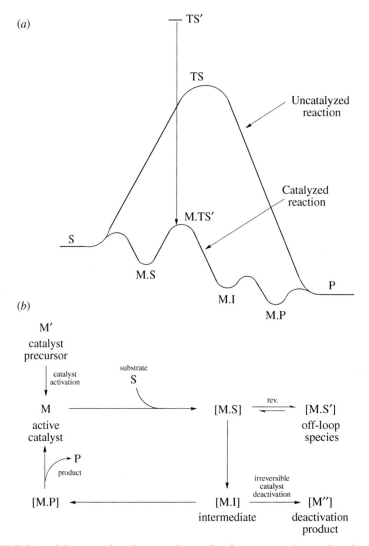

FIGURE 9.1 (*a*) A catalyst lowers the activation energy for a chemical reaction. Here, the uncatalyzed conversion of substrate S to product P passes by way of the high-energy transition state TS. In this case, the metal-catalyzed version goes via a different transition state TS′, which is very unstable in the free state but becomes viable on binding to the catalyst as M.TS′. The arrow represents the M–TS′ binding energy. (*b*) Typical catalytic cycle in schematic form.

Kinetics

In a simple $A \Rightarrow B$ reaction, each catalytic turnover corresponds to one mole of B being formed per mole of catalyst. The catalytic rate is often given as a *turnover frequency* (TOF), the number of catalytic cycles completed per unit time (usually h^{-1}). Catalyst lifetime is measured by the *turnover number* (TON), the number of cycles before deactivation, assuming excess substrate still remains. The TON and TOF depend on the conditions, which must therefore be stated.[8] Since the TOF continually varies with the elapsed time, the maximum TOF during the catalytic run is often cited. This often occurs at the outset of the reaction, and we often see the *initial rate* reported as a TOF. Comparison of the TOFs can tell us which catalyst has the best rate, while comparison of the TONs tells us which catalyst is the most robust. Conversion (%) measures how much substrate has been converted at a given point, typically after the reaction has come to a halt. Yield (%) measures the amount of any one product relative to the theoretical maximum yield derived from the chemical equation, given the conversion achieved. Selectivity (%) measures the amount of the desired product relative to the theoretical maximum yield. This means the yield is the conversion times the selectivity.

The catalyzed pathway is usually completely different mechanistically from the uncatalyzed one. As shown in Fig. 9.1a, instead of passing through the high-energy uncatalyzed transition state TS, the catalyzed reaction normally goes by a multistep mechanism in which the metal stabilizes intermediates and transition states that are accessible only when metal bound. One such transition state M.TS' is shown in Fig. 9.1. The TS' structure in the absence of the metal would be extremely unstable, but the energy of binding is so high that M.TS' is now much more favorable than TS and the reaction all passes through the catalyzed route. Different metal species may stabilize other transition states TS'' that lead to entirely different products from the same starting materials — hence different catalysts can give different products and the catalysis products can be different than the ones accessible without catalysis.

In a stoichiometric reaction, the passage through M.TS' would be the slow, or rate-determining, step. In a catalytic reaction, the cyclic nature of the system means that the rates of all steps are identical. On a circular track, the same number of trains must pass each point per unit time. The equivalent to the slow step is called the *turnover limiting step*. Any change that lowers the barrier for this step will increase the TOF, but changes elsewhere will not affect the TOF. For a high TOF, we require that none of the intermediates be bound too strongly — otherwise they may be too stable and not react further — and that none of the transition states be prohibitively high in energy. Indeed, the whole reaction profile must not stray from a rather narrow range of free energies, accessible at the reaction temperature. This is why catalysts can be hard to tune—a change in a ligand designed to lower the turnover limiting transition state energy may also lower the energy of the preceding intermediate, resulting in no net change in the reaction barrier and thus in the rate.

A catalyst may cycle only a few times and then "die." Such deactivation is a serious problem for practical applications of homogeneous catalysts, but this area still attracts few studies.[9] There are many ways in which a catalyst can fail, so we have to look hard for the right metal, ligand set, solvent, temperature range, and conditions. In the selectivity determing step of the cycle, which may or may not be turnover limiting, a choice is made between two possible pathways that lead to different products, such as between linear or branched aldehydes in hydroformylation.

Excellent quantitative accuracy is now available from computationally derived energy diagrams of the type shown in Fig 9.1a, and a computed TOF value can even be obtained for direct comparison with experiment.[10]

Mechanism, Intermediates, and Kinetic Competence

In a catalytic cycle (Fig. 9.1b), the *active catalyst* M is often rather unstable and is only formed in situ from the catalyst precursor (or precatalyst), M'. If we monitor the system, for example, by NMR, we normally see only the disappearance of S and the appearance of P, not the transient catalytic intermediates. We may still see only M' because only a small fraction of the metal is likely to be on the loop at any given time. Even if we appear to see an intermediate, we cannot be sure it is not M·S', an off-loop species. If a species builds up steadily during the reaction, it might be a *catalyst deactivation product* M", in which case, the catalytic rate will fall as [M"] rises. Excellent reviews are available on the determination of mechanism in catalytic reactions.[1b]

Catalysis is a kinetic phenomenon, so activity may rely on a minor, even minuscule, catalyst component. This emphasizes the danger of relying too heavily on spectroscopic methods. The fact that a series of plausible intermediates can all be seen spectroscopically in the catalytic mixtures does not mean these are the true intermediates. Instead, we need to show that each of the proposed intermediates reacts sufficiently fast to account for the formation of products, that is, that each is *kinetically competent*.

Cooperative Catalysis

Cooperative catalysis combines two or more catalysts, such as organo-catalysts with organometallics, to produce multistep, one-pot transformations, possible because the catalysts act independently to mediate separate steps of an overall process. Organocatalysts are simple organic compounds, such as amino acids, oligopeptides, Brønsted acids, Lewis acids, or nitrogen heterocycles that catalyze a wide variety of organic reactions, often with a high level of asymmetric induction.[11] For instance, an alkene hydroformylation (Section 9.4) with a racemic Rh catalyst to give an aldehyde can be followed by a Mannich or aldol procedure, organocatalyzed by L-proline or a proline derivative, that not only builds molecular complexity, but is also responsible for the asymmetric outcome.[12]

Huff and Sanford[13] have a case where three different homogeneous catalysts, $RuCl(PMe_3)_4(OAc)$, $Sc(OTf)_3$, and $RuH(PNN)(CO)(H)$, operate in sequence to promote the reduction of CO_2 to MeOH via the intermediates HCOOH and HCOOMe. This is clearly an area with big future possibilities.

Deactivation

There are several possible reasons why a catalyst for A + B = C may stop at, say 50% conversion of A. If equilibrium has been reached, addition of C may reverse the process. If we have run out of coreagent B, addition of more B may restart the process. Catalyst deactivation is often the culprit,[14] in which case addition of fresh catalyst may restart it. Deactivation is a key reason for poor catalyst performance, and in such a case, identification of the failure mode(s) can greatly help catalyst optimization.[15]

Choice of Metal and Ligands

The choice of metals tends to be governed by preexisting work on the particular catalytic reaction of interest. Early successes achieved with Ti (polymerization), Co and Rh (hydrogenation and hydroformylation), and Pd (C–C coupling) continued to influence researchers for a long time. Related metals, including Zr, Ir, and Ni, only later gained attention. Orphan elements, such as Re, seem to have been left behind, whether from neglect or from their systematically poorer reactivity remains unclear. The main group is almost untouched and offers a tempting future target for catalytic chemists.

Phosphines and NHCs provide useful ligands because they are tunable electronically and sterically, thus permitting optimization of catalyst properties in a systematic way, for example using the Tolman map. NHCs can be much stronger donors than PR_3, for example, in one case, the acidity of a metal hydride was reduced by 7 pK_a units on moving to the NHC version.[16] NHCs have the disadvantage that they can sometimes reductively eliminate with a hydride to give the free imidazolium salt and thus be cleaved from the metal during catalysis.[17]

Interest is growing in multifunctional or noninnocent ligands that do more than merely bind to the metal.[18] Some can gain or lose electrons,[19] others can gain or lose protons, but in either case, the ligand changes its properties and may also cooperate more effectively with the metal to facilitate reactions. One promising class of ligand contains molecular recognition groups that orient the substrate via hydrogen bonding so as to enhance selectivity. In one case, four N–H⋯O hydrogen bonds from a ligand hold a carboxylate of the substrate so as to affect the regioselectivity in hydroformylation of ω-unsaturated carboxylates.[20]

9.2   ALKENE ISOMERIZATION

A 1,3-migration of hydrogen substituents in alkenes moves the C=C group along a linear chain (Eq. 9.1). This is often a side reaction in other catalyzed reactions—sometimes useful, sometimes not. Two mechanisms apply: one goes via alkyl intermediates (Fig. 9.2a), the other by an η^3-allyl (Fig. 9.2b). Since all steps are reversible, a nonthermodynamic ratio of cis/trans **9.2** can form here as early-stage kinetic products, the thermodynamic ratio eventually being formed if the catalyst does not "die." In many other catalytic reactions, where product formation is essentially irreversible, nonthermodynamic products are common. In asymmetric catalysis, for example, the two product enantiomers have the same energy, so the one that dominates is automatically a kinetic product.

Alkyl Mechanism

This requires an M–H bond and a vacant site so the alkene can bind then insert to give the alkyl. 1-butene can give the 1° or the 2° alkyl, but β elimination of the 1° alkyl can merely give back 1-butene. Only the 2° alkyl can give both 1- and cis- and trans-2-butene. The initial 2-butene cis/trans ratio depends on the catalyst, cis often being favored—the final thermodynamic ratio usually favors trans. Such is the mechanism for $RhH(CO)L_3$ (L = PPh_3),[21] a coordinatively saturated

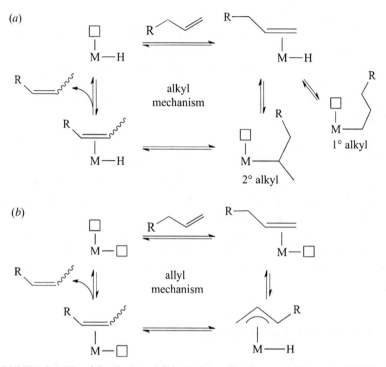

FIGURE 9.2 The (*a*) alkyl and (*b*) allyl mechanisms of alkene isomerization. The open box represents a 2e vacancy or potential vacancy in the form of a labile 2e ligand.

18e species that must lose a PPh$_3$ to form a coordinatively unsaturated intermediate before alkene can bind.

Allyl Mechanism

This needs two 2e vacant sites and has been established for Fe$_3$(CO)$_{12}$ as precatalyst. Fe(CO)$_3$, formed on heating, is believed to be the active species,[22] but as a 14e fragment, Fe(CO)$_3$ may always be tied up with substrate or product. The open square in Fig. 9.2 thus represents a vacant site or a labile ligand. In this mechanism, the C–H bond at the activated allylic position of the alkene undergoes an oxidative addition to the metal to give an η^3-allyl hydride. We only need a reductive elimination to give back the alkene. Again, we have nonproductive cycling if the H returns to the same site it left, rather than to the opposite end of the allyl group.

The two routes can be distinguished by a crossover experiment (Section 6.6) on a mixture of C$_5$ and C$_7$ alkenes (Eq. 9.2).[23] On the allyl mechanism, the D in **9.3a** should end up only in the C_7 product **9.3b** after an intramolecular 1,3 shift, as in fact seen. For the hydride mechanism, the D would be transferred to the catalyst that would in turn transfer it by crossover to the C_5 product.

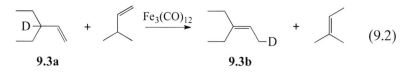

$$\text{9.3a} \qquad\qquad\qquad\qquad \text{9.3b} \qquad\qquad\qquad (9.2)$$

9.3   HYDROGENATION

Hydrogenation catalysts[1c] add H_2 to an unsaturated C=X bond (X = C, N, O). Substrates range from alkenes and alkynes to the more challenging cases of arenes, nitriles, and esters.[24] Three general types of catalyst differ in the way they activate H_2. This can happen by (1) oxidative addition, (2) heterolytic activation, and (3) homolytic activation. One further type, (4) outer sphere, is distinguished by the substrate never becoming bound to the metal.

Oxidative Addition Pathway

One catalytic cycle for Wilkinson's catalyst, $RhCl(PPh_3)_3$ (**9.4**) is shown in Fig. 9.3. Hydrogen addition to give a dihydride leads to labilization of one of the PPh_3 ligands to give a site at which the alkene binds.

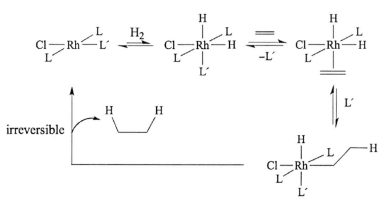

FIGURE 9.3   One mechanism for the hydrogenation of alkenes by Wilkinson's catalyst (L = PPh_3). Other pathways also operate, however, notably involving prior dissociation of PPh_3 before H_2 binding in which case $L' = $ solvent.

The alkene inserts into the Rh–H bond, as in isomerization, but the intermediate alkyl is irreversibly trapped by RE with the second hydride to give an alkane. This is an idealized mechanism[2] because **9.4** can also lose PPh_3 to give $RhCl(PPh_3)_2$, as well as dimerize via Cl bridges with each of these species having their own catalytic cycles. Indeed, the majority of the reaction goes through $RhClL_2$ because it reacts so much faster with H_2 than does **9.4**. Reversibility arguments see rapid OA of H_2 to the three-coordinate d^8 $RhClL_2$ to give five-coordinate d^6 RhH_2ClL_2 relative to the corresponding four-coordinate → six-coordinate conversion in **9.4** as consistent with the tendency for faster RE from five-coordinate d^6 species discussed in Section 6.6. In a key study by Tolman,[25] the dihydride was detected by ^{31}P NMR under H_2 and the reversible loss of the PPh_3 trans to H inferred from a broadening of the appropriate resonance, as discussed in Section 10.5. As predicted by the mechanism of Fig. 9.3, hydrogen gives syn addition to the alkene, although it is only possible to tell this in cases such as Eq. 9.3.

Figure 9.3 represents the *hydride mechanism* in which H_2 adds before the olefin. Sometimes, the olefin adds first (the *olefin mechanism*), as proposed for the Schrock-Osborn catalyst, $[Rh(dpe)(MeOH)_2]BF_4$, formed by hydrogenation in MeOH of the placeholder cyclooctadiene (cod) ligands of the catalyst precursor, $[(cod)Rh(dpe)]BF_4$.

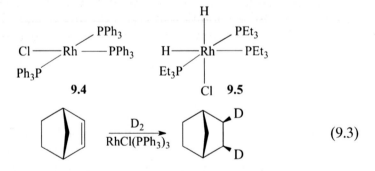

$$(9.3)$$

To bind two hydrides and the alkene, the 16e catalyst $RhCl(PPh_3)_3$ needs to dissociate PPh_3 first. The PEt_3 analog of **9.4** reacts with H_2 to give a stable dihydride $RhH_2Cl(PEt_3)_3$, **9.5**. The small PEt_3 does not dissociate, so **9.5** is catalytically inactive. An active PEt_3 analog is possible if an RhH_2ClL_2 intermediate is formed in situ by hydrogenating $[(nbd)RhCl(PEt_3)_2]$. Under H_2, the norbornadiene (nbd) is removed by hydrogenation, to give the active $RhH_2Cl(PEt_3)_2$ system.

Isomerization is often a minor pathway in a hydrogenation catalyst if the intermediate alkyl β-eliminates before it has a chance to reductively eliminate. The more desirable catalysts, such as **9.4**, tend to give little isomerization. Unhindered alkenes are preferred: monosubstituted > disubstituted > trisubstituted > tetrasubstituted = 0. This means that **9.4** reduces the triene **9.6** largely to the octalin **9.7** (Eq. 9.4). Heterogeneous catalysts give none of this product, but only the fully saturated decalin (**9.9**), and the isomerization product, tetralin (**9.8**) (Eq. 9.4). The C=O and C=N double bonds of ketones and imines are successfully reduced only by certain catalysts. Other functional groups that can be reduced by heterogeneous catalysts, such as –CN, –NO$_2$, –Ph, and –CO$_2$Me, are rarely reduced by homogeneous catalysts.

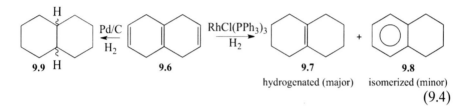

9.9   9.6   9.7   9.8
hydrogenated (major)   isomerized (minor)

$$(9.4)$$

IrCl(PPh$_3$)$_3$, the iridium analog of **9.4**, is inactive because IrH$_2$Cl(PPh$_3$)$_3$ fails to lose PPh$_3$ as a result of the stronger M–L bond strengths for the third-row metals. Using the same general strategy we saw for Rh, more useful catalysts are obtained by moving to **9.10** and **9.11**.[26] On hydrogenation, these tend to bind a solvent, such as EtOH, to give the isolable species **9.12** (solv = acetone, ethanol, or water). As a result, the catalyst can be used in CH$_2$Cl$_2$, a much more weakly coordinating solvent than EtOH, where **9.11** is unusual in reducing even highly hindered alkenes. The high activity of **9.10** at first escaped attention because it was initially tested in EtOH, the conventional solvent for hydrogenation at that time. Screening a new catalyst under a variety of conditions is therefore advisable. An Ir(III)/(V) cycle may apply to **9.11** as opposed to the Rh(I)/(III) cycle accepted for **9.4**.[16]

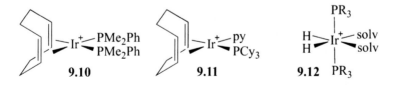

9.10   9.11   9.12

Directing Effects

9.11 shows strong directing effects, meaning that H$_2$ is added to one face of the substrate if there is a directing group (e.g., –OH, –COMe,

and –OMe) on that face (Eq. 9.5). The net positive ionic charge makes
the metal hard enough to bind to the directing group and, as $\{IrLL'\}^+$
is a 12e fragment, it has enough sites to bind all the needed ligands in
the key iridium dihydride intermediate in Eq. 9.5. Of the four possible
geometrical isomers of the saturated ketone, only one is formed, H_2
having been added syn to the directing group.

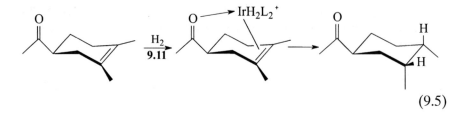

$$(9.5)$$

Asymmetric Catalysis

The importance of this area was emphasized by the award of the 2001
Nobel Prize to William S. Knowles, Ryoji Noyori and K. Barry Sharpless. In a typical case, the Schrock-Osborn $[(cod)RhL_2]^+$ catalysts,
equipped with homochiral ligands, can give asymmetric reduction of a
prochiral alkene **9.13**. Although achiral, **9.13** can give homochiral hydrogenation products by favoring either **9.14** or **9.15** (Eq. 9.6).[27] The hard
part is finding a ligand and conditions that can give a practically useful
bias in favor of a desired enantiomer.

$$
\underset{R'}{\overset{R}{}}C{=}CH_2 \longrightarrow \underset{R'}{\overset{R}{}}H{\blacktriangleright}C{-}CH_3 + \underset{R'}{\overset{R}{}}H_{\cdot\cdot}C{-}CH_3 \qquad (9.6)
$$

$$\text{9.13} \qquad\qquad \text{9.14} \qquad\qquad \text{9.15}$$

   In **9.13**, asymmetric hydrogenation is possible if H_2 prefers to add to
one face over the other. Equation 9.7 shows how a prochiral alkene
binds to an achiral metal to give two enantiomers; that is, the complex
is chiral even though neither the free ligand nor the metal were chiral.
We can regard the "chiral carbon" indicated by the asterisk as having
four different substituents, one of which is the metal.

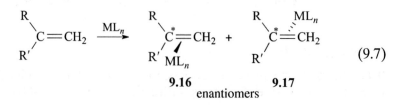

$$(9.7)$$

9.16   **9.17**

enantiomers

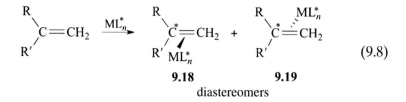

$$(9.8)$$

9.18   **9.19**

diastereomers

An ML_n catalyst with a homochiral L can indeed bias the H_2 addition to one face of a prochiral alkene. In Eq. 9.8, instead of forming two enantiomeric complexes **9.16** and **9.17**, which react at equal rates to give a racemic mixture of products, we will now have diastereomeric alkene–catalyst complexes, **9.18** and **9.19**, where we now have two asymmetric centers, the coordinated alkene and the ML_n groups. Since diastereomers can have different chemical properties, **9.18** and **9.19** can have different rates of hydrogenation. This bias can give us one of the pair of enantiomers **9.14** or **9.15** over the other. Each enantiomer of the catalyst should ideally give us one enantiomer of the hydrogenation product. This is valuable in that a large amount of a product enantiomer comes from a small amount of resolved ligand L. This is also the natural route to pure enantiomers from enzyme catalysis, where the selectivity is near-perfect.

In asymmetric hydrogenation, 95–99% enantiomeric excess [e.e. $= 100 \times \{$amount of major isomer $-$ amount of minor isomer$\}/$ $\{$total of both isomers$\}$] can be obtained in favorable cases. The first alkenes to be reduced with high e.e. contained a coordinating group: for example, **9.20** and **9.21**.

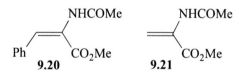

These bind to the metal via the amide carbonyl just as in directed hydrogenation. This improves the rigidity of the alkene–catalyst complex, which in turn increases the chiral discrimination of the system. As in directed hydrogenation, a 12e catalyst fragment is required, as is indeed formed by hydrogenation of the Schrock–Osborn precatalyst.

Some of the best chiral ligands, such as BINAP (**9.22**), have a C_2 axis that result in the symmetry of a propellor, which can either have a left-handed or a right-handed twist. The chiral centers impose a twist on the conformation of the BINAP–metal complex that in turn leads to a chiral, propeller-like arrangement of the PPh_2 groups (**9.23**). These groups transmit the chiral information from the asymmetric centers to the binding site for the alkene on the opposite side of the catalyst. The advantage of a C_2 symmetry is that the substrate sees the same chirality however it binds.

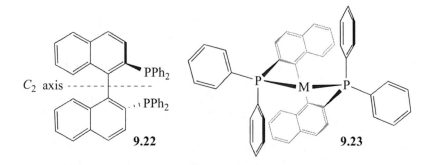

C_2 axis

9.22   **9.23**

It was once thought that the binding preference—one face of the substrate often binds better to the catalyst than the other—always determines the sense of asymmetric induction, thus preferential binding of the pro-S face was thought to lead to a preference for the S product. In a classic study, Landis and Halpern[28] showed that in a catalyst that gives the S product, however, the metal is preferentially bound to the "wrong" pro-R face (**9.24**). A kinetic study showed that the minor isomer had to react at $\sim 10^3$ times the rate of the other (Eq. 9.9) to produce the S product. The hydrogenation rapidly depletes the minor isomer, but **9.24** and **9.25** interconvert even more rapidly, so the deficit is immediately made up. This common behavior is termed Halpern selectivity.[29]

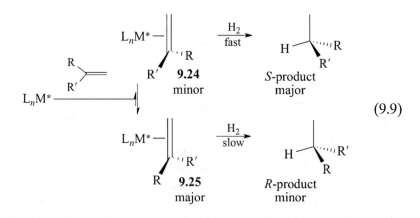

(9.9)

Asymmetric alkene hydrogenation was used in the successful commercial production of the pain reliever, naproxen, and of the Parkinson's disease drug l-DOPA, formed by hydrogenation of the alkene **9.26**.[30]

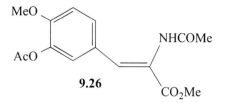

9.26

Another commercial success, this time for Novartis, was the Ir-catalyzed asymmetric synthesis of the herbicide, (S)-metolachlor, from an imine precursor. The key advantage of iridium is the extremely high rate ($>200{,}000$ TOF h^{-1}) and catalyst lifetime ($\sim 10^{6}$ TON) despite a substantially lower e.e. than with Rh. This shows both that C=N bonds can be hydrogenated, and that in commercial applications, it is not just high e.e. that counts but also productivity per unit reactor volume per unit time.

Reversibility

In hydrogenation, the final step, the reductive elimination of the product, is irreversible. This contrasts with the reversibity of alkene isomerization. In a reversible cycle, the products can equilibrate among themselves, and a thermodynamic mixture is always eventually obtained if the catalyst remains active. This is not the case in asymmetric hydrogenation; if it were, the R and S products would eventually come to equilibrium and the e.e. would go to zero with time. Only an irreversible catalytic cycle with an irreversible last step can give a nonthermodynamic final product ratio. This means we can obtain different kinetic product ratios with different irreversible catalysts. Reversible catalysts can give a nonthermodynamic product ratio initially, but the final ratio will be thermodynamic.

Heterolytic H₂ Activation

Catalysts such as $[RuHCl(PPh_3)_3]^{31}$ activate H_2 heterolytically (Eq. 9.10) by σ-*bond metathesis*,[32] a reaction that has the general form of Eq. 9.11, in which Y is often a hydrogen atom.

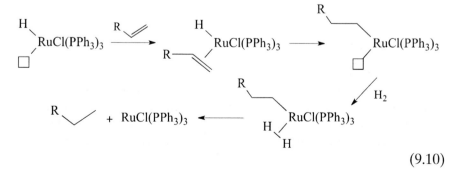

$$(9.10)$$

$$
\begin{array}{c}
\text{M} \longrightarrow \text{X} \\
+ \\
\text{Y} \longrightarrow \text{Z}
\end{array}
\quad \longrightarrow \quad
\begin{array}{cc}
\text{M} & \text{X} \\
| & + & | \\
\text{Y} & \text{Z}
\end{array}
\tag{9.11}
$$

By going via a $Ru^{II}(H_2)$ intermediate, the metal gives the same products that would have been obtained from OA-RE sequence, but by avoiding the OA step, the metal avoids becoming Ru(IV), not very stable for Ru. Otherwise, these catalysts act very similarly to the OA group. As a 16e hydride complex, $RuH_2(PPh_3)_3$ can coordinate the alkene, undergo insertion to give the alkyl, then liberate the alkyl by a heterolytic activation of H_2, in which the alkyl group takes the proton and the H^- goes to the metal to regenerate the catalyst.

Homolytic H_2 Activation

Iguchi's paramagnetic d^7 $[Co^{II}(CN)_5]^{3-}$, a very early (1942) homogeneous hydrogenation catalyst, is an example of a rare group of catalysts that activate hydrogen homolytically by a binuclear oxidative addition. This is not unreasonable for this Co(II) complex ion, a metal-centered radical that has a very stable oxidation state, Co(III), one unit more positive. Once $[HCo^{III}(CN)_5]^{3-}$ has been formed, a hydrogen atom is transferred to the substrate in the second step, an outer sphere reaction that therefore does not require a vacant site at the metal, but does require the resulting organic radical to be moderately stable—hence the Iguchi catalyst will reduce only activated alkenes, such as the cinnamate ion, in which the radical is benzylic and therefore stabilized by resonance. Finally, in a second outer sphere step, the organic radical abstracts H• from a second molecule of the cobalt hydride to give the saturated product.

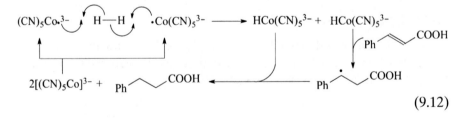

$$
\tag{9.12}
$$

Outer Sphere Hydrogenation

Noyori's[33a] highly effective asymmetric hydrogen transfer (Eq. 9.13) catalysts go by an outer sphere route, a mechanism of rising importance.[33b] The metal donates a hydride to the substrate C=O carbon while the adjacent Ru–NH_2R group simultaneously donates a proton to the C=O oxygen. H_2 then binds to the now 16e metal as an H_2 complex, leading to proton transfer to the ligand to restore the system to its original state. A point of particular interest is that the oxidation state of the metal remains unchanged throughout.

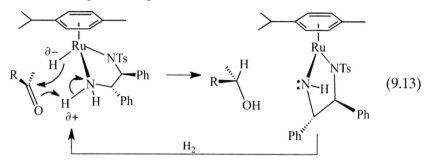

$$(9.13)$$

An outer sphere hydrogenation of quinolines involves a nonconcerted transfer, first of H^+ from the dihydrogen complex to the quinoline N. This polarizes the C=N bond to facilitate the subsequent H^- transfer from the resulting hydride.[34] Again, there is no change of oxidation state throughout the cycle. OA/RE mechanisms require a $\Delta(OS)$ of two units, appropriate for the precious metals (e.g., Ir(I),(III),(V)), but not for the inexpensive ones (e.g., Co(I),(II),(III)). This suggests that $\Delta(OS) = 0$ mechanisms might be well adapted for the cheaper metals—indeed, in Eq. 16.34, we will see that [Fe] hydrogenase carries out a reaction rather like Eq. 9.14 by a $\Delta(OS) = 0$ mechanism.

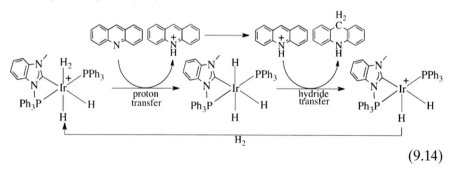

$$(9.14)$$

Transfer Hydrogenation

Transfer hydrogenation avoids free H_2 by using a liquid, typically isopropanol, both as solvent and as reductant that can donate $(H^+ + H^-)$ to the substrate.[35] The ease of handling iPrOH makes transfer hydrogenation a good choice for industrial applications.

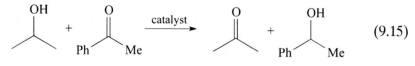

$$(9.15)$$

The reaction is particularly good for reducing ketones and imines, and for asymmetric catalysis (Section 14.4). Bäckvall and coworkers have shown how $RuCl_2(PPh_3)_3$ is effective at 80°C with added base as catalyst promoter. The role of the base is no doubt to form the isopropoxide ion, which presumably coordinates to Ru and by β elimination forms a hydride and acetone.

Nanoparticles

Since catalysis can arise from a small, highly active component of the reaction mixture, it is easy to misassign the observed catalytic activity to the major component of the catalyst that is seen spectroscopically.[36] Few homogeneous catalysts hydrogenate benzene, but heterogeneous catalysts such as metallic Rh do so readily. This means that hydrogenation of arenes is a "red flag" for the possible intervention of metal nanoparticles formed by partial decomposition of the ostensible homogeneous catalyst. Their 1–100 nm size range means the nanoparticles may stay suspended in the solvent and mimic a homogeneous catalyst. To complicate this question, a few truly homogeneous arene hydrogenation catalysts have been found, such as $[Cp^*RhCl_2]_2$, but proving homogeneity required heroic efforts.[37] Ultimately, there is no sharp division between homogeneous and heterogeneous catalysis when the possibility of the formation of small clusters is taken into account. If an M_4 cluster were active, assignment as a homogeneous catalyst would follow, an M_{400} cluster would be considered heterogeneous, but an M_{40} cluster would fall in between.

Asymmetric catalysis might be thought of as a guarantee of catalyst homogeneity since this outcome has traditionally been associated with homogeneous catalysis. In fact, nanoparticles can bind asymmetric ligands and give high levels of asymmetric catalysis even when formed in situ during a catalytic run from a homogeneous catalyst precursor.[38]

9.4   ALKENE HYDROFORMYLATION

In the late 1930s, Otto Roelen discovered the hydroformylation, or *oxo* process, one of the first commercially important homogeneous catalytic reactions. He found that a C_n alkene can be converted to a C_{n+1} aldehyde by the addition of H_2 and CO, catalyzed by $Co_2(CO)_8$; further reduction to the alcohol can occur, depending on conditions (Eq. 9.16).

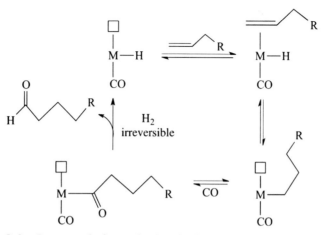

FIGURE 9.4 One catalytic cycle for hydroformylation with HCo(CO)$_4$. Alkene insertion also takes place in the opposite direction to give the secondary alkyl, which goes on to the branched or *iso*-aldehyde RCH(Me)CHO, but this parallel and less important side-reaction is not shown.

Many million tons of aldehydes and alcohols are now made annually in this way.

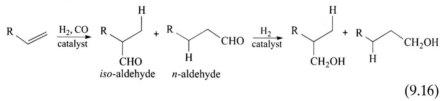

$$(9.16)$$

In the pathway of Fig. 9.4, the Co$_2$(CO)$_8$ first gives a binuclear oxidative addition with H$_2$ to form HCo(CO)$_4$ as the active catalyst. CO dissociation then generates the vacant sites required for alkene binding. The first steps resemble hydrogenation in that an alkyl is formed by alkene insertion. Since there is no second Co–H, the Co–R cannot reductively eliminate with H, as in hydrogenation, so the alkyl undergoes migratory insertion to M–CO to give the acyl, M–COR, followed by a heterolytic H$_2$ cleavage (e.g., Eq. 9.10) to give RCHO and regenerate the catalyst.

Depending on the direction of insertion, the 1° and 2° alkyls can be formed, corresponding to 1° or 2° aldehydes. The linear 1° aldehyde is much more valuable commercially, so catalysts that favor it are preferred. The product regiochemistry is normally decided not by the direction of alkene insertion — a reversible step — but by the

rate at which the 1° and 2° alkyls are irreversibly trapped by migration to CO; this is the selectivity determining step of the cycle.
Slaugh and Mullineaux made the commercially important discovery
that the addition of P(n-Bu)$_3$ to Co$_2$(CO)$_8$ gives a catalyst that is not
only much more active—5–10 atm H$_2$/CO are required versus
100–300 atm for Co$_2$(CO)$_8$—but which also shows a greater preference for the 1° over the 2° aldehyde (n:iso = 8:1 vs. 4:1 for
Co$_2$(CO)$_8$). The steric bulk of P(n-Bu)$_3$ both helps formation of the
less hindered 1° alkyl and, more importantly, speeds up migratory
insertion. The rhodium complex, RhH(CO)(PPh$_3$)$_3$, is an even more
active catalyst, operating at 1 atm H$_2$/CO pressure and 25°, and it is
also even more selective for the 1° product.

Both Co and Rh catalysts are also very active for alkene isomerization and so almost the same mixture of aldehydes is formed from 1- and
2-butene. This implies that commercially valuable n-aldehydes can still
be obtained from the cheaper internal alkenes. The catalyst first isomer-
izes 2-butene, to a mixture including 1-butene. The latter is hydroformylated much more rapidly, accounting for the predominant n-aldehyde
product. 1-butene is always a minor component of the alkene mixture,
but the n-aldehyde is formed from it, providing another example of a
catalytic process in which the major product is formed from a minor
intermediate and leads to the general principle that what we see in the
catalytic solution may have little or no relation to the active cycle.

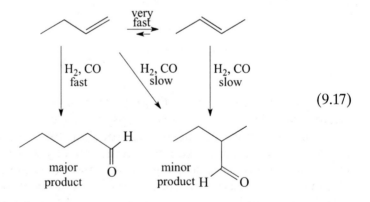

$$(9.17)$$

Chelating and Phosphite Ligands

In Rh-catalyzed hydroformylation, the n:iso ratio increases with the
bite angle[39] = (preferred P–M–P angle) of a chelate phosphine, probably because these ligands facilitate the RE step in the mechanism. The
Rh complex (**9.27**) of the wide bite angle ligand, BISBI, has proved
particularly useful.[40]

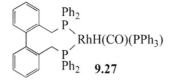

9.27

9.5   ALKENE HYDROCYANATION

That the great strength of protein biopolymers, such as spider web material, relies on N–H···O=C hydrogen bonding suggested to Carothers at du Pont that the peptide link, –NHCO–, might also be useful in artificial polymers. Out of this work came nylon-6,6 (**9.28**), one of the first useful petroleum-based polymers.

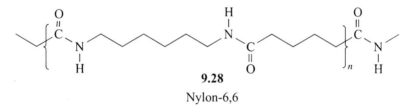

9.28

Nylon-6,6

The polymer itself is made from adipoyl chloride and hexamethylene diamine, both obtained from adiponitrile. Now that the original patents have long expired, the key to making nylon-6,6 commercially is therefore having the least expensive source of adiponitrile. Originally made commercially by the chlorination of butadiene, this old route involves Cl_2 and thus generates much toxic waste as well as causing corrosion problems (Eq. 9.18). Homogeneous catalysis provided the means to improve the adiponitrile synthesis by a new route, hydrocyanation, discovered at duPont by Drinkard. In this reaction, two equivalents of HCN give anti-Markovnikov addition to butadiene with a NiL_4 catalyst to obtain adiponitrile directly (Eq. 9.19).

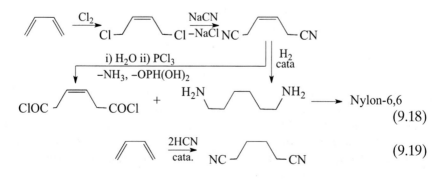

(9.18)

(9.19)

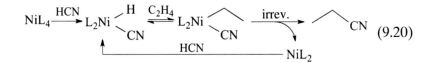

Ethylene hydrocyanation by [Ni{P(O-o-tolyl)₃}₄] follows the cycle of Eq. 9.20. Oxidative addition of HCN to the metal gives a 16e nickel hydride that undergoes ethylene insertion to give an ethyl complex, followed by reductive elimination to give the EtCN product. The reaction with butadiene is more complex but goes by a closely related route not discussed in detail here. The best ligands are bulky, π-acceptor P-donors, such as tri-o-tolyl phosphite.

9.6   ALKENE HYDROSILYLATION AND HYDROBORATION

Hydrosilylation

The addition of R_3Si–H across a C=C bond to give the R_3Si–C–C–H unit is commercially important for the synthesis of such products as the self-curing silicone rubber formulations in common domestic use.[41] The preferred catalysts, $H_2[PtCl_6]$, Speier's catalyst,[42] and **9.29**, Karstedt's catalyst, are active even at 0.1 ppm. Speier's catalyst requires an induction period before hydrosilylation begins, attributed to reduction to the active Pt(0) state. Careful work was needed to determine the homogeneity of the catalyst.[43] The Chalk–Harrod mechanism[41] of Eq. 9.22, was accepted for many years. The true catalyst may be platinum nanoparticles of 10–1000 Å diameter. If so, this implies that the active form of Speier's catalyst is heterogeneous. Other hydrosilylation catalysts, such as $Co_2(CO)_8$, $Ni(cod)_2$, $NiCl_2(PPh_3)_2$, and $RhCl(PPh_3)_3$, do seem to be authentically homogeneous, however.

As in hydroformylation, both linear and branched products can be obtained from $RCH=CH_2$. The *dehydrogenative silylation* product, $RCH=CHSiR'_3$, is often present and can even predominate under some conditions (Eq. 9.21). The dehydrogenative path can only be explained on the modified Chalk–Harrod mechanism of Eq. 9.23, in which the alkene first inserts into the M–Si bond. β elimination of the intermediate alkyl now leads directly to the vinylsilane, the two H atoms thus released go on to hydrogenate the substrate leading to coproduction of alkane. As in hydrogenation, syn addition is generally observed. Apparent anti addition is due to isomerization of the intermediate metal vinyl, as we saw in Eq. 7.21, also a reaction in which initial insertion of alkyne into the M–Si bond must predominate (>99%). Progress has been made in eliciting good hydrosilylation activity from nonprecious metals, such as Fe.[44]

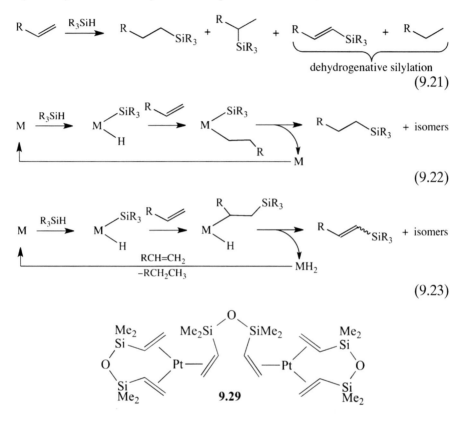

$$(9.21)$$

$$(9.22)$$

$$(9.23)$$

Hydroboration

$RhCl(PPh_3)_3$ catalyzes the addition of the B–H bond in catecholborane {HB(cat)} to alkenes (Eq. 9.24).[45] The uncatalyzed reaction also goes forward, although more slowly, but the catalytic reaction has usefully different chemo-, regio-, and stereoselectivities. An oxidative workup of the product, R'B(cat), is normally adopted and leads to R'OH. The catalytic cycle is complex, with more than one species contributing to activity, and the results depend on whether aged or freshly prepared catalyst is used. Improved catalysts, including asymmetric catalysts are now available.[45]

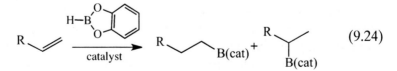

$$(9.24)$$

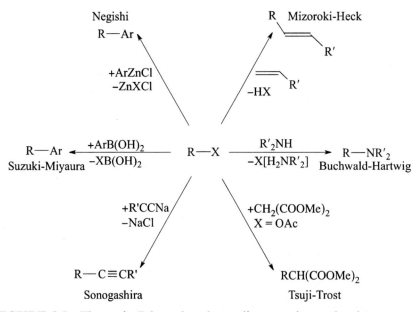

FIGURE 9.5 The main Pd-catalyzed coupling reactions take the names of their discoverers, identified above. RX is typically an aryl or vinyl halide.

9.7 COUPLING REACTIONS

The development of this reaction class led to the award of the 2011 Nobel Prize to Heck, Negishi, and Suzuki.[46] Palladium catalysts, very extensively used for carbon–carbon and carbon–heteroatom coupling reactions,[47] now occupy a central place in synthetic organic methodology and in the pharmaceutical industry. For example, the Mizoroki–Heck reaction is used to make the tricyclic ring system at the heart of the antitumor drug, Taxol. Among the most useful coupling reactions, shown in Fig. 9.5, bear the names of their developers.

Often catalyzed by a variety of palladium complexes or simply by a mixture of $Pd(OAc)_2$ and PR_3, these involve initial reduction of a Pd(II) precursor to Pd(0), normally stabilized by a single 2e ligand, L, typically a phosphine or an NHC. Subsequent oxidative addition of RX generates an R–Pd(II)(L)X intermediate. Basic, bulky phosphines, such as $P(t\text{-}Bu)_3$ or X-phos (**4.11**) facilitate the OA by favoring the formation of this highly reactive zerovalent, 1:1 complex, PdL, in line with the idea that the microscopic reverse, RE from Pd(II), often takes place from a three-coordinate LPd(R)(X) intermediate (Section 6.6). R is often aryl or vinyl to avoid the β elimination that would be likely if R were an alkyl. This restriction is not universal, however, because in a growing number of cases, coupling of simple alkyl groups can be achieved.[48] In such cases, either β elimination, although kinetically accessible, is uphill thermodynamically,[49] or else reductive elimination may be faster than β elimination.

In the Tsuji–Trost reaction, an allylic acetate first oxidatively adds to the Pd(0) catalyst to give a π-allyl complex, which undergoes nucleophilic attack by the carbanion derived from the deprotonated active methylene compound; allyl alcohols and aldehydes can be coupled by a related procedure.

In the Mizoroki–Heck reaction,[50] an alkene inserts into the Pd^{II}–R bond, followed by β elimination to give the product and LPd(H)(X). A base such as NaOAc reduces Pd^{II} to Pd^0 by removing HX in the last RE step. The electron-withdrawing group (EWG), R′, on the alkene cosubstrate ensures that the insertion step takes place in the direction shown, to give R′CH=CHR, not CH_2=CRR′. If the Pd–R bond is polarized as Pd^+–R^-, the R group can attack the more positive, terminal carbon of the C^+=C-EWG⁻ group. Many catalysts require temperatures over 120°, in which case catalyst decomposition to Pd nanoparticles, the true catalyst, is possible.[51] $Pd_2(dba)_3$, a common Pd(0) catalyst precursor, has been shown to contain up to 40% Pd NPs, themselves potent catalysts, thus complicating interpretation of work carried out with this material.[52]

In the other coupling reactions,[47] the anionic X group of the R–Pd(II)-X intermediate is then replaced by the nucleophilic group from the cosubstrate, either aryl or NR_2. In the final step, reductive elimination gives the product. Other nucleophiles also work, for example, C–O coupling to form aryl ethers is possible with aryl halides and phenolates.[53] Equation 9.25 and Equation 9.26 indicate the main steps of these reactions, where Nu–E is the generalized reaction partner for the ArX. The classic Ullman coupling of R_2NH with ArI to give R_2NAr, mediated by Cu(I) at >180° is not a mild procedure, but use of a cheap metal is attractive and photolytic activation avoids the need for heating by inducing a radical coupling pathway.[54]

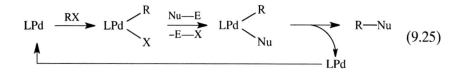

$$(9.25)$$

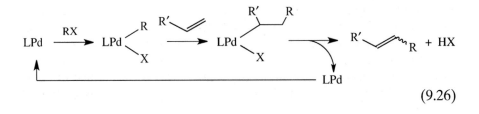

$$(9.26)$$

9.8   ORGANOMETALLIC OXIDATION CATALYSIS

Organometallic catalysts are more common in reduction than oxidation, in part because air and oxidants can cleave organometallic ligands from the metal, resulting in deactivation or formation of a coordination catalyst.[55-57] Coordination compounds, having more oxidatively robust ligands and preferring higher oxidation states, are much better known in oxidation catalysis, as in the Wacker process[58] (Section 8.3). Nevertheless, organometallic precatalysts or intermediates can be involved in oxidation and interest in the area is on the rise.

Oxidase Reactions

Stahl[59] and Sheldon[60] have shown how oxidations can be driven by air as *primary oxidant*, or source of stochiometric oxidizing power. Like the catalysts in this subsection, biological oxidases are enzymes that use O_2 but do not incorporate its O atoms into the substrate. For example, Pd(OAc)$_2$-pyridine is active for alcohol oxidation, intramolecular hetero- and carbocyclization of alkenes, intermolecular O–C and C–C coupling reactions[55] with alkenes, and oxidative C–C bond cleavage of tertiary alcohols. A pathway for alcohol oxidation is shown in Eq. 9.27. Normally a 4e process, reduction of O_2 can be hard to couple with oxidation of the catalytic intermediates, processes that often proceed in 2e steps. In this case, intermediate η^2-peroxo Pd(II) complexes can be formed from reaction of Pd(0) intermediates with O_2, which thus acts as a 2e oxidant.

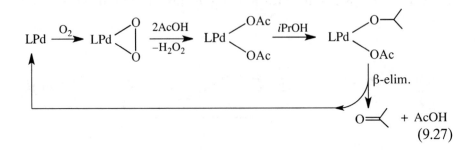

$$(9.27)$$

Water Splitting and C–H Oxidation

Water splitting is a component of many artificial photosynthesis schemes since it converts a photogenerated electrochemical potential into a fuel, H_2, as well as releasing O_2.[61] Although nickel and iridium oxides have long been used as heterogeneous electrocatalysts for water oxidation, and a few Mn and Ru coordination catalysts are known, some organometallic Cp*Ir^{III} compounds have recently proved active.[62] [Cp*Ir^{III}(OH$_2$)$_3$]SO$_4$ is a precursor to a highly active Ir oxide-based heterogeneous catalyst, while **9.30** gives a homogeneous catalyst. Both lose the Cp* ligand, but the catalyst from **9.30** retains the N,O chelate. The primary oxidant can either be an electrode or a chemical oxidant, such as Ce(IV) or NaIO$_4$. The catalysts are believed to operate via a proposed Ir^V=O intermediate that undergoes nucleophilic attack by H_2O or periodate to generate the O–O bond. The Ir^V=O can both oxidize water and also hydroxylate C–H bonds with retention of stereochemistry (Eq. 9.28). Retention suggests that no radical species are involved, since the 9-decalyl radical rapidly ($\sim 10^8$ s^{-1}) loses stereochemistry and in that case, trans-9-decalol would be the predominant product.[63] The water reduction (WR) component of water splitting is discussed in relation to hydrogenases in Section 16.4.

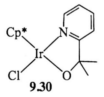

9.30

Many electrocatalytic reactions involve proton transfers, as in $2H_2O \Rightarrow O_2 + 4H^+ + 4e^-$. In an important general principle, concerted loss (or gain) of a proton and an electron together is often much easier than any sequential two-step process because moving the proton and electron together avoids costly charge separation. This important pathway, termed proton-coupled electron transfer or PCET, is very common in both synthetic and enzyme catalysts.[64]

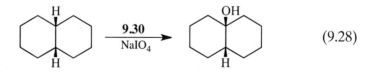

(9.28)

9.9 SURFACE, SUPPORTED, AND COOPERATIVE CATALYSIS

Organometallic complexes can be supported in a variety of ways to give heterogenized catalysts that are more readily separable from the soluble products of the reaction. For example, polystyrene (P) beads can be functionalized with –CH$_2$PPh$_2$ groups, allowing attachment of a variety of catalysts, including (P–CH$_2$PPh$_2$)RhCl(PPh$_3$)$_2$. The bead swells in organic solvents to admit substrate and the catalytic cycle proceeds normally.[65a] Leaching of metal from the support is often a problem, however.

A catalyst can also be supported in a separate liquid phase if the catalyst is made soluble in that liquid by appending solubilizing groups, SG, to a ligand, as in P(C$_6$H$_4$(SG))$_3$. Solubility in water can be induced with –SO$_3$Na solubilizing groups and in fluorocarbons with –CH$_2$CH$_2$(CF$_2$)$_n$CF$_3$. The reaction is run in a mixed solvent such that the substrate and products concentrate in the organic phase and the catalyst in the water or fluorocarbon layer; in the case of mixed fluorocarbon–hydrocarbon solvents, the two layers become miscible on heating but separate on cooling.[65b]

Surface Organometallic Chemistry

Catalysts functionalized with siloxane anchors can be attached to SiO$_2$ nanoparticles (NPs) via [Si]–O–M links involving surface silanol groups, denoted [Si]OH. They thus benefit from their high surface area of NPs and relatively easy separability. Similarly, catalysts supported on magnetic Fe$_3$O$_4$ nanoparticles can be magnetically separated from the reaction medium for reuse.[66] Other advantages accrue: catalysts that are insoluble in a given solvent become viable when supported on NPs; two different catalysts that might otherwise mutually interfere in conventional cooperative catalysis can be kept out of contact on separate NPs.

A variety of organometallics has been covalently anchored to a silica surface at single sites by [Si]–O–M links involving [Si]OH groups. The oxophilic early metals are particularly well suited to this approach. Once bound to the surface, many of the usual solution characterization methods no longer apply. A combination of EXAFS (extended X-ray absorption fine structure: see Chapter 16), solid-state NMR, and IR spectroscopy, however, can often give sufficient information. Unusual reactivity can be seen, probably as a result of *site isolation*, which prevents the formation of inactive M(μ-OR)$_n$M dimers.[67] Many such species are catalytically active. Cp*ZrMe$_3$ on Al$_2$O$_3$ gives an ethylene polymerization catalyst in the presence of the usual MAO activator ([MeAlO]$_n$; see Section 12.2); [([Si]O)Re($\equiv$CtBu)(=CHtBu)(CH$_2$tBu)] is active in alkene metathesis. Remarkably, a number of these species carry out alkane conversion reactions unknown in heterogeneous and very rare in homogeneous catalysis. For example, ([Si]–O)$_3$TaH causes disproportionation of acyclic alkanes into lower and higher homologs, such as of ethane into methane and propane. A number of commercially important catalysts consist of organometailic compounds covalently attached to surfaces. In the Phillips alkene polymerization catalyst,[68] for example, CrCp$_2$ is supported on silica.

Cooperative Catalysis

If two or more catalysts operate within the same reactor to bring about a tandem reaction that relies on them both, we have cooperative or tandem catalysis. Common cases involve a metal complex and an organocatalyst, the latter often supplying the asymmetric aspect.[69a] In another example, light alkanes were first dehydrogenated to alkenes with a pincer Ir catalyst (Section 12.4) and the resulting olefins were then upgraded to heavier hydrocarbons by a Cp*TaCl$_2$(C$_2$H$_4$) alkene dimerization catalyst (Section 12.2).[69b]

Hidden Acid Catalysis

If a reaction is catalyzed by a proton acid, a metal-catalyzed version may also be possible. Such is the case for addition of alcohols or carboxylic acids to alkenes and alkynes catalyzed by silver salts such as AgOTf. In hidden acid catalysis,[70] the metal may liberate free protons that are the true catalyst. Careful control experiments are needed to test this possibility.

- Catalysis, a key organometallic application, goes by the steps discussed in Chapters 6–8.
- Directed and asymmetric catalysis (Section 9.3) can lead to high selectivity.
- Intermediates must be kinetically competent (Section 9.3); apparent intermediates may in fact be off-loop species (Fig. 9.1).

REFERENCES

1. (a) D. Steinborn, *Fundamentals of Organometallic Catalysis*, Wiley-VCH, Weinheim, 2012; B. Cornils, ed., *Multiphase Homogeneous Catalysis*, Wiley-VCH, Chichester, 2005; (b) B. Heaton, ed., *Mechanisms in Homogeneous Catalysis: A Spectroscopic Approach*, Wiley-VCH, Weinheim, 2005. (c) P. W. N. M. van Leeuwen, *Homogeneous Catalysis: Understanding the Art*, Kluwer, Dordrecht, 2004.

2. C. A. Busacca, D. R. Fandrick, J. J. Song, and C. H. Senanayakea, *Adv. Synth. Catal.*, **353**, 1825, 2011.

3. J.-M. Savéant, *Chem. Rev.*, **108**, 2348, 2008.

4. R. H. Crabtree, *Chem Rev.*, **112**, 1536, 2012.

5. S. Jansat, M. Gomez, K. Philippot, G. Muller, E. Guiu, C. Claver, S. Castillon, and B. Chaudret, *J. Am. Chem. Soc.*, **126**, 1592, 2004.

6. N. Castellanos-Blanco, M. Flores-Alamo, and J. J. García, *Organometallics*, **31**, 680, 2012.

7. P. F. Cook and W. W. Cleland, *Enzyme Kinetics and Mechanism*, Garland Science, New York, 2007.

8. S. Kozuch and J. M. L. Martin, *ACS Catal.*, **2**, 2787, 2012; G. Lente, *ACS Catal.*, **3**, 381, 2013.

9. P. W. N. M. van Leeuwen, *Appl. Catal., A*, **212**, 61, 2001.

10. S. Kozuch and S. Shaik, *Acc. Chem. Res.*, **44**, 101, 2011.

11. S. Bertelsen and K. A. Jørgensen, *Chem. Soc. Rev.*, **38**, 2178, 2009.

12. Z. T. Du and Z. H. Shao, *Chem. Soc. Rev.*, **42**, 1337, 2013.

13. C. A. Huff and M. S. Sanford, *J. Am. Chem. Soc.*, **133**, 18122, 2011.

14. X. Solans-Monfort, C. Copéret, and O. Eisenstein, *Organometallics*, **31**, 6812, 2012.

15. T. J. Collins, *Acc. Chem. Res.*, **35**, 782, 2002.

16. Y. Zhu, Y. Fan, and K. Burgess, *J. Am. Chem. Soc.*, **132**, 6249, 2010.

17. C. Gandolfi, M. Heckenroth, A. Neels, G. Laurenczy, and M. Albrecht, *Organometallics*, **28**, 5112, 2009.

18. R. H. Crabtree, *New J. Chem.*, **35**, 18, 2011.

19. V. Lyaskovskyy and B. de Bruin, *ACS Catal.*, **2**, 270, 2012; V. K. K. Praneeth, M. R. Ringenberg, T. R. Ward, *Angew. Chem. Int. Ed.*, **51**, 10228, 2012; O. R. Luca, and R. H. Crabtree, *Chem. Soc. Rev.*, **42**, 1440, 2013.

20. P. Dydio, W. I. Dzik, M. Lutz, B. de Bruin, and J. N. H. Reek, *Angew. Chem. Int. Ed.*, **50**, 396, 2011.

21. S. Krompiec, M. Pigulla, M. Krompiec, B. Marciniec, and D. Chadyniak, *J. Mol. Catal. A*, **237**, 17, 2005.

22. D. Seyferth, *Organometallics*, **22**, 2, 2003 and references cited.

23. C. P. Casey and C. R. Cyr, *J. Am. Chem. Soc.*, **95**, 2248, 1973.

24. P. A. Dub and T. Ikariya, *ACS Catal.*, **2**, 1718, 2012.

25. P. Meakin, C. A. Tolman, and J. P. Jesson, *J. Am. Chem. Soc.*, **94**, 3240, 1972.

26. R. H. Crabtree, K. Fujita, Y. Takahashi, and R. Yamaguchi, (1,5-Cyclooctadiene) (tricyclohexylphosphine)(pyridine)iridium(I) Hexafluorophosphate, in the *Encyclopedia of Reagents for Organic Synthesis*, Wiley, Hoboken, NJ, 2007.

27. W. J. Tang and X. M. Zhang, *Chem. Rev.*, **103**, 3029, 2003.

28. C. R. Landis and J. Halpern, *J. Am. Chem. Soc.*, **109**, 1746, 1987.

29. J. S. T. Henriksen, P. -O. Norrby, P. Kaukoranta, and P. G. Andersson, *J. Am. Chem. Soc.*, **130**, 10414, 2008.

30. H. U. Blaser, *Top. Catal.*, **53**, 997, 2010 and *Chimia* **64**, 65, 2010 and references cited.

31. S. Siegel, Ruthenium catalysts, in the *Encyclopedia of Reagents for Organic Synthesis*, Wiley, Hoboken, NJ, 2007.

32. R. N. Perutz and S. Sabo-Etienne, *Angew Chem. -Int. Ed. Engl.*, **46**, 2578, 2007.

33. (a) T. Ohkuma, N. Utsumi, K. Tsutsumi, K. Murata, C. Sandoval, and R. Noyori, *J. Am. Chem. Soc.*, **128**, 8724, 2006; (b) [R Noyori and S Hashiguchi, *Acc. Chem. Res.*, **30**, 97, 1997.

34. G. E. Dobereiner, A Nova, N. D. Schley, N Hazari, S. J. Miller, O Eisenstein, and R. H. Crabtree, *J. Am. Chem. Soc.*, **133**, 7547, 2011.

35. C. Wang, X. Wu, and J. Xiao, *Chem. Asian J.*, **3**, 1750, 2008; G. E. Dobereiner and R. H. Crabtree, *Chem. Rev.*, **110**, 681, 2010.

36. R. H. Crabtree, *Chem Rev.*, **112**, 1536, 2012.

37. E. Bayram, J. C. Linehan, J. L. Fulton, J. A. S. Roberts, N. K. Szymczak, T. D. Smurthwaite, S. Ozkar, M. Balasubramanian, and R. G. Finke, *J. Amer. Chem. Soc.*, **133**, 18889, 2011.

38. J. F. Sonnenberg, N. Coombs, P. A. Dube, and Robert H. Morris, *J. Amer. Chem. Soc.*, **134**, 5893, 2012.

39. M. N. Birkholz, Z. Freixa, and P.W. N. M. van Leeuwen, *Chem. Soc. Rev*, **38**, 1099, 2009.

40. T. A. Puckette, *Top. Catal.*, **55**, 421, 2012.

41. B. Marciniec (ed.), *Hydrosilylation*, Springer, Dordrecht, The Netherlands, 2009.

42. M. P. Sibi, *Hydrogen Hexachloroplatinate(IV)* in the *Encyclopedia of Reagents for Organic Synthesis*, Wiley, Hoboken, NJ, 2007.

43. J. Stein and L. N. Lewis, *J. Am. Chem. Soc.*, **121**, 3693, 1999.

44. A. M. Tondreau, C. C. Hojilla Atienza, K. J. Weller, S. A. Nye, K. M. Lewis, J. G. P. Delis, and P. J. Chirik, *Science*, **335**, 567, 2012.

45. J. Ramirez, V. Lillo, A. M. Segarra, and E. Fernandez, *Curr. Org. Chem.*, **12**, 405, 2008.

46. C. C. C. Johansson Seechurn, M. O. Kitching, T. J. Colacot, and V. Snieckus, *Angew. Chem. Int. Ed.*, **51**, 5062, 2012.

47. R. Martin and S. L. Buchwald, *Acct. Chem. Res.*, **41**, 1461, 2008; J. F. Hartwig, *Acct. Chem. Res.*, **41**, 1534, 2008; E. Negishi and L. Anastasia, *Chem. Rev.*, **103**, 1979, 2003; P. Beletskaya and A. V. Cheprakov, *Chem. Rev.*, **100**, 3009, 2000.

48. B. Saito and G. C. Fu, *J. Am. Chem. Soc.*, **130**, 6694, 2008; Z. Li, Y. -Y. Jiang, and Y. Fu, *Chem. Eur. J.*, **18**, 4345, 2012; S. L. Zultanski and G. C. Fu, *J. Am. Chem. Soc.*, **135**, 624, 2013.

49. J. Breitenfeld, O. Vechorkin, C. Corminboeuf, R. Scopelliti, and X. Hu, *Organometallics*, **29**, 3686, 2010.

50. J. Le Bras and J. Muzart, *Chem. Rev.*, **111**, 1170, 2011.

51. J. G. de Vries, *Dalton Trans.*, **2006**, 421.

52. S. S. Zalesskiy and V. P. Ananikov, *Organometallics*, **31**, 2302, 2012.

53. R. Chinchilla and C. Najera, *Chem. Rev.*, **107**, 1979, 2003.

54. S. E. Creutz, K. J. Lotito, G. C. Fu, and J. C. Peters, *Science*, **338**, 647, 2012.

55. C. S. Yeung and V. M. Dong, *Chem. Rev.*, **111**, 1215, 2011.

56. F. Meyer and C. Limberg, Eds., *Organometallic Oxidation Catalysis*, Springer, New York, 2007; S. S. Stahl, *Science*, **309**, 1824, 2005.

57. U. Hintermair and R. H. Crabtree, *J. Organometal. Chem.*, 2014, in press.

58. K. M. Gligorich and M. S. Sigman, *Chem. Commun.*, **2009**, 3854.

59. M. M. Konnick, N. Decharin, B. V. Popp, and S. S. Stahl, *Chem. Sci.*, **2**, 326, 2011; Y. Izawa, D. Pun, and S. S. Stahl, *Science*, **333**, 209, 2011.

60. R. A. Sheldon, I. W. C. E. Arends, G. J. Ten Brink, and A. Dijksman, *Acct. Chem. Res.*, **35**, 774, 2002.

61. N. S. Lewis and D. G. Nocera, *Proc. Nat. Acad. USA*, **103**, 15729, 2006.

62. J. D. Blakemore, N. D. Schley, G. Olack, C. D. Incarvito, G. W. Brudvig, and R. H. Crabtree, *Chem. Sci.*, **2**, 94, 2011; U. Hintermair, S. M. Hashmi, M. Elimelech, and R. H. Crabtree, *J. Am. Chem. Soc.*, **134**, 9785, 2012, and **135**, 10837, 2013.

63. M. Zhou, D. Balcells, A. Parent, R. H. Crabtree, and O. Eisenstein, *ACS Catalysis*, **2**, 208, 2012.

64. C. J. Gagliardi, A. K. Vannucci, J. J. Concepcion, Z. F. Chen, and T. J. Meyer, *Energy Environ. Sci.*, **5**, 7704, 2012.

65. (a) N. E. Leadbeater and M. Marco, *Chem. Rev.*, **102**, 3217, 2002; (b) Special Issue on Recoverable Reagents: *Chem. Rev.*, **102**, 3215–3892, 2002.

66. V. Polshettiwar and R. S. Varma, *Green Chem.*, **12**, 743, 2010.

67. J. M. Basset, C. Copéret, D. Soulivong, M. Taoufik, and J. T. Cazat, *Acc. Chem. Res.*, **43**, 323, 2010.

68. E. Groppo, C. Lamberti, S. Bordiga, G. Spoto, and A. Zecchina, *Chem. Rev.*, **105**, 115, 2005.

69. (a) L. Stegbauer, F. Sladojevich and D. J. Dixon, *Chem. Sci.*, **3**, 942, 2012; (b) D. C. Leitch, Y. C. Lam, J. A. Labinger, and J. E. Bercaw, *J. Am. Chem. Soc.*, **135**, 10302, 2013.

70. T. T. Dang, F. Boeck, and L. Hintermann, *J. Org. Chem.*, **76**, 9353, 2011.

PROBLEMS

It can be useful to work backwards from the product by identifying reactant-derived fragments to see how they might be assembled by *standard* organometallic steps.

9.1. Compound **9.31** is hydrogenated with a number of homogeneous catalysts. The major product in all cases is a ketone, $C_{10}H_{16}O$, but small amounts of an acidic compound $C_{10}H_{12}O$, **9.32**, are also formed. What is the most reasonable structure for **9.32**, and how could it be formed?

9.31

9.2. Would you expect Rh(triphos)Cl to be a hydrogenation catalyst for alkenes (triphos = $Ph_2PCH_2CH_2CH_2PPhCH_2CH_2CH_2PPh_2$)? How might the addition of BF_3 or $TlPF_6$ affect the result?

9.3. Predict the steps in the hydrocyanation of 1,3-pentadiene to 1,5-pentanedinitrile with HCN and $Ni\{P(OR)_3\}_4$.

9.4. Write out a mechanism for arene hydrogenation with $(\eta^3$-allyl)-$Co\{P(OMe)_3\}_3$, invoking initial propene loss. Why do you think arene hydrogenation is so rare for homogeneous catalysts? Do you think that diphenyl or naphthalene would be more or less easy to reduce than benzene? Explain your answer.

9.5. Suggest plausible mechanisms for the reactions shown below, which are catalyzed by a Rh(I) complex, such as $RhCl(PPh_3)_3$.

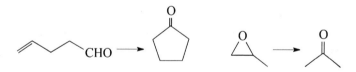

9.6. Comment on the possibility of finding catalysts for each of the following:

$CO_2 \longrightarrow CO + 0.5O_2$

$CH_4 \xrightarrow{0.5O_2} CH_3OH$

9.7. What do you think is the proper structural formulation for H_2PtCl_6? Why do you think the compound is commonly called chloroplatinic *acid*? Make sure that your formulation gives a reasonable electron count and oxidation state.

9.8. In some homogeneous alkyne hydrosilations, a second product (**B**) is sometimes found in addition to the usual one (**A**). How do you think **B** is formed? Try to write a balanced equation for the reaction, assuming an **A/B** ratio of 1:1 and you will see that **A** and **B** cannot be the only products. Suggest the most likely identity for a third *organosilicon* product **C**, which is always formed in equimolar amounts with **B**.

$$RC \equiv CH \xrightarrow{Me_3SiH} \quad R \diagdown \diagdown SiMe_3 \quad + \quad RC \equiv C - SiMe_3$$
$$\qquad\qquad\qquad\qquad\qquad A \qquad\qquad\qquad\qquad B$$

9.9. The following reaction, catalyzed by $(\eta^6\text{-}C_6H_6)$-$Ru(CH_2{=}CHCO_2Et)_2/Na[C_{10}H_8]$ ($Na[C_{10}H_8]$ is simply a reducing agent), has been studied by workers at du Pont as a possible route to adipic acid, an important precursor for Nylon. Suggest a mechanism. How might you use a slightly modified substrate to test your suggestion?

$$\diagup\diagdown COOMe \quad\longrightarrow\quad MeOOC \diagup\diagdown\diagdown COOMe$$

9.10. $(\eta^6\text{-}C_6H_6)Mo(CO)_3$ is a catalyst for the reduction of 1,3-dienes to cis monoenes with H_2; suggest how this might work, why the cis product is formed, and why the alkene is not subsequently reduced to alkane.

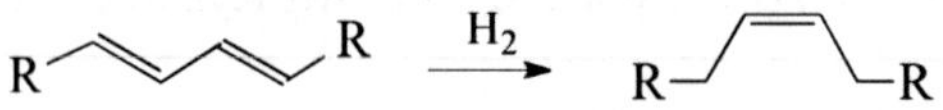

$$R \diagdown\diagdown\diagup R \xrightarrow{H_2} R \diagdown\diagup\diagdown R$$

9.11. A Pd(II) precatalyst with $tBuPPh_2$ as supporting ligand gives a catalyst that, with trimethylsilyl iodide and NEt_3 as coreactants, converts styrene to $PhCH{=}CH(SiMe_3)$. Propose a mechanism and explain the role of the amine. (R. McAtee, S. E. S. Martin, D. T. Ahneman, K. A. Johnson, and D. A. Watson, *Angew. Chem. Int. Ed.*, **51**, 3663, 2012.)

10

PHYSICAL METHODS

We now look at spectroscopic and crystallographic methods for identifying a new complex, assigning its stereochemistry, and learning about its properties.*

10.1 ISOLATION

Isolation and purification procedures closely resemble those of organic chemistry. Most organometallics are solids at 20°, although some are liquids, for example, $CH_3C_5H_4Mn(CO)_3$, or even volatile liquids, such as $Ni(CO)_4$. Numerous organometallics are air and water stable and can be handled exactly like organic compounds, but inert atmosphere work is sometimes required, notably for the electropositive f-block, and early d-block metals. In those cases, air and water must be completely absent. Typical methods involve flasks and filter devices fitted with ground joints for making connections and vacuum taps for removing air or admitting

---

*Undergraduates taking this course may not have had a physical chemistry course. The material on spectroscopy has therefore been gathered together here, so that instructors have the option of omitting all or part of it without losing the narrative flow of the rest of the book.

The Organometallic Chemistry of the Transition Metals, Sixth Edition.
Robert H. Crabtree.

nitrogen. In this Schlenk glassware, all operations can be carried out under an inert atmosphere on an ordinary benchtop. As an alternative, operations can be carried out in a N_2-filled inert atmosphere box. Details of these techniques are available in comprehensive monographs.[1]

10.2 ¹H NMR SPECTROSCOPY

Of all spectroscopic techniques,[2] organometallic chemists tend to rely heavily, perhaps too heavily, on NMR spectroscopy. The commonest situation involves observing $I = \frac{1}{2}$ nuclei with sufficient isotopic abundance, such as ^{1}H (~100% abundance), ^{13}C (~1%), ^{31}P (100%), and ^{19}F (100%). Each chemically different nucleus in a molecule normally gives a distinct signal. Any J coupling to adjacent inequivalent $I = \frac{1}{2}$ nuclei can provide evidence about the local environment of the atom in question. Beyond identifying the organic ligands, the ^{1}H NMR technique[3] is particularly useful for metal hydrides, which resonate in an otherwise empty spectral region (0 to -40δ). This unusual chemical shift is ascribed to shielding by the metal d electrons, and the shifts indeed become more negative for higher d^n configurations. The number of hydrides present may be determined by integration or, if phosphines are also present, from $^{2}J(P,H)$ coupling in the ^{31}P NMR spectrum (Section 10.4), where the term $^{n}J(X,Y)$ refers to the coupling of nucleus X and Y through n bonds. For the $^{2}J(P,H)$ coupling of M–H to adjacent PR_3 groups, the fact that trans couplings (90–160 Hz) are larger than cis (10–30 Hz) often allows full stereochemical assignment, as seen in Fig. 10.1 for some Ir(III) hydrides. Similar cis < trans coupling relationships hold for other pairs of NMR-active donor atoms. The 5-, 7-, 8-, and 9-coordinate hydrides are often *fluxional* so that the ligands exchange positions within the complex sufficiently fast to become equivalent on the NMR timescale (~10^{-2} s). We look at some consequences of fluxionality later (Section 10.5).

^{2}H NMR spectroscopy is useful for following the fate of deuterium in mechanistic experiments. Even though D is an $I = 1$ nucleus, the ^{2}H spectrum is still obtainable but has broader resonances than for ^{1}H. The chemical shifts are essentially identical to those seen in the ^{1}H NMR spectrum, however, which greatly simplifies the interpretation, but all J coupling to ^{2}H are reduced by a factor of 6.5 versus ^{1}H because of the lower gyromagnetic ratio for ^{2}H.

Virtual Coupling

Virtual coupling in the ^{1}H NMR spectrum can help geometry assignments for complexes involving phosphines such as PMe_3 or PMe_2Ph. If

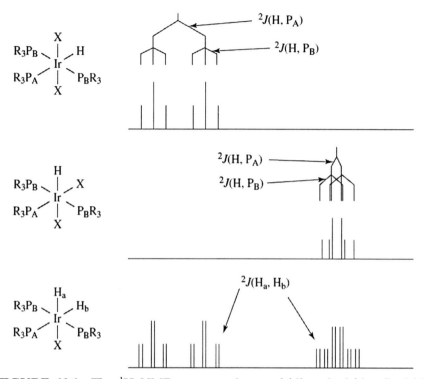

FIGURE 10.1 The ¹H NMR spectra of some iridium hydrides (hydride region). Each stereochemistry gives a characteristic coupling pattern.

two such phosphines are cis, they behave independently, and we see a $^2J(P,H)$ doublet for P–*Me*. If they are trans, the $^2J(P,P')$ coupling coupling becomes so large that the ¹H NMR of the P–*Me* unit is affected. Instead of a simple doublet, we see a distorted triplet with a broad central peak giving the appearance that the P–*Me* is coupled to P and P' about equally (Fig. 10.2*a*). Intermediate P–M–P angles between 90° and 180° give intermediate patterns (Fig. 10.2*b* and 10.2*c*).

Diastereotopy

The ¹H NMR spectrum of a PMe₂Ph ligand in **10.1** and **10.2** can provide stereochemical assignments from symmetry (Fig. 10.3). In **10.1**, a mirror plane containing M, X, Y, and the PMe₂Ph phosphorus reflects one P–Me group into the other and makes them equivalent; **10.2** lacks such a plane of symmetry, and the inequivalent Me' and Me" groups are termed *diastereotopic*.[3a] In general, two groups will be inequivalent if no symmetry element of the molecule exchanges one with the other.

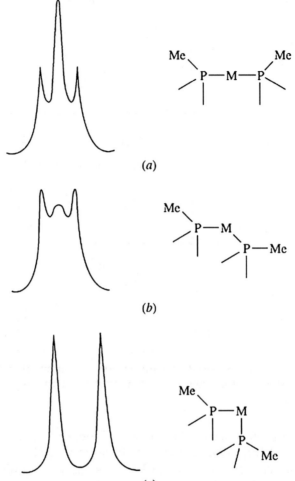

FIGURE 10.2   Virtual coupling in the P*Me* proton resonance of methylphos-phine complexes. Each methyl group shows coupling both to P and to P′ as long as $^2J(P, P')$ is large enough. As the *Me*P–M–P′ angle decreases from 180°, the virtual coupling decreases, until at an angle of 90°, we see a simple doublet, owing to coupling of the P*Me* protons only to P, not to P′. At intermediate angles the spectrum takes up a ghostly appearance (case *b*).

Diastereotopic groups are inequivalent and generally resonate at dif-ferent chemical shifts. We will therefore see a $^2J(P,H)$ doublet for **10.1** and a pair of $^2J(P,H)$ doublets for **10.2**. The appearance of the spectrum changes on moving to a higher field spectrometer (Fig. 10.3) because the diastereotopic resonances differ by a certain chemical shift in parts

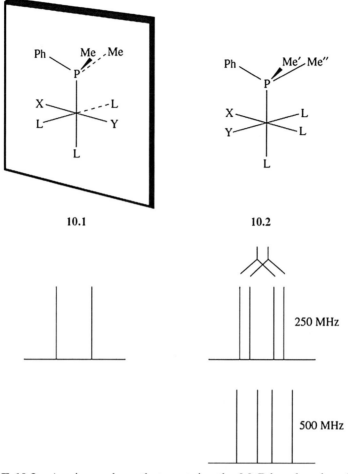

FIGURE 10.3 A mirror plane that contains the M–P bond makes the P*Me* groups in **10.1** equivalent by ¹H NMR so they appear as a single ²*J*(P, H) doublet. P*Me* and P*Me′* groups in **10.2** are inequivalent (diastereotopic) and so resonate at different frequencies. The two distinct doublets that result do not appear the same at a higher field and so are distinguishable from a doublet of doublets due to coupling, the appearance of which would be invariant with field.

per million (ppm), while ²*J*(P,H) coupling has a constant value in hertz; the pattern therefore changes at higher field, where there are more hertz per ppm. The same inequivalence is found for any compound (e.g., **10.3**) in which no element of symmetry can make the two groups equivalent. The inequivalence is independent of M–P or C–C bond rotation—only one rotamer need show the effect.

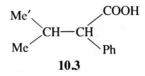

10.3

Chemical Shifts

In organic compounds, certain chemical shift ranges are diagnostic for certain groups, but in organometallic chemistry, the shifts are much more variable. For example, the vinyl protons of a coordinated alkene can resonate anywhere from 2 to 5δ (free alkene: $5-7\delta$). In the MCP (X_2) extreme (Section 5.1), the shifts are at the high-field end of the range, closer to those in cyclopropane, but in the opposite D–C (L) extreme, they are closer to those in the free alkene, near 5δ. Hydride resonances are even more variable. In Ir(III) complexes, they tend to depend on the nature of the trans ligand and can range from -10δ, for high-trans-effect ligands, (e.g., H) to -40δ, for low-trans-effect ligands (e.g., H_2O). Structural assignments based on coupling constants tend to be more secure than ones based on chemical shifts, however. Signals from common impurities need to be identified to avoid misleading interpretations.[4]

Paramagnetic NMR

Metal complexes can be paramagnetic, and this can lead to large shifts in the NMR resonances;[3b] for instance, $(\eta^6-C_6H_6)_2V$ (**10.4**) has a ^1H NMR resonance at 290δ. The Cr–Cr bonded dimer $[CpCr(CO)_3]_2$ has a somewhat broadened proton NMR spectrum because the dimer partially dissociates to give the paramagnetic, 17e $[CpCr(CO)_3]$ monomer. Assignments of resonances in paramagnetic complexes is becoming easier thanks to computational and experimental advances.[5] These resonances can be broadened to such an extent that they become effectively unobservable, however, so a featureless NMR spectrum does not necessarily mean that no organometallic complexes are present.

10.3 ^{13}C NMR SPECTROSCOPY

M–C resonances in alkyl complexes appear from -40 to $+20\delta$, π-bonded carbon ligands such as alkenes, Cp, and arenes from $+40$ to $+120\delta$, carbonyls around $150-220\delta$ (terminal) and $230-290\delta$ (bridging), and carbenes come in the range $200-400\delta$.[6] Relaxation (Section 10.7) of the ^{13}C nuclei, especially in M–CO, may be slow, which makes them difficult to observe unless a relaxation reagent such as paramagnetic, d^3 Cr(acac)$_3$ is present. Since the dynamic range of the method greatly exceeds that of ^{1}H NMR, the ^{13}C peaks for different carbons in a complex are normally much farther apart in frequency (hertz) than the corresponding ^{1}H peaks. This means that the spectra of complicated molecules are much easier to assign because overlapping of peaks is less likely and also that slower fluxional processes (Section 10.5) can be studied. Coupling is transmitted by the σ bonds of a molecule—the higher the s character of a bond, the higher the coupling. Thus ^1J(C,H) values depend on the C–H hybridization: sp^3, ~125 Hz, sp^2, ~160 Hz, and sp, ~250 Hz. Trans couplings are larger than cis ones, for example, in, ^2J(C,P) is ~100 Hz for trans Me–M–PR$_3$ groups, but only ~10 Hz for the analogous cis couplings.

By off-resonance decoupling, the ^{13}C spectrum shows only ^1J(C,H) couplings, that is, couplings to H atoms directly bound to the carbon. This procedure allows a distinction to be made between CH$_3$, CH$_2$ or CH groups, which give a quartet, a triplet, or a doublet, respectively. The structure can often be deduced in this way even when the ^{1}H NMR spectrum is too complex to decipher, as was the case for **10.5** and **10.6**, although these were only obtainable as an inseparable mixture. Beyond the PPh$_3$ resonances, each complex showed two quartets, two triplets, two doublets, and a singlet in the off-resonance ^{1}H-decoupled ^{13}C spectrum. These were uniquely assigned as shown.

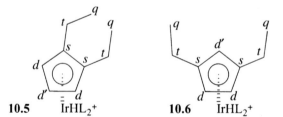

10.5 $\overline{\text{I}}$rHL$_2^+$ **10.6** IrHL$_2^+$

Integration of carbon spectra can be unreliable for carbons lacking H substituents, because of their long relaxation times. This means that the nuclei are easily saturated and intensities are low, but a relaxation reagent or a relaxation delay (e.g. 5 s) can be introduced in acquiring the spectrum. If the sample is concentrated enough, it is sometimes possible to obtain usable spectra with a single pulse, where relaxation is no longer a problem.

In polyene and polyenyl complexes, carbons directly attached to the metal tend to be more shielded on binding, and a coordination shift (i.e., relative to the free ligand) of ~25 ppm to high field is common. Metal nuclei with $I = ½$ show coupling to the metal in ^{1}H and ^{13}C spectra. Diastereotopy also applies in ^{13}C spectra and is seen for the diastereotopic P–*Me* carbons in **10.2**.

10.4 ^{31}P NMR SPECTROSCOPY

In ^{31}P NMR studies of phosphine complexes, the ligand protons are normally all decoupled to simplify the spectra.[6] Different types of ligand normally resonate in different chemical shift ranges, so that phosphines and phosphites can be reliably distinguished, for example. Free and bound P-donors also show large coordination shifts that are useful in characterizing cyclometallated phosphines and mono-dentate diphosphines, otherwise hard to do other than by crystal-lography. If the phosphorus is part of a four-, five-, or six-membered ring, as in a cyclometalation product, chelation shifts of -50, $+35$, or -15 ppm are seen relative to a similar, coordinated but nonche-lating phosphine ligand because the ring size affects the hybridiza-tion at phosphorus.

Mechanistic Study of Wilkinson Hydrogenation

Tolman[7] was able to look by ^{31}P NMR at events related to the mecha-nism of Wilkinson hydrogenation (Fig. 10.4, Eq. 10.1 and Section 9.3). Spectrum A shows the ^{1}H-decoupled ^{31}P NMR of RhCl(PPh$_3$)$_3$ itself. Two types of phosphorus are seen in a 2:1 ratio, P$_a$ and P$_b$ in **10.7**, each showing coupling to Rh ($I = ½$, 100% abundance). P$_a$ also shows a cis coupling to P$_b$, and P$_b$ shows two cis couplings to the two P$_a$s. On adding H$_2$ (spectrum B), the starting material almost disappears and is replaced by a new species, **10.8**, in which only P$_a$ now couples cleanly to Rh, and P$_b$ is a broad hump. Slowing the exchange by cooling to $-25°$ (spectrum B$'$) restores the coupling pattern for static **10.8**. The change from B to B$'$ is the result of P$_b$ dissociating at a rate that is slow at $-25°$ but comparable with the NMR timescale at $+30°$ (Section 10.5). In spectrum B, P$_a$ retains clean coupling to Rh and must remain bound, while P$_b$ does not, so P$_b$ must be dissociating. The reason for the loss of coupling is that two coupled nuclei that stay together during fluxionality retain their mutual J-coupling, but with fast dis-sociation, we have crossover of PR_3 between Rh centers so $^1J(P,Rh)$ coupling is lost by averaging. Each of the two peaks of P$_a$ doublet in spectrum B comes from a different population of molecules, one with the Rh spin α and the other with β spin. When P$_b$ moves from mole-cule to molecule, it samples α and β Rh spins equally and so the whole population of P atoms ends up resonating at an averaged chemical shift. The amount of *free* PPh$_3$ always remains very small—the arrows show where free PPh$_3$ would appear. Passing N$_2$ partially reverses the reaction by sweeping out H$_2$ and a mixture of **10.7** and **10.8** results (spectrum C).

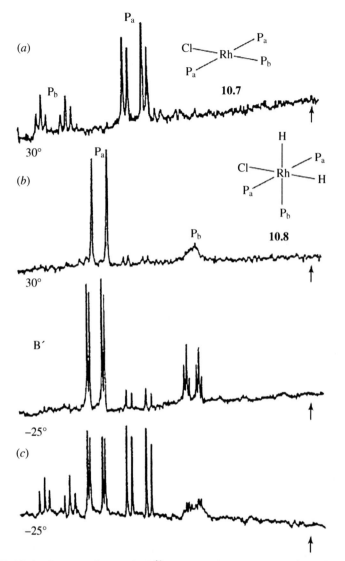

FIGURE 10.4 Proton-decoupled ^{31}P NMR data for RhCl(PPh$_3$)$_3$: (A) dissolved in CH$_2$Cl$_2$; (B) after addition of H$_2$ at 30°; (B′) after addition of H$_2$ and cooling to −25°; (C) after sweeping solution B with N$_2$. The different P nuclei in the complex are seen, together with ^1J(P,Rh) coupling and ^2J(P$_a$,P$_b$) couplings (small). In spectrum B, the loss of ^1J(P$_b$,Rh) coupling indicates that R$_3$P$_b$ is reversibly dissociating. Free PPh$_3$ (arrow) is absent. Source: From Meakin et al., 1972 [7]. Reproduced with permission of the American Chemical Society.

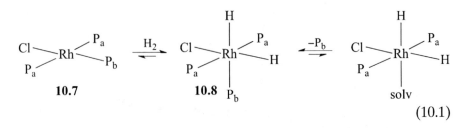

$$(10.1)$$

10.5 DYNAMIC NMR

When organometallic species give fewer NMR resonances than would be predicted from their static structures, molecular nonrigidity may be the cause. If the nuclei concerned are exchanging places at a rate much faster than the NMR timescale ($\sim 10^{-1}$–10^{-6}s), then a sharp averaged resonance results. For example, $Fe(CO)_5$ gives only one carbon resonance at 25°, and yet its IR spectrum—a technique with the much faster timescale of $\sim 10^{-12}$—indicates a TBP structure with two types of carbonyl. Axial and equatorial carbonyls exchange easily by the *Berry pseudorotation* mechanism of Eq. 10.2. Ligands 1–4 become equivalent in the square pyramidal intermediate, and 1 and 4, which were axial in TBP, become equatorial in TBP′.

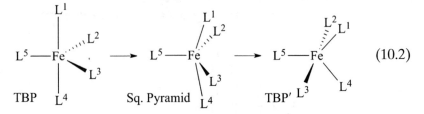

$$(10.2)$$

Rate of Fluxionality

Sometimes an exchange takes place at a rate that is comparable with the NMR timescale. When this happens, we can slow the exchange by cooling the sample until we see the static spectrum at the *low-temperature limit*. On the other hand, if we warm the sample, the rate of exchange can rise so as to give the fully averaged spectrum at the *high-temperature limit*. In between these extremes, broadened resonances are seen. Take a molecule with two sites A and B that are equally populated: on warming, we will see the sequence of spectra illustrated in Fig. 10.5. The two sharp peaks broaden as the temperature rises. If we measure this initial broading at half peak height in units of hertz, and subtract out the natural linewidth that was present before broadening

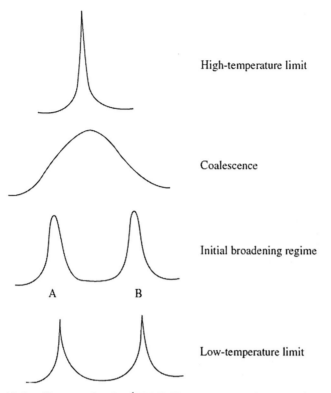

FIGURE 10.5 Changes in the ¹H NMR spectrum of a two-site system on warming of the H_A and H_B protons begin to exchange at rates comparable with the NMR timescale.

set in, then we have $W_{1/2}$, a measure by Eq. 10.3 of the rate at which the nuclei leave the site during the exchange process.

$$\text{Rate} = \pi(W_{1/2}) \qquad (10.3)$$

As we continue to warm the sample, the broadening increases until the two peaks *coalesce*. According to Eq. 10.4, the exchange rate required to do this depends on $\Delta\nu$, the separation of the two resonances of the static structure.

$$\text{Rate} = \frac{\pi\Delta\nu}{\sqrt{2}} \qquad (10.4)$$

On further warming, the single coalesced peak gets narrower according to Eq. 10.5, and we finally reach a point at which the signal is sharp once more.

$$\text{Rate} = \frac{\pi(\Delta v)^2}{2(W_{1/2})} \tag{10.5}$$

This happens because the exchange is now much faster than the NMR timescale and only an averaged resonance is seen. Note that Eq. 10.4 and Eq. 10.5 contain $\Delta \nu$, the separation of the two resonances in Hz. Since this is different at different magnetic fields, the coalescence temperature and the high-temperature limit are field dependent. A $\Delta \delta$ of 1 ppm translates to 400 Hz at 400 MHz, but to 600 Hz at 600 MHz. On cooling, *decoalescence* causes the same changes to occur in reverse. The position of the averaged resonance at the high-temperature limit is simply the weighted average of the resonance positions at the low-temperature limit. For example, if we have n_1 nuclei resonating at δ_1 and n_2 at δ_2, then at the high-temperature limit, the resonance position will be the weighted average δ_{av}, given by Eq. 10.6.

$$\delta_{av} = \frac{n_1\delta_1 + n_2\delta_2}{n_1 + n_2} \tag{10.6}$$

Dynamic NMR is a very powerful method for obtaining kinetic information about processes that occur at a suitable rate, typically ones having a barrier in the 12–18 kcal/mol range.

Ligand Fluxionality

Beyond fluxionality of the metal geometry, the ligand can also be fluxional. The classic example, $CpFe(CO)_2(\eta^1\text{-}C_5H_5)$ (Fig. 10.6), shows only two proton resonances at room temperature, one for the $\eta^5\text{-}C_5H_5$ and one for the fluxional $\eta^1\text{-}C_5H_5$. The iron atom migrates around the $\eta^1\text{-}C_5H_5$ ring at a rate sufficient to average all the proton environments of the ring. On cooling, separate resonances appear for the three different proton environments in the low-temperature limiting spectrum of the static $\eta^1\text{-}C_5H_5$. On warming, each signal broadens, but in a different way depending on whether the fluxionality involves 1,2 or 1,3 shifts. Since the H_C protons are adjacent, a 1,2 shift—equivalent to a 1,5 shift—will result in one of the H_C nuclei staying in an H_C site after the shift; in contrast, all the H_B nuclei will end up in non-H_B sites. The exchange rate for H_C will therefore be one-half of the exchange rate for H_B, and thus show less initial broadening. Conversely, since H_B nuclei are three carbons apart, 1,3 shifts will result in the H_B signal showing less initial broadening. Experimentally, a 1,2 shift in fact takes place.[13a] To do this analysis, however, we first need to assign H_B and H_C signals correctly— often a hard step.

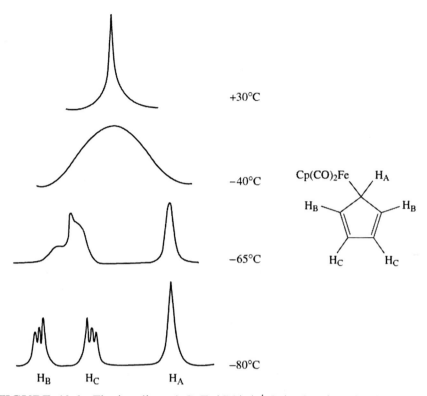

FIGURE 10.6 Fluxionality of $CpFe(CO)_2(\eta^1\text{-}Cp)$, showing the faster collapse of the H_B resonance, indicating a 1,2 rather than a 1,3 fluxional shift. Only the resonances for the η^1 Cp group are shown.

For Cp, it is impossible to distinguish between a Woodward–Hoffmann orbital symmetry-allowed 1,5 shift and a least-motion 1,2 shift because they are equivalent. In an $\eta^1\text{-}C_7H_7$ system, the two cases are distinguishable, Woodward–Hoffman giving a 1,4 and least motion a 1,2 shift. A similar analysis shows that $(\eta^1\text{-}C_7H_7)Re(CO)_5$ follows a least motion and $\eta^1\text{-}C_7H_7SnMe_3$ a Woodward–Hoffmann path.

Another important case of fluxionality is bridge-terminal exchange in carbonyl complexes. The classic example is $[CpFe(CO)_2]_2$, which shows separate Cp resonances for cis and trans CO-bridged isomers in the ^1H NMR below $-50°C$, but one resonance at room temperature owing to fast exchange.

10.6 SPIN SATURATION TRANSFER

When fluxional exchange is too slow to detect by the methods of Section 10.5, we may still be able to use *spin saturation transfer*.[6,8] To do this, we irradiate one of the resonances of the two exchanging species or sites and watch for the effects on the signal for the other species or site. For example, if we irradiate the Me_A protons in **10.9a** in Eq. 10.7, we see a diminution in the intensity of the resonance for Me_B in **10.9b**. This shows that Me_A in **10.9a** becomes Me_B in **10.9b** in the course of a slow exchange; likewise, irradiation of H_C affects the intensity of the H_D. In this way, we can obtain mechanistic information about this alkene isomerization process.

$$(10.7)$$

10.9a $\quad ReH_3L_2$ $\qquad\qquad$ **10.9b** $\quad ReH_3L_2$

By irradiating the Me_A protons, we equalize their α and β spin populations. If Me_A protons now become Me_B protons by exchange, then they carry the memory of the equalized populations. Since we need unequal α and β populations in order to observe a spectrum, the newly arrived Me_B protons do not contribute their normal amount to the intensity of the resonance. These new Me_B protons begin to lose their memory of the original, artificially equalized α- and β-spin populations over a few seconds by *relaxation*. The initially equal populations in the newly arriving protons relax back to the equilibrium population ratio with a rate $1/T_1(B)$, where $T_1(B)$ is the spin lattice relaxation time, or T_1, of the Me_B site; the T_1 data must be measured independently. The exchange rate has to be faster than $\sim 10 T_1$, or >0.1 s^{-1}, to give a measurable effect. If the initial intensity of the B resonance is I_0, the relaxation time of the B protons is $T_1(B)$, and the final intensity of the B resonance on irradiating the A resonance is I_f, then the exchange rate k is as given by Eq. 10.8.

$$\frac{I_0}{I_f} = \frac{\{T_1(B)\}^{-1}}{k + \{T_1(B)\}^{-1}} \tag{10.8}$$

By learning which protons exchange with which, we can solve some difficult mechanistic problems. The nuclear Overhauser effect (NOE) (Section 10.7) can affect the experimental outcome and must also be taken into account.

10.7 T_1 AND THE NUCLEAR OVERHAUSER EFFECT

To determine the T_1 for any signal, we first put our sample in the magnetic field of the NMR spectrometer.[6] If z is the direction of the applied magnetic field, then the nuclei will line up with and against the field. The ΔE between these two states being small, the excess of the more stable α spins is very slight, but enough to give a net sample magnetization along $+z$ (Fig. 10.7). If a $90°$ radio-frequency pulse is now applied, the magnetization vector rotates precisely into the xy plane. We can only measure the magnetization in the xy plane where the vector now rotates around the z axis at the Larmor frequency; the oscillating magnetic field due to the rotating magnetization generates a signal in the receiver coil of the instrument. This is the conventional Fourier transform (FT) NMR experiment.

In the inversion/recovery method for determining T_1, we apply a $180°$ pulse that inverts the spins and moves the magnetization from the $+z$ to the $-z$ direction. The original slight excess of α spins is now converted into a slight excess of β spins. We now wait for a variable time, t, to allow relaxation to convert the new nonequilibrium distribution favoring β back to the old one favoring α. In separate experiments, we can sample the spins with a $90°$ pulse after different times, t, to put the magnetization back into the xy plane, where we can follow the path to recovery (Fig. 10.7). The negative peaks at short times reflect the inverted spin populations at those times; at longer times, the resonances become positive and the populations eventually completely recover by a first-order process with rate constant $1/T_1$. The spectrometer software automatically calculates T_1 on request.

T_1 and H_2 Complexes

T_1 data helps distinguish between molecular hydrogen complexes, **10.10**, and classical dihydrides, **10.11**. Two protons that are very close together can relax one another very efficiently by the dipole–dipole mechanism. Dipole–dipole couplings are several orders of magnitude larger than the usual J couplings we see as splitting in the normal NMR spectrum. We do not see dipole–dipole splittings in the normal spectrum, however, because they average exactly to zero with the tumbling of the molecule in solution. Although we cannot see the effects of dipole–dipole coupling directly, it is nevertheless the principal mechanism for spin relaxation in most cases. The random tumbling of the molecule in solution causes one nucleus, say, H_A, to experience a randomly fluctuating magnetic field due to the magnetic field of a nearby nucleus, H_B, that is rotating around H_A with the tumbling of the molecule. If these fluctuations happen to occur at the Larmor frequency, then H_A can undergo a spin flip, and the α and β spins are eventually brought to thermal equilibrium, or relaxed. Relaxation is important because to see an NMR signal we need a difference in the populations

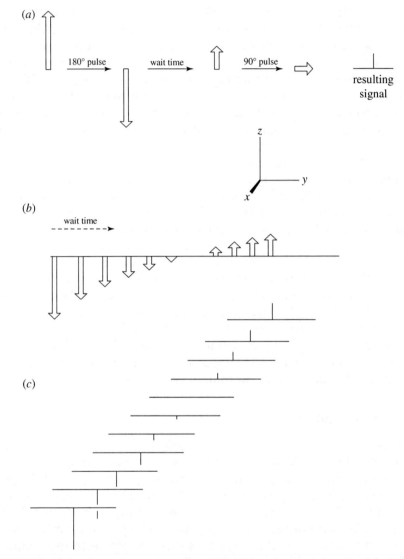

FIGURE 10.7 Inversion recovery method for determining T_1. (*a*) A 180° pulse inverts the spins. They partially recover during the wait time and are sampled by a 90° pulse. (*b*) Varying the wait time allows us to follow the time course of the recovery process, as seen in a stacked plot of the resulting spectra (*c*).

of α and β spins—when the populations are equal in Fig. 10.7, there is no signal. Observing the signal pumps energy into the spins and tends to equalize their populations—relaxation drains energy from the spins and tends to reestablish the population difference. Careful analysis of the T_1 data at variable temperature can give the H–H distance and thus distinguish **10.10**, typically having a T_1 of ~30 ms, as in **10.12**, from **10.11**, with a value of 300 ms or more.

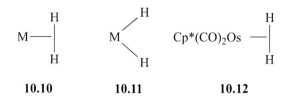

10.10	**10.11**	**10.12**

PHIP and SABRE

To see PHIP,[10] or para-hydrogen-induced polarization, we first cool a sample of H_2 with a catalyst so that the H_2 becomes enriched in the slightly more stable p-H_2 in which the two nuclear spins are aligned.[9] If a hydrogenation reaction is now carried out in an NMR tube with p-H_2 enriched gas, the two hydrogens may be transferred together to a substrate. Their spin alignment in p-H_2 is also transferred to the hydrogenation product, which results in an extremely nonthermal distribution of spins in the product, and this in turn leads to very large enhancements of the resonances.

The effect decays with rate $1/T_1$, so the T_1 of the protons in the product must not be very short. Traces of a metal dihydride in equilibrium with H_2, even if undetectable by standard NMR, may be seen using PHIP. In a related technique, signal amplification by reversible exchange, SABRE, signal amplifications of 1000-fold are possible.[11]

Nuclear Overhauser Effect

NOE spectroscopy is an NMR technique for determining the conformation of a molecule in solution.[6] NOE is observed for any two nuclei, say, H_A and H_B, that are close enough to relax each other by the dipole–dipole mechanism. For this, the two nuclei need to be <3 Å apart. Distance is the only criterion—no bonds are needed.

Irradiating H_A, while observing H_B, can ideally lead to an increase in the intensity of the H_B resonance by as much as 50% with NOE, but usually only by 5–10%. In a typical application, NOE is expected in only one of two related isomers. For example, H_A and H_B in **10.13**, but not in **10.14**, show NOE, leading to the structural assignments shown and later confirmed crystallographically (R = C_2H_5).

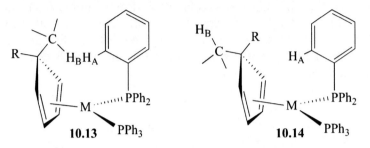

10.13   PPh₃   **10.14**   PPh₃

By irradiating H_A, we equalize its α and β spin populations. Dipole–dipole relaxation then transfers some of the increased spin population in the upper β state of H_A to the lower α state of H_B and consequently increases the intensity of the H_B resonance. The enhancement is measured by the NOE factor, η, given by Eq. 10.9, where I_0 and I_f are the initial and NOE-enhanced intensities, respectively.

$$\eta = \frac{I_f - I_0}{I_0} \tag{10.9}$$

10.8   IR SPECTROSCOPY

Bands in the IR spectrum correspond to vibrational modes of a molecule.[12] The position of the band, v, depends (Eq. 10.10, where c = the velocity of light) on the strength of the bond(s) involved as measured by a force constant k, and on the reduced mass of the system, m_r. Equation 10.11 shows the reduced mass calculated for a simple diatomic molecule, where m_1 and m_2 are the atomic weights of the two atoms:

$$v = \frac{1}{2\pi c}\left\{ \sqrt{\frac{k}{m_r}} \right\} \tag{10.10}$$

$$m_r = \frac{m_1 m_2}{m_1 + m_2} \tag{10.11}$$

The band intensity (I) depends on the dipole moment change during the vibration, $d\mu/dr$. This is big for polar bonds such as OH, NH, or $R_2C{=}O$, smaller for lower polarity bonds such as C–H, but zero for free H_2 or N_2, where no signal is seen.

The most intense, high energy IR bands therefore arise from light atoms being bound together in strong, polar bonds, for example, HF, H_2O or MeC≡N.

Carbonyls

Infrared spectroscopy is especially useful for metal carbonyls because the intense C=O stretching vibration at 1700–2100 cm^{-1} appears a region that is relatively free of other bands. The intensity is large thanks to the polarization on binding (M–C$^{\partial+}$=O$^{\partial-}$) and consequent large $d\mu/dr$. In polycarbonyls, the $\nu(CO)$ bands are coupled in a way that depends on the symmetry of the M(CO)$_n$ fragment.

In an octahedral trans dicarbonyl, coupling leads to the spectrum of Fig. 10.8a. The COs may vibrate in phase, in which case they both

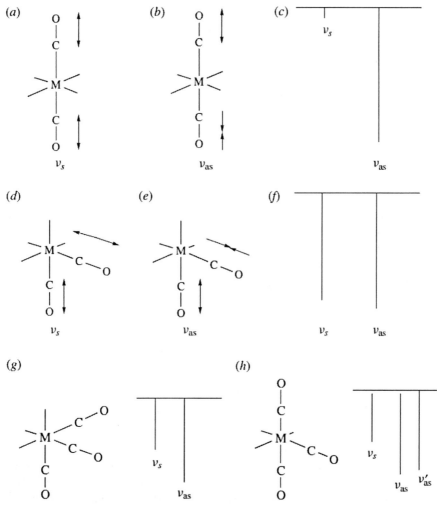

FIGURE 10.8 Effect of the structure of a metal carbonyl on the IR absorption pattern observed.

stretch simultaneously (ν_s, Fig. 10.8a), or they may vibrate out of phase (ν_{as}, 10.8b), in which case one stretches when the other compresses. Although there are two CO bands and two COs, both COs contribute to both bands.

The in-phase, or symmetric, vibration, ν_s, appears at higher frequency than ν_{as} because it is harder to stretch both COs at once than in alternation. On stretching, each CO becomes a better π acceptor, so it is easier for the metal to satisfy each CO by back bonding if they stretch alternately, rather than simultaneously. The intensity of the in-phase vibration is low because the dipoles of the two COs are opposed. The absorption does not have zero intensity because of mixing with other, allowed vibrations. The spectrum (Fig. 10.8a) therefore has an intense band at lower energy and a weak band at higher energy. A cis dicarbonyl shows the same two bands, but now with approximately equal intensity, because ν_s now has a large $d\mu/dr$. The relationship between the ratio of the intensities and θ, the angle between the two COs, is shown in Eq. 10.12.

$$\frac{I_s}{I_{as}} = \cot^2 \theta \qquad (10.12)$$

Octahedral tricarbonyls can be facial (*fac*) or meridional (*mer*); tetracarbonyls can be cis or trans, where these labels now refer to the geometry of the noncarbonyl ligands; only one isomer occurs for penta- and hexacarbonyls. In each case, a characteristic pattern of IR bands allows us to identify the isomer; Fig. 10.8g and h show the spectra expected for the two tricarbonyl isomers.

The pattern moves to higher or lower frequency with change of net ionic charge, noncarbonyl ligands, or of the metal. For example, a net negative charge, or more strongly donor ligands, or a more π-basic metal give more back bonding and so weaken the C=O bond. This shifts the IR frequencies to lower energy, which means to lower wavenumber (Table 2.10).

Other Ligands

Hydrides often show $\nu(M–H)$ bands, but the intensities can be very low as the polarity of the bond is usually small. Carboxylates can be chelating or nonchelating, and the IR data helps distinguish the two cases. Complexes of CO_2, SO_2, NO, and other oxygen-containing ligands give intense bands that are often useful in their identification. Oxo ligands give very intense bands around $500–1000 \text{ cm}^{-1}$, but the usual polyenes and polyenyls do not give very characteristic absorptions. In an agostic

C–H system, the bond is sometimes sufficiently weakened to give a band at $\sim$2800 cm^{-1}. Dihydrogen complexes sometimes give a similar band at 2300–2700 cm^{-1}, but in this case, we again rely on mixing to gain intensity and the band is completely absent in some cases.

Band Identification by Isotope Substitution

We may need to assign a given IR band as arising from a specific bond. For example, a weak band at 2000 cm^{-1} might be a ν(M–H), or there might be a small amount of a CO complex present. This kind of problem is solved by isotopic substitution. If we repeat the preparation with deuteriated materials, then we will either see a shift of the band to lower frequency, in which case we have a ν(M–H,D) stretch, or not, in which case ν(CO) is likely; if so, the band should then shift appropriately in the ^{13}CO analog. The shift can be estimated by calculating the reduced masses of the normal and isotopically substituted systems from Eq. 10.11, assuming that L$_n$M can be assigned infinite mass, and deducing the shift from Eq. 10.10. In the case of a ν(M–H) at 2000 cm^{-1}, the ν(M–D) will come at $2000/\sqrt{2} = 1414$ cm^{-1}.

Raman Spectroscopy

This is rarely applied to organometallic species in part because laser irradiation can cause complexes to decompose, but the method is in principle useful for detecting nonpolar bonds, which do not absorb, or absorb only weakly in the IR.[13] The intensity of the Raman spectrum depends on the change of polarizability of the bond during the vibration. One of the earliest uses was to detect the Hg–Hg bond in the mercurous ion [ν(Hg–Hg) = 570 cm^{-1}], a case where the polarizability change is large as a result of multielectron atoms being involved. Unlike IR spectroscopy, the method is compatible with measurements in aqueous solution.

10.9 CRYSTALLOGRAPHY

Crystal structure determination is important, particularly for identifying features, such as cyclometalation, that are hard to detect otherwise or in characterizing new ligand binding modes.[14] The three-dimensional structure of a crystal is built from a repetitive arrangement of the simplest structural unit, called the *unit cell*, just as a single tile is often a unit cell for a two-dimensional ceramic tiling pattern. Depending on the nature of the unit cell, Bragg's law is satisfied only at certain orientations of the crystal, and a beam of X-rays will then emerge from the crystal at a certain angle to the incident beam. Bragg's law (Eq. 10.13, where λ is the wavelength of the radiation, 2θ is the angle between the incident and diffracted ray, n is an integer, and d is the spacing of the cells) requires that the diffracted radiation from different layers of unit cells be in phase. The positions and intensities of the diffracted beams are measured by an automated diffractometer. Their positions relative to the incident beam carry the information about the arrangement of the unit cells in space, while the diffraction intensities depend on the nature and arrangement of the atoms in the unit cell.

$$2d \sin \theta = n\lambda \qquad (10.13)$$

Limitations

The X-rays are diffracted by the electrons around each atom. This means that the diffraction pattern is often dominated by the metal in a complex because it usually has a far greater number of electrons than the other atoms present. Conversely, hydrogen atoms may not appear at all because they have so few electrons. Where it is important to know the hydrogen positions, as in metal hydrides, dihydrogen complexes, or in determining the bond angles at carbon in ethylene complexes, neutron diffraction is preferred. Neutrons are diffracted from the nuclei of the atoms and so give precise internuclear distances. All elements have broadly similar ability to diffract neutrons, so that the resulting intensities are not dominated by any one atom, and the positions of all the atoms can therefore be obtained. The hydrogens, and even the heavy atoms in BH_3NH_3 were correctly located only by neutron diffraction.[15] Only a few laboratories are equipped to carry out neutron work, however, and an added inconvenience is the much larger crystal size often required to obtain good data. In contrast, most chemistry departments have an X-ray facility, and a substantial fraction of papers in organometallic chemistry include one or more X-ray structures.

An X-ray structure is often represented in a diagram showing the positions of all the atoms in the molecule (e.g., Fig. 5.7). These have a deceptively persuasive appearance so we have to be aware of the potential pitfalls. Is the crystal representative of the bulk? Minor impurities can crystallize while the major species do not. As a check, an IR spectrum of the specific crystal used for the structure can be compared with the bulk sample. The more difficult question is whether the structure in the solid state is really the same as the structure of the same material in solution, to which the solution reactivity and NMR data correspond. Some organometallics exist as one isomer in solution but as another in the solid state.[16] If several isomers are interconverting, any crystals that form will consist of the tautomer that is least soluble or kinetically fastest to crystallize, not necessarily the most stable. Surprisingly large forces are present within the lattices, especially of ionic crystals; these *packing forces* may change the details of the structure compared with solution. This makes NMR methods of structure determination in solution useful. IR spectroscopy also helps because we can obtain a spectrum both in solution and in the solid state to see if there are any significant differences. NMR spectra on solid-state samples, obtained by "magic angle" spinning, can also show if any changes take place on going from the solution to the solid. Co-crystallization with impurities can also lead to highly deceptive artifacts, such as erroneous bond lengths, as in the misconceived attribution of "bond stretch isomerism" to several series of oxometal halide complexes. Sophisticated detective work showed that the apparently variable M–O bond length that came from the X-ray work was in fact the result of $L_nM=O$ and L_nM-Cl cocrystallizing in different proportions in the lattice.[17]

10.10   ELECTROCHEMISTRY AND EPR

The increasing interest in redox events in organometallics has raised the profile of cyclic voltammetry (CV). Electrochemistry is too extensive a field to do more than mention it here and the interested reader is referred to the standard text.[18] In this technique, the voltage applied to a solution of a sample complex is continuously ramped back and forward in a sawtooth manner. If a reversible oxidation occurs, an oxidation wave is seen in the plot of current passed; if the oxidation product is sufficiently long-lived, this should be accompanied by a reduction wave having reverse current flow on the reverse scan. The redox potential for the complex is then obtained from the average potentials corresponding to the peak currents for oxidation and reduction waves. A good example from Betley[19] (Fig. 10.9) shows five 1e redox events for a hexanuclear iron cluster, $[X_{12}Fe_6(NCMe)_6]^{n+}$ as n changes from -1 to $+4$. Seeing well-defined waves in the CV relies on the redox event being reversible and fast relative to the scan rate of the CV experiment. Many redox events are poorly reversible or slow, however, and analysis of such data requires specialized knowledge. With this and related methods, the redox potentials and estimates of the lifetimes of the oxidized or reduced species can be determined.[20a] A stable species can sometimes be isolated either by chemical oxidation using an oxidant adapted to the redox potential of the complex or

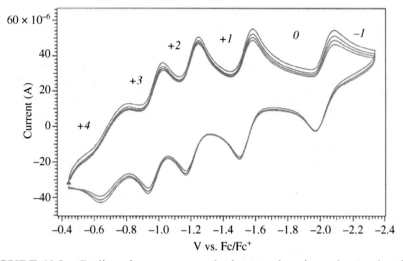

FIGURE 10.9 Cyclic voltammogram of a hexanuclear iron cluster showing five well-defined, reversible 1e redox events (Fc = ferrocene). Source: From Zhao et al., 2011 [19]. Reproduced with permission of the American Chemical Society.

by preparative scale electrochemistry. In some cases, the oxidized or reduced form of the starting complex has some useful reactivity toward an organic substrate in which case electrocatalytic oxidation or reduction can result.[20b]

Electrochemical processes often involve production of paramagnetic species for which electron paramagnetic resonance (epr) spectroscopy[3d] helps in characterizing the symmetry, and in determining how the unpaired electron is delocalized. Hyperfine coupling to $I \neq 0$ ligand atoms can sometimes also be seen, as for ^1H and ^{31}P in [Rh^{II}H(CO)-(PPh$_3$)$_3$]$^+$.[21] The principles resemble the situation for NMR spectroscopy, but in EPR, the electron spin is involved, rather than the nuclear spin. Irradiation with microwaves while varying the magnetic field brings the electron spin into resonance. The intensity of the signal increases as the temperature is lowered, because the population difference between the α and β spins is then enhanced, hence most EPR spectra are collected at liquid He temperature. The equivalent of the chemical shift in NMR is the g factor, with the free electron and simple organic radicals resonating near $g = 2$.

Paramagnetic complexes may also give usable NMR spectra, but the resonance positions may be strongly shifted and broadened compared to a diamagnetic complex. If we oxidize a Ni(II) complex, LNi, we may make a paramagnetic species LNi^+. Sometimes, the EPR of the product gives a resonance near $g = 2$ appropriate for an organic radical, in which case we assign the complex as $Ni(II)(L^{.+})$, with the oxidation having taken place at the ligand. In other cases, the EPR shows $g \neq 2$ in which case a $Ni(III)L$ formulation may be considered more appropriate. Assignment of the oxidation or reduction to M or L can be a contentious issue, however, because the real structure may not be purely $Ni(II)$ $(L^{.+})$ or $Ni(III)L$. In other cases, $Cp*Ni(acac)$, for example,[22] diamagnetic and paramagnetic spin states can be in a temperature-dependent equilibrium, resulting in the appearance of strongly temperature-dependent chemical shifts.

10.11 COMPUTATION

Molecular orbital (MO) theory[23] includes a series of quantum mechanical (QM) methods for describing the behavior of electrons in molecules by combining the familiar s, p, d, and f atomic orbitals (AOs) of the individual atoms to form MOs that extend over the molecule as a whole. The accuracy of the calculations critically depends on the way the interactions between the electrons (electron correlation) are handled. More exact treatments generally require more computer time, so the problem is to find methods that give acceptable accuracy for systems of chemical interest without excessive use of computer time. For many years, the extended Hückel (EH) method was widely used in organometallic chemistry, largely thanks to the exceptionally insightful contributions of Roald Hoffmann. The EH method allowed structural and reactivity trends to be discussed in terms of the interactions of specific MOs but is not able to give good energetic information.

Advances in computing power and computational methods since the late 1990s have allowed improved implementation for organometallic molecules. These methods make fewer assumptions and are based more directly on the physics of the system. Once again, the critical issue is handling electron correlation—very important in transition metals. A major step forward has been the widespread adoption of the present standard method, density functional theory (DFT), in which the energy of a molecule is calculated from an expression involving the electron density distribution, the potential of the atomic nuclei, and a mathematical device called a *functional*. By replacing the inner electrons, not involved in bonding, with a potential drastically reduces the number of electrons that have to be considered and allows good calculations on molecules containing heavy atoms.

A geometry optimization process looks for a minimum in the total energy of the molecule and provides a structure with the corresponding energy. Typical errors are ± 0.02 Å for bond lengths and ± 5 kcal/mol for energies, but in a series of similar molecules, systematic errors cancel, so trends can be more reliable than would appear from the errors quoted above. DFT methods are very versatile but may not be very accurate for paramagnetic, open-shell species and should be used with caution. DFT methods do allow prediction of spectroscopic information, such as NMR and IR spectra, often with good accuracy. With all these quantitative methods, the simple molecular orbital analysis is lost, but the natural bond orbital (NBO) analysis can advantageously replace it.[24]

By using classical mechanics, not QM, molecular mechanics (MM)[25] provides a very large gain in computational time, allowing big systems to be treated, such as proteins. In MM, the molecules are considered as if they contained classical atoms connected by springs. The quality of the results depends on proper parametrization of all the force constants (stretching, bending, electrostatic, van der Waals, etc.). No single satisfactory parametrization has proved possible for transition metals, which need to be represented at the QM level. For example, four-coordinate carbon is ideally tetrahedral, but four-coordinate nickel can be square planar or tetrahedral, and one thus needs QM methods to resolve the problem.

Combining QM and MM methods so that the metal and immediate ligand sphere is described by QM methods and the outer, purely organic part of the ligand by the much less expensive MM technique is also possible. Other ways to integrate MO and MM are now available.[26] As computing power continues to increase, the fraction of the molecule described by QM has also grown larger. Ideally, the modern preference is to carry out full QM calculations.

A very great advantage of computational methods is that structures can be obtained for postulated transient intermediates and even for transition states, where experimental methods are unavailing. The accuracy of the computational results is often sufficient to rule out a postulated intermediate or decide between two competing mechanisms or structures even where there is no convincing experimental method for making the distinction.

In molecular dynamics computations,[27] a molecular system, including reactants and solvent, is allowed to evolve for some length of time, typically a few picoseconds, to give a "movie" of the course of events. Clearly, only low barrier processes can occur spontaneously in such a short time, although certain stratagems can be employed to encourage higher barrier processes to occur within the short observation time.

10.12 OTHER METHODS

Kinetic isotope effects come from the measurement of the k_H/k_D rate ratio for a given reaction where an X-(H,D) bond is present in the substrate. The types of experiment involved and the intepretations that are permitted by the data have been discussed in detail in an important paper by Simmons and Hartwig.[28]

The UV–visible spectrum of an organometallic complex is most commonly obtained when photochemical experiments are carried out, to help decide at which wavelength to irradiate the sample (see Section 4.7). A detailed interpretation of the spectrum has been carried out for few organometallic complexes, a situation that contrasts with that in coordination chemistry, where UV–visible spectroscopy and the ligand field interpretation of the results has always been a strong focus of attention.

When considering ligand designs for future synthesis, it is always best to model the system, either computationally or even with a physical model set, to identify problems of steric clash or incompatibility with the metal's geometric preferences, to avoid the problems that can otherwise arise before investing time in experimental work.[29]

Microanalysis of purified products is standard practice, and the values found for C and H are normally acceptable if they fall within +0.3% of the calculated figure. Solvent of crystallization can be present in the lattice and can alter the percentages observed; the presence of any such solvent should be confirmed by another method, such as NMR or IR, so the calculated analytical data can be suitably adjusted. The molecular weight of suitable complexes can be obtained by electrospray mass spectroscopy.[30]

Mass Spectroscopy

Some volatile organometallic compounds can also be studied by mass spectrometry, or electrospray mass spectrometry (ESI-MS) for involatiles.[31] Mass spectrometry often allows the molecular weight of a complex to be measured directly, if the molecular ion can be seen. Some ligands such as CO may so easily dissociate in the spectrometer that true molecular ions may be lacking. The isotopic distribution for many of the heavier elements (e.g., Mo, Cl, Br, Pd, and Ru) is distinctive, and so the nature and number of these elements can usually be unambiguously identified both in the molecular ion and in other fragments. Thermodynamic data about the strength of bonds within the complex can sometimes be approximately estimated from the appearance potentials of certain fragments in the spectrum.[32] In another variant of the method, *ion cyclotron resonance spectroscopy*, the vapor-phase reactions of metal ions or of metal fragment ions with organic molecules can be studied. For example, ESI-MS shows that bare $[VPO_4]^+$ ions react readily with alkanes.[31] ESI-MS data can identify some of the metal-containing intermediates present in a catalytic reaction. The results have helped throw light on the mechanism of the Wacker process, for example.[33] MS has also been applied to analyzing transfer hydrogenation by η^6-arene Ru complexes where intermediates having lifetimes in the submillisecond to millisecond range were detected by desorption ESI (DESI).[34] A limitation of simple MS is that the observed mass may correspond to any of a number of possible isomers of the molecule under study. It has now proved possible to obtain infrared spectra from gas phase organometallics buried in a cluster of inert molecules such as H_2 by an indirect method that monitors the evaporative loss of H_2 as the sample is irradiated at IR wavelengths; at the appropriate irradiation frequencies, absorption takes place, the sample is heated and H_2 is lost from the cluster. This IR data helps differentiate between the possible isomers.[35]

Single-Molecule Imaging

Methods discussed up to now involve average measurements on a large *ensemble* of molecules—what if single molecules behave very differently from one another? For example, suppose we want to determine what percentage of a given catalyst is in the active form at any one time. Are all the molecules active or does the activity come from just a small percentage of the molecules? Single-molecule fluorescence microscopy, SMFM, has the potential to tell us because it makes it possible to detect chemical events at the single-molecule level.[36] High-resolution AFM microscopy now has the resolution to show the detailed structure of molecules, at least in favorable cases.[37]

Interpretation of Results

Care always needs to be taken with interpreting physical data because Nature has a thousand ways to mislead. An approach to test your conclusion is to ask if there is any combination of events that could falsify it. Devising good control experiments is critical for testing alternate explanations of the data.

- NMR is useful for diamagnetic complexes.
- IR spectroscopy and crystallography also apply to paramagnetic complexes.
- Computational information plays a critical role in understanding organometallic structure and reactivity.

REFERENCES

1. G. S. Girolami, T. B. Rauchfuss, and R. J. Angelici, *Synthesis and Technique in Inorganic Chemistry*, University Science Books, Mill Valley, CA, 1999; D. F. Shriver, *The Handling of Air-Sensitive Compounds*, McGraw-Hill, New York, 1969.

2. E. A. V. Ebsworth, D. W. H. Rankin, and S. Cradock, *Structural Methods in Inorganic Chemistry*, Blackwell, Oxford, 1987.

3. (a) P. J. Hore, J. A. Jones, and S. Wimperis, *NMR, the Toolkit*, Oxford University Press, New York, 2000, is a good nonmathematical introduction. (b) G. N. Lamar, W. D. Horrocks, and R. H. Holm, *NMR of Paramagnetic Molecules*, Academic, New York, 1973; (c) M. Gielen, R. Willem, and B. Wrackmeyer, *Advanced Applications of NMR to Organometallic Chemistry*, Wiley, New York, 1996. (d) R. A. Scott and C. M. Lukehart (eds.), *Applications of Physical Methods to Inorganic and Bioinorganic Chemistry*, Wiley, Hoboken, NJ, 2007.

4. G. R. Fulmer, A. J. M. Miller, N. H. Sherden, H. E. Gottlieb, A. Nudelman, B. M. Stoltz, J. E. Bercaw, and K. I. Goldberg, *Organometallics*, **29**, 2176, 2010.

5. P. Roquette, A. Maronna, M. Reinmuth, E. Kaifer, M. Enders, and H. -J. Himmel, *Inorg. Chem.*, **50**, 1942, 2011.

6. P. S. Pregosin, *NMR in Organometallic Chemistry*, Wiley-VCH, Weinheim, 2012.

7. P. Meakin, J. P. Jesson, and C. A. Tolman, *J. Am. Chem. Soc.*, **94**, 3240, 1972.

8. R. L. Jarek, R. J. Flesher, and S. K. Shin, *J. Chem. Ed.*, **74**, 978, 1997.

9. L. T. Kuhn and J. Bargon, *NMR Methods Catal.*, **276**, 25, 2007.

10. R. Sharma and L. -S Bouchard, *Scientific Reports*, **2**, Article 277, 2012.

11. R. W. Adams, J. A. Aguilar, K. D. Atkinson, M. J. Cowley, P. I. P. Elliott, S. B. Duckett, G. G. R. Green, I. G. Khazal, J. López-Serrano, and D. C. Williamson, *Science*, **323**, 1708, 2009.

12. H. Günzler, *IR Spectroscopy*, Wiley-VCH, Weinheim, 2002.

13. D. J. Gardiner, *Practical Raman spectroscopy*, Springer-Verlag, Berlin, 1989.

14. M. F. C. Ladd, *Structure Determination by X-Ray Crystallography*, Kluwer, New York, 2003.

15. W. T. Klooster, T. F. Koetzle, P. E. M. Siegbahn, T. B. Richardson, and R. H. Crabtree, *J. Am. Chem. Soc.*, **121**, 6337, 1999.

16. T. Gottschalk-Gaudig, J. C. Huffman, K. G. Caulton, H. Gerard, and O. Eisenstein, *J. Am. Chem. Soc.*, **121**, 3242, 1999.

17. G. Parkin, *Chem. Rev.*, **93**, 887, 1993.

18. A. J. Bard and L. R. Faulkner, *Electrochemical Methods: Fundamentals and Applications*, 2nd ed., Wiley NewYork, 2000.

19. Q. Zhao, T. D. Harris, and T. A. Betley, *J. Am. Chem. Soc.*, **133**, 8293, 2011.

20. (a) M. Fourmigué, *Accts. Chem. Res.*, **37**, 179, 2004; (b) D. Serra, M. C. Correia, and L. McElwee-White, *Organometallics*, **30**, 5568, 2011.

21. D. Menglet, A. M. Bond, K. Coutinho, R. S. Dickson, G. G. Lazarev, S. A. Olsen, and J. R. Pilbrow, *J. Am. Chem. Soc.*, **120**, 2087, 1998.

22. M. E. Smith and R. A. Andersen, *J. Amer. Chem. Soc.*, **118**, 11119, 1996.

23. T. A. Albright, J. K. Burdett, and M. -H. Whangbo, *Orbital Interactions in Chemistry*, Wiley, New York, 1985.

24. F. Weinhold and C. R. Landis, *Discovering Chemistry with Natural Bond Orbitals*, Wiley New York, 2012.

25. N. L. Allinger, *Molecular Structure: Understanding Steric and Electronic Effects from Molecular Mechanics*, Wiley, Hoboken, NJ, 2010.

26. C. N. Rowley and T. K. Woo, *Organometallics*, **30**, 2071, 2011.

27. A. Pavlova and E. J. Meijer, *Chem. Phys. Chem.*, **13**, 3492, 2012. A. Comas-Vives, A. Stirling, A. Lledos, and G. Ujaque, *Chem. Eur. J.*, **16**, 8738, 2010.

28. E. M. Simmons and J. F. Hartwig, *Angew. Chem. Int. Ed.*, **51**, 3066, 2012.

29. J. -Y. Shao, J. Yao, and Y. -W. Zhong, *Organometallics*, **31**, 4302, 2012.

30. A. T. Lubben, J. S. McIndoe, and A. S. Weller, *Organometallics*, **27**, 3303, 2008.

31. N. Dietl, T. Wende, K. Chen, L. Jiang, M. Schlangen, X. Zhang, K. R. Asmis, and H. Schwarz, *J. Am. Chem. Soc.*, **135**, 3711, 2013.

32. M. T. Rodgers and P. B. Armentrout, *Mass Spectr. Rev.*, **19**, 215, 2000.

33. D. Harakat, J. Muzart, and J. Le Bras, *RSC Advances*, **2**, 3094, 2012.

34. R. H. Perry, K. R. Brownell, K. Chingin, T. J. Cahill, R. M. Waymouth, and R. N. Zare, *Proc. Nat. Acad. Sci. USA*, **109**, 2246, 2012.

35. E. Garand, J. A. Fournier, M. Z. Kamrath, N. D. Schley, R. H. Crabtree, and M. A. Johnson, *Phys. Chem. Chem. Phys.*, **14**, 10109, 2012.

36. (a) N. M. Esfandiari, Y. Wang, J. Y. Bass, and S. A. Blum, *Inorg. Chem.*, **50**, 9201, 2011; (b) T. Cordes and S. A. Blum, *Nature Chem*, **5**, 993, 2013.

37. D. G. de Oteyza, P. Gorman, Y. -C. Chen, S. Wickenburg, A. Riss, D. J. Mowbray, G. Etkin, Z. Pedramrazi, H. -Z. Tsai, A. Rubio, M. F. Crommie, and F. R. Fischer, *Science*, **340**, 1434, 2013.

PROBLEMS

10.1. Sketch the ^{1}H NMR spectrum of (i) *cis-* and (ii) *trans-*$OsH_2(PMe_3)_4$. How could we go about finding the value of a trans $^2J(H,H)$ coupling by looking at the spectra of an isotopic modification of one of these complexes?

10.2. *trans-*$OsH_2(PMe_3)_4$ reacts with HBF_4 to give $[OsH_3(PMe_3)_4]^+$. What structures should we consider for this species, and how might ^{1}H NMR spectroscopy help you decide which structure is in fact adopted?

10.3. (Indenyl)$_2$W(CO)$_2$ is formally a 20e species. How might it achieve a more reasonable 18e configuration, and how could you use ^{13}C NMR spectroscopy to test your suggestion?

10.4. How could we distinguish between an [(η^6-benzene)ML$_n$] and an [(η^4-benzene)ML$_n$] structure for a given diamagnetic complex?

10.5. Two chemically inequivalent hydrides, H$_a$ and H$_b$ in a metal dihydride complex at 50°C, resonate at -5δ and -10δ, respectively, and are exchanging so that each resonance shows an initial broadening of 10 Hz at a field corresponding to 500 MHz. What is the rate of exchange?

10.6. Which of the methods (a–e) would be suitable for solving parts 1–6? (a) X-ray crystallography, (b) ^{1}H NMR spectroscopy, (c) ^{31}P NMR spectroscopy, (d) IR spectroscopy, or (e) magnetic moment determination: (1) Characterizing a cyclometallated Ph$_2$PC$_6$H$_4$ complex, (2) characterizing a dihydrogen complex, (3) characterizing a CO$_2$ complex, (4) determining the stereochemistry of M(CO)$_2$(dppe)$_2$, (5) comparing the relative donor properties of a series of ligands L in LNi(CO)$_3$, and (6) finding out whether a given complex NiCl$_2$L$_2$ were square planar or tetrahedral in solution and how would you interpret the data. If you cite more than one method, be sure to state which method you would use first.

10.7. IrCl(CO)$_2$(PMe$_3$)$_2$ has two solution IR bands in the CO region, for which I_{sym}/I_{asym} is 0.33. What is the preferred geometry of this complex in solution?

10.8. Why are the CO stretching bands of a bridging carbonyl at lower frequency in the IR spectrum than those of a terminal CO? What would you expect for a μ_3-CO?

10.9. How can a complex having an apparent formulation [IrHCl(CO)-(acetate)(PR$_3$)$_2$], as judged from analytical and NMR measurements, be formulated with (a) an κ^1-acetate, (b) an κ^2-acetate in solution? For each of your suggested formulations, state what methods of characterization would be useful to test your suggestions.

10.10 [Ir(cod)(PMe$_2$Ph)(2-methylpyridine)]$^+$ shows a *pair* of doublets for the PMePh protons in the ^{1}H NMR; explain (Coupling to the metal is not responsible; Ir does not have an $I = \frac{1}{2}$ nucleus.)

10.11. Vibrational spectra are obtainable in aqueous solution only by Raman spectroscopy, not by IR. Why do you think this is?

11

M–L MULTIPLE BONDS

We now look in detail at compounds with multiple bonds to C, N, and O, particularly carbene and carbyne complexes, $L_nM=CR_2$ and $L_nM\equiv CR$, which at least formally contain M=C or M≡C multiple bonds.

11.1 CARBENES

A free carbene such as CH_2 has two spin states, singlet ($\downarrow\uparrow$) and triplet ($\uparrow\uparrow$) that are distinct spin isomers with different reactivities and structures.[1] In the singlet, the electrons are paired up in the sp^2 lone pair, but the triplet has one electron in each of the sp^2 and p orbitals (Fig. 11.1a). Unlike many of the ligands discussed previously, carbenes are rarely stable in the free state. Methylene, $:CH_2$, for example, is a transient intermediate that even reacts with alkanes. This instability—both thermodynamic and kinetic—contributes to its very strong metal binding by disfavoring carbene dissociation.

Fischer versus Schrock Carbenes

The two main types of coordinated carbene are named after their discoverers: Fischer[1] and Schrock.[2] Each represents an extreme formulation

The Organometallic Chemistry of the Transition Metals, Sixth Edition.
Robert H. Crabtree.
© 2014 John Wiley & Sons, Inc. Published 2014 by John Wiley & Sons, Inc.

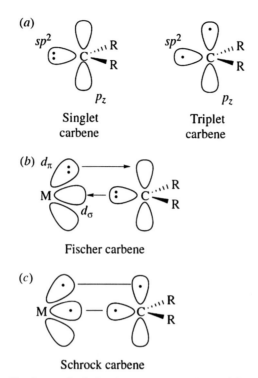

FIGURE 11.1 Singlet and triplet forms of a carbene (*a*) can be considered as the parents of the Fischer (*b*) and Schrock (*c*) carbene complexes. In the Fischer case, direct C→M donation predominates, and the carbon tends to be positively charged. In the Schrock case, two covalent bonds are formed, each polarized toward the carbon, giving it a negative charge.

of the bonding of the CR$_2$ group to the metal, reminiscent of the Dewar–Chatt (D-C) and metalacyclopropane (MCP) models for metal alkene complexes. Carbenes, L$_n$M=CR$_2$, have Fischer character for low oxidation state, late transition metals, having π-acceptor ligands on the metal, and π-donor substituents, R, such as –OMe or –NMe$_2$, on the carbene carbon. A Fischer carbene receives reduced back donation from the metal and is electrophilic, reminiscent of D-C alkene complexes. As an L-type ligand, it is counted as a 2e donor from the filled lone pair of the singlet carbene. It can be considered as a metal-stabilized singlet carbene (Fig. 11.1*b*). By combining with a triplet metal fragment, the triplet carbene gives a diamagnetic complex, just as two triplet CH$_2$ groups can combine to give diamagnetic C$_2$H$_4$.

Schrock carbenes are usually found in high oxidation-state, early-transition metal complexes stabilized by strong donor ligands such as

Cp. In addition, H or alkyl substituents are typically found on the carbene carbon, which acts as a nucleophilic, ∂^- center. The carbene itself is often counted as an X_2 ligand, formally derived from the triplet carbene (Fig. 11.1c), leading to an increase in the metal oxidation state by two units on binding.

Intermediate between the two extremes are carbenes, such as $L_nM=C(Hal)_2$, the halide being intermediate in π-donor strength between –H and –OMe; neither model is satisfactory in this intermediate zone and we encounter another L/X_2 oxidation state ambiguity.

The reactivity of the carbene carbon is controlled by the bonding. A Fischer carbene is predominantly an L-type σ donor via the lone pair, but the empty p orbital on carbon is also a weak acceptor for π back donation from the $M(d_\pi)$ orbitals (Fig. 11.1b). This leads to an electrophilic carbene carbon because direct C→M donation is only partly compensated by M→C back donation; nucleophilic attack at the carbene carbon is thus favored. A Schrock carbene acts as an X_2 ligand by forming an M=C double bond via interaction of the two electrons of triplet CR_2 with any metal fragment that also has two unpaired electrons (Fig. 11.1c). The M–C bonds are polarized toward carbon because C is more electronegative than M, leading to a nucleophilic carbene carbon. Electrophilic attack at the carbene carbon is thus favored. A change in oxidation state can alter the situation: for example, $RuCl_2COL_2(=CF_2)$ is predominantly Fischer type and $Ru(CO)_2L_2(=CF_2)$, with its higher-energy $M(d_\pi)$ orbitals and enhanced back donation, is borderline Schrock type.

The electron-deficient Fischer carbene carbon receives π donation from the lone pair(s) of the π-donor substituents, denoted OR(lp). Structure **11.1** shows how the $M(d_\pi)$ and OR(lp) orbitals compete for π donation to the carbene carbon. This can be described in valence bond (VB) language by resonance between **11.2** and **11.3**. The real structure often resembles **11.3** rather than **11.2**, as shown by the long M–C and short C–O bonds found by X-ray studies. For electron counting purposes, we regard the Fischer carbene as an L-type ligand like CO. The true M=C bond order is much less than 2, thanks to the contribution of **11.3**.

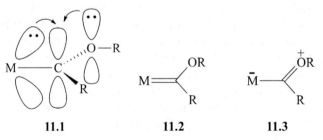

        11.1   **11.2**   **11.3**

Structures **11.4** and **11.5** show how formal oxidation states are assigned differently for the two types. Binding of a Fischer (singlet) carbene does not alter the oxidation state of the metal, but as an X_2 diyl ligand, a Schrock carbene is counted as raising the oxidation state of the metal by two units. Alkenes are all conventionally taken to be L for oxidation state calculations, but for historical reasons, the same choice was not made for carbenes, where the two extremes are treated differently and intermediate cases can be treated either way.

$$(CO)_5W \longleftarrow :C \begin{array}{c} \diagup OR \\ \diagdown R \end{array} \qquad Cp_2(Me)Ta = C \begin{array}{c} \diagup H \\ \diagdown H \end{array}$$

11.4 W(0), 18e   **11.5** Ta(V), 18e

An *alkylidene* is a carbene, CR_2, with alkyl substituents; for example, $MeCH=ML_n$ is an ethylidene complex. "Alkylidene" was sometimes used as a synonym for "Schrock carbene" in the older literature because the first alkylidenes were of the Schrock type. Electrophilic Fischer alkylidenes as well as nucleophilic Schrock ones are now known, however, so the terms should be kept separate. For example, $[Cp_2W(=CH_2)Me]^+$ and $Cp_2Ta(=CH_2)Me$ are isoelectronic, but the former is electrophilic (Fischer) and the latter nucleophilic (Schrock);[1] the net positive charge on the tungsten complex must stabilize the $M(d_\pi)$ levels, leading to much weaker back donation. Schrock carbenes with aryl substituents, such as $[Cp^*(Me_3P)(ArN)Nb=CHPh]$,[2] cannot be called alkylidenes. A small third class of carbene ligand beyond Fischer and Schrock, having electron withdrawing substituents at the carbene carbon, is beginning to attract attention in natural product synthesis in connection with metal catalyzed C–H functionalization by reaction with precursors to free carbenes, such as $N_2CH(COOMe)$, a precursor to $:CH(COOMe)$.[3]

Fischer Carbenes

Fischer made the first recognized carbene complexes in 1964 by treatment of Mo or W carbonyl with RLi then MeI (Eq. 11.1). On the Fischer bonding picture, the methoxy substituent helps stabilize the empty p orbital on the carbene carbon by π donation from one of the lone pairs on oxygen. Resonance form **11.3** is probably dominant in the heteroatom-stabilized Fischer carbenes. The multiple character of the C–OR bond is responsible for the restricted rotation often observed in NMR work and results in a reduced bond order in the M–C bond, often closer to single than double. A C–OR multiple bond in **11.3** implies cis-trans isomerism: isomers **11.6** and **11.7** indeed exist but rapidly interconvert at room temperature (Eq. 11.2) and only decoalesce below −40°C in the ^{1}H NMR spectrum. Another important type of Fischer carbene is the *N*-heterocyclic carbene or NHC, **11.8**, dealt with in Sections 4.3 and 11.4.

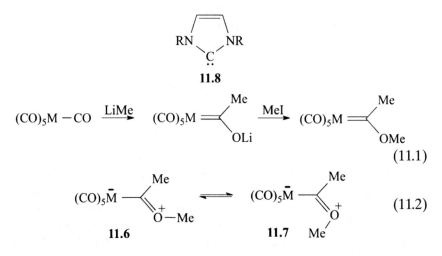

Preparation of Fischer Carbenes The key synthetic routes are illustrated by Eq. 11.1–Eq. 11.5. In Eq. 11.1, we see the alternation of nucleophilic and electrophilic attack, in this case, via an acyl. Eq. 11.3 and Eq. 11.4 contrast abstraction of an H$^-$ (Eq. 11.3) with an electrophile (e.g., Ph$_3$C$^+$) to give a Fischer carbene with abstraction of an H$^+$ (Eq. 11.4) with a base (e.g., Me$_3$P=CH$_2$) to give a Schrock carbene, abstraction taking place in each case from the α C–H of an alkyl. In Eq. 11.5, a classical carbene precursor transfers CH$_2$ to the metal. In Eq. 11.4 and Eq. 11.5, the L$_n$M fragment must be able to accept an extra pair of electrons during the reaction, and so the starting material must be <18e or else lose a ligand.

$$L_nM-CHR_2 \xrightarrow{\ E^+\ } L_nM^+{=}CR_2 \ + \ EH \qquad (11.3)$$

$$L_nM-CHR_2 \xrightarrow{\ Nu^-\ } L_nM^-{=}CR_2 \ + \ NuH \qquad (11.4)$$

$$L_nM + CH_2N_2 \rightarrow L_nM = CH_2 + N_2 \qquad (11.5)$$

Isonitrile complexes are more liable to nucleophilic attack than carbonyls, and a wide range of bisheteroatom-stabilized carbenes can be obtained.[4]

$$[\text{Pt(CNMe)}_4]^{2+} \xrightarrow{\;4\text{MeNH}_2\;} \left[\text{Pt} \Bigl(\substack{\text{NHMe} \\ \diagdown \\ \text{NHMe}} \Bigr)_4 \right]^{2+} \qquad (11.6)$$

Chugaev's[5] carbene complexes (Eq. 11.6) of 1915 escaped structural assignment with the methods then available. Unexpectedly good bases via their resonance form $L_nM^+=C=C^-R$, acetylides $L_nM–C{\equiv}CR$ can react with in acidic alcohol to give the carbenes (Eq. 11.7) via an intermediate vinylidene cation that undergoes nucleophilic attack by the alcohol. In this case, the order of attack followed in Eq. 11.1 (Nu^-, then E^+) is inverted.

$$\text{ClL}_2\text{Pt} - C{\equiv}CR \xrightarrow{\;H^+\;} \text{ClL}_2\text{Pt}^+ = C = CHR \xrightarrow{\;\text{EtOH}\;} \text{ClL}_2\text{Pt}^+ = C \begin{smallmatrix} \diagup \text{OEt} \\ \diagdown \text{CH}_2\text{R} \end{smallmatrix}$$

$$(11.7)$$

Electrophilic abstraction from an alkyl complex (Eq. 11.3) is illustrated by Eq. 11.8.

$$\text{Cp(CO)}_2\,\text{FeCH}_2\text{OMe} + H^+ \rightarrow \underset{\text{reactive transient}}{\text{Cp(CO)}_2\,\text{Fe} = \text{CH}_2} \rightarrow \text{other products}$$

$$(11.8)$$

Carbenes can sometimes be made from organic carbene precursors, such as diazo compounds, from 1,1-diphenylcyclopropene (Eq. 11.9),[6] or from rearrangement of an alkynyl complex, as in the first step of Eq. 11.7 to form a vinylidene. NHC (**11.8**) syntheses are discussed in Section 4.3.

$$\text{RuCl}_2(\text{PPh}_3)_3 \xrightarrow{\;\overset{\displaystyle \overset{\text{Ph}_2}{\underset{\text{C}}{\triangle}}}{\;}\;} \text{Cl}_2(\text{PPh}_3)_2\text{Ru}{=}\!\!\diagup\!\!\diagdown\!\!{=}\text{CPh}_2 \qquad (11.9)$$

Spectroscopy ^{13}C NMR data is very valuable for detecting carbene complexes because their very deshielded carbene carbon resonates at $\sim$200–400 ppm to low field of TMS. An M=C–*H* gives a ^{1}H NMR resonance in the range +10 to +20δ.

Reactions of Fischer Carbenes Thermal decomposition of carbene complexes usually leads to one or both of two types of alkenes: one type is formed by the 1,2-shift of a hydride, and the other by dimerization of the carbene (Eq. 11.10). Neither pathway goes via the free organic carbene because cyclobutanone, known to be formed in the rearrangement of the free carbene, was not seen.

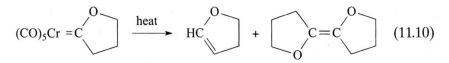

$$(11.10)$$

Fischer carbenes without a heteroatom substituent are much more reactive. The protonation of vinyl complexes[7] gives one such type.

$$(11.11)$$

The ethylidene intermediate readily gives a 1,2 shift of the β proton to give the thermodynamically more stable alkene complex. Even carbenes that lack β hydrogens can be unstable: $[Cp(CO)_2Fe=CH-CMe_3]^+$ and $[Cp(CO)_2Fe=CH-CMe_2Ph]^+$ both rearrange by a 1,2 shift of a methyl or a phenyl anion, respectively, to the electron-deficient carbene carbon (Eq. 11.12). This reaction, analogous to the Wagner–Meerwein rearrangement in carbonium ions, is fast because of the electron-deficient character of the carbene carbon, which could be considered a metal-stabilized carbonium ion.

$$(11.12)$$

$[Cp(Ph_2PCH_2CH_2PPh_2)Fe=CH-CMe_3]^+$ does not rearrange, however, probably because the increased back donation to the carbene by the more basic phosphine-substituted iron decreases the electron deficiency at the carbene carbon.

Where the resulting carbene is sufficiently stabilized, an alkene can even rearrange to the corresponding carbene, the reverse of Eq. 11.12, as in Eq. 11.13. In Eq. 11.14, alkene and carbene forms are in equilibrium.[8]

$$(11.13)$$

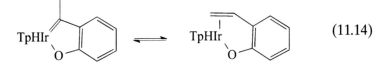

$$(11.14)$$

Oxidative cleavage of a carbene ligand[9] can be achieved with oxidants such as Ce(IV) salts, pyridine N-oxide, or DMSO, or even with air. The product is normally the ketone corresponding to the starting carbene. This reaction is useful for helping to characterize the original carbene (e.g., Eq. 11.15):

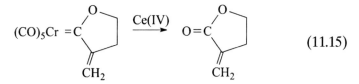

$$(11.15)$$

The synthesis of **11.9** illustrates another useful reaction of Fischer carbenes, the abstraction of a proton β to the metal by a base such as an organolithium reagent. The resulting negative charge can be delocalized onto the metal as shown in Eq. 11.16 and is therefore stabilized. The anion can be alkylated by carbon electrophiles as shown.

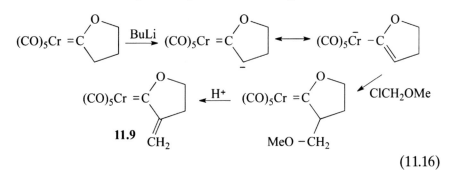

$$(11.16)$$

Fischer carbenes readily undergo nucleophilic attack at the carbene carbon,[10] as shown in Eq. 11.17. The attack of amines can give the zwitterionic intermediate shown, or by loss of methanol, the aminocarbene. If we mentally replace the $(CO)_5Cr$ group with an oxygen atom, we can see the relation of this reaction to the aminolysis of esters to give amides.

$$(11.17)$$

The addition of alkenes can lead to the formation of metalacycles. These can break down to a carbene and an alkene (Eq. 11.18a), or reductive elimination may take place to give a cyclopropane (Eq. 11.18b). Equation 11.18a is the key step in alkene metathesis (Section 12.1).[6]

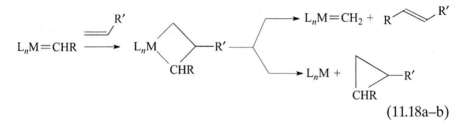

$$(11.18a\text{–}b)$$

Schrock Carbenes

High-valent metal alkyls, especially of the early metals, can undergo proton abstraction at the α carbon to give nucleophilic Schrock carbenes. The first high oxidation-state carbene was formed in an attempt to make TaNp$_5$ (Np = CH$_2$CMe$_3$, or neopentyl), by the reaction of TaNp$_3$Cl$_2$ with LiNp.* In fact, the product is Np$_3$Ta=CH(t-Bu) (Eq. 11.19). The reaction may even go via TaNp$_5$, which then loses neopentane by an α-proton abstraction from one Np ligand—probably agostic—by another. With R = Me$_3$SiCH$_2$, less bulky from the longer bonds to silicon, TaR$_5$ could be isolated at −80°C.

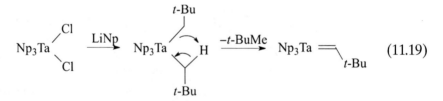

$$(11.19)$$

A requirement for α elimination is that the molecule be crowded. Substitution of a halide in Np$_2$TaCl$_3$ with a Cp group (Eq. 11.20) is enough to do this, for example, as is addition of a PMe$_3$ (Eq. 11.21).[2] The corresponding benzyl complexes require one of the more bulky pentamethylcyclopentadienyls, Cp* (Eq. 11.22), or two plain Cp groups (Eq. 11.23).

$$Np_2Ta\begin{smallmatrix}Cl\\\blacktriangleleft Cl\\Cl\end{smallmatrix} \xrightarrow{TlCp} CpCl_2Ta = \diagdown_{t\text{-Bu}} + \; t\text{-BuMe} \qquad (11.20)$$

*Interestingly, Wittig was trying to make Ph$_3$PMe$_2$ when he discovered Ph$_3$P=CH$_2$.

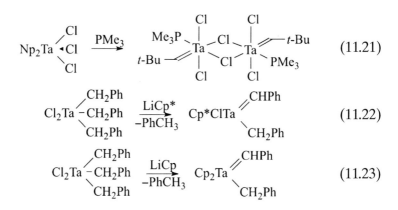

$$Np_2Ta \overset{Cl}{\underset{Cl}{\overset{|}{\longleftarrow}}}Cl \xrightarrow{PMe_3} \quad (11.21)$$

Addition of two PMe$_3$ ligands is enough to give α proton abstraction from a benzylidene to form a benzylidyne in Eq. 11.24.

The methyl group is so sterically undemanding that it does not α-eliminate under the same conditions (Eq. 11.25). The synthesis of a methylene complex requires a deprotonation of a methyl complex by a strong base. A net positive charge on the complex can activate the methyl for this reaction. Equation 11.26 shows how this can be done by an electrophilic abstraction of Me$^-$. Had this been a low-valent, late metal, Ph$_3$C$^+$ might have abstracted H$^-$ to give a Fischer methylene complex.

$$TaMe_3Cl_2 \xrightarrow{LiCp} CpTaMe_3Cl \xrightarrow{LiCp} Cp_2TaMe_3 \quad (11.25)$$

$$Cp_2TaMe_3 \xrightarrow{Ph_3C^+} CpTaMe_2^+ \xrightarrow{base} Cp_2Ta(=CH_2)Me \quad (11.26)$$

Structure and Spectra Few of these early metal complexes are 18e: TaMe$_3$Cl$_2$ is ostensibly 10e, for example. This is not unusual for high oxidation-state complexes, especially in the early metals, where the *d* orbitals are not as strongly stabilized as in lower oxidation states or for later metals (Chapter 15). The halide has lone pairs that might π donate to the empty d_π orbitals, and the alkyl C–H bonds might become agostic. Indeed, Schrock carbene complexes with <18e commonly have agostic C–H bonds. When this happens, the proton on the carbene carbon bends back toward the metal, the M=C bond becomes shorter, and the C–H bond becomes longer (**11.10**). In contrast, in late metals, these d_π orbitals are usually full and the complex is often 18e and lacks agostic C–H bonds.

$$M=C \begin{smallmatrix} H \\ \\ R \end{smallmatrix}$$

11.10

Agostic binding leads to a high-field NMR shift for the C–*H*–M proton and a reduction of the $^1J(C,H)$ coupling constant, together with a lowering of $v(C-H)$ in the IR. In 18e carbene complexes, such protons are not agostic and usually appear at 12δ with a $^1J(C,H)$ of 105–130 Hz; in complexes with <18e, if the CH binds, the agostic proton can come as high as −2δ with a $^1J(C,H)$ of 75–100 Hz. The $v(C-H)$ band in the IR indicates a weakened CH bond, for example, $v(C-H) = 2510$ cm^{-1} in CpTa{CH(*t*-Bu)}Cl$_2$. Crystal structures show that the M=C–R angle can open up to as much as 175°, while the M=C–H angles fall to as little as 78°. The M=C bond length is always short (at least 0.2 Å shorter than an M–C single bond) in all cases, but is even shorter in the complexes with <18e. The oxo alkylidene Cl$_2$(PEt$_3$)$_2$W(=O)(=CHCMe$_3$) has a much less distorted alkylidene group probably because the oxo lone pairs are more basic and so more available for the metal than the C–H bonding pair.

A countersteric conformation is usually a sign of an electronic factor at work. In the structure of Cp$_2$Ta(CH$_2$)Me by neutron diffraction, for example, the CH$_2$ is oriented at right angles to the mirror plane of the molecule, contrary to the conformation predicted on steric grounds which would have the CH$_2$ lying in the plane. The experimental structure is adopted because it allows the filled CH$_2$ p_z orbital to interact with one of the empty metal orbitals lying in the mirror plane of the molecule (see Section 5.4), thus fixing the countersteric out-of-plane CH$_2$ orientation (Fig. 11.2). The larger CHR alkylidenes, having the same orientation as CH$_2$, make the two Cp groups inequivalent at 25°,

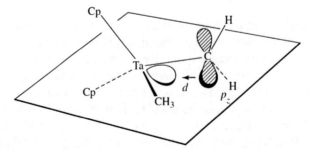

FIGURE 11.2 The orientation of the methylene group in Cp$_2$Ta(CH$_2$)Me is contrary to our expectation from steric effects and is electronically controlled by the overlap of the C(p_z) with a metal d orbital that lies in the plane shown. Filled orbital hatched following the (CH$_2$)$^-$ model.

but fluxionality makes them equivalent at elevated temperature. Since the fluxional process is rotation about the M=CHR bond, the alkylidene must lie in the mirror plane in the transition state, with no M–C π interaction. The $\Delta G^{\ddagger}$ deduced from the data, 25 kcal/mol, therefore gives an estimate of the strength of the Ta=C π bond.

Reactions The reactions of Schrock carbenes illustrate their nucleophilic character. For example, they form adducts with the Lewis acid AlMe$_3$ (Eq. 11.27) and also react with ketones as would a Wittig (Ph$_3$P=CH$_2$) reagent (Eq. 11.28).

$$\text{Cp}_2\text{Ta} \overset{\text{CH}_2}{\underset{\text{Me}}{}} \xrightarrow{\text{AlMe}_3} \text{Cp}_2\text{Ta}^+ \overset{\text{CH}_2-\text{AlMe}_3}{\underset{\text{Me}}{}} \qquad (11.27)$$

$$\text{Np}_3\text{Ta} =\text{CH}(\textit{t}\text{-Bu}) \xrightarrow{\text{Me}_2\text{CO}} \text{Me}_2\text{C}=\text{CH}(\textit{t}\text{-Bu}) + [\text{Np}_3\text{TaO}]_x \qquad (11.28)$$

Carbenes react with alkenes to give metalacycles, which can subsequently react in several ways, either by reversal of the formation reaction to give alkene and a carbene (Eq. 11.18a), by RE to give a cyclopropane (Eq. 11.18b), or by β elimination to give an allyl hydride. The first route is the most important. Each time the RCH=ML$_n$ complex encounters an external alkene, it can exchange alkylidene (RCH=) groups between itself and the alkene. The final result is that alkylidene groups are catalytically exchanged between all the alkenes present. This alkene metathesis reaction[6] (Eq. 11.29) has proved to be of remarkably wide applicability in both organic and polymer chemistry and is discussed in detail in Sections 12.1 and 14.2.

$$\overset{\text{R}}{=\!\!\!/} \; \overset{\text{L}_n\text{M}=\text{CH}_2}{\rightleftharpoons} \; =\!\!\! \; + \; \overset{\text{R}}{\underset{\text{R}}{=\!\!\!/}} \qquad (11.29)$$

M=C multiple bonds can also undergo addition of X–H bonds to give an X–M–C–H unit for X = C, N, and O, as in Eq. 11.30.

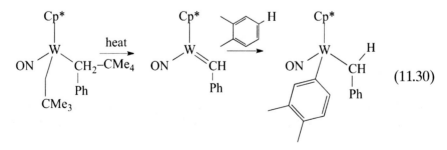

$$(11.30)$$

Intermediate Cases

The Os complex of Eq. 11.31 contains a carbene with character intermediate between the Fischer and Schrock extremes because it reacts both with electrophiles such as SO_2 (Eq. 11.31a) or H^+ and with nucleophiles such as CO (Eq. 11.31b) or CNR.[11] This is consistent with our bonding picture: the osmium has π-donor (Cl) as well as π-acceptor (NO) ligands, the metal is in an intermediate oxidation state (Os(II) if we count the carbene as L, Os(IV) if X_2), and the carbene carbon has non-π-donor substituents (H). Such carbenes cannot be securely classed as either Fischer or Schrock forms, leading to an oxidation state ambiguity, since the convention differs for the two forms.

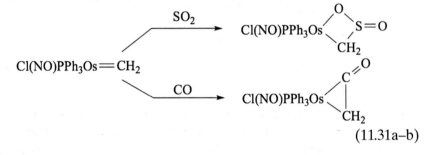

$$(11.31a{-}b)$$

Boryls

The $[BR_2]^-$ group is isoelectronic with CR_2 and several boryl complexes are known, including $Cp_2WH(B\{cat\})$ (**11.11**), $CpFe(CO)_2(B\{cat\})$ and $RhHCl(B\{cat\})(PPh_3)_2$ (cat = catecholate), which is one of the products formed from the oxidative addition of H–B(cat) with Wilkinson's catalyst.[12] As in a carbene, an M=B multiple bond seems to be present; for example, in $Cp_2WH(B\{cat\})$, the B(cat) group is aligned in the least sterically favorable conformation, shown below, so the empty p orbital on boron can π bond with the filled metal d orbital shown. The π bond is not particularly strong, however, because the NMR spectrum shows that the B(cat) group is rapidly rotating.

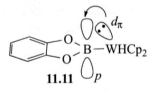

11.11

11.2 CARBYNES

Carbynes M≡CR also have extreme bonding formulations analogous to Fischer and Schrock carbenes, although the distinction is less marked

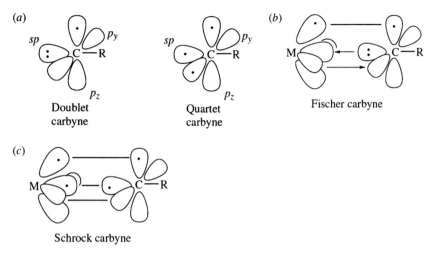

FIGURE 11.3 Doublet and quartet forms of (*a*) a carbyne can be considered as the parents of the (*b*) Fischer and (*c*) Schrock carbyne complexes.

than for M=CR$_2$.[13] In one bonding model, the free carbyne can be considered as doublet for Fischer and quartet for Schrock forms (Fig. 11.3*a*). A doublet carbene is a 2e donor via its *sp* lone pair and forms an additional covalent π bond (Fig. 11.3*b*). One *p* orbital on carbon remains empty and is able to receive back donation from the filled M($d_π$) orbital. We therefore have an LX ligand, 3e on the covalent model (ionic model: 4e). A quartet carbene can form three covalent bonds to a metal having three unpaired electrons, giving an X$_3$ ligand (Fig. 11.3*c*); this is also a 3e ligand on the covalent model (ionic model: 6e).

Oxidation state assignments again depend on the carbyne type. For example, the Fischer carbyne, Br(CO)$_4$W≡CR, is considered as W(II), and the Schrock carbyne, Br$_3$L$_2$W≡CR, as W(VI). Once again, we have ambiguity in intermediate cases.

Synthesis

Fischer first prepared carbyne complexes (1973) by electrophilic abstraction of methoxide ion from a methoxy methyl carbene.

$$L(CO)_4M =C \genfrac{}{}{0pt}{}{\nearrow Me}{\searrow OMe} \xrightarrow[-BX_2(OMe)]{2BX_3} [L(CO)_4M \equiv C-Me]BX_4 \xrightarrow[-L,-BX_3]{} [X(CO)_4M \equiv C-Me]$$

$$(11.32)$$

If L is CO, then the halide ion (Cl, Br, or I) displaces the CO trans to the carbyne in the intermediate cationic complex, showing the high trans effect of the carbyne. On the other hand, if L is PMe_3, then the cationic species is the final product.

By carefully controlled oxidation, it is possible to remove the carbonyl ligands in a Fischer carbyne to give a Schrock carbyne, thus making a direct link between the two types. In Eq. 11.33, we can think of the Br_2 oxidizing the metal by two units. This destabilizes the metal d_π orbitals relative to the carbon p orbitals, and so switches the polarity of the metal–carbon multiple bond. The coligands change from soft carbonyls in the W(II) reactant to the hard O-donor dme in the d^0 W(VI) product. Schrock carbynes are nearly always d^0 if the carbyne is counted as an X_3 ligand.

$$[Br(CO)_4W \equiv CMe] \xrightarrow[\ \ \ \ \ \ \ \ \ ]{Br_2} [Br_3W \equiv CMe] \quad (11.33)$$

Schrock carbynes can be made (i) by deprotonation of an M=CHR group and (ii) by an α elimination, in which this CH bond in effect oxidatively adds to the metal (Eq. 11.34).

$$[Cp*Br_2Ta = CH t Bu] \xrightarrow[Na/Hg]{} [Cp*HTa \equiv C t Bu] \quad (11.34)$$

Structure and Spectra

The carbyne ligand is linear, having sp hybridization, and the M≡C bond is very short (first row, 1.65–1.75 Å; second and third rows, 1.75–1.90 Å). The ^{13}C NMR shows a characteristic low-field resonance for the carbyne carbon at +250 to +400 ppm.

Reactions

Two carbynes can couple to give an alkyne or alkyne complex.[14] For instance, $Br(CO)_4Cr\equiv CPh$ reacts with Ce(IV) to give free PhC≡CPh. In the Fischer series, the carbyne carbon is electrophilic and subject to nucleophilic attack,[15] for example, by PMe_3, pyridine, RLi, or isonitrile (=Nu) to give a carbene of the type $L_nM=CR(Nu)$. Alternatively, the nucleophile may attack the metal in $L_n(CO)M\equiv CR$ and produce a ketenyl complex $L_n(Nu)M(\eta^2\text{-}OC{=}CR)$ or $L_n(Nu)_2M(\eta^1\text{-}OC{=}CR)$. On the other hand, Schrock carbynes are nucleophilic and subject to attack by electrophiles, for instance, $(t\text{-}BuO)_3W{\equiv}C(t\text{-}Bu)$ reacts with HCl to give $(t\text{-}BuO)_2Cl_2W{=}CH(t\text{-}Bu)$. In one case, a carbyne complex, $[(\mu^3\text{-}MeC)_2Mo_3(OAc)_6(OH_2)_3]^{2+}$, is believed to spontaneously release free carbyne radicals in solution. These give a variety of reactions including forming $MeC{\equiv}CMe.$[16]

11.3 BRIDGING CARBENES AND CARBYNES

Like CO, CR_2 can act not only as a terminal but also as a bridging ligand. On the traditional model, when CO or CR_2 bridge, a metal–metal bond is usually present (**11.12** and **11.13**). In bridging, the carbene carbon moves from tricoordinate sp^2 toward tetracoordinate sp^3. Fischer methylenes are rare, while the bridged form is better known and less reactive. Bridging carbenes can be made from diazomethane (Eq. 11.35).

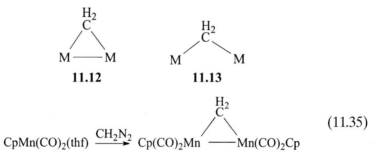

$$\text{CpMn(CO)}_2\text{(thf)} \xrightarrow{\text{CH}_2\text{N}_2} \text{Cp(CO)}_2\text{Mn}\underset{\displaystyle}{\overset{\displaystyle}{\triangle}}\text{Mn(CO)}_2\text{Cp} \qquad (11.35)$$

Diazomethane adds not only to monomeric metal complexes but also to compounds containing metal–metal double bonds, a reaction somewhat analogous to the addition of a free carbene to a C=C double bond to give a cyclopropene (Eq. 11.36):

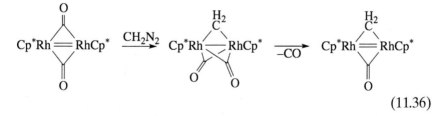

$$(11.36)$$

Structure and Spectra

^{13}C NMR data for carbenes reflect the higher sp^3 character on moving from terminal (250–500δ) to bridging with an M–M bond (100–210δ) to bridging without an M–M bond $(0-10\delta)$; for comparison, metal alkyls resonate at -40 to 0δ.

Reactions

Hydride abstraction from a bridging carbene can give an unsaturated and very reactive μ^2-bridging carbyne, having pronounced carbonium ion character. The bonding scheme resembles the one we saw for Fischer carbenes, except that this is a bis-metal-stabilized carbonium ion, **11.14**. Carbynes can also bridge three metals, as in the long-known and very stable tricobalt complex **11.15**; these are much less reactive than the unsaturated μ^2-carbynes discussed earlier.

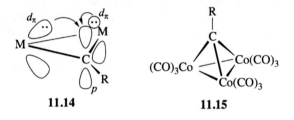

11.14   **11.15**

11.4 N-HETEROCYLIC CARBENES

The majority of the carbene complexes described up to now are reactive. They are generally actor ligands because the M=C bonds are easily broken in typical reactions. N-heterocyclic carbenes,[17] or NHCs, are an exception–their M=C bonds are so stable that NHCs are normally reliable spectator ligands. Although discovered in the 1960s, they languished in relative obscurity for many years before emerging in the last decade to rival phosphines in importance. The signature NHC series **11.16** is derived from N,N'-diaryl or -dialkyl imidazolium compounds by replacing the C–H bond at C2 by a C–ML$_n$ group. These NHCs are sometimes given a shorthand designation, for example, IMes for **11.16** (R = mesityl). They can be considered as Fischer carbenes on steroids because they are doubly flanked by two strongly π donor nitrogens.

NHCs are strong σ donors with some back bonding, but the adjacent N lone pairs donate into the carbene carbon p orbital sufficiently strongly that back bonding from the metal is not required for the stability of the carbene. This permits main-group NHCs to be stable even though these elements do not back bond,[18] for example, the pure σ-bonding group, H^+ is bound much more tightly in an imidazolium salt (H–IPr$^+$ pK$_a$ = 24) than in Ph$_3$P–H$^+$ (pK$_a$ = 2.7). Back bonding is variable, depending on the substituents and transition metal, although its extent is still a matter of debate.[19] Like phosphines, NHCs are electronically and sterically tunable and promote a wide series of catalytic reactions, and homochiral NHCs are also available for asymmetric catalysis.[20] NHCs differ from PR$_3$ in important ways, however. Not only are NHCs such as **11.16** considerably stronger donors but also higher trans effect ligands than any PR$_3$. Chelate formation in bis-NHCs is hampered by the thermodynamic instability of free NHCs that strongly disfavors M–NHC dissociation. Unlike M(H)PR$_3$, irreversible reductive elimination of the imidazolium salt can occur from M(H)(NHC) (Eq. 11.47).[21] Many catalysts containing NHCs are nevertheless stable for thousands of turnovers, so productive chemistry can be much faster than decomposition via Eq. 11.37.

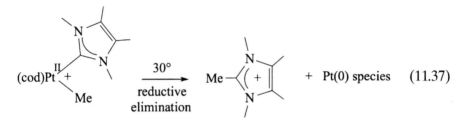

$$(11.37)$$

The Tolman electronic parameters (TEP) for typical NHCs show the higher donor power (lower energy $\nu(CO)$) than phosphines: PMe$_3$, 2064 cm^{-1}; **11.16**, 2054 cm^{-1}; **11.17**, 2050 cm^{-1}. Unlike PR$_3$, the nature of the R groups at N1 and N3 has less effect on the TEP than a change in the nature of the heterocyclic ring.[22] The R groups do influence the steric effect of the NHC, but the ligand is fan shaped, not cone shaped like PR$_3$, and rotation about the formal M=C bond usually allows the NHC to orient so as to avoid steric clashes, thus making the NHC less bulky than might appear. NHCs are variously represented in the literature, for example, as **11.18a–c**; all these refer to the same ligand. By the *Wanzlick equilibrium*, **11.17** is in equilibrium with its dimer, **11.19**.

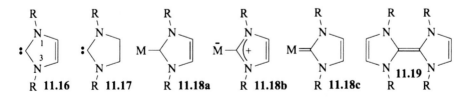

Other NHCs, such as **11.20, 11.21, 11.22**, and **11.23**, are readily accessible by similar routes, starting from the corresponding azoles: **11.20**, deriving from 1,2,4-triazole, and **11.21**, from thiazole. **11.23** is an abnormal or mesoionic carbene because no resonance structure can be written with all-zero formal charges.[23] Perhaps as a result, these aNHCs tend to be even more donor than the normal NHCs, but are more easily lost by protonation to form the free imidazolium salt. Abnormal NHC complexes go back almost to the earliest work on carbene complexes.[24]

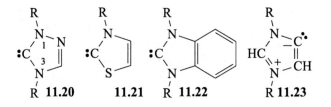

Synthesis of NHC Complexes

The commonest route goes via the free NHC, **11.16**, formed via deprotonation of the parent imidazolium salt with a strong base, such as BuLi (Eq. 11.38).[25] Bulky R groups such as mesityl prevent the free carbene from dimerizing to **11.19**, but the need for BuLi forbids the presence of functional groups with labile protons in the NHC structure. These limitations have led to the development of milder routes that avoid the free carbene.

Simplest among these is direct oxidative addition (Eq. 11.39), where the outcome can be complicated by subsequent reactions of the hydride formed in the oxidative addition step. Direct metallation can be assisted by weak bases such as acetate because agostic binding of C(2)-H makes it easier to deprotonate the imidazolium ion.[26] A very useful method[27] is the initial formation of a silver carbene **11.21** from Ag_2O, followed by transmetallation to give the final product (Eq. 11.50). NHC carboxylate **11.22** and its esters are also useful NHC transfer agents (Eq. 11.51).[28]

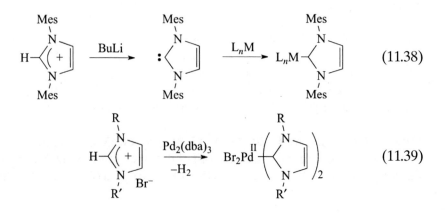

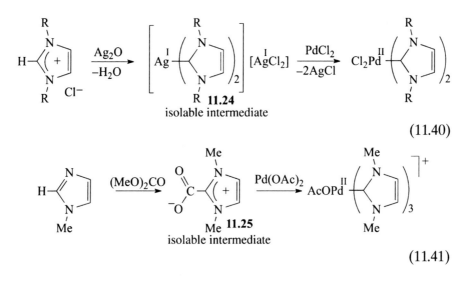

(11.40)

(11.41)

Polydentate NHCs

In an example that shows the strong donor character of NHCs, the tripodal polydentate NHC ligand in **11.26** stabilizes Fe(V) as an organometallic nitride.[29] Not all potentially chelating bis-NHC ligands in fact form chelates, however, even when the chelate would be thermodynamically favored. Each NHC often binds to a separate metal in a 2:1 complex as kinetic product because, unlike M–PR$_3$, M–NHC bond formation is not reversible, so the 'error' cannot be remedied by reversible dissociation (Eq. 11.42).

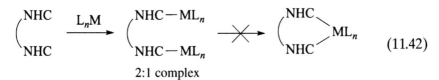

(11.42)

2:1 complex

Applications

After initial activity in the 1960s and 1970s had slowed, Arduengo drew attention back to the area in 1991 with the isolation of the first NHC in the free state, where bulky R groups stabilize the carbene center. From 1994, Herrmann developed NHCs as spectator ligands in homogeneous catalysis. Perhaps the most dramatic success came from modification of the original Grubbs[30] alkene metathesis catalyst **11.27** by replacing one phosphine with an NHC to give a much improved "second generation" version, **11.28**, with rates 10^2–10^3 faster than in the original **11.27**. **11.28** also features a saturated NHC with a CH$_2$CH$_2$ backbone — these are more donor than the standard ligand **11.16**.

There are numerous catalytic applications[31] of NHC complexes (hydrogenation, hydrosilation, metathesis, coupling chemistry, etc.) in which they show advantages over phosphines. Rates can be faster, and the catalysts usually do not need protection from air during catalysis. Imidazoles are also more readily synthesized in a variety of structural modifications, although subsequent formation of the M–C bond can be somewhat more difficult than in the case of PR_3. Ruthenium NHCs can even be stable under intensely oxidative and acidic conditions in catalytic water oxidation driven by Ce(IV).[32] Since free NHCs would be easily oxidized, this emphasizes the kinetic inertness of M–NHC bonds and contrasts with the ease of oxidation of many M–PR_3 to give O=PR_3.

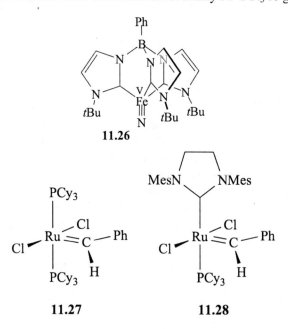

11.26

11.27   **11.28**

11.5   MULTIPLE BONDS TO HETEROATOMS

Related to M=CR_2 and M≡CR are oxo M=O, nitrido M≡N, and imido M=NR. Their high electronegativity gives such ligands "Schrock" character so they can be regarded as O^{2-}, NR^{2-}, and N^{3-}. They form stable complexes with metals located along a periodic table diagonal from V to Os, with a maximum at Mo. Oxo groups have a high tendency to form M–O–M bridges.

For M=O in an octahedral complex, there are strong interactions between two of the M d_π orbitals and the O lone pairs (Fig. 11.4). When the two d orbitals are empty (d^0 to d^2), the interaction is bonding, and

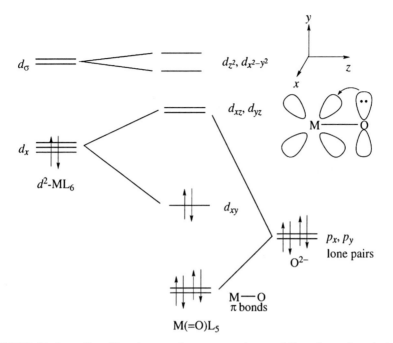

FIGURE 11.4 π Bonding in metal oxo complexes. After the σ bonds have been considered, a d^2 ML_6^{2+} species has a two-above-three orbital pattern characteristic of an octahedron. As long as they remain empty, two of the three d_π orbitals (xz and yz) can accept electrons from the O^{2-} lone pairs; one of these interactions is shown at the top right. This is a special case of the situation shown in Figure 1.10. With one σ bond and two π bonds, the net $M^-\equiv O^+$ bond order is three.

the M=O group has triple-bond character **11.29** with the LX_2 O atom as a 6e donor. This can be represented as **11.29a** or **11.29b**.

$$\bar{M}\equiv\overset{+}{O} \qquad M\overset{\longleftarrow}{=}\overset{..}{O}$$

11.29a   **11.29b**

Oxo Wall

Many metal oxos have electron configurations from d^0 to d^2. The "oxo wall" is often invoked to explain the lack of isolable octahedral d^6 oxo complexes, particularly noticeable for the later transition metals. On this idea,[33] M=O groups are only stabilized by six-coordinate metal centers with an oxidation state of no less than 4+ and a d electron count no higher than d^4 or d^5. This is ascribed to destabilizing electron–electron repulsion between M d_π orbitals, only completely filled in d^6, and O lone pairs. Lower coordination numbers than 6 can free up orbitals to participate in stabilizing M-O π bonding—examples are four coordinate, d^4 OIr^V(mesityl)$_3$ or d^6 (pincer)Pt^{IV}=O. The d^4 oxo species, Re(=O)-X(RC≡CR)$_2$, adopts a tetrahedral structure, thus avoiding the destabilization that would arise in an octahedral ligand field.

Similar ideas hold for M$^-$≡NR$^+$ and M≡N, where M$^-$≡NR$^+$ is linear at nitrogen, as expected for an M≡N triple bond. The d^6 (η^6-C$_6$H$_4$(i-Pr)Me)-Os$^-$≡NAr$^+$ and (η^5-C$_5$Me$_5$)Ir$^+$≡NAr$^+$ avoid the "azo wall" by being linear. A rare bent M=NR double-bonded structure is found in **11.30**, where the M=NR bond length of 1.789 Å can be compared with the adjacent $^-$M≡NR$^+$ at 1.754 Å. The reason for the unusual structure is that since =NR is an X$_2$ and ≡NR is an LX$_2$ ligand, if both imides were linear the Mo would have 20e.

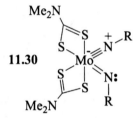

11.30

Synthesis

The complexes are often formed by oxidation, hydrolysis, or aminolysis (Eq. 11.43–Eq. 11.45).

$$[Os^{III}(NH_3)_6]^{3+} \xrightarrow{Ce^{IV}} [N≡Os^{VI}(NH_3)_5]^{3+} \quad (11.43)$$

$$WCl_6 \xrightarrow{tBuNH_2} [tBu\overset{+}{N}≡\overset{-}{W}^{VI}(NHtBu)_4] \quad (11.44)$$

$$[Np_3W≡CtBu] \xrightarrow[-NpH]{H_2O} Np_2\overset{-}{W}\overset{\overset{+}{O}}{\underset{CHtBu}{\diagdown}} \quad (11.45)$$

Spectra and Structure

The $^-$M≡O$^+$ band at 900–1100 cm^{-1} in the IR spectrum is characteristic of the terminal oxo group; $^-$M≡NR$^+$ appears at 1000–1200 cm^{-1} and M≡N at 1020–1100 cm^{-1}. The assignment can be confirmed by ^{18}O or ^{15}N substitution. An exception is Cp$_2$M=O (M = Mo, W), with ν(M–O) frequencies below 880 cm^{-1}; electron counting shows that these must be M=O, not $^-$M≡O$^+$, however, as is indeed consistent with the long

M=O bond length of 1.721 Å in $(MeC_5H_4)_2Mo=O$. Low frequencies are also seen in bis-oxo species where the two oxo groups probably compete for bonding with the $M(d_\pi)$ orbitals. Both nuclei having $I = \frac{1}{2}$, ^{17}O- and ^{15}N-NMR spectra for isotopically substituted complexes can greatly help assigning bridging or terminal bonding modes and distinguishing between M-OH and M-OH$_2$.[34]

- Carbenes form a series between Fischer and Schrock extremes (Table 11.1).
- N-heterocyclic carbenes (Section 11.4) are a rising class of spectator ligand.

TABLE 11.1 Fischer versus Schrock Carbenes, $L_nM=CR_2$

Property	Fischer	Schrock
Reactivity of carbene carbon	Electrophilic	Nucleophilic
Typical R groups on carbon	π Donor (e.g., –OR)	Alkyl, H
Typical metal	Mo(0), Fe(0)	Ta(V), W(VI)
Typical ligands on the metal	π Acceptor (e.g., CO)	Cl, Cp, alkyl
Electron count (covalent model)	2e (L)	2e (X$_2$)
Electron count (ionic model)	2e	4e
ΔOS	0	+2
Back bonding to carbene	Weak	Strong

Note: ΔOS = change in oxidation state of the metal on binding CR$_2$ to L$_n$M.

REFERENCES

1. (a) G. Bertrand (ed.), *Carbene Chemistry: From Fleeting Intermediates to Powerful Reagents*, Marcel Dekker, New York, 2002. (b) P. de Fremont, N. Marion, and S. P. Nolan, *Coord. Chem. Rev.*, **253**, 862, 2009.

2. R. R. Schrock, *Chem. Rev.*, **102**, 145, 2002.

3. H. M. L. Davies and J. R. Denton, *Chem. Soc. Rev.*, **38**, 3061, 2009.

4. J. Barluenga, A. de Prado, J. Santamaria, and M. Tomas, *Organometallics*, **24**, 3614, 2005.

5. A. I. Moncada, M. A. Khan, and L. M. Slaughter, *Tet. Lett.*, **46**, 1399, 2005.

6. T. M. Trnka and R. H. Grubbs, *Acct. Chem. Res.*, **34**, 18, 2001.

7. C. W. Cheng, Y. C. Kuo, S. H. Chang, Y. C. Lin, Y. H. Liu, and Y. Wang, *J. Am. Chem. Soc.*, **129**, 14974, 2007.

8. M. Paneque, M. L. Poveda, L. L. Santos, E. Carmona, A. Lledos, G. Ujaque, and K. Mereiter, *Angew Chem. -Int. Ed. Engl.*, **43**, 3708, 2004.

9. R. Stumpf, M. Jaeger, and H. Fischer, *Organometallics*, **20**, 4040, 2001.

10. Y. S. Zhang and J. W. Herndon, *Org. Lett.*, **5**, 2043, 2003.

11. M. A. Esteruelas, A. I. Gonzalez, A. M. Lopez, and E. Oñate, *Organometallics*, **22**, 414, 2003.

12. J. F. Hartwig, K. S. Cook, M. Hapke, C. D. Incarvito, Y. B. Fan, C. E Webster, and M. B. Hall, *J. Am. Chem. Soc.*, **127**, 2538, 2005.

13. J. W. Herndon, *Coord. Chem. Rev.*, **256**, 1281, 2012.

14. A. Mayr, C. M. Bastos, N. Daubenspeck, and G. A. McDermott, *Chem. Ber.* **125**, 1583, 1992.

15. J. Chen and R. Wang, *Coord. Chem. Rev.*, **231**, 109, 2002.

16. B. Bogoslavsky, O. Levy, A. Kotlyar, M. Salem, F. Gelman, and A. Bino, *Angew. Chem. Int. Ed.* **51**, 90, 2012.

17. O. Schuster, L. R. Yang, H. G. Raubenheimer, and M. Albrecht, *Chem. Rev.*, **109**, 3445, 2009.

18. S. Budagumpi and S. Endud, *Organometallics*, **22**, 1537, 2013.

19. M. Alcarazo, T. Stork, A. Anoop, W. Thiel, and A. Fürstner, *Angew. Chem. Int. Ed.*, **49**, 2542, 2010.

20. F. J. Wang, L. J. Liu, W. F. Wang, S. K. Li, and M. Shi, *Coord. Chem. Rev.*, **256**, 804, 2012.

21. D. S. McGuinness, N. Saendig, B. F. Yates, and K. J. Cavell, *J. Am. Chem. Soc.*, **123**, 4029, 2001.

22. D. G. Gusev, *Organometallics*, **28**, 6458, 2009; R. Tonner and G. Frenking, *Organometallics*, **28**, 6458, 3901.

23. R. H. Crabtree, *Coord. Chem. Rev.*, **257**, 755, 2013.

24. J. Müller, K. Öfele, G. Krebs, *J. Organomet. Chem.*, **82**, 383, 1974.

25. L. Benhamou, E. Chardon, G. Lavigne, S. Bellemin-Laponnaz, and V. Cesar, *Chem. Rev.*, **111**, 2705, 2011.

26. M. Albrecht, J. R. Miecznikowski, A. Samuel, J. W. Faller, and R. H. Crabtree, *Organometallics*, **21**, 3596, 2002.

27. J. B. Lin and C. S. Vasam, *Coord. Chem. Rev.*, **251**, 642, 2007.

28. A. M. Voutchkova, M. Feliz, E. Clot, O. Eisenstein, and R. H. Crabtree, *J. Am. Chem. Soc.*, **129**, 12834, 2007.

29. J. J. Scepaniak, C. S. Vogel, M. M. Khusniyarov, F. W. Heinemann, K. Meyer, and J. M. Smith, *Science*, **331**, 1049, 2011.

30. G. C. Vougioukalakis and R. H. Grubbs, *Chem. Rev.*, **110**, 1746, 2010. Trnka; B. K. Keitz and R. H. Grubbs, *J. Amer. Chem. Soc.*, **133**, 16277, 2011 and references cited.

31. N. Marion and S. P. Nolan, *Acct. Chem. Res.*, **41**, 1440, 2008; *Chem. Soc. Rev.*, **37**, 1776, 2008. J. C. Lewis, R. G. Bergman and J. A. Ellman, *Acct. Chem. Res.*, **41**, 1013, 2008. S. P. Nolan, *Acct. Chem. Res.*, **44**, 91, 2011.

32. M. R. Norris, J. J. Concepcion, D. P. Harrison, R. A. Binstead, D. L. Ashford, Z. Fang, J. L. Templeton, and T. J. Meyer, *J. Am. Chem. Soc.*, **135**, 2080, 2013.

33. R. H. Holm, *Chem. Rev.*, **87**, 1401, 1987; J. R. Winkler and H. B. Gray, *Struct. Bonding (Berlin)*, **142**, 17, 2011.

34. D. T. Richens, *Chem. Rev.*, **105**, 1961, 2005.

PROBLEMS

11.1. Cp_2TiCl_2 reacts with $AlMe_3$ to give $Cp_2Ti(\mu\text{-}Cl)(\mu\text{-}Me)AlMe_2$. Suggest a mechanism.

11.2. Provide a plausible mechanism for Eq. 11.46 and suggest experimental mechanistic tests for your mechanism.

$$(OC)_5\,W = \begin{array}{c} \nearrow \\ Ar \end{array} \xrightarrow{OMe} \longrightarrow (OC)_5\,W = \begin{array}{c} OMe \\ H \end{array} + \begin{array}{c} Ar \\ \end{array} \quad (11.46)$$

11.3. Can you suggest a mechanism for the reactions of Eq. 11.13 and Eq. 11.14?

11.4. (a) We can view $Ph_3P{=}CH_2$ as a carbene complex of a main-group element. Does it show Fischer- or Schrock-like behavior? Using arguments of the type shown in Fig. 11.1, explain why it behaves as it does. (b) Metal oxo complexes, such as $Re({=}O)Cl_3(PPh_3)_2$, might also be regarded as carbene-like if we make the isoelectronic substitution of O for CH_2. Do the same arguments of Fig. 11.1 give any insight into whether an M=O group will have greater or lesser nucleophilic character than the corresponding $M{=}CH_2$ species?

11.5. Propose a mechanism for Eq. 11.47.

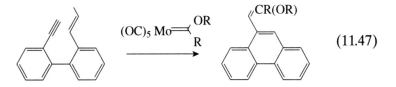

$$(11.47)$$

11.6. Would you expect changes in the formal orbital occupation to effect the orientation of a CH_2 group? Given the orientation shown in Fig. 11.2, draw the appropriate diagram for the isoelectronic $[Cp_2W({=}CH_2)Me]^+$, which has an electrophilic methylene.

What about the hypothetical $[Cp_2W(=CH_2)Me]^-$? What would be the CH_2 orientation, and would you expect the complex to be stable?

11.7. Why is an NHC ligand, such as **11.31**, regarded as a 2e neutral donor L ligand even though its M–C bond resembles that for the undoubted anionic X ligand M–Ph? Is **11.32** an L or an X ligand? What happens if it is deprotonated to give **11.33**?

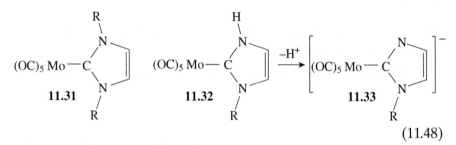

$$(11.48)$$

11.8. The anionic PNP pincer complex **11.34** shown below reacts with ethane at 21° to give a Ti(IV) intermediate **11.35** that is formed without loss of any ligands from **11.34** and that on reaction with RN_3 gives ethylene, N_2, and complex **11.36**. Suggest a pathway, including identifying plausible structures for the intermediate **11.35**. (V. N. Cavaliere, M. G. Crestani, B. Pinter, M. Pink, C. -H. Chen, M. -H. Baik, and D. J. Mindiola, *JACS*, **133**, 10700, 2012.)

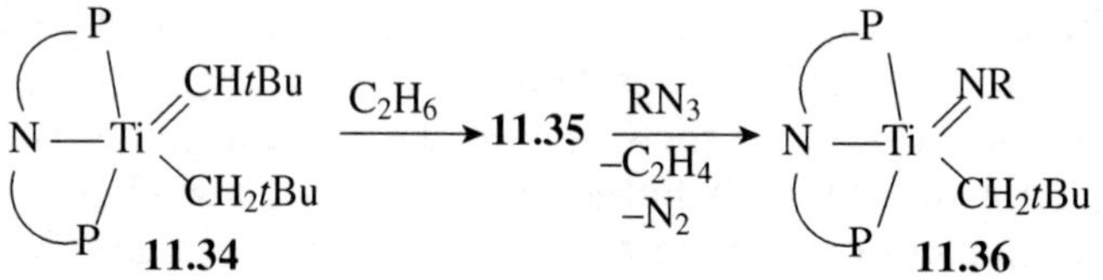

$$(11.49)$$

12

APPLICATIONS

Organometallic catalysts saw early practical applications in hydroformylation (Section 9.4) and the Wacker process (Section 8.3). Here, we continue this industrial theme with alkene metathesis, now widely applied in organic and polymer synthesis, and alkene polymerization, where catalysis provides an exceptional level of control over the molecular structure and the resulting polymer properties. CO chemistry is illustrated by the water–gas shift reaction that is of commercial importance in providing a route to H_2. Other areas show practical promise but are not yet perfected. These include catalytic C–H bond functionalization, an area that has seen rapid growth in connection with green chemistry aspirations, because we start with a cheap hydrocarbon and introduce C–X functionality with minimal waste formation. The rising interest in alternative energy has begun to provide a new set of potential applications in energy capture and storage. Materials and organic synthetic applications are deferred to Chapters 13 and 14.

12.1 ALKENE METATHESIS

In this transformation, the C=C bond of an alkene such as RCH=CHR′ is broken with the resulting RHC and R′HC fragments being redistributed (Eq. 12.1).[1] Originally developed in industry,[2] metathesis could at first

The Organometallic Chemistry of the Transition Metals, Sixth Edition.
Robert H. Crabtree.
© 2014 John Wiley & Sons, Inc. Published 2014 by John Wiley & Sons, Inc.

only be applied to simple alkenes because the early catalysts were intolerant of functionality. The key development of much more tolerant and versatile catalysts, together with wider diffusion of knowledge of the reaction, has led to numerous applications to functionalized alkenes in pharmaceutical, organic, and polymer synthesis. Its importance was emphasized by the award of the 2005 Nobel Prize to Chauvin, Grubbs, and Schrock for their work in the area.

$$
\begin{array}{ccc}
\text{RHC} & \text{CHR}' & \\
\| & + & \| \\
\text{R'HC} & \text{CHR} &
\end{array}
\xrightleftharpoons{\text{cata.}}
\begin{array}{c}
\text{RHC}=\text{CHR}' \\
+ \\
\text{R'HC}=\text{CHR}
\end{array}
\qquad (12.1)
$$

To make catalysts more tolerant of functionality, a move to the right in the periodic table became necessary. The early titanium catalysts are least tolerant because they react preferentially with heteroatom functionalities in the order:

$$RCOOH > ROH > R_2CO > RCO_2R > C=C,$$

in line with the highly oxophilic, hard character of early metals. Mo and W catalysts are intermediate in character, while soft Ru catalysts prefer C=C bonds over heteroatoms in the order:

$$C=C > RCOOH > ROH > R_2CO > RCO_2R'.$$

Rhodium is too far to the right and fails to give metathesis—the key carbene intermediate instead undergoes RE to give cyclopropanes. Grubbs' Ru catalysts[1] (**12.1**) have proved to be the easiest to handle, but some applications require Schrock's more reactive Mo catalysts (**12.2**).[3] Both contain the critical metal carbene unit required for catalysis.

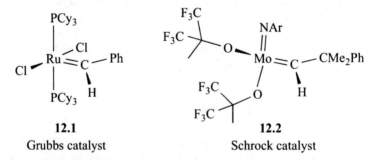

12.1	**12.2**
Grubbs catalyst	Schrock catalyst

Metatheses naturally divide into types, depending on the substrates and products. Beyond *simple metathesis* (Eq. 12.1) involving a single alkene as reactant, comes *cross metathesis* (CM, Eq. 12.2), where two different alkenes react. In a common variant of CM, one product is removed, such as volatile C_2H_4 in Eq. 12.2, to drive the reaction to the right. With some choices of R and R′, the cross product RCH=CHR′ can be strongly favored kinetically. This happens in Eq. 12.3, where R is a electron donor alkyl or aryl and R′ is an electron withdrawing group, stabilizing the mixed product by a push-pull effect.

$$
\begin{array}{c}
RHC \quad CHR' \\
\| \; + \; \| \\
H_2C \quad CH_2
\end{array}
\xrightleftharpoons{cata.}
\begin{array}{c}
RHC = CHR + RHC = CHR' + R'HC = CHR' \\
+ \\
H_2C = CH_2 \uparrow
\end{array}
$$

$$(12.2)$$

$$
\begin{array}{c}
PhHC \quad CHCOOEt \\
\| \; + \; \| \\
H_2C \quad CH_2
\end{array}
\xrightleftharpoons{cata.}
\begin{array}{c}
PhHC = CHCOOEt \\
+ \\
H_2C = CH_2 \uparrow
\end{array}
$$

$$(12.3)$$

With an unconjugated diene, ring-closing metathesis (RCM) is possible (Eq. 12.4), a reaction that is particularly good for forming medium and large rings.[1,4] The reverse of Eq. 12.4 is ring-opening metathesis (ROM), favored by ring strain or a large excess of C_2H_4. The outcome is governed by the thermodynamics of Eq. 12.4, together with the possibility of driving off the volatile C_2H_4 in RCM.

$$
(CH_2)_n
\;
\overset{\substack{\text{ring-closing}\\ \text{metathesis}\\ \text{(RCM)}}}{\underset{\substack{\text{ring-opening}\\ \text{metathesis}\\ \text{(ROM)}}}{\rightleftharpoons}}
\;
(CH_2)_n \; \| \; + \; \|
$$

$$(12.4)$$

The efficiency of the best catalysts is high enough for polymer formation. The two best known cases[1] are acyclic diene metathesis (ADMET, Eq. 12.5) and ring-opening metathesis polymerization, or ROMP (Eq. 12.6), driven by ring strain (e.g., ~15 kcal/mol for norbornene). These reactions are considered living polymerizations because the catalyst remains fully active in the resting state [Cl_2L_2Ru=CH–{P}], where {P} is the polymer chain. This means that once one monomer, A, is used up, a second monomer, B, can be added to form a block copolymer (. . . AAAABBBB . . .). Such a material has very different physical properties from a mixture of homopolymers A_n and B_n or a random copolymer (. . . AABABB . . .). Once again, the reaction can be very tolerant of functional groups.

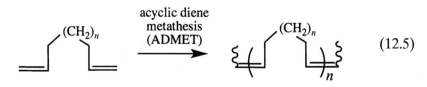

$$(12.5)$$

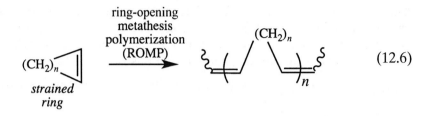

$$(12.6)$$

Mechanism

After the initial discovery, several early mechanistic suggestions appeared in the mid-1970s, shown for ethylene in Eq. 12.7.[5] A cyclobutane metal complex was considered, but cyclobutanes were not formed in the reaction and added cyclobutane did not participate. A tetracarbene complex, $M(=CHR)_4$, seemed possible. A metallocyclopentane might have been formed from oxidative coupling of two alkenes, but how could it rearrange as required? All these mechanisms proved misconceived. In an earlier (1971) article that had completely escaped the attention of the organometallic community—no doubt because it was published in a polymer journal—Hérisson and Chauvin[6] suggested the correct solution. A series of well-chosen "double-cross" experiments ruled out *pairwise mechanisms* in which the two alkenes simultaneously bind to the metal, as is the case in all the previously suggested mechanisms, in favor of a *nonpairwise mechanism* in which the alkenes are converted one by one. The specific nonpairwise mechanism they suggested, shown in Eq. 12.8, is now known as the *Chauvin mechanism*. A metalacyclobutane is formed from an initial carbene reacting with an incoming alkene and then cleaving in a different direction to give the new alkene and a different carbene. The tendency of R and R' to occupy different faces of the metalacyclobutane as a result of mutual steric repulsion translates into preferential formation of trans (E) alkenes.

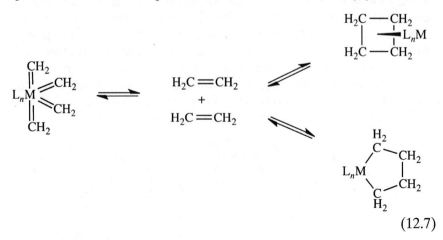

$$(12.7)$$

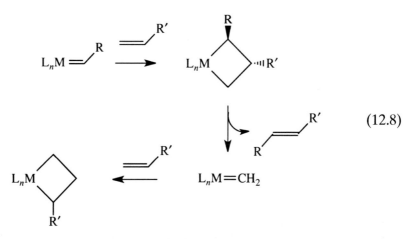

(12.8)

The critical experiment, the double cross shown in Eq. 12.9, is a more elaborate form of the crossover experiment. In a pairwise case, we will see initial products from only two of the alkenes (e.g., the C_{12} and C_{16} products in Eq. 12.9), not the double-cross product with fragments from all three alkenes. The double cross C_{14} product would only form initially in a nonpairwise mechanism. Later on, double-cross products are bound to form, whatever the mechanism, by subsequent metathesis of C_{12} with C_{16}.

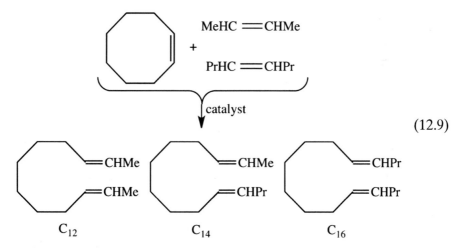

(12.9)

The production of C_{12}, C_{14}, and C_{16} was followed over time, and the $[C_{14}]/[C_{12}]$ and $[C_{14}]/[C_{16}]$ ratios extrapolated back to time zero. These ratios should be zero for a pairwise pathway, since no C_{14} should be formed initially. The results instead showed that a nonpairwise mechanism operates because $[C_{14}]/[C_{12}]$ extrapolated to 0.7 and, more impressively, $[C_{14}]/[C_{16}]$ was 8.35 for a standard catalyst, $MoCl_2(NO)_2(PPh_3)_2/Me_3Al_2Cl_3$.

Reminiscent of the Werner–Jorgensen disputes (Sec 1.3), staunch adherents of the pairwise mechanism suggested the "sticky olefin" hypothesis in which the alkene is strongly retained by the metal, where it undergoes multiple metathesis events, leading to the C_{14} product being released and detected even at the earliest times. Testing this required a more sophisticated test, involving a case in which the initial metathesis products do not themselves metathesize, so that we can be sure that we see the *initial* reaction products. In Eq. 12.10, **12.3** is converted into ethylene and phenanthrene, neither of which undergo further metathesis with the Mo catalyst chosen, so the initial isotopic distributions will then truly reflect the outcome of a single catalytic cycle. This reverse double cross gave a purely statistical 1:2:1 mixture of d^0, d^2, and d^4 ethylene, confirming the nonpairwise mechanism. Only at the end of the 1970s was a consensus in favor of the Chauvin mechanism finally established, however.

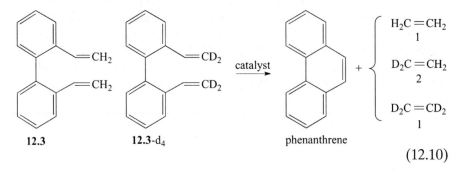

12.3	**12.3**-d_4	phenanthrene	

$$(12.10)$$

Selectivity in Cross Metathesis

Olefins fall into different classes[5a] according to their ease of metathesis. Type I substrates give facile homodimerization and the homodimers are themselves reactive. Type II substrates are less reactive and the homodimers show little or no reactivity. Type III substrates show no homodimerization, although they can still participate in cross metathesis with a more reactive olefin. Type IV are unreactive. The classes depend on the catalyst: for example, 1,1-disubstituted olefins are type IV with the first-generation catalyst, $RuCl_2(PCy_3)_2(CHPh)$, but type III with the more reactive second-generation catalyst, **11.28**; similarly, styrene is type II with the first and type I with the second. Selective cross metathesis can be encouraged by choosing reactants belonging to different types, for example, styrene and $Me_2C=CH_2$ are expected to selectively cross-metathesize with the second-generation catalyst. Olefins in the same class are expected to give near-statistical mixtures of all possible products.

Commercial Applications

The neohexene process starts with the acid-catalyzed dimerization of isobutene, followed by metathesis with ethylene, to give neohexene, an intermediate in the manufacture of synthetic musk, and regenerate isobutene.[7]

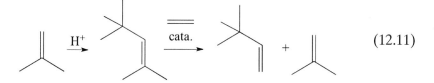

$$(12.11)$$

The commercial synthesis of the housefly pheromone **12.4** illustrates the technique of driving the metathesis reaction by removing the more volatile alkene product, in this case, ethylene; undesired noncross products can easily be separated by distillation. Unfortunately, the presence of the alkylaluminum co-catalyst severely limits the range of functional groups tolerated by this system.

$$Me(CH_2)_7CH=CH_2$$
$$+ \xrightarrow{\text{cata.}} Me(CH_2)_7CH=CH(CH_2)_{12}Me + H_2C=CH_2$$
$$Me(CH_2)_{12}CH=CH_2 \qquad\qquad \textbf{12.4}$$

$$(12.12)$$

Commercial synthesis of unusual polymers has also been possible with the Grubbs metathesis catalyst. Polydicyclopentadiene can be formed from dicyclopentadiene by ROMP. In the reaction, the strained C=C bond indicated by the arrow in **12.5** initially polymerizes; the presence of the second C=C bond allows some cross-linking to occur, giving exceptional strength to the material, which can even stop bullets within a few centimeters! Direct reaction injection molding has proved possible in which the monomer and catalyst are injected into the heated mold and the item formed in place. The polymer is being used to fabricate sports equipment and several other applications are being considered.

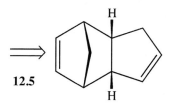

12.5

Alkynes can be cross-metathesized by complexes such as $(t\text{-BuO})_3W\equiv C(t\text{-Bu})$ (**12.6**), apparently via the tungstenacyclobutadiene species **12.7** in Eq. 12.13.[8]

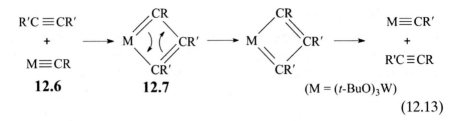

$$(M = (t\text{-}BuO)_3W)$$

$$(12.13)$$

Alkene metathesis also plays a key role both in the SHOP process, discussed in the next section, and in ROMP polymerization (Eq. 12.6).

12.2 DIMERIZATION, OLIGOMERIZATION, AND POLYMERIZATION OF ALKENES

The title reactions are related in relying on chain extension by repeated 1,2-insertion of an alkene into the catalyst M–C bond, but the extension proceeds to different extents (Eq. 12.14) depending on the catalyst and conditions. Dimerization requires one such insertion, oligomerization up to 50, and beyond that point, the product is considered a true high polymer.

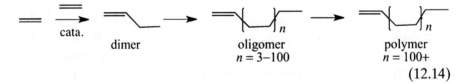

$$(12.14)$$

Alkene polymerization is one of the most important catalytic reactions in commercial use and was an important advance in polymer and materials science. The Ziegler–Natta catalysts, for which they won the Nobel Prize in 1963, account for more than 50 million tons of polyethylene and polypropylene annually. Whether fully homogeneous or supported heterogeneously, these catalysts are all believed to act similarly. In contrast with the 200°C and 1000 atm conditions required for thermal polymerization, a Ziegler–Natta catalyst such as $TiCl_3/Et_2AlCl$ is active at 25°C and 1 atm. Not only are the conditions milder, but the product also shows much less branching. This led to the commercialization of linear low-density polyethylene (LLDPE). Propylene, which does not form useful polymers thermally, now gives highly crystalline stereoregular polymer, **12.8**.

Better defined, homogeneous versions of the catalysts often have the general form $[LL'MCl_2]$ (M = Ti, Zr, or Hf), where L and L' are a series of C- or N-donor ligands. Initially, L and L' were Cp groups, hence the term *metallocene catalysts*. Later improvements involved a much wider range of ligands, and so the term *single-site catalyst* is now also used. The Phillips catalyst, consisting of Cr supported on Al_2O_3, behaves similarly.

These catalysts have had a revolutionary impact on the polymer industry because the variation of L and L′ allows delicate control over the microstructure of the polymer—how the atoms are connected in the chains—and over the polydispersity—the distribution of chain lengths. The catalyst structure controls the physical properties of the final polymer, affecting how it can be of practical use. Metallocene polymers can be designed to be very tough, or act as elastomers, or be easily heat-sealed, or have excellent optical properties, or have easy processability, and they have therefore displaced higher-cost polymers, such as polyurethanes, in many applications. Their economic advantage comes from the low cost of ethylene and propylene. Syndiotactic polypropylene (**12.9**), unobtainable in pure form before metallocene catalysis, is softer but tougher and optically more transparent than other forms. It is used in films for food storage and in medical applications.

Catalyst Activation

Kaminsky showed that Cp_2ZrCl_2 must first be activated with methylalumoxane (MAO, $[MeAlO]_n$), formed by partial hydrolysis of $AlMe_3$. Initial methylation by MAO gives Cp_2ZrMe_2, followed by Me^- abstraction by MAO to form the active 14e species, $[Cp_2ZrMe]^+$, stabilized by the "noncoordinating" $[Me\{MeAlO\}_n]^-$ counterion. Mass spectral studies have thrown some light on the structure of MAO; one component is $[(MeAlO)_{21}(AlMe_3)_{11}Me]^-$.[9]

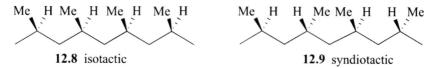

12.8 isotactic   **12.9** syndiotactic

Microstructure

Metallocenes produce polyethylene that is strictly linear, without side branches, termed LLDPE (linear low-density polyethylene). Other processes tend to produce branches and hence a lower quality product. If shorter chains are needed, H_2 can be added to cleave them via heterolysis (Eq. 12.15).

$$M{-}H \Longrightarrow M{\Large\{}\!\!\bigwedge\!\!{\Large\}}_n^H \xrightarrow{H_2} M{-}H + H{\Large\{}\!\!\bigwedge\!\!{\Large\}}_n^H \qquad (12.15)$$

Polypropylene has an almost perfectly regular head-to-tail structure when produced with metallocenes. The arrangement of the methyl groups in isotactic polypropylene (**12.8**) gives the polymer chain a helical rod structure. The rods are chiral, and catalysts that form isotactic polypropylene are also chiral. Since both hands of the catalyst are normally present, rods of both left- and right-handed forms are present in equal amounts.

Syndiotactic polypropylene has no chirality and is formed by catalysts lacking chirality. It tends to adopt a planar zigzag conformation (**12.9**) of the main chain.

Mechanism

Dimerization, oligomerization, and polymerization all rely on the Cossee–Arlman mechanism that consists of repeated alkene 1,2-insertion into the M–C bond of the growing polymer chain (Fig. 12.1).[10] The three types only differ in their k_g/k_t ratio, that is in their relative rates of chain growth by insertion (k_g) to termination by β elimination (k_t). If chain termination is very efficient, k_g/k_t is small and we may see dimerization; if k_g/k_t is somewhat greater, oligomerization, as in the SHOP process discussed later; and if k_g/k_t is very large, true polymerization will result, as in Ziegler–Natta and metallocene catalysis. Although discussed separately, they are nevertheless closely related mechanistically (Eq. 12.16).

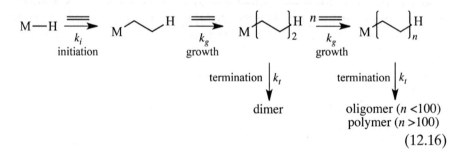

$$(12.16)$$

Unlike the conversion of ethylene to linear polyethylene (PE), propylene polymerization to polypropylene (PP) introduces stereochemical complexity because we can obtain **12.8, 12.9** or a random atactic product. Surprisingly, selective formation of syndiotactic propylene (**12.9**) is seen for many metallocene polymerization catalysts. To see why, we need to know that d^0 [Cp$_2$ZrR]$^+$ is pyramidal (**12.12** in Fig. 12.1)[11] for much the same reasons that made d^0 WMe$_6$ prismatic (Section 3.1). We next have to assume that the pyramidality inverts after each insertion step, transferring the polymer chain from one side to the other like a windshield wiper. The nth alkene to insert therefore occupies the opposite binding site from the $(n − 1)$th and $(n + 1)$th alkene—once the insertion takes place, the newly formed M–C bond automatically finds itself in the other binding site (Fig. 12.1).

In catalyst **12.10**, each binding site is locally chiral, but because the whole molecule has C_2 symmetry, both sites have the same local symmetry. The propylene monomers insert in the same way, leading to isotactic product **12.8**. In catalyst **12.11**, each binding site is again locally chiral, but because the whole molecule has a plane of symmetry, each site has the opposite local symmetry. The propylene monomers insert in the two possible ways with alternation between the two on each successive insertion, leading to syndiotactic product **12.9**.

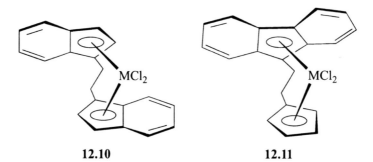

12.10   **12.11**

Computational work indicates the probable structures for the key intermediate propylene complexes in the two classes of catalyst. In the chiral isotactic catalyst, **12.10**, the methyl group tends to be located as shown in Fig. 12.2 (*upper*), so that successive propylenes enter with the same chiralities and bind via the same face (*re* in the figure). In the achiral syndiotactic catalyst, **12.11**, in contrast, successive propylenes enter with opposite chiralities and bind via alternating faces (*re* then *si*).

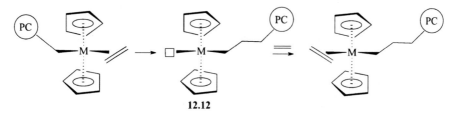

12.12

FIGURE 12.1   Windshield wiper model for alkene polymerization by metallocene catalysts. The insertion causes the M–C bond to the polymer chain (PC) to move alternately from one side to the other in the pyramidal $[Cp_2ZrR]^+$ intermediate as each insertion occurs. The open box represents a vacant site in $[Cp_2ZrR]^+$ where the next alkene can bind.

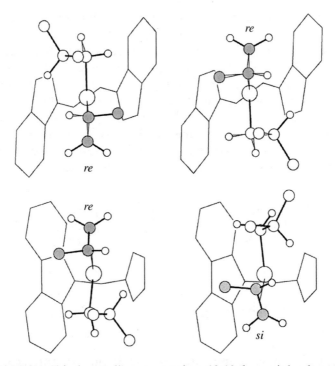

FIGURE 12.2   Chiral metallocene catalyst **12.10** (*upper*) leads to alternate propylenes (shaded) binding via the same *re*-face to give isotactic polymer. The achiral catalyst **12.11** (*lower*) leads to alternate propylenes binding via the opposite faces, *re* then *si*, to give syndiotactic polymer. Source: From Brintzinger et al., 1995 [64]. Reproduced with permission of Wiley-VCH.

The Cossee–Arlman mechanism involving C=C insertion into the M–C bond of the growing polymer chain seems to apply generally. The insertion is much faster in the Ziegler–Natta catalysts than in many isolable 18e alkyl olefin complexes because the reaction is strongly accelerated by coordinative unsaturation in the key intermediate, such as 16e $[Cp_2ZrMe(C_2H_4)]^+$. The alkyl can become agostic and rotate to direct the lone pair of the R^- ligand toward the alkene, facilitating insertion (modified Green-Rooney mechanism). Theoretical work has indicated that in the model intermediate $[Cp_2ZrMe(C_2H_4)]^+$, the CH_3 group is agostic (Fig. 12.3, *left*), as allowed by the formally 16e count for this species. The principal axis (C_3 axis) of the methyl group is indeed rotated by 40°, turning the CH_3 sp^3 hybrid orbital toward the alkene. At the transition state for insertion (Fig. 12.3, *right*), this value has increased to 46°.

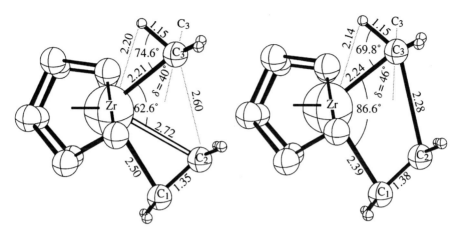

FIGURE 12.3   Structures of a model intermediate $[Cp_2ZrMe(C_2H_4)]^+$ (*left*), showing the agostic methyl. The methyl leans over even more at the transition state (*right*). The results were obtained by Ziegler and coworkers by density functional theoretical calculations. Source: From Fan et al., 1995 [65]. Reproduced with permission of the American Chemical Society.

In the *f*-block metals, successive alkene insertions into a Lu–R bond can be observed stepwise (Eq. 12.17). Not only do the alkenes insert but the reverse reaction, β elimination of an alkyl group, as well as the usual β elimination of a hydrogen, are both seen. For the *d* block, a β elimination of an alkyl group would normally not be possible, but the greater M–R bondstrengths in the *f* block makes the alkyl elimination process sufficiently favorable to compete with β elimination of H.

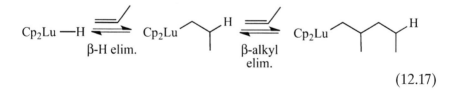

$$(12.17)$$

SHOP Oligomerization

Most late *d* block metals favor β elimination, thus their higher k_t often leads to dimerization or oligomerization, rather than polymerization. The Shell higher olefins process (SHOP) is based on homogeneous nickel catalysts (Fig. 12.4) discovered by Keim.[12] These oligomerize ethylene to give 1-alkenes of various chain lengths (e.g., C_6–C_{20}). Insertion is therefore considerably but not overwhelmingly faster than β

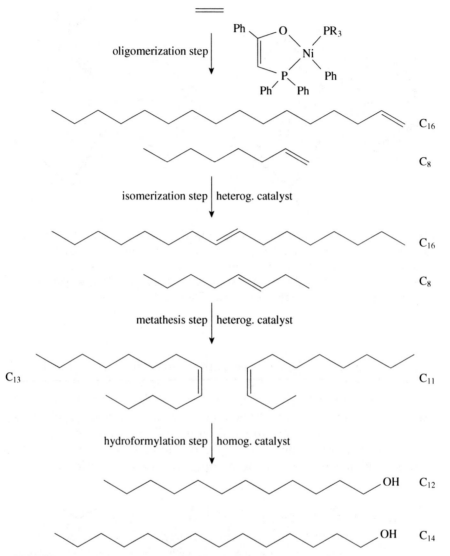

FIGURE 12.4 In the Shell higher olefins process (SHOP), Keim's nickel catalyst gives 1-alkenes of various chain lengths. The subsequent steps allow the chain lengths to be manipulated to maximize the yield of C_{10}–C_{14} products. Finally, SHOP alkenes are often hydroformylated, in which case, the internal alkenes largely give the linear product, as discussed in Chapter 9.

elimination. The C_{10}–C_{14} fraction is a desirable feedstock; for example, hydroformylation gives C_{11}–C_{15} alcohols that are useful in detergent manufacture. The broad chainlength distribution from SHOP means that there is a big non-C_{10}–C_{14} fraction with longer ($>C_{14}$) and shorter ($<C_8$) chain lengths. Figure 12.4 shows how this process minimizes waste by design via isomerization and metathesis steps that manipulate the chain lengths so as to produce more C_{10}–C_{14} material from the longer and shorter chains. The fact that internal C_{10}–C_{14} alkenes are formed does not matter because hydroformylation gives linear alcohols even from internal alkenes, as discussed in Section 9.4. Homogeneous catalysts were strong contenders for the isomerization and metathesis steps of SHOP, but in practice, heterogenized catalysts were adopted. Several plants are now operating with a production of $>10^7$ tons/y.

Another commercially important reaction is du Pont's synthesis of 1,4-hexadiene. This is converted to synthetic rubber by copolymerization with ethylene and propylene, which leaves the polymer with unsaturation. Unsaturation is also present in natural rubber, a 2-methylbutadiene polymer **12.13**, and is necessary for imparting elastomer properties and permitting vulcanization, a treatment with S_8 that cross-links the chains via C–S–C units and greatly hardens the material.

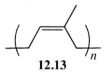

12.13

The 1,4-hexadiene is made by codimerization of ethylene and butadiene, with a $RhCl_3$/EtOH catalyst (Eq. 12.18). The catalyst is about 80% selective for *trans*-1,4-hexadiene, a remarkable figure considering all the different dimeric isomers that could have been formed. The catalyst is believed to be a rhodium hydride formed by reduction of the $RhCl_3$ with the ethanol solvent (Section 3.2). This must react with the butadiene to give mostly the *anti*-methylallyl (crotyl) intermediate, which selectively inserts an ethylene at the unsubstituted end. The cis/trans ratio of the product probably depends on the ratio of the two isomers of the crotyl intermediate. Adding ligands such as HMPA to the system greatly increases the selectivity for the trans diene. By increasing the steric hindrance on the metal, the ligand probably favors the syn isomer of the crotyl ligand over the more hindered anti isomer. The rhodium hydride is also an isomerization catalyst, and so the 1,4-hexadiene is also converted to the undesired conjugated 1,3 isomers. The usual way around a problem like this is to run the reaction only to low conversion, so that the side product is kept to a minimum. The substrates, which are more volatile than the products, are easily recycled.

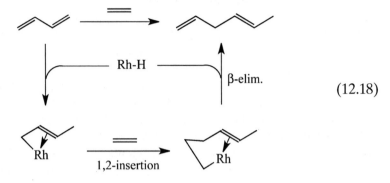

(12.18)

12.3   ACTIVATION OF CO AND CO$_2$

Most organic commodity chemicals are currently made commercially from ethylene, a product of oil refining. In the next several decades, we may see a shift toward other carbon sources for these chemicals. Either coal or natural gas (CH$_4$) can be converted with steam into CO/H$_2$ mixtures called "water–gas" or "synthesis gas" and then on to methanol or to alkane fuels with various heterogeneous catalysts (Eq. 12.19). In particular, the Fischer–Tropsch reaction converts synthesis gas to a mixture of long-chain alkanes and alcohols using heterogeneous catalysis.[13]

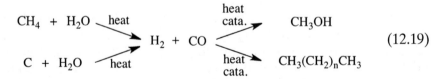

(12.19)

Water–Gas Shift

The H$_2$:CO ratio in synthesis gas depends on the conditions of its formation, but the initial ratio obtained is often ~1:1, insufficiently high for a number of applications. For example, conversion of CO to CH$_3$OH requires a 2:1 H$_2$:CO ratio. If so, we can change the ratio via the water–gas shift reaction (Eq. 12.20), catalyzed either heterogeneously (Fe$_3$O$_4$ or Cu/ZnO) or by a variety of homogeneous catalysts, such as Fe(CO)$_5$. The reagents and products in Eq. 12.20 have comparable free energies so the reaction can be run in either direction but H$_2$ production from CO and H$_2$O is the usual goal.[14]

$$H_2O + CO \rightleftharpoons H_2 + CO_2 \qquad (12.20)$$

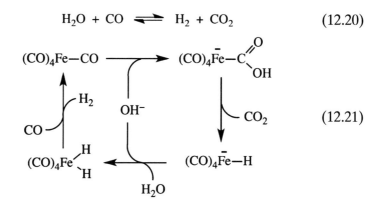

$$(12.21)$$

In the mechanism proposed for $Fe(CO)_5$ (Eq. 12.21), CO bound to Fe becomes activated for nucleophilic attack by OH^- at the CO carbon. Decarboxylation of the resulting metalacarboxylic acid probably does not take place by β elimination because this would require prior loss of CO to generate a vacant site; instead, deprotonation may precede loss of CO_2, followed by reprotonation at the metal to give $[HFe(CO)_4]^-$. Protonation of this anionic hydride liberates H_2 and regenerates the catalyst.

Monsanto Acetic Acid Process

Over 8 million tons of acetic acid derivatives a year are produced in >99% selectivity by carbonylation of methanol with a Rh(I) catalyst, $[RhI_2(CO)_2]^-$ (Eq. 12.22).[15] The process is 100% atom economic since all the reactant atoms appear in the acetic acid. The net effect is the cleavage of the methanol $H_3C{-}OH$ bond and insertion of a CO. The methanol substrate requires activation with HI to produce an equilibrium concentration of MeI, which can oxidatively add to the metal in the turnover limiting step (Fig. 12.5).

$$MeOH + CO \xrightarrow[{[Rh(CO)_2I_2]^-}]{MeI} MeCOOH \qquad (12.22)$$

Once the rhodium methyl is formed, migratory insertion with CO gives an acetylrhodium iodide. Reductive elimination of the acyl iodide is followed by hydrolysis to give acetic acid and HI, which is recycled. The Monsanto process for making acetic acid is replacing the older route that goes from ethylene by the Wacker process to acetaldehyde, followed by oxidation to acetic acid in a second step. An improved process based on iridium (Cativa process) has been developed by BP-Amoco,[15] and a biological analog of this reaction is discussed in Section 16.4.

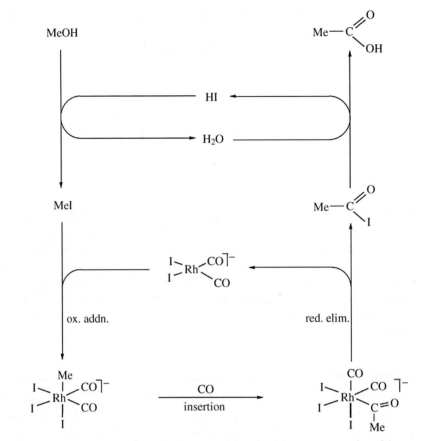

FIGURE 12.5   Catalytic cycle proposed for the Monsanto acetic acid process that converts MeOH and CO to MeCOOH with a Rh catalyst.

CO₂ Activation

A related process, CO_2 activation, has attracted much attention in the hope of producing useful chemicals from a cheap starting material.[16,17] CO_2 is so thermodynamically stable, however, that few potential products can be made from CO_2 by exothermic processes. With $\sim 10^{12}$ tons of excess CO_2 already in the atmosphere and $\sim 2.4 \times 10^{9}$ tons being added per year,[16b] CO_2 conversion to chemical products cannot have a significant impact on mitigating the climate change problem, but it at least goes in the right direction.

Catalytic reduction of CO_2 with H_2 to give HCOOH involves CO_2 insertion into M–H bonds. Although this is "uphill" thermodynamically ($\Delta G = +8$ kcal/mol), the reaction becomes favorable under gas pressure or in the presence of base to deprotonate the formic acid. One of the best homogeneous catalysts to date is **12.14**, which gives 150,000 turnovers per hour at 200°.[18] As an 18e catalyst, a hydride is likely to attack an outer sphere CO$_2$ to give HCOO$^-$ ion that can now coordinate to Ir via O, so this is an unusual type of insertion, greatly favored because a hydride trans to another hydride is particularly hydridic, consistent with the sd^n model of Section 1.8. This step is followed by RE of HCOOH and OA of H$_2$ to close the cycle (Eq. 12.23).

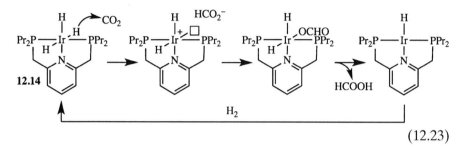

$$(12.23)$$

Formic acid can easily be further converted, for example, to CH$_3$OH + CO$_2$ by disproportionation using [Cp*Ir(dipy)(OH$_2$)][OTf]$_2$[19] or to H$_2$ + CO$_2$ with [{P(CH$_2$CH$_2$PPh$_2$)$_3$}FeH] as catalyst.[20]

Carbon–carbon bond formation from CO$_2$ is illustrated by the Pd catalyzed conversion of CH$_2$=CHCH$_2$SnR$_3$ to CH$_2$=CHCH$_2$CO$_2$SnR$_3$ by a series of [(η^3-allyl)PdL(OOCR)] complexes (L = phosphine or NHC). The stannane transfers the substrate allyl to Pd, followed by the attack of the resulting η^1-allyl terminal =CH$_2$ group on CO$_2$ in the key C–C bond-forming step.[21]

The most important CO$_2$ activation process is photosynthesis in green plants, in which solar photons drive a reaction that would otherwise be uphill thermodynamically: the reduction of CO$_2$ to carbohydrates coupled to water oxidation to O$_2$. Many metalloenzymes are involved in these processes, such as ribulose diphosphate carboxylase that "fixes" CO$_2$ via nucleophilic attack on an enolate anion from a sugar. Artificial photosynthesis[22] takes the natural version as inspiration and seeks to photochemically reduce CO$_2$ to fuels such as MeOH. Naturally, a catalyst is needed–Re(CO)$_3$(bpy)X holds promise in this regard by converting CO$_2$ to CO and HCOOH.[23]

Assigning mechanisms in electrocatalysis is hard, as illustrated by what was initially considered a "metal-free" electroreduction of CO$_2$ to HCOOH with pyridinium ion as the electrocatalyst, where direct interaction of the 1e-reduced [C$_5$H$_5$NH] radical with CO$_2$ was proposed. An alternative mechanism involving the Pt electrode, thought to form surface Pt hydride that attacks CO_2 in a way reminiscent of Eq. 12.23 has now been suggested on computational grounds.[24] This illustrates the difficulty of securely identifying the catalytically active species in a case where a very small fraction of the material may be the active component.

12.4 C–H ACTIVATION

C–H activation refers to any reaction step in which a metal complex cleaves a C–H bond.[25] The aim is a subsequent functionalization step that converts the resulting C–M unit into a C–X group, where X is any of a wide range of useful functionality (OH, NH_2, aryl, etc.). The ultimate goal of the field is the replacement of activated reactant molecules such as RBr and ROTs in synthesis by simple, greener, and less expensive RH. RH being relatively unreactive, metal catalysis is now required. Waste and toxicity problems are both reduced and reaction steps eliminated by starting from RH instead of RBr. This problem also relates to biology in that many enzymatic C–H functionalization reactions, such as C–H hydroxylation or desaturation of $-CH_2-CH_2-$ to $-CH=CH-$ proceed by C–H activation.[26]

Methane (natural gas) will become a more common feedstock for the chemical and energy industries in future, in which case methane activation will be needed. Some natural gas is found at geographically remote sites, where transport to consumers is hampered by methane being a permanent gas that cannot be liquefied at ambient temperature. A goal is to convert methane on-site to more easily transported materials such as MeOH or Me_2O.

Organometallic complexes often activate the C–H either by oxidative addition (Fig. 6.3, path *a*) or σ-bond metathesis, or σ-CAM (path *b*). These reactions favor attack at a terminal C–H bond, leading to subsequent terminal functionalization (e.g., PrH → *n*-PrX), or at an arene C–H bond (e.g., ArH → ArX). This selectivity usefully contrasts with standard organic reactions via radicals or carbonium ions that are selective for the most highly substituted or benzylic CH bonds (e.g., PrH → *i*-PrX; ArMe → $ArCH_2X$). Species such as *i*-Pr· or *i*-Pr$^+$ are more stable and more rapidly formed than *n*-Pr· or *n*-Pr$^+$. Numerous organic synthetic applications of C–H activation continue to be found (Chapter 14).

Catalysis by coordination compounds also plays an important role in the field because high valent Fe and Mn oxo complexes can abstract a hydrogen atom from a C–H bond, leading to fast "rebound" of the newly formed OH group to the C radical to give the alcohol (Eq. 12.24), as in the P-450 enzymes that have an oxoiron porphyrin (por) active site. In the oxo form, the porphyrin is oxidized to a cation radical, illustrating the use of a redox active ligand to store part of the oxidizing power of the system.

$$(12.24)$$

In suitable cases, desaturation can occur by double H atom abstraction: $CH–CH + M=O \rightarrow C=C + M(OH_2)$.[27] Both in enzymes and even in some synthetic catalysts, the resulting radical type selectivity can be modified by molecular recognition between the catalyst and substrate, so that the substrate is held in an orientation that dictates the selectivity.[28]

Shilov Chemistry

Alexander Shilov[29] was the first to see a preference for terminal reactivity in alkane reactions with transition metal through H/D exchange catalyzed by Pt(II) in $D_2O/DOAc$. This was the first indication of the special organometallic reactivity pattern that favors the 1 position of n-alkanes, as distinct from standard organic reactivity in which tertiary. and secondary positions are preferred because they give more stable radicals and carbonium ions. This meant that a new mechanism was at work—one that leads to an intermediate n-alkylplatinum complex. With $[Pt^{IV}Cl_6]^{2-}$ as oxidant and the same Pt(II) catalyst, alkanes, RH, were converted to a mixture of ROH and RCl, the same linear product always being preferred. This suggested that the Pt(IV) intercepts the same intermediate alkyl that led to RD in the deuteration experiments. With methane as substrate, it was even possible to detect a methylplatinum intermediate. Labinger and Bercaw[30] applied a series of mechanistic probes that confirmed and extended Shilov's main points. Figure 12.6 shows the current mechanistic view. An alkane complex either leads to oxidative addition of the alkane and loss of a proton, or the alkane σ complex loses a proton directly (Eq. 12.25). In isotope exchange, the resulting alkyl is cleaved by D^+ to give RD. In the alkane functionalization, oxidation of the Pt(II) alkyl by Pt(IV) gives a Pt(IV) alkyl by electron transfer. The Pt(IV) now becomes a good leaving group, and Cl^- or OH^- can nucleophilically attack the R–Pt(IV) species with departure of Pt(II) to regenerate the catalyst. With the usual organic mechanisms, CH_3OH is much more reactive than CH_4, and so rapid overoxidation of CH_3OH to CO_2 prevents buildup of the desirable

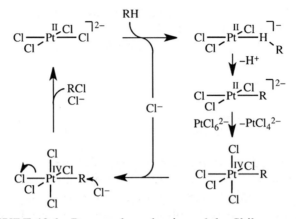

FIGURE 12.6 Proposed mechanism of the Shilov reaction.

product, CH_3OH. With the Shilov system, the reactivity of the CH bonds of CH_3OH is not very different from that of methane and methanol is not overoxidized to the same extent.

Periana and coworkers[25] made Shilov-like chemistry much more efficient. In early work with Hg(II) salts as catalyst in H_2SO_4 at 180°, the acid acts both as solvent and as oxidant for the Hg(0) $\Rightarrow$ Hg(II) step that regenerates the catalyst (Eq. 12.26). Methane was converted to the methanol ester, methyl bisulfate, $MeOSO_3H$, in which the $-OSO_3H$ provides a powerful deactivating group to prevent overoxidation of the methyl group. At a methane conversion of 50%, 85% selectivity to methyl bisulfate (ca. 43% yield) was achieved with the major side product being CO_2 from overoxidation. The expected intermediate $MeHg^+$ cation was seen by NMR spectroscopy, and a Shilov-like mechanism proposed. Since Hg(II) is not expected to give oxidative addition, Hg(IV) being unknown, the initial activation step must occur via deprotonation of a σ complex. Similar selectivity was seen for Pt(II) in H_2SO_4 at 180° (Eq. 12.26).[25]

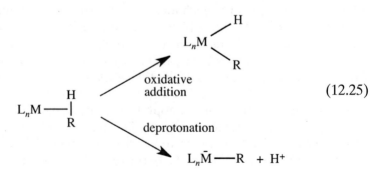

$$(12.25)$$

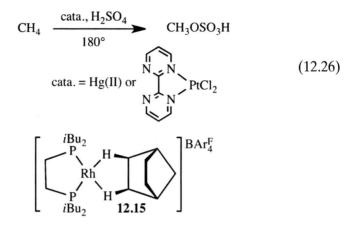

$$CH_4 \xrightarrow[180°]{cata., H_2SO_4} CH_3OSO_3H$$

cata. = Hg(II) or

(12.26)

12.15

In line with the proposed intermediacy of alkane CH σ complexes, several such complexes are now known, one of which, **12.15**, is even stable in the solid state.[31]

Other Routes

Alkane dehydrogenation[32,33] has proved possible by reversing transition metal catalyzed hydrogenation of alkenes to alkanes (Eq. 12.27), but since the thermodynamics are now "uphill," special strategies are needed to drive the reaction. The H_2 formed can be continuously removed, either by introducing a sacrificial acceptor such as tBuCH= CH_2, or by refluxing the solvent to sweep out gaseous hydrogen. Of all alkenes, tBuCH=CH$_2$ has one of the highest affinities for H_2 because of the relief of strain on hydrogenation. In photochemical dehydrogenation, the photon energy supplies the required driving force (Eq. 12.28).

Equation 12.27 shows how the reaction goes via an oxidative addition of a terminal alkane CH bond followed by β elimination.[33] The reaction often requires heating to dissociate some of the monodentate ligands and provide a site for the alkane to bind, so finding a ligand to stabilize the complex is hard. Pincers[32] have worked well (e.g., **12.16**); some have even been able to tolerate 200° (Eq. 12.27),[34] a reaction temperature that normally decomposes organometallic compounds. As in the case of **12.14**, we once again see a special feature of pincers—the ability to stabilize complexes at elevated temperatures. Since 1-octene is the kinetic product, β elimination of an n-octyliridium intermediate is proposed, consistent with the finding that, contrary to radical or electrophilic CH activation in organic chemistry, organometallics typically favor attack at the least hindered position of the alkane.

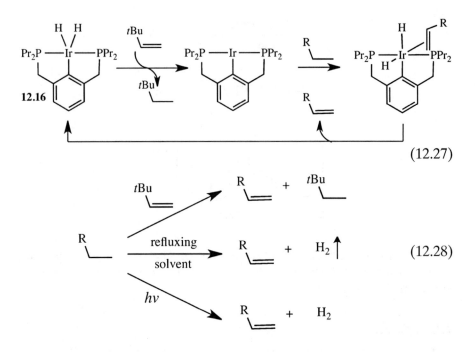

$$(12.27)$$

$$(12.28)$$

A similar pincer catalyst is the key component of the alkane metathesis system[35] of Fig. 12.7. This consists of three steps run in tandem with two catalysts present. The Ir pincer catalyst first dehydrogenates the alkane, with selective formation of the terminal alkene. This selectivity is expected from the usual organometallic selectivity pattern of initial terminal CH oxidative addition followed by β elimination. A Schrock

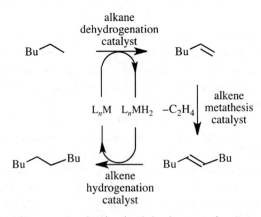

FIGURE 12.7 Alkane metathesis via dehydrogenation/metathesis/hydrogenation. The same catalyst brings about both the first and last steps.

alkene metathesis catalyst then takes over and preferentially converts the alkene to decene and ethylene. The Ir catalyst, being also a hydrogenation catalyst can use the H atoms abstracted from the alkane in the first step to hydrogenate the alkenes in the last step. The result is the formation of n-decene from n-pentane (Fig. 12.7). Similar principles operate here as in alcohol activation by the "hydrogen-borrowing" pathway (Section 12.5).

Catalytic dehydrogenation can also apply to functionalized compounds, such as primary amines, which can be dehydrogenated to nitriles.[36]

Transition metal-catalyzed terminal borylation of linear alkanes with $Cp*Rh(\eta^4-C_6Me_6)$ gives linear alkylboranes from commercially available borane reagents under thermal conditions in high yield (Eq. 12.29).[37]

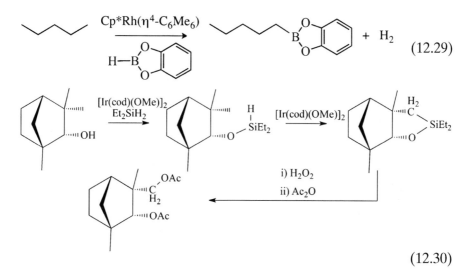

$$(12.29)$$

$$(12.30)$$

Alkanes can also be functionalized with silanes in a similar way; this step can be followed by oxidation to provide alcohols or esters. Simmons and Hartwig[38] have treated fenchol with Et_2SiH_2 in the presence of $[Ir(cod)(OMe)]_2$ catalyst. In the first step, the alcohol is silated to so as to direct a subsequent CH activation to the adjacent methyl group. After oxidative cleavage of the C–Si bond, the diol ester is formed (Eq. 12.30).

On treatment with $NaIO_4$, precatalyst **12.17** gives rise to a homogeneous catalyst that hydroxylates alkanes with retention of configuration so that cis-decalin gives cis-9-decalol (Eq. 12.31). The Cp* in **12.17** acts as placeholder ligand by being oxidatively released from the metal, and the resulting active oxidant is believed to be an Ir(V) oxo species. This contrasts with the classic metal oxo H atom abstraction "rebound"

mechanism of Eq. 12.24 in which the 9-decalyl radical would rapidly
($\sim10^8$ s^{-1}) lose its cis stereochemistry and give trans-9-decalol. Unlike
the Cp*, the alkoxy function in **12.17** is stable to the oxidative conditions because it lacks a β H atom.[39]

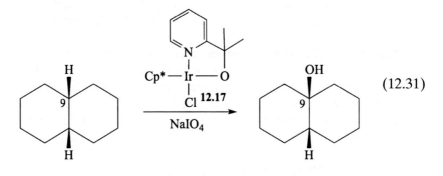

$$(12.31)$$

C–C Bond Formation and Cleavage

Breaking the C–C bonds of alkanes is worse both thermodynamically
and kinetically than breaking the C–H bond because we make two
relatively weak M–C bonds (together worth ~70 kcal/mol), for the loss
of a C–C bond (~85 kcal/mol) and a C–C bond is also less sterically
accessible than a C–H bond. Direct alkane C–C bond breaking is seen
only for very strained alkanes where relief of strain drives the reaction
(Eq. 12.32).[40]

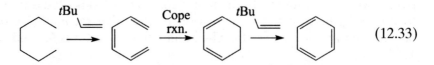

Conversely, C–C bond making from an alkane is seen in Goldman's
conversion of n-hexane to benzene via dehydrogenation followed by a
Cope reaction.

$$(12.33)$$

   In spite of these advances, the development of a series of robust and
selective catalysts for selective CH to CX conversion reactions at unactivated positions with a variety of X functionalization has remained a
continuing challenge to organometallic chemists from the 1970s right
up to the present.

12.5   GREEN CHEMISTRY

Green or sustainable chemistry brings together a constellation of ideas aimed at minimizing environmental impacts of chemicals and chemical processes.[41] Points of emphasis include minimizing inputs and maximizing outputs, as well as designing products to enhance sustainabilty and minimize environmental impact from cradle to grave. Green procedures have gained increasing attention in the pharmaceutical industry, where production methods have seen much recent improvement.[42]

Of prime relevance for us, organometallic catalysis plays a key role in realizing many green aspirations, such as atom economy (Eq. 12.34),[43] which measures the efficiency of incorporation of reactant atoms into products in the theoretical chemical equation. For example, the Monsanto process (Section 12.3) has 100% atom economy (MeOH + CO $\Rightarrow$ MeCOOH) but requires catalysis to activate the reactants. With their high selectivity, catalysts often avoid the need for separations and for protection/deprotection steps.

$$\text{Atom economy } (\%) \quad = \quad \frac{\text{mass of desired products}}{\text{mass of total products}} \times 100 \qquad (12.34)$$

Alcohol Activation and Hydrogen Borrowing

Atom economy can be improved by dispensing with conventional activating groups, such as iodide in RI, because they lead to waste formation.[44] In the absence of such groups, the relatively unreactive alternative reagents employed, such as ROH, need activation from another source, hence the need for catalysis. For example, we can avoid using RCH_2I for amine alkylation by catalytically activating the corresponding alcohol, RCH_2OH. In this case, dehydrogenating the alcohol to the aldehyde, RCHO, provides a much more reactive species (Fig. 12.8, left). After condensation with an amine, $R'NH_2$, to form the imine, $R'N{=}CHR$, the hydrogen removed in the alcohol activation step is returned to the imine to give the amine, $R'NHCH_2R$, as final product in a hydrogen-borrowing process.[45] $[Cp^*IrCl_2]_2$ is one of the many catalysts for this reaction.[46–48] The reaction of a primary alcohol and a primary amine can be selective for the secondary amine product. Under suitable conditions, the secondary amine initially produced does not form an imine, so no overalkylation occurs. This constitutes a selectivity advantage over standard alkylation of RNH_2 with RI, where polyalkylation occurs to give NR_3 or even $[NR_4]^+$.

Milstein[49] modified this pathway with a catalyst that dehydrogenates the hemiaminal intermediate shown in Fig. 12.8 (right) to give the

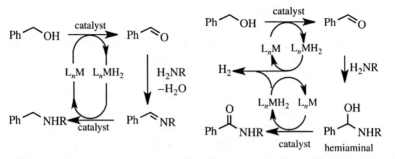

FIGURE 12.8 Two types of alcohol activation catalysis: alcohol amination (*left*) and alcohol amidation (*right*).

highly desirable amide product. Since there is no amide hydrogenation step, H_2 is released and no H borrowing occurs. If the hemiaminal remains metal-bound, β elimination can give the amide; if the hemiaminal dissociates, loss of water gives the imine that can be hydrogenated to the amine.[50] The Milstein catalyst also converts $RCH_2OH/NaOH$ to RCOONa, and by liberating H_2, thus avoids waste formation with standard oxidants.[51]

Catalysts such as $[Cp*IrCl_2]_2$ can help recovery of materials that would otherwise become waste. For example, after the desired enantiomer is removed from the mixture in a resolution step or enzymatically, the undesired isomer left behind can be catalytically racemized back to a 50–50 mixture of enantiomers via the sequence of Eq. 12.35. Dehydrogenation destroys the initial chirality so the hydrogenation step produces a 50–50 racemic mixture, from which more of the desired isomer can be extracted as before.

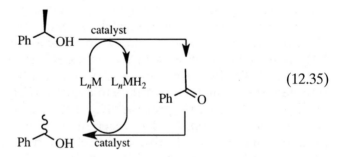

(12.35)

12.6   ENERGY CHEMISTRY

The rising field of alternative energy[52] has gained attention in connection with concerns about climate change. This is a highly interdisciplinary area, but organometallic chemistry can play a role in providing catalysts for key transformations. For example, low carbon footprint energy production via solar, wind, or even nuclear methods provides electric power but not a storable fuel for automotive and air transport. Electrocatalysis may therefore be an important area for development. Electrocatalytic reduction of CO_2 to storable fuels such as MeOH or MeOMe would be useful; so far, Re(bipy)(CO)$_3$Cl complexes can at least reduce CO_2 as far as HCOOH in this way.[53]

Making biofuels, such as EtOH from corn, is contentious since they can compete with food production and put a strain on water resources.[54] The inedible parts of plants are a promising alternative and organometallic catalysts such as MeReO$_3$/H$_2$O$_2$ have been suggested as a way to obtain useful products from lignins,[55] which are complex polymeric aromatic ethers typically making up 15–30% of biomass. The oxygen-rich three-dimensional structure that gives lignin its strength consists of numerous aromatic ethers. Depolymerization therefore requires cleavage of these ethers, hard to do while maintaining the integrity of the aromatic rings. This has been achieved with a Ni catalyst formed in situ from Ni(cod)$_2$ and a saturated NHC precursor as applied to the lignin model compound shown in Eq.12.36 (Ar = 2,6-iPrC$_6$H$_4$).[56]

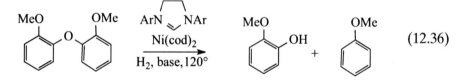

$$(12.36)$$

Hydrogen is considered a useful potential fuel for a future "hydrogen economy" but difficult problems of H_2 production and storage need to be solved. Classical electrochemical H_2 production from water requires Pt electrodes, so efforts have been made to replace these with cheap metal catalysts. A number of very active Ni catalysts, such as **12.18**, have been reported that on electroreduction can undergo protonation to give a nickel hydride; **12.18** also incorporates pendant bases that in the protonated form can deliver H^+ to a metal-bound hydride to generate H_2.[57]

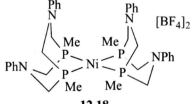

12.18

Numerous suggestions have been made for efficient, reversible hydrogen storage,[58] including Ti-doped $NiAlH_4$, where the Ti catalyzes the release and storage steps, as well as Ti nanoparticles. Amine-borane adducts, such as H_3NBH_3, are very attractive in %H content and ease of H_2 release by simple heating, but regeneration procedures are far less simple.[59] Reversible hydrogenation-dehydrogenation of organic heterocycles is another case where organometallic catalysis could play a role.[60] Methanol[61] from CO and H_2 and ammonia[62] from N_2 and H_2, can be storable forms of H_2, since both can be directly employed in internal combustion engines. Equation 12.37 shows catalyzed H_2 release from MeOH.

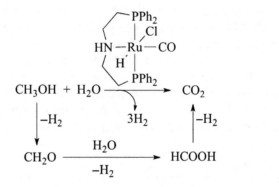

$$(12.37)$$

Beyond catalysis, organometallics can be effective photosensitizers for Grätzel cells, where sunlight photoinjects an electron from a dye into a semiconductor electrode in the key electric power-producing step. The oxidized dye is then reduced by iodide ion in the electrolyte to reset the system for the next injection event. Polypyridyl thiocyanate complexes of Ru(II) are highly effective in these cells with the added advantage that substitution—R in **12.19**—can easily be incorporated to tune the photophysical properties.[63] The COOH groups serve to attach the dye to the TiO_2 semiconductor.

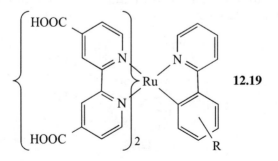

12.19

- Catalytic alkene metathesis (Section 12.1) is the most important application of carbene complexes.
- Catalytic alkene polymerization is one of the most important applications of organometallic chemistry.
- C–H activation (Section 12.4) is rising in importance.

REFERENCES

1. R. H. Grubbs, *Handbook of Metathesis*, Wiley-VCH, Hoboken, NJ, 2003; G. C. Vougioukalakis and R. H. Grubbs, *Chem. Rev.*, **110**, 1746, 2010.

2. R. L. Banks, *Chemtech.*, **16**, 112, 1986.

3. R. R. Schrock and A. H. Hoveyda, *Angew. Chem. -Int. Ed. Engl.*, **42**, 4592, 2003.

4. T. A. Moss, *Tet. Lett.*, **54**, 993, 2013.

5. (a) R. H. Grubbs, *Tetrahedron*, **60**, 7117, 2004; (b) H. -C. Yang, Y. -C. Huang, Y. -K. Lan, T. -Y. Luh, Y. Zhao, and D. G. Truhlar, *Organometallics*, **30**, 4196, 2011.

6. J. L. Herisson and Y. Chauvin, *Makromol. Chem.*, **141**, 161, 1971.

7. V. Dragutan, A. Demoncean, I. Dragutan, and E. S. Finkelshtein (eds.), *Green Metathesis Chemistry*, Springer, Dordrecht, 2010.

8. A. Fürstner, *Angew. Chem., Int. Ed.*, **52**, 2794, 2013.

9. T. K. Trefz, M. A. Henderson, M. Y. Wang, S. Collins, and J. S. McIndoe, *Organometallics*, **32**, 3149, 2013.

10. J. I. Martinez-Araya, R. Quijada and A. Toro-Labbe, *J. Phys. Chem. C*, **116**, 21318, 2012.

11. M. Kaupp, *Chem. Eur. J.*, **4**, 1678, 1998.

12. J. C. Mol, *J. Mol. Catal. A*, **213**, 39, 2004.

13. O. R. Inderwildi, S. J. Jenkins, and D. A. King, *J. Phys. Chem. C*, **112**, 1305, 2008.

14. J. D. Holladay, J. Hu, D. L. King, and Y. Wang, *Catal. Today*, **139**, 244, 2009.

15. I. Omae, *Appl. Organometal. Chem.*, **23**, 91, 2009.

16. (a) E. E. Benson, C. P. Kubiak, A. J. Sathrum, and J. M. Smieja, *Chem. Soc. Rev.*, **38**, 89, 2009; (b) M. Mikkelsen, M. Jørgensen, and F. C. Krebs, *Energy Environ. Sci.*, **3**, 43, 2010.

17. I. Omae, *Coord. Chem. Rev.*, **256**, 1384, 2012.

18. R. Tanaka, M. Yamashita, and K. Nozaki, *J. Am. Chem. Soc.*, **131**, 14168, 2009.

19. A. J. M. Miller, D. M. Heinekey, J. M. Mayer, and K. I. Goldberg, *Angew. Chem. Int. Ed.*, **52**, 3981, 2013.

20. M. Nielsen, E. Alberico, W. Baumann, H. -J. Drexler, H. Junge, S. Gladia, and M. Beller, *Science*, **333**, 1733, 2011.

21. J. Wu and N. Hazari, *Chem. Comm.*, **47**, 1069, 2011.

22. D. Gust, T. A. Moore, and A. L. Moore, *Acc. Chem. Res.*, **42**, 1890, 2009.

23. A. J. Morris, G. J. Meyer, and E. Fujita, *Acct. Chem. Res.*, **42**, 1983, 2009.

24. M. Z. Ertem, S. J. Konezny, C. M. Araujo, and V. S. Batista, *J. Phys. Chem. Lett.*, **4**, 745, 2013.

25. B. G. Hashiguchi, S. M. Bischof, K. M. Konnick, and R. A. Periana, *Acct. Chem. Res.*, **45**, 885, 2012 and references cited.

26. P. R. Ortiz de Montellano, *Chem. Rev.*, **110**, 932, 2010.

27. A. Conde, L. Vilella, D. Balcells, M. M. Díaz-Requejo, A. Lledós, and P. J. Pérez, *J. Am. Chem. Soc.*, **135**, 3887, 2013.

28. S. Das, C. D. Incarvito, R. H. Crabtree, and G. W. Brudvig, *Science*, **312**, 1941, 2006.

29. See G. B. Shul'pin, Oxidations of C–H compounds catalyzed by metal complexes, in *Transition Metals for Organic Synthesis*, eds. M. Beller and C. Bolm, 2nd ed., Wiley-VCH, Weinheim, 2004, Chapter 2.2, pp. 215–241.

30. J. A. Labinger and J. E. Bercaw, *Top Organometal. Chem.*, **35**, 29, 2011.

31. S. D. Pike, A. L. Thompson, A.G. Algarra, D. C. Apperley, S. A. Macgregor, and A. S. Weller, *Science*, **337**, 1648, 2012.

32. G. van Koten, *J. Organometal. Chem.*, **730**, 156, 2013.

33. J. Choi, A. H. R. MacArthur, M. Brookhart, and A. S. Goldman, *Chem. Rev.*, **111**, 1761, 2011.

34. K. E. Allen, D. M. Heinekey, A. S. Goldman, and K. I. Goldberg, *Organometallics*, **32**, 1579, 2013.

35. M. C. Haibach, S. Kundu, M. Brookhart, and A. S. Goldman, *Acc. Chem. Res.*, **45**, 947, 2012.

36. W. H. Bernskoetter and M. Brookhart, *Organometallics*, **27**, 2036, 2008.

37. I. A. I. Mkhalid, J. H. Barnard, T. B. Marder, J. M. Murphy, and J. F. Hartwig, *Chem. Rev.*, **110**, 890, 2010.

38. E. M. Simmons and J. F. Hartwig, *Nature*, **483**, 70, 2012.

39. M. Zhou, D. Balcells, A. Parent, R. H. Crabtree, and O. Eisenstein, *ACS Catal.*, **2**, 208, 2012.

40. M. E. van der Boom and D. Milstein, *Chem. Rev.*, **103**, 1759, 2003.

41. P. T. Anastas and J. C. Warner, *Green Chemistry: Theory and Practice*, Oxford University Press, New York, 1998; M. Lancaster, *Green Chemistry: An Introductory Text*, Royal Society of Chemistry, Cambridge, 2002; R. A. Sheldon, *Chem. Commun.*, 3352, 2008; P. Anastas and N. Eghbali, *Chem. Soc. Rev.*, **39**, 301, 2010.

42. P. J. Dunn, A. S. Wells, and M. T. Williams (eds.), *Green Chemistry in the Pharmaceutical Industry*, Wiley-VCH, Weinheim, 2010.

43. B. M. Trost, *Acct. Chem. Res.*, **35**, 695, 2002; G. J. ten Brink, I. W. C. E. Arends, and R. A. Sheldon, *Science*, **287**, 1636, 2000.

44. G. E. Dobereiner and R. H. Crabtree, *Chem. Rev.*, **110**, 681, 2010. A. J. A. Watson, A. C. Maxwell, and J. M. J. Williams, *Org. Biomol. Chem.*, **10**, 240, 2012.

45. T. D. Nixon, M. K. Whittlesey, and J. M. J. Williams, *Dalton Trans.*, **753**, 2009.

46. T. Suzuki, *Chem. Rev.*, **111**, 1309, 2011.

47. F. Li, C. L. Sun, H. X. Shan, X. Y. Zou, and J. J. Xie, *CHEMCATCHEM*, **5**, 1543, 2013.

48. B. Royo and E. Peris, *Eur. J. Inorg. Chem.*, **2012**, 1309.

49. C. Gunanathan, Y. Ben-David, and D. Milstein, *Science*, **317**, 790, 2007.

50. D. Balcells, A. Nova, E. Clot, D. Gnanamgari, R. H. Crabtree, and O. Eisenstein, *Organometallics*, **27**, 2529, 2008.

51. E. Balaraman, E. Khaskin, G. Leitus, and D. Milstein, *Nature Chem.*, **5**, 122, 2013.

52. T. R. Cook, D. K. Dogutan, S. Y. Reece, Y. Surendranath, T. S. Teets, and D. G. Nocera, *Chem. Rev.*, **110**, 6474, 2010; M. Z. Jacobson, *Energy Environ. Sci.*, **2**, 148, 2009.

53. J. M. Smieja and C. P. Kubiak, *Inorg. Chem.*, **49**, 9283, 2010.

54. D. Tilman, R. Socolow, J. A. Foley, J. Hill, E. Larson, L. Lynd, S. Pacala, J. Reilly, T. Searchinger, C. Somerville, and R. Williams, *Science*, **325**, 270, 2009.

55. TJ Korstanje, RJMK Gebbink, and R. J. M. Klein, *Topics Organomet. Chem.*, **39**, 129, 2012; J. C. Hicks, *J. Phys. Chem. Lett.*, **2**, 2280, 2011.

56. A. G. Sergeev and J. F. Hartwig, *Science*, **332**, 439, 2011.

57. S. Wiese, U. J. Kilgore, D. L. DuBois, and R. M. Bullock, *ACS Catal.*, **2**, 720, 2012.

58. U. Eberle, M. Felderhoff, and F. Schüth, *Angew. Chem. -Int. Ed.*, **48**, 6608, 2009.

59. A. D. Sutton, A. K. Burrell, D. A. Dixon, E. B. Garner, J. C. Gordon, T. Nakagawa, K. C. Ott, P. Robinson, and M. Vasiliu, *Science*, **331**, 1426, 2011.

60. R. H. Crabtree, *Energy Environ. Sci.*, **1**, 134, 2008.

61. M. Nielsen, E. Alberico, W Baumann, H. -J. Drexler, H. Junge, S. Gladiali, and M. Beller *Nature*, **495**, 85, 2013.

62. C. Zamfirescu and I. Dincer, *Fuel Proc. Technol.*, **90**, 729, 2009.

63. K. C. D. Robson, P. G. Bomben, and C. P. Berlinguette, *Dalton Trans.*, **41**, 7814, 2012.

64. H. H. Brintzinger. D. Fischer, R. Mulhaupt, B. Rieger, and R. M. Waymouth, *Angew. Chem. -Int. Ed. Engl.*, **34**, 1143, 1995.

65. L. Y. Fan, D. Harrison, T. K. Woo, and T. Ziegler, *Organometallics*, **14**, 2018, 1995.

PROBLEMS

12.1. Given the reaction of Fig. 12.6, what can you say about the mechanism of a related reaction in the reverse sense, starting from MeI and $[PtCl_4]^{2-}$? What OA product is expected and what geometry would it have?

12.2. The attempted metathesis of ethyl vinyl ether, $EtOCH=CH_2$, with Grubbs's catalyst $[RuCl_2(PCy_3)_2(=CHPh)]$, gives only a stable metal complex and one equivalent of a free alkene as product. Predict the structures of these products and explain why the reaction is only stoichiometric, not catalytic.

12.3. The reaction shown below appears to be a cyclometallation, but is there anything unusual about it that might excite suspicion that it does not go by a conventional oxidative addition mechanism? Suggest an alternative.

$$Pt(PMe_2Ph)_2Cl_2 \xrightarrow{\text{LiMe}} \begin{array}{c} PhMe_2P \quad\quad Me \\ \diagdown \diagup \\ Pt \\ \diagup \diagdown \\ PhMeP \text{———} CH_2 \end{array} \qquad (12.38)$$

12.4. Suppose that you were about to study the following complexes to see if any of them bind CO_2. Describe what type(s) of product you would anticipate in each case: $[Re(PMe_3)_5]^-$, $(\eta^5\text{-Indenyl})\text{-}Ir(PMe_3)_2$, and $CpMo(CO)_3H$. Given that you had samples of all three, which would you try first as the most likely to bind CO_2?

12.5. Suggest a plausible mechanism for Eq. 12.39.

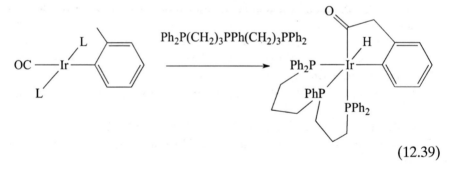

$$(12.39)$$

12.6. Suggest a plausible mechanism and experimental tests for your mechanism:

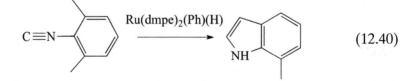

$$(12.40)$$

12.7. Suggest a plausible mechanism and experimental tests for your mechanism:

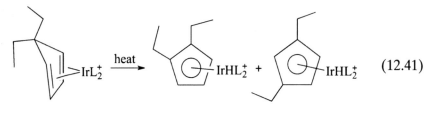

50:50 mixture

12.8. Suggest a plausible mechanism for Eq. 12.42 and some ways of testing your suggestion:

$$P_2Pt\overset{\displaystyle\square}{\underset{Ph}{}} \rightleftharpoons P_2Pt\overset{\displaystyle\square}{}\!-\!Ph \tag{12.42}$$

12.9. Suggest a plausible mechanism for Eq. 12.43 and some ways of testing your suggestion:

$$[(cod)IrL(thf)]^+ \xrightarrow{\text{HCOONa}} [(cod)IrHL]^+ \tag{12.43}$$

12.10. Account for the product formed in Eq. 12.44:

$$L_3Fe\underset{}{\overset{}{\Big]}} \xrightarrow[\text{(ii) oxidize}]{\text{(i) } CO_2} HO_2C\!\!\diagdown\!\!\diagup\!\!\diagdown\!\!\diagup\!\!\diagdown\!\!\diagup\!\!\diagdown_{CO_2H} \tag{12.44}$$

12.11. Hydrosilation (shown below) is mediated by a variety of catalysts, both homogeneous and heterogeneous. Write a plausible mechanism for a generalized homogeneous catalyst L_nM.

$$RC\equiv CH \xrightarrow[\text{cata.}]{R_3SiH} RHC\!=\!CH(SiR_3) + \overset{R}{\underset{R_3Si}{\diagup}}C\!=\!CH_2 \tag{12.45}$$

12.12. If methanol/HI is carbonylated in a system resembling the Monsanto acetic acid process, but with $[(dpe)RhI(CO)]$ as catalyst and H_2 present, ethanol is formed from methanol. Provide two reasonable mechanisms and suggest an experimental test to distinguish between them.

12.13. A small amount of acetic anhydride, $(MeCO)_2O$, is sometimes formed in the acetic acid process of Fig. 12.5. How would this be formed and how could one enhance the rate of anhydride formation? The Eastman process for the synthesis of acetic anhydride converts MeCOOMe and CO into (MeCO)$_2$O with the same [RhI$_2$(CO)$_2$]$^-$ catalyst as in the Monsanto process but with LiI as additive. Suggest a mechanism given that the first step is: LiI + MeCOOMe $\Rightarrow$ MeI + MeCOOLi (J. R. Zoeller, V. H. Agreda, S. L. Cook, N. L. Lafferty, S. W. Polichnowski, and D. M. Pond, *Catal. Today,* **13**, 73, 1992).

12.14. Why is the methane activation step of Eq. 12.46 proposed to be a σ-bond metathesis rather than oxidative addition/reductive elimination, and why does β elimination not occur to give Me$_2$C=CH$_2$ as final product?

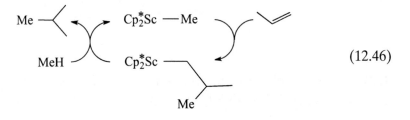

$$(12.46)$$

13

CLUSTERS, NANOPARTICLES, MATERIALS, AND SURFACES

We now go beyond mononuclear metal complexes to look at larger structures. Beginning with metal clusters, typically consisting of 2–50 metal atoms, we move to metal nanoparticles (NPs) with 50 to ca. 10^8 atoms and finally to bulk organometallic materials.

Unsaturated (<18e) ML_n fragments can combine to form $[ML_n]_m$ clusters, held together by M–M bonds or bridging ligands.[1] Unlike elements of groups 14–16 that form chains and rings (e.g., C_6H_{12}, and S_8), transition metals, as electron-deficient elements, prefer to form clusters that maximize the number of M–M bonds, and thus maximize sharing of the limited number of available electrons. They thus resemble boron, a main-group, electron-deficient element that also forms numerous clusters.

A cluster is a small fragment of metal, often surrounded by stabilizing ligands. This picture is best adapted to understanding NPs, where a somewhat larger core is involved than in a cluster. More labile ligands than CO are often chosen for NPs to favor dissociation to reveal the surface metal atoms for enhanced reactivity, for example, when a substrate for a catalytic reaction is present. The term nanocluster is sometimes used for structures having ~50 atoms that are intermediate in size between clusters and NPs.

The Organometallic Chemistry of the Transition Metals, Sixth Edition.
Robert H. Crabtree.
© 2014 John Wiley & Sons, Inc. Published 2014 by John Wiley & Sons, Inc.

Reactive organometallic groups can be grafted to surfaces of materials, such as SiO_2, so as to give unusual reactivity patterns. We therefore conclude this survey with a study of some typical cases.

13.1 CLUSTER STRUCTURES

The term *cluster* is used in a very broad sense in chemistry to indicate an assembly of similar units. In the organometallic field, it generally means a complex having a number of metal–metal bonds, but in bioinorganic or coordination chemistry usage (Chapter 16), the term refers to any multimetal unit where the metals are merely held together by bridging ligands without metal–metal bonds.

Being small and high field, carbon monoxide is the most common ligand in low valent organometallic metal clusters, for example, $Ru_3(CO)_{12}$ (**13.1**). CO is small enough to bind in sufficient numbers to electronically saturate each metal of the cluster without causing a steric clash with neighboring COs.

Higher valent clusters such as $[Re_2Cl_8]^{2-}$ (**13.2**) also exist.[2a] As expected for hard, high valent Re(III), the preferred ligand sets include halides and N and O donors, rarely CO.

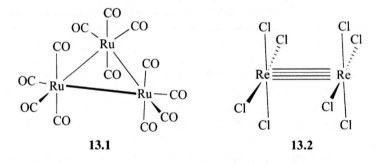

13.1   **13.2**

Cluster chemistry usually requires X-ray crystallography for characterization, so structural aspects have received the most attention. The M–M single-bond lengths are often comparable with those found in the elemental metal, and the metals can also be bridged by ligands such as CO. Not all M–M bonds are bridged; $[(CO)_5Mn–Mn(CO)_5]$ is a relatively rare unbridged example, but this bond is weak (28 ± 4 kcal/mol) and unusually long, at 2.93 Å versus 2.46 Å in $[(CO)_3Fe(\mu\text{-}CO)_3Fe(CO)_3$. With a bond strength of only 17 kcal/mol, the unsupported M–M bond of $[CpCr(CO)_3]_2$ reversibly dissociates even at 25°. We use the traditional bonding model in this chapter, not the new one[2b] mentioned on pp. 104–105.

Effective Atomic Number (EAN) Rule

Only the simpler clusters are best described in terms of the 18e rule. For example, each 16e $Os(CO)_4$ group in $Os_3(CO)_{12}$, **13.3**, attains 18e by forming two M–M bonds, each bond adding 1e to the count for each metal. Since the metals have the same electronegativity, the M–M bond is considered as contributing nothing to the oxidation state — the complex thus contains 18e, Os(0).

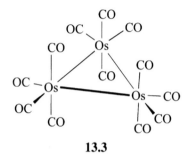

13.3

It is more common to count the electrons for the cluster as a whole, rather than attempt to assign electrons individually to each metal. On this convention, $Os_3(CO)_{12}$ is a $3 \times 8e$ (Os is in group 8) $+ 12 \times 2e = 48e$ cluster. We might think $3 \times 18e = 54e$ would be the right number to expect, because we have $3 \times 9 = 27$ orbitals. The difference arises because in summing the totals from each metal, we count the M–M bonding electrons twice over. In counting $Os^{(1)}$, we count 1e "originating" from a bookkeeping point of view from $Os^{(2)}$. In counting $Os^{(2)}$, we would count the same M–M bonding electrons again. Six M–M bonding electrons are involved in the double count, so we expect $54 - 6 = 48e$ to adjust for the overcount arising from the three M–M bonds. Since cluster electron counts are always $>18e$, we use an alternative name, the effective atomic number, or EAN, rule. In **13.3**, Os attains the same electron count as radon and is therefore said to have the same effective atomic number as Rn.

The EAN electron count for a cluster of nuclearity x and having y metal–metal bonds is given by Eq. 13.1.

$$\text{EAN count} = 18x - 2y \qquad (13.1)$$

For $Mo(CO)_6$, for example, y is 0 and we expect an 18e count from Eq. 13.1. For $(CO)_5Mn–Mn(CO)_5$, y is 1 and we expect a count of $(2 \times 18) - 2 = 34e$. This is indeed the case because 2 Mn contribute 14e and 10 COs contribute 20e, so this is an EAN cluster. The 48e $Os_3(CO)_{12}$ case ($y = 3$) was discussed above. For tetrahedral $Rh_4(CO)_{12}$, $y = 6$

$Re_4H_4(CO)_{12}$	$4 \times Re$	= 28		$Fe_6C(CO)_{16}^{2-}$	$6 \times Fe$	= 48
	$4 \times H$	= 4			$1 \times C$	= 4
	$12 \times CO$	= $\underline{24}$			$16 \times CO$	= 32
		56			$2 \times e^-$	= $\underline{2}$
						86
$Os_3H_2(CO)_{10}$	$3 \times Os$	= 24		$Fe_3(\mu\text{-}CO)_2(CO)_{10}$	$3 \times Fe$	= 24
	$2 \times H$	= 2			$2 \times \mu\text{-}CO$	= 4
	$10 \times CO$	= $\underline{20}$			$10 \times CO$	= $\underline{20}$
		46				48

FIGURE 13.1   Electron counting in clusters.

(**13.4**) and we expect 60e from Eq. 13.1, as are indeed present. For TBP $Os_5(CO)_{16}$, $y = 9$ (**13.5**) and we expect 72e for an EAN cluster, as again found. The bridging CO adds 2e to the cluster count as a whole, just like a terminal CO, so we cannot tell by counting electrons whether to expect terminal or bridging COs. In the isoelectronic Group 8 $M_3(CO)_{12}$ series, only the iron analog, **13.6**, has bridging COs–the others have none (e.g., **13.3**). These ideas are extended to more complex clusters in Fig. 13.1.

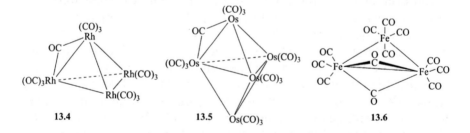

The 46e cluster, $Os_3H_2(CO)_{10}$, lacks 2e from the EAN count of 48e and is therefore an unsaturated cluster and much more reactive than $Os_3(CO)_{12}$. On the traditional view, it contains a Os=Os "double bond" because the EAN count for a system with four M–M bonds in a three-atom cluster is 46e. We would then regard an Os=Os double bond, like a C=C double bond, as being unsaturated. Structure **13.7** shows that there are in fact two Os–H–Os bridges that can be considered as protonated M–M bonds just as we saw for M(μ-H)M bridges in Section 3.3. For greater clarity in **13.7**, a single unlabeled line drawn from the metal denotes a terminal carbonyl and a bent line denotes a bridging CO; only non-CO ligands are always shown explicitly.

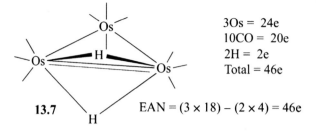

$$3Os = 24e$$
$$10CO = 20e$$
$$2H = 2e$$
$$Total = 46e$$

13.7

$$EAN = (3 \times 18) - (2 \times 4) = 46e$$

This means that the Os=Os "double bond" is really a reflection of the presence of the two hydride bridges, each of which can open to generate a vacant site. This makes the dihydride far more reactive than $Os_3(CO)_{12}$ and therefore a very useful starting material in triosmium cluster chemistry.

The tetranuclear group 9 clusters $M_4(CO)_{12}$ have 60e. By Eq. 13.1, six M–M bonds are present as required by the EAN rule. We can either assign the EAN count if we know how many M–M bonds are present, or we can assume an EAN structure and deduce the number of M–M bonds expected.

Face (μ_3) bridging is a bonding mode unique to polynuclear complexes. If we have a face bridging CO (**13.8**), we count only the 2e of the carbon lone pair as contributing to the cluster. On the other hand, some ligands have additional lone pairs they can bring into play. A Cl ligand is 1e when terminal, **13.9**, but 3e when edge (μ_2) bridging, **13.10**, and has 5e to donate to the cluster if it is face bridging (**13.11**), as two of its lone pairs now come into play.

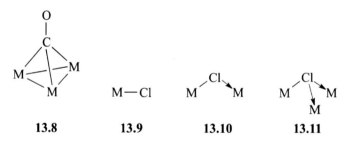

| **13.8** | **13.9** | **13.10** | **13.11** |

Wade–Mingos Rules

With six or more metals, the EAN picture can start to fail. For example, $[Os_6(CO)_{18}]^{2-}$, **13.12**, has 86e. Assuming there are 12 M–M bonds, Eq. 13.1 predicts that an EAN structure should have 84e. Yet the cluster shows no tendency to lose electrons or expel a ligand. $Os_6(CO)_{18}$, **13.13**, an authentic 84e cluster, does not adopt the same Os_6 octahedron framework but does have 12 M–M bonds.

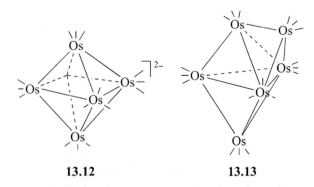

13.12   **13.13**

The cluster-counting model that applies to these non-EAN clusters is the polyhedral skeletal electron pair theory, known as the *Wade–Mingos rules* (W-M).[3] On this picture, an analogy is drawn between the metal cluster and the corresponding boron hydride cluster. Elements such as C and H, that have the same number of electrons and orbitals, can form closed-shell molecules, as in CH_4. Elements to the right of carbon, such as N, have more electrons than orbitals and so give molecules with lone pairs, as in NH_3. Like transition metals, boron has fewer electrons than orbitals, and so it forms compounds in which the BH_x units cluster together to try and share out the few electrons that are available by using 2e, three-center bonds, as in B_2H_6. The higher borane hydride anions ($n = 6$–12) form polyhedral structures, some of which are shown in Fig. 13.2, that form the basis for the polyhedral structures adopted by all W-M clusters. The shape of a Wade cluster is decided purely by the number of cluster electrons, called "skeletal" electrons.

To assign skeletal electrons for $[B_nH_n]^{2-}$, we note that each B has one nonbridging B–H bond that is a normal 2e covalency, requiring 1e from H and 1e from B. Boron now has two of its three valence electrons left to contribute to the cluster, and this in turn means that $[B_nH_n]^{2-}$ has $2n + 2$ cluster electrons, $2n$ of which come from the n BH groups, and

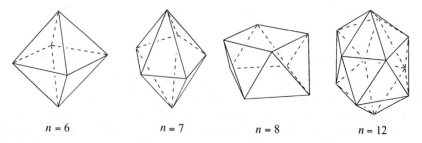

$n = 6$ $n = 7$ $n = 8$ $n = 12$

FIGURE 13.2   Some common polyhedral structures adopted by boranes $[B_nH_n]^{2-}$.

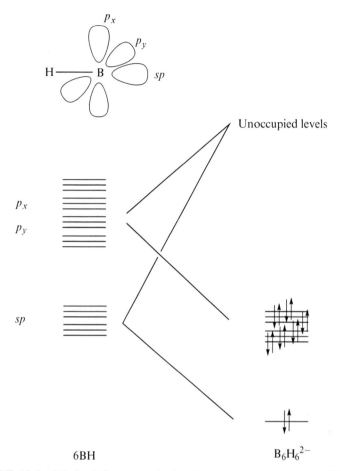

FIGURE 13.3 Wade–Mingos analysis of a closo borane cluster $[B_6H_6]^{2-}$.

the remaining two from the 2− net ionic charge. In order to see where these $2n + 2$ electrons go, we consider that each BH unit has an sp orbital pointing directly toward the center of the cluster, and a p_x and a p_y orbital, pointing along the surface (Fig. 13.3). The MO analysis predicts that the radial sp orbitals all contribute in phase to a single low-lying orbital. The lateral p orbitals, $2n$ in number, combine to give n filled bonding MOs and n empty antibonding MOs. This picture provides a total of $n + 1$ bonding orbitals, which offer a home for $2n + 2$ skeletal electrons.

Since the cluster shape depends only on the numbr of skeletal electrons, we can remove a vertex as a BH^{2+} group without changing the cluster structure, because we leave behind the two skeletal electrons provided by that vertex. From $[B_6H_6]^{2-}$, we can therefore remove BH^{2+} to get the hypothetical $[B_5H_5]^{4-}$ (Eq. 13.2). This retains the original polyhedral structure and if we add four H^+ to neutralize the 4 ion charge, we get the known neutral borane, B_5H_9. The protons bridge the faces of the polyhedron that include the missing vertex, attracted by the electron density left behind on removal of the BH^{2+}. As a species with one empty vertex, B_5H_9 is given the descriptor *nido*. Molecules that have every vertex occupied are designated *closo*. In general, a species $[B_xH_y]^{z-}$ has the number of skeletal electron pairs, e_{skel}, and number of vertices, v, given by Eq. 13.3 and Eq. 13.4.

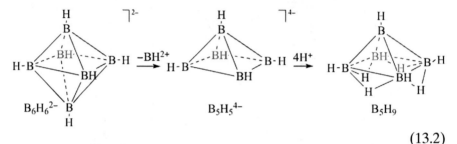

$$(13.2)$$

$$e_{skel} = \tfrac{1}{2}(x+y+z) \tag{13.3}$$

$$v = \tfrac{1}{2}(x+y+z)-1 \tag{13.4}$$

Say the number of BH groups that exist in a given compound and for which we have to find vertices is x. Then if the number of vertices v called for by W-M rules also happens to equal x, each vertex will be occupied and we will have a closo structure. On the other hand, if x is $(v-1)$, one vertex will be empty and a nido structure will result. If x is $(v-2)$ or $(v-3)$, then the resulting *arachno* and *hypho* structures have two or three empty vertices. In such a case, the empty vertices prefer to be adjacent.

W-M rules can also apply to other main group clusters. For example, $[Sn_6]^{2-}$ is an octahedral cluster with 14 skeletal electrons because Sn, with four valence electrons, uses up $2e$ in the terminal lone pair analogous to the exo B–H bonds of $[B_6H_6]^{2-}$ and has two left to act as skeletal electrons. We can also replace BH groups by such L_nM groups that also provide the same number of skeletal electrons. Since transition metals have nine orbitals but only three are required for cluster bonding on the W-M picture, we first have to fill the six orbitals not required for cluster bonding to see how many electrons remain for the cluster-bonding orbitals. For the $Os(CO)_3$ fragment, we assign the nine orbitals as follows: (1) three orbitals are filled with the three CO lone pairs; (2) three more orbitals are filled with six electrons out of the eight

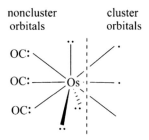

FIGURE 13.4 Applying Wade–Mingos rules to a transition metal fragment. The three CO groups of Os(CO)₃ supply 6e, and these electrons occupy three of the metal's nine orbitals. Six of the eight metal electrons occupy the d_π orbitals and back-bond to the CO groups. Two metal electrons are left to fill the three cluster-bonding orbitals shown to the right of the dotted line.

electrons appropriate for d^8Os(0)–these electrons back-bond to the COs; and (3) two metal electrons are now left for the remaining three orbitals, the ones that bond to the cluster (Fig. 13.4). This implies that Os(CO)₃ contributes the same number of skeletal electrons—two—as does a BH group. We can therefore replace all the BHs in $[B_6H_6]^{2-}$ with Os(CO)₃ groups without altering the W-M structure. We end up with $[Os_6(CO)_{18}]^{2-}$, **13.12**, the very cluster we could not rationalize on the EAN model.

In *metalaboranes*, some of the vertices have a boron atom and others a transition metal, as in *closo*-(CpCo)₂(BH)₄(μ₃-H)₂ **(13.14)**. For the fragment MX_aL_b, the W-M analysis leads us to predict that the cluster electron contribution, F, of that fragment from Eq. 13.5, where N is the group number of metal.

$$F = N + a + 2b - 12 \qquad (13.5)$$

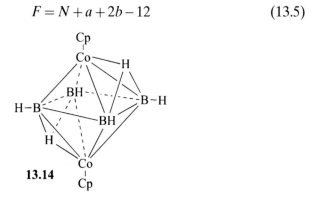

13.14

To find the total number, T, of cluster electrons, we then sum the contribution from all the fragments in the cluster, add the sum of the contributions from the bridging ligands (ΣB) to account for any electrons donated to the cluster by edge bridging, face bridging, or encapsulated atoms (see example below), and adjust for the total charge, z^-, on the cluster as a whole by Eq. 13.6, where $B = 1$ for bridging H, 2 for bridging CO, 3 for μ^2-Cl, and so on. The number of vertices, v, in the cluster will then be given by Eq. 13.7.

$$T = \Sigma F + \Sigma B + z \qquad (13.6)$$

$$v = \left(\frac{T}{2}\right) - 1 \qquad (13.7)$$

We have seen what happens in a borane cluster if there are not enough BH fragments to fill the vertices—we get a nido structure with an empty vertex. The same is true for transition metal clusters; for example, in $Fe_5(CO)_{15}C$, the carbon atom is encapsulated within the cluster and gives all its four valence electrons to the cluster and so this carbon is not considered a vertex atom. The $Fe(CO)_3$ fragment contributes two cluster electrons as it is isoelectronic with $Os(CO)_3$. The total Wade count is therefore $(5 \times 2) + 4 = 14$, and the number of vertices is $(14/2) - 1 = 6$. This requires the structure shown as **13.15**, as is indeed observed for this and the analogous Ru and Os species, where one vertex is empty.

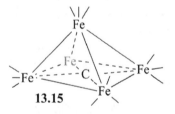

13.15

What happens when there are more atoms than vertices into which they can fit? For example, $Os_6(CO)_{18}$ is a $(6 \times 2) = 12$ cluster electron species. This means that the number of vertices required by W–M rule is $(12/2) - 1 = 5$. The structure found for the molecule, **13.13**, shows that the extra metal atom bridges to a face of the five-vertex base polyhedron, and so is able to contribute its electrons to the cluster, even though it cannot occupy a vertex.

Only when we move up to clusters of nuclearity 6–12, do the EAN and Wade predictions become different. Often the W–M structure is the one observed, but sometimes we find that both a W–M and an EAN cluster are stable. Adams and Yang have shown how in such situations there can be facile interconversion between the two forms by gain or loss of a ligand (Eq. 13.8).

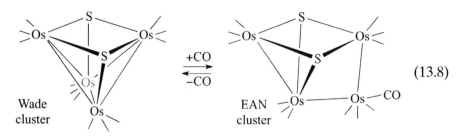

$$(13.8)$$

Some structures, such as $[Ni_2Sn_7Bi_5]^{3-}$, escape classification under either rule, however.[4]

M–M Multiple Bonds and Formal Shortness Multiply bonded species, such as $[Re_2Cl_8]^{2-}$ (**13.2**) are seen for the middle transition elements, the same ones that give strong $M\equiv O$ multiple bonds (Section 11.5). For two L_xM units to form a bond of order n, the L_xM fragment has to have a d^n or higher configuration because it needs a minimum of n electrons in n orbitals to form n M–M bonds, just as the CH fragment needs three available electrons in three orbitals to form $HC\equiv CH$. In **13.2**, two square planar fragments face each other in the unusual eclipsed geometry. Taking the M–M direction as z, the quadruple bond is formed from overlap of the d_{z^2} (a σ bond), the d_{xz} and d_{yz} (two π bonds), and of the d_{xy} on each Re, which forms the δ bond. The δ bond can only form in the eclipsed geometry where $d_{xy}-d_{xy}$ overlap is possible (**13.16**). The M–M bonding in **13.2** is $\sigma^2\pi^4\delta^2$, while $(RO)_3Mo\equiv Mo(OR)_3$ has an $\sigma^2\pi^4$ M–M triple bond in a staggered geometry, **13.17**.

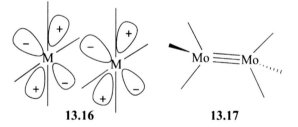

13.16   **13.17**

M–M multiple bonds are short—typical values for Mo are 2.1 Å for bond order 4; 2.2 Å for order 3; 2.4 Å for 2 and 2.7 Å for a single Mo–Mo bond; for comparison Mo-Mo is 2.78 Å in elemental Mo. Bond strengths are known for few systems—for the ReRe bond of order 4 in **13.2** it is $85 + 5$ kcal/mol, of which only ~ 6 kcal/mol is assigned to the δ bond, making it comparable to a hydrogen bond.

The length of an M–M multiple bond is judged by the formal shortness ratio, or FSR, defined as the M–M distance divided by the sum of the appropriate Pauling atomic radii. $[Re_2Cl_8]^{2-}$ has an FSR of 0.87, for example. A quintuple CrCr bond (1.835 Å) is known in ArCrCrAr.[5] Quintuple bonds have been proposed, for example, for the dimolybdenum complex, **13.18**.[6] Known only in the gas phase, diatomic Cr_2, a d^6 Cr(0) dimer, could in principle have a bond order of six, and indeed its FSR is even lower, 0.71, but the dissociation energy being a mere 33 cal/mol^{-1}, a sextuple bond may not be the best model.

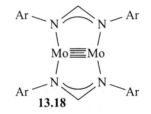

13.18

M–M Dative Bonds   If a reduced, 18e Lewis basic metal is adjacent to a 16e Lewis acidic one, a weak dative bond can form between them, with the Lewis base formally acting as a 2e donor L ligand to the Lewis acid. Structure **13.19** has an Os(0) donor and a Cr or W(0) acceptor lacking any bridging ligands to provide any additional help. Other cases are proposed for bioinorganic metal clusters (Chapter 16).[7]

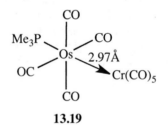

13.19

Metallophilicity   The heavy d^{10} metals can exhibit weak homonuclear M⋯M interactions, best known for Au⋯Au (aurophilicity), and appearing in crystal structures as short contacts of 2.8–3 Å versus the van der Waals Au Au nonbonding contact distance of 3.3 Å; interaction energies in the range of 5–15 kcal/mol have been estimated.[8] The preferred linear AuL_2 coordination for Au(I) permits close Au⋯Au approach by minimizing interligand steric effects. [Pt(NH$_2$R)$_4$][PtCl$_4$] (**13.20**) derivatives form linear M⋯M⋯M chains both in the solid and in solution, indicating that d^8 metals can show similar effects.[9]

13.2   THE ISOLOBAL ANALOGY

Hoffmann's[10] isolobal analogy is a unifying principle that identifies analogies between organic and inorganic structures in terms of their bonding pattern. Different L_nM groups are considered isolobal with

CH_3, CH_2, or CH when they have similar orbital symmetry and occupation that give rise to similar bonding preferences. A methyl radical is univalent because of its singly occupied sp^3 orbital. This fragment has one orbital and one "hole" to give a 7e configuration. As far as the X group in Me–X is concerned, a methyl radical can be considered as providing a hole and an orbital. Hoffmann points out that any fragment like $\cdot Mn(CO)_5$ with a half-filled orbital of a σ type may show analogy with the methyl group. Indeed, the $\cdot Mn(CO)_5$ radical can replace one methyl group in ethane to give $MeMn(CO)_5$, or both of them to give $(CO)_5Mn-Mn(CO)_5$. The two fragments, $\cdot CH_3$ and $\cdot Mn(CO)_5$, are not isoelectronic because $\cdot Mn(CO)_5$ has far more electrons than $\cdot CH_3$, but the key orbitals involved in forming bonds are the same in number, symmetry, and occupancy. The isolobal analogy is expressed by a double-headed twisted arrow, as in Eq. 13.9.

$$Me-Me \quad \longleftrightarrow \quad Me-Mn(CO)_5 \quad \longleftrightarrow \quad (CO)_5Mn-Mn(CO)_5$$

$$(13.9)$$

$Cr(CO)_5$ with one empty orbital having two holes is isolobal with CH_3^+. Just as $Cr(CO)_5$ reacts with CO to give $Cr(CO)_6$, CH_3^+ reacts with CO to give the acetyl cation CH_3CO^+, a Friedel–Crafts intermediate, formed from MeCOCl and $AlCl_3$.

The CH_2 fragment can use its two orbitals and two holes quite flexibly by changing its hybridization. If CH_2 binds two H atoms to give CH_4, sp^3 hybridization now applies, but if it binds to a second CH_2 to give C_2H_4, then sp^2 hybrids form the σ bonds and a p orbital forms the π bond. $Mo(CO)_5$ is isolobal with CH_2 because it also has two orbitals and two holes. The empty orbital is the 2e vacancy at the metal and the other is a d_π orbital normally involved in back bonding. Fischer carbenes $(CO)_5Mo=CR_2$ show how the $Mo(CO)_5$ fragment replaces one CR_2 in the alkene $R_2C=CR_2$.

Table 13.1 shows how the analogy works. We need to calculate n_H, the number of holes in our metal fragment via Eq. 13.10, where N is the group number of the metal, shows this explicitly for the $[MX_aL_b]^{c+}$ ion.

$$n_H = 18 - N - a - 2b + c \qquad (13.10)$$

This shows us which organic fragments are isolobal with the organometallic fragment in question. The most direct analogy will be with the organic fragment that has the same number of orbitals. For the metal fragments, the number of orbitals, n_o, is calculated on the basis of an octahedral model. If there are three ligands in the fragment, three orbitals of the octahedron are available.

TABLE 13.1 Isolobal Relationships^a

Inorganic Fragment	n_H	n_o	Organic Fragment	Complex	Isolobal with
$Mn(CO)_5$	1	1	CH_3	$Me–Mn(CO)_5$	$Me–Me$
$Mo(CO)_5$	2	1	$\{CH_3\}^+$	$Me_3P–Mo(CO)_3$	$Me_3P–Me^+$
$Mo(CO)_5$	2	2^b	CH_2	$OC=Mo(CO)_5$	$OC=CH_2$
$Mo(CO)_5$	2	3^b	CH^-	–	–
$Fe(CO)_4$	2	2	CH_2	$(C_2H_4)–Fe(CO)_4$	Cyclopropane
$Cp(CO)_2Mo$	3	2^b	$\{CH_2\}^+$	–	–
$Cp(CO)_2Mo$	3	3^b	CH	$Cp(CO)_2Mo\equiv CR$	Acetylene
$CpRh(CO)$	2	2	CH_2	$\{CpRh(CO)\}_2(\mu\text{-}CH_2)$	Cyclopropane
$\{PtCl_3\}^-$	2^c	1^d	$\{CH_3\}^+$	$[Cl–PtCl_3]^{2-}$	Cl–Me
$\{PtCl_3\}^-$	2^c	$2^{b,d}$	CH_2	$[(\eta^2\text{-}C_2H_4)–PtCl_3]^-$	Cyclopropane

$^a n_H$ and n_o are the number of holes and of orbitals.
^{b}After rehybridizing to include one or more d_π orbitals. Note that on the deprotonation analogy, CH_3, CH_2^- and CH^{2-} are isolobal, as are CH_3^-, CH_2, and CH^- and CH_3^{2+}, CH_2^-, and CH.
^{c}On the basis of a 16e closed shell.
^{d}On a square planar basis.

$$n_O = 6 - a - b + \pi \qquad (13.11)$$

Metal fragments can make up to three more orbitals available by using their d_π set; these are denoted as π in Eq. 13.11, where π can go from zero to three. Reference to Table 13.1 will show how we often have to resort to using the d_π set. For example, $Mo(CO)_5$ is isolobal with CH_3^+ by Eq. 13.10 and Eq. 13.11 ($n_H = 2$, $\pi = 0$, $n_o = 1$). If we bring in an extra filled d_π orbital, we move to ($n_H = 2$, $\pi = 1$, $n_o = 2$), which makes the fragment isolobal with CH_2. This means that the $Me_3P–Mo(CO)_5$ or $Me–Mn(CO)_5$ bonds are formed without a significant contribution from a d_π orbital, while the $OC=Mo(CO)_5$ double bond with its strong Mo-to-CO π back-bonding component requires a strong contribution from a d_π orbital.

Because CH has three orbitals and three holes, the most direct analogy ($\pi = 0$) is therefore with the Group 9 $M(CO)_3$ fragments, such as $Co(CO)_3$. Figure 13.5 shows the stepwise conceptual conversion of the hydrocarbon tetrahedrane into a tetrahedral $M_4(CO)_{12}$ cluster by the isolobal replacement of $M(CO)_3$ groups by CH. $Co_4(CO)_{12}$ has a bridged structure, and only the Rh and Ir analogs are all-terminal; since the all-terminal structure can only be unstable with respect to the real structure by a few kilocalories per mole for Co, we must not hold it against the isolobal analogy, or any counting rule for not being able to

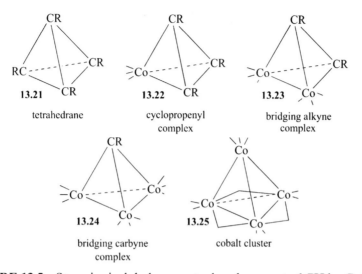

FIGURE 13.5 Stepwise isolobal conceptual replacement of CH by $Co(CO)_3$ in tetrahedrane. $Co_4(CO)_{12}$ has the CO bridged structure shown.

predict the pattern of CO bridges. Structure **13.24**, best known for Co, is normally considered as μ_3-carbyne cluster. Structure **13.23** is usually considered as a bridging alkyne complex of $Co_2(CO)_8$, and **13.22** as a cyclopropenyl complex of $Co(CO)_3$. The all-carbon compound, **13.21**, is unstable and reverts to two molecules of acetylene, but stable tetrahedranes C_4R_4 have been made by using very bulky R groups.

For those metals that prefer to be 16e, such as Pt(II), the number of holes is determined on the basis of a closed shell of 16e, not 18e, so 16 replaces 18 in Eq. 13.10. The argument is that the fifth d orbital, although empty, is too high in energy to be accessible, and so its two holes do not count. For example, the 14e $[PtCl_3]^-$ fragment is considered as having two holes, not four. The number of orbitals is also calculated on the basis of a square planar structure, so 4 replaces 6 in Eq. 13.11 and $[PtCl_3]^-$ has one orbital and is therefore isolobal with CH_3^+. Both species form a complex with NH_3, for example, $[H_3NPtCl_3]^-$ and $[H_3NCH_3]^+$. An extra nonbonding orbital on Pt can also be considered to contribute ($\pi = 1$), giving two orbitals and two holes, which makes $[PtCl_3]^-$ isolobal with CH_2. Both fragments form complexes with ethylene, $[(\eta^2\text{-}C_2H_4)PtCl_3]^-$, and cyclopropane, respectively.

Any bridging hydrides can be removed as protons leaving the M–M bond that accompanies bridging; for example, the dinuclear hydride in Eq. 13.12 is isolobal with acetylene because the 15e $[IrHL_2]^+$ fragment has three holes and three orbitals. CO ligands contribute in the same way whether they are bridging or terminal (e.g., Eq. 13.12), but the isolobal (Eq. 13.13) rhodium dimer (Eq. 13.14) has bridging CO groups.

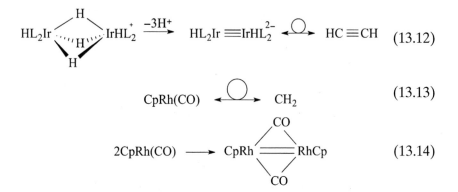

We must not expect too much from such a simple model. Many molecules isolobal with stable organic compounds have not been made. Finally, the isolobal analogy is structural—we cannot expect it to predict reaction mechanisms, for example. The isolobal concept has been greatly developed in recent years, notably to the case of gold.[10b]

13.3 NANOPARTICLES

Nanoparticles (NPs), termed colloids in the older literature, occur in a wide variety of practical contexts, such as the gold or silver NPs that give stained glass its deep red or light yellow color, respectively, or TiO_2 NPs that form the white base material of paint. Surprisingly, bulk silver and copper metal release metal NPs under conditions of high humidity, implying that humans have been exposed to these nanomaterials since prehistoric times.[11,12]

NPs have attracted intense attention and have numerous recent applications, notably in biomedicine and energy conversion.[13] They can be formed from almost any substance but most relevant to organometallic chemistry are transition metal NPs. These are small particles in the 1 to 100 nm (10–1000 Å) range, the smallest particles sometimes being termed nanoclusters. Their key characteristic is having a large surface area per unit mass of material—a significant fraction of the atoms present are at the surface. The most desirable synthetic routes control the size and size distribution as well as the nature of the surface, which is often stabilized by surfactants or ligands. Because small amounts of impurity can affect the NP growth, irreproducibility can be a problem in the area.

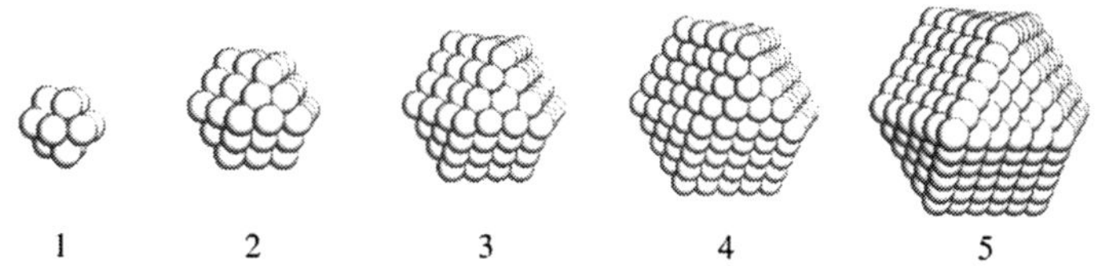

Full-Shell "Magic Number" Clusters				
Number of shells	1	2	3	4	5
Number of atoms in cluster	M_{13}	M_{55}	M_{147}	M_{309}	M_{561}
Percentage surface atoms	92%	76%	63%	52%	45%

FIGURE 13.6 Idealized nanoclusters of close-packed atoms with one to five shells, together with the numbers of atoms (magic numbers) in these nanoclusters. Source: From Aiken and Finke, 1999 [49]. Reproduced with permission of Elsevier.

NPs can also be formed inadvertently in supposedly homogeneous catalysts, where they can contribute to the catalytic activity.[14] For example, Pd NPs can catalyze all of the C–C coupling reactions of Section 9.7. Metal NPs can also be deliberately applied to catalysis.[15]

NPs maximize the fraction of metal at the surface. A metal powder, such as freshly precipitated palladium metal, for example, typically has a particle diameter on the order of 10^5 Å with only ca. 0.05% of the Pd on the surface. In contrast, a small NP of diameter 12 Å has $\sim$50% of the Pd atoms on the surface (Fig. 13.6), each potentially able to act as a catalytic center. Varying the size and shape of NPs change the properties of the material itself. As the particle size increases, typical bulk metallic properties develop incrementally, so NPs have different physical properties from the bulk material, such as the deep red color seen for gold NPs. Figure 13.6 shows the "magic numbers" of metal atoms associated with globular particles, following equation Eq. 13.15, where the $(10n^2 + 2)$ term represents the number of atoms in the nth shell of the perfect close packed NP, although such perfection is only an ideal.

$$\text{No. of atoms} = 1 + \sum_{1}^{n} (10n^2 + 2) \qquad (13.15)$$

Syntheses start from inorganic salts or organometallic precursors, with the size, shape, and surface properties of NPs controlled by the conditions.[16] Particle growth follows initiation. If initiation is fast relative to growth, the particles will be small and many, if slow, large and few. Additives such as polyvinyl alcohol $(CH_2–CH_2OH)_n$ bind to the surface of the particles, slowing growth and inhibiting agglomeration. The particles may have the same close-packed structure as the bulk metal, with specific faces of the nanocrystal being structurally distinct. If an additive binds selectively at a particular face of a growing nanocrystal, particle growth can be selectively suppressed at that face so that the shape of the nanocrystal can be affected. Recent advances in sample preparation and instrumentation have allowed the details of the growth of Pt NPs to be directly imaged at atomic level resolution by transmission electron microscopy; in particular, nanocrystals were found to coalesce by attaining a specific orientation such that their crystal lattices matched.[17] Clusters as small as Os_3 can also be successfully imaged.[18]

Magnetic NPs of Fe or Co, suspended in a hydrocarbon, can act as magnetoresponsive fluids and can form liquid seals when held in place by a magnetic field. The needed Co NPs can be obtained from $Co_2(CO)_8$ and AlR_3.

Chaudret[19] has shown that the metal–organic precursor $Fe[N(SiMe_3)_2]_2$ can be reduced by H_2 in the presence of n-$C_{16}H_{33}NH_2$ to give iron NPs of very similar shape and size—cubes of 7 Å edge length. They even "crystallize" into a cubic superlattice.

Gold NPs stabilized with PhC_2H_4SH are isolable as $Au_{38}(SR)_{24}$.[20] When two different metals are reduced, alloy or "onion" structures can be formed. In the latter case, a colloid of one metal is used as the seed particles for growing a second metal: Au encapsulated by Pt is an example. Bare Au_{19} and Au_{20} clusters (**13.26–13.27**) have been structurally characterized in the gas phase by comparison of their experimental IR spectra with those predicted by DFT computations.[21] The catalytic activity of small Au NPs, surprising since bulk Au is inert, has been associated with the more reactive, low coordinate "corner" Au atoms. Au NPs in the 3–10 atom size range can be extremely active catalysts for the hydration of alkynes with very high TOFs in the range 10^5–10^7 h^{-1}.[22] Unlike the thiol-stabilized Au NPs mentioned earlier, in this case, the NPs are substrate stabilized so no other groups are present that might limit substrate access to the catalyst. The very small size of the clusters and the solution environment mean that these catalysts fall in between the homogeneous and heterogeneous realms—we can best consider them as operationally homogeneous.

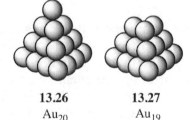

13.26	13.27
Au_{20}	Au_{19}

Gold NPs have also attracted biomedical attention. These particles show enhanced light scattering via a quantum effect that permits localization of cancers if suitable ligands are attached to the NPs that bind specifically to cancer cells. Au NPs can also convert incident light into heat, suggesting the possibility of thermal destruction of cancer cells.[23]

Beyond elemental metals, metal oxide NPs are gaining attention. Grätzel's solar cells rely on TiO_2 NPs for supporting the photosensitizer, a Ru polypyridyl complex.[24] One role of the TiO_2 is to provide a high surface area for the surface-bound complex to intercept the maximal amount of sunlight. Another is to accept electrons photoinjected from the photosensitizer into the TiO_2 conduction band. Unlike a molecule, with discrete, well-separated bonding and antibonding levels, NPs have a band structure, consisting of continuous energy ranges that have filled energy levels, the valence band, higher energy ranges that have unfilled energy levels, the conduction band, and a range between the two that is empty, the band gap. The ~3.2 eV band gap of TiO_2 corresponds to ~390 nm so the gap is just big enough to make the material white in color by preventing light absorption in the visible range (400–800 nm). IrO_2 and RuO_2 NPs are among the best known water oxidation catalysts and evolve O_2 when driven electrochemically or with chemical oxidants such as Ce(IV).[25]

13.4 ORGANOMETALLIC MATERIALS

Inorganic materials from concrete to silicon chips are indispensable to modern life. The need for designed materials with special properties will continue to grow in the future, and organometallic chemistry is beginning to contribute in several ways. For example, in metal organic chemical vapor deposition (MOCVD),[26] a volatile metal compound is decomposed on a hot surface to deposit a film of metal. A typical example is the pyrolysis of $Cr(C_6H_6)_2$ to deposit Cr films. Atomic layer deposition (ALD) is a more controlled form of this process in which much thinner layers are deposited, even on surfaces with complicated shapes. $[(MeC_5H_4)PtMe_3]$, Cp_2Ru, and lanthanide N,N-dimethylaminodiboranates are established ALD precursors for Pt, Ru, and lanthanide thin films, for example.[27]

Bulk Materials

Bulk elemental metals have close-packed atoms in which layers such as **13.28** are stacked such that each metal has 12 neighbors, 6 in the same layer and 3 each from the layers above and below. This produces voids between the atoms that can be filled by other atoms that are small enough. In each layer, there is a set of three-coordinate trigonal holes, but these are too small for most atoms to fit. Between the layers are larger four-coordinate tetrahedral holes (the central T in **13.29**) and even larger six-coordinate octahedral holes (the central O in **13.30**) that are suitable for larger atoms. For a metal of radius r_M, the tetrahedral hole has a radius $0.414\ r_M$ and the octahedral hole has a radius $0.732\ r_M$. A number of metals can take up H_2 to give interstitial metal hydrides in which the H_2 is dissociated into atoms that are typically located in tetrahedral holes. For example, the alloy $LaNi_5$ both absorbs and releases H_2 readily, accounting for its use in batteries, where it can reversibly store H_2. With increasing interest in hydrogen storage in alternative energy applications, hydrides such as MgH_2 are being considered as storage materials.[28]

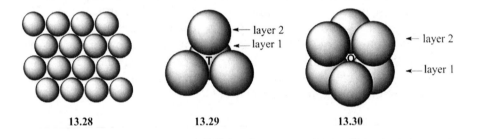

| 13.28 | 13.29 | 13.30 |

SrMg$_2$FeH$_8$, formed from SrMg alloy with Fe powder under H_2 at 500°C, contains an $[FeH_6]^{4-}$ unit best considered as an 18e d^6 Fe(II) polyhydride anion. After its discovery in this insoluble material, THF-soluble salts of $[FeH_6]^{4-}$ were also prepared, such as $[MgBr(thf)_2]_4[FeH_6]$. A large number of similar hydride materials exist with 16e ($[Li_3(RhH_4)]$) or 18e polyhydride anions, ($[Mg_3(H)(MnH_6)]$ or $[Mg_2(NiH_4)]$).[29]

Carbide (C^{4-}) has a big radius because the high anionic charge considerably expands the ion; this fits it for octahedral holes. These carbides are extremely high melting and very hard: tungsten carbide (WC) has a melting point of 2870°C and is almost as hard as diamond. HfC has the astonishingly high melting point of 3890°C. Other useful properties appear: NbC is a superconductor below 10 K. Several carbides, including WC, are catalytically active.[30]

Polymers of metal-containing monomers have applications from sensors to catalysis.[31] For example, numerous catalysts have been attached to polymer supports for easier recovery and reuse. Ferrocenyl units have even replaced the native Cu ion in the electron transfer protein, azurin, with retention of redox activity. Conjugated polymers containing $-C{\equiv}C\text{-}PtL_2\text{-}C{\equiv}C-$ units (L = PBu$_3$) have shown promise as the light-absorbing component in organic photovoltaics for the conversion of solar to electrical energy.[32]

Organometallic compounds such as $[Cp*Ir(OH_2)_3]SO_4$ can be precursors for electrodeposition of a heterogeneous hydrated iridium oxide material that is an excellent electrocatalyst for water oxidation to O_2. The deposit is not formed from standard Ir salts and may consist of small $(IrO_2)_n$ clusters ($n = 2,3$) held together by carboxylate ligands formed in the oxidative degradation of $Cp*$.[33]

Porous Materials

Porous materials,[34] with well-defined structures having voids in the interior, have proved exceedingly valuable as catalysts. For example, *zeolites* have aluminosilicate lattices. Each relacement of Si(IV) in SiO_2 materials by an Al(III) leads to an additional unit negative charge. This is compensated by the presence of an acidic proton in the pores. Being poorly stabilized by the environment, this proton has superacid properties and acts as an acid catalyst, even able to protonate such weak bases as alkanes. Reaction with substrate only happens in the interior of the structure within a small cavity having small access channels of defined size, so only compounds having certain sizes can enter or leave, depending on the exact zeolite structure. Exxon-Mobil's acidic ZSM-5 zeolite catalyst, for example, converts MeOH to gasoline range hydrocarbons and water. Main-group elements predominate in this area, but hybrid materials with transition metal catalytic sites are becoming more frequently seen.

Metal Organic Frameworks (MOFs)

MOFs are three-dimensional coordination polymers.[35] In a typical example[36] bridging ligands such as 4,4'-dipyridine act as rigid rods to connect metal ions or small clusters to form open lattices that possess zeolite-like cavities with access channels; these *mesoporous* materials can be crystallized and their structures determined. So far, none is organometallic, but they have very impressive absorptive power for guest molecules including H_2 and CH_4. Another type of organometallic material is formed by crystallizing organometallic precursors that have hydrogen bonding groups capable of establishing a network of hydrogen bonds throughout the lattice.[37] This is sometimes called crystal engineering.

Porous Organic Polymers (POPs)

Similar to MOFs but constructed of organic monomer units are POPs. These promise to be more thermally and hydrolytically stable than

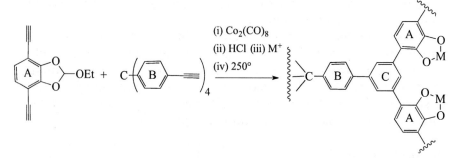

FIGURE 13.7 A porous organic polymer synthesis via Co-catalyzed alkyne trimerization.[38] The newly formed arene ring is labeled *C*.

MOFs but should otherwise show similar absorptive characteristics for small molecules. The organometallic aspect in this case is the $Co_2(CO)_8$ catalyst that brings about the needed polymerization by catalyzing alkyne trimerization. For example, Fig. 13.7 shows how the ethynyl groups of the monomers combine under the influence of the catalyst to produce the POP.[38]

Organometallic Polymers

Organometallic chemistry has contributed strongly to the polymer industry by providing polymerization catalysts (Section 12.2), but polymers derived from organometallic monomers are also attracting attention,[39] although for the moment they remain laboratory materials. Among the best studied polymers of this type are the poly(ferrocenylsilanes), formed by ring-opening polymerization (ROP) of ferrocenophanes (ferrocenes with bridges between the rings, e.g., **13.31**). The strain of ca. 16–20 cal/mol present in the bridge, evident from the ring tilt angle of 16–21°, serves to drive the polymerization. The thermal route of Eq. 13.16 gives very high-molecular-weight material (polymer from **13.31** M_w ca. 10^5–10^6). The polymer is processable and films can be formed by evaporating a solution on a flat surface. The nature of the R groups can also be readily changed (e.g., OR, NR_2, alkyl), allowing a range of materials to be accessed.

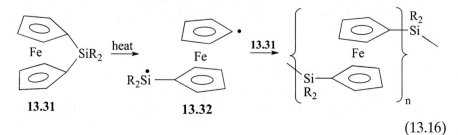

$$(13.16)$$

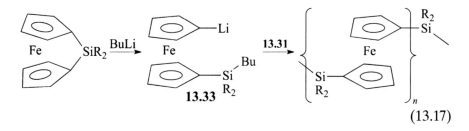

$$(13.17)$$

The thermal ROP presumably goes via the diradical **13.32**, but the polymerization can also be initiated by BuLi in THF (Eq. 13.17), in which case the intermediate is likely to be **13.33**. This route is more easily controllable and gives better polydispersity (less deviation of molecular weights of individual chains from the average molecular weight). The polymer is also *living*, meaning that the Li remains at the end of the chain, allowing chain-end functionalization or the introduction of a second monomer to form a new block of a second polymer. Transition metal-catalyzed ROP of **13.31** is also possible, and this has the advantage that the process is less affected by impurities than the BuLi-initiated version.

Molecular Wires

A number of conjugated 2D polymers are electrically conducting and offer promise for application in molecular electronics, a field that pushes the goal of miniaturization to the ultimate extreme. Self-assembled monolayers (SAMs) can be assembled on a gold electrode and the other end of the molecular wire probed with a Au tip of a conductive atomic force microscope (CAFM) for the conductivity measurement. An organometallic example is shown in Fig. 13.8, where conductivity is much enhanced by the presence of the redox-active ferrocene units.[40]

Molecular Electronics

Electrically conductive organic polymers, for which the Nobel Prize was awarded in 2000, rely on doping with electron donor or acceptor additives. By adding or subtracting electrons from an initially essentially nonconducting, conjugated, spin-paired polymer, a mobile unpaired electron is either introduced into the LUMO (*n*-type) or is left in the HOMO (*p*-type) after electron abstraction. Organometallics have proved useful in generating *n*-type polymers. For example, the C–C coupled dimer (**13.34**) of the 19e [Cp*RhCp] monomer can be doped into a number of organic polymers. On interaction with the polymer, the dimer dissociates into [Cp*RhCp]$^+$ monomer units that transfer electrons into the polymer, leading to *n*-type electrical conductivity.[41]

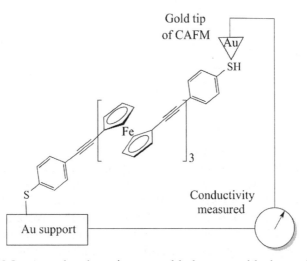

FIGURE 13.8   A molecular wire assembled on a gold electrode with the gold tip of a conductive atomic force microscope (CAFM) for a conductivity measurement.

13.34

Beyond simple conduction, a number of structures have been shown to act as diodes, conductors that permit current flow in one direction, while inhibiting it in the other. Diodes are required for constructing electronic circuits of any complexity. Large rectification effects ($>10^2$) have been seen in self-assembled monolayers (SAMs), but only where ferrocene units are present to act as redox active units. Figure 13.9 shows an idealized version of such a rectifying SAM in which the alkyl-thiolate chain is anchored to an Ag contact and the redox-active ferrocene end is in contact with a gallium oxide layer. Electrons are thought to tunnel through the nonconducting alkyl chains. Biasing the voltage across the junction such that the ferrocene units are oxidized by the Ga_2O_3 contact allows electron tunneling over a shorter distance from the Ag electrode, than with the opposite bias, when tunneling has to occur across the whole junction to the Ag, ferrocene being an e donor $(FcH \Rightarrow [FcH]^+ + e^-)$ but not an e acceptor.[42]

Nonlinear Optical (NLO) Materials

Most materials respond to light in a linear way, so that the polarization induced in the material by the electric field component of the light

FIGURE 13.9   Idealized representation of a SAM-rectifier.[42]

depends linearly on the light intensity. In NLO materials, this no longer holds true, which leads to useful effects, such as frequency doubling, by which two photons of incident light of wavelength λ are converted into one photon of emitted light of wavelength $\lambda/2$. Many NLO materials are simple salts, such as $LiIO_3$ or $KNbO_3$, but organic materials such as L-arginine maleate dihydrate have advantages for some applications. Organometallic compounds have also shown useful NLO effects because of the presence of polarizable metals and extended conjugated π systems, as in the case of **13.35**.[43]

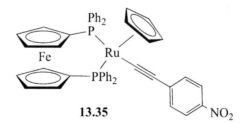

13.35

Organic Light-Emitting Diodes (OLEDs)

OLEDs, sometimes just called LEDs, emit light in response to a voltage.[44] Although many OLEDs involve organometallics, the term "organic" is preferred because some OLEDs have no metals. They are found in cell phones, TVs, and other displays because they are resistant to bending and shock and are very thin. Over 300 million OLED telephone displays have now been shipped by Samsung for a total of $7B in global revenue, and the first TV screens are already being shipped. An applied voltage injects electrons from one side and holes from the other side of the material, and when these find each other in the same molecule — electron in the LUMO and hole in the HOMO– energy is emitted that corresponds to the HOMO–LUMO gap as the electron falls from the LUMO to fill the hole in the HOMO. The 1 : 3 ratio of singlet to triplet electron–hole pairs (called excitons), dictated by the quantum physics of the device, limits the emission quantum efficiency to 25% because only the singlet can emit. In the absence of a spin flip, the triplet cannot emit and its energy can then go into degrading the device. Only heavy metals such as Ir have a spin-orbit coupling strong enough to facilitate this spin flip, resulting in ~100% emission efficiency. The key advance in the OLED field was the adoption of cyclometalated iridium(III) complexes,[45] such as the one shown nearby, that can be tuned by variation of the structure to vary the emission wavelength. Commercial production of cell phone displays incorporating iridium OLEDs has been linked with the recent price rise of this metal (2003: ~$100/oz. 2009: ~$400/oz. 2013: ~$1000/oz.).

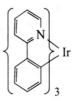

A related cell-permeable, cyclometalated iridium(III) complex, [Ir(2-C_6H_4-quinoline)$_2$(H_2O)$_2$]OTf, is a phosphorescent probe for cell imaging by preferentially staining the cytoplasm of both live and dead cells with a bright luminescence.[46]

Sensors

Sensors, analytical devices for detecting variations in the levels of specific compounds over time, are attracting increasing attention for environmental monitoring.[47] One organometallic application follows the degree of ripening of fruit by monitoring levels of the fruit ripening hormone, ethylene. This is done by doping a carbon nanotube (NT) array with a TpCu(I) derivative that binds to the NTs in the absence of C_2H_4. As the ethylene level builds up, the Cu(I) becomes detached from the NT, instead binding reversibly to C_2H_4 and thus affecting the electrical resistance of the NT array to give a continuous readout of the C_2H_4 level.[48]

- Cluster structures can follow the EAN or Wade's rules (Section 13.1).
- Isolobal relationships (Section 13.2) draw analogies between organic and inorganic structures.
- Nanoparticles (Section 13.3) and materials (Section 13.4) offer many new directions for future development of organometallic chemistry.

REFERENCES

1. P. J. Dyson, *Coord. Chem. Rev.*, **248**, 2443, 2004.
2. (a) A. Krapp, M. Lein, and G. Frenking, *Theor. Chem. Acct.*, **120**, 313, 2008; (b) See Reference 3 in Chapter 4.
3. D. M. P. Mingos, *Acc. Chem. Res.*, **17**, 311, 1984.
4. F. Lips and S. Dehnen, *Angew. Chem. Int. Ed.*, **50**, 955, 2011.
5. G. La Macchia, G. L. Manni, T. K. Todorova, M. Brynda, F. Aquilante, B. O. Roos, and L. Gagliardi, *Inorg. Chem.*, **49**, 5216, 2010 and references cited.
6. M. Carrasco, N. Curado, C. Maya, R. Peloso, A. Rodríguez, E. Ruiz, S. Alvarez, and E. Carmona, *Angew. Chem. -Int. Ed.*, **52**, 3227, 2013.
7. P. A. Lindahl, *J. Inorg. Biochem.*, **106**, 172, 2012.
8. H. Schmidbaur and A. Schier, *Chem. Soc. Rev.*, **37**, 1931, 2008.
9. L. H. Doerrer, *Dalton Trans.*, **39**, 3543, 2010.
10. (a) E. M. Brzostowska, R. Hoffmann, and C. A. Parish, *J. Am. Chem. Soc.*, **129**, 4401, 2007; (b) H. G. Raubenheimer and H. Schmidbaur, *Organometallics*, **31**, 2507, 2012, and references cited.
11. G. Schmid (ed.), *Nanoparticles: From Theory to Application*, Wiley-VCH, Weinheim, 2004.
12. R. D. Glover, J. M. Miller, and J. E. Hutchison, *ACS Nano.*, **5**, 8950, 2011.
13. S. E. Lohse and C. J. Murphy, *J. Am. Chem. Soc.*, **134**, 15607, 2012.
14. R. H. Crabtree, *Chem. Rev.*, **112**, 1536, 2012.
15. D. Astruc (ed.), *Nanoparticles and Catalysis*, Wiley-VCH, Weinheim, 2008; V. Polshettiwar and R. S. Varma, *Green Chem.*, **12**, 743, 2010.
16. P. X. Zhao, N. Li, and D. Astruc, *Coord. Chem. Rev.*, **257**, 638, 2013.
17. J. M. Yuk, J. Park, P. Ercius, K. Kim, D. J. Hellebusch, M. F. Crommie, J. Y. Lee, A. Zettl, and A. P. Alivisatos, *Science*, **336**, 61, 2012.
18. S. Mehraeen, A. Kulkarni, M. Chi, B. W. Reed, N. L. Okamoto, N. D. Browning, and B. C. Gates, *Chem. Eur. J.*, **17**, 1000, 2011.
19. F. Dumestre, B. Chaudret, C. Amiens, P. Renaud, and P. Fejes, *Science*, **303**, 821, 2004.

20. R. L. Donkers, D. Lee, and R. W. Murray, *Langmuir*, **20**, 1945, 2004.

21. P. Gruene, D. M. Rayner, B. Redlich, A. F. G. van der Meer, J. T. Lyon, G. Meijer, and A. Fielicke, *Science*, **321**, 674, 2008.

22. J. Oliver-Meseguer, J. R. Cabrero-Antonino, I. Domínguez, A. Leyva-Pérez, A. Corma, *Science*, **338**, 1452, 2012.

23. P. K. Jain, I. H. El-Sayed, *Nanotoday*, **2**, 18, 2007.

24. M. Grätzel, *Acc. Chem. Res.*, **42**, 1788, 2009.

25. Y. Lee, J. Suntivich, K. J. May, E. E. Perry, and Y. Shao-Horn, *J. Phys. Chem. Lett.*, **3**, 399, 2012.

26. S. A. Chambers, *Adv. Mater.*, **22**, 219, 2010.

27. A. J. M. Mackus, N. Leick, L. Baker, and W. M. M. Kessels, *Chem. Mater.*, **24**, 1752, 2012; T. Aaltonen, A. Rahtu, M. Ritala, and M. Leskelä, Electro-chem. *Solid-State Lett.*, **6**, C130, 2003; S. R. Daly, D. Y. Kim, Y. Yang, J. R. Abelson, and G. S. Girolami, *J Am Chem Soc.*, **132**, 2106, 2010.

28. J. Graetz, *Chem. Soc. Rev.*, **38**, 73, 2009.

29. R. B. King, *Coord. Chem. Rev.*, **200–202**, 813. 2000.

30. A.-M. Alexander and S. J. Justin, *Chem. Soc. Rev.*, **39**, 4388, 2010.

31. J. -C. Eloi, L. Chabanne, G. R. Whittell, and I. Manners, *Mater. Today*, **11**, 28, 2008.

32. W. Y. Wong and C. -L. Ho, *Acc. Chem. Res.*, **43**, 1246, 2010.

33. N. D. Schley, J. D. Blakemore, N. K. Subbaiyan, C. D. Incarvito, F. D'Souza, R. H. Crabtree, and G. W. Brudvig, *J. Am. Chem. Soc.*, **133**, 10473, 2011; U. Hintermair, S. W. Sheehan, A. R. Parent, D. H. Ess, D. T. Richens, P. H. Vaccaro, G. W. Brudvig, and R. H. Crabtree, *J. Am. Chem. Soc.*, **135**, 10837, 2013.

34. J. Perez-Ramirez, C. H. Christensen, K. Egeblad, C. H. Christensen, and J. C. Groen, *Chem. Soc. Rev.*, **37**, 2530, 2008.

35. J. Lee, O. K. Farha, J. Roberts, K. A. Scheidt, S. T. Nguyen, and J. T. Hupp, *Chem. Soc. Rev* **38**, 1450, 2009; H. C. Zhou, J. R. Long, and O. M. Yaghi, *Chem. Rev.*, **112**, 673, 2012.

36. M. Eddaoudi, D. B. Moler, H. L. Li, B. L. Chen. T. M. Reineke, M. O'Keeffe, and O. M. Yaghi, *Acct. Chem. Res.* **34**, 319, 2001.

37. D. Braga and F. Grepioni, *Acct. Chem. Res.*, **33**, 601, 2000.

38. M. H. Weston, G. W. Peterson, M. A. Browe, P. Jones, O. K. Farha, J. T. Hupp, and S. T. Nguyen, *Chem. Comm.*, **49**, 2995, 2013.

39. I. Manners, *J. Organometal. Chem.*, **696**, 1146, 2011.

40. C. -P. Chen, W. -R. Luo, C. -N. Chen, S. -M. Wu, S. Hsieh, C. -M. Chiang, and T. -Y. Dong, *Langmuir*, **29**, 3106, 2013.

41. S. Guo, S. K. Mohapatra, A. Romanov, T. V. Timofeeva, K. I. Hardcastle, K. Yesudas, C. Risko, J. -L. Brédas, S. R. Marder, and S Barlow, *Chem. Eur. J.*, **18**, 14760, 2012.

42. C. A. Nijhuis, W. F. Reus, and G. M. Whitesides, *J. Am. Chem. Soc.*, **132**, 18386, 2010.

43. B. A. Babgi, A. Al-Hindawi, G. J. Moxey, F. I. A. Razal, M. P. Cifuentes, E. Kalasekera, R. Stranger, A. Teshome, I. Asselberghs, K. Clays, and M. G. Humphrey, *J. Organometal. Chem.*, **730**, 108, 2013.

44. J. A. G. Williams, *Chem. Soc. Rev.*, **38**, 1783, 2009; A. Extance, *Chem. World*, June 2013, 51.

45. A. F. Rausch, M. E. Thompson, and H. Yersin, *Inorg. Chem.*, **48**, 1928, 2009 and references cited.

46. D. L. Ma, H. J. Zhong, W. C. Fu, D. S. H. Chan, H. Y. Kwan, W. F. Fong, L. H. Chung, C. Y. Wong, and C. H. Leung, *PLOS ONE*, **8**, e55751, 2013, DOI: 10.1371/journal.pone.0055751; E. Baggaley, J. A. Weinstein and J. A. G. Williams, *Coord. Chem. Rev.*, **256**, 1762, 2012.

47. J. W. Steed, *Chem. Soc. Rev* **38**, 506, 2009.

48. B. Esser, J. M. Schnorr, and T. M. Swager, *Angew. Chem. Int. Ed.*, **51**, 5752, 2012.

49. J. D. Aiken and R. G. Finke, *J. Mol. Catal.*, **145**, 1, 1999.

PROBLEMS

13.1. Given the existence of cyclopropenone, suggest two cluster complexes that are isolobal with this species, and how you might try to synthesize them.

13.2. Give the cluster electron counts (see Fig. 13.1) of the following: $Cp_3Co_3(\mu_3\text{-}CS)(\mu_3\text{-}S)$; $Fe_3(CO)_9(\mu_3\text{-}S)_2$; $Fe_3(CO)_{10}(\mu_3\text{-}S)_2$. In deciding how to count the S atoms, take account of the fact that these seem to have one lone pair not engaged in cluster bonding, as shown by their chemical reactivity in methylation with Me_3O^+, for example.

13.3. For the species listed in Problem 13.2, how many M–M bonds would you expect for each? Draw the final structures you would predict for these species.

13.4. $Co_4(CO)_{10}(EtC{\equiv}CEt)$ has structure **13.36** shown below. What is the cluster electron count? Does it correctly predict the number of M–M bonds? How would you describe the structure on a Wade's rule approach?

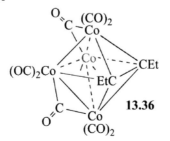

13.5. What light do the isolobal ideas throw on structures **13.37** and **13.38**?

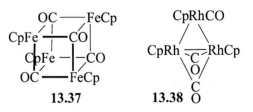

          13.37   **13.38**

13.6. What structures would you predict for $[Fe_4(CO)_{13}]^{2-}$, $[Ni_5(CO)_{12}]^{2-}$, and $[Cr_2(CO)_{10}(Ph_2PCH_2PPh_2)]$?

13.7. Pt(0) forms an RC≡CR complex $Pt(C_2R_2)_n$. Predict the value of n based on an isolobal relationship with structure **13.39** (below). Why are the two W–C vectors orthogonal in **13.39**?

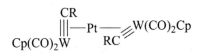

13.39

13.8. Predict the structure of **13.40**, making it as symmetric as possible. With what organoiron complex is **13.40** isolobal?

$$Fe(CO)_3\{B_4H_4\}$$

13.40

13.9. Why do boron and transition metal hydrides tend to form clusters, when carbon and sulfur hydrides tend to form open-chain hydrides $Me(CH_2)_nMe$, and $HS(S)_nSH$? Why is sulfur able to form clusters in the compounds mentioned in Problem 13.2?

13.10. $Os_3(CO)_{10}(\mu_2\text{-}CH_2)(\mu_2CO)$ reacts with CO to give structure **13.41**, which reacts with H_2O to form acetic acid. Suggest a structure for **13.41**.

$$Os_3(CO)_{12}\{CH_2CO\}$$

13.41

14

ORGANIC APPLICATIONS

In the earliest period of complex natural product synthesis, from Robinson's[1] 1917 tropinone synthesis to Eschenmoser and Woodward's[2] 1973 coenzyme B_{12} synthesis, metal-catalyzed reactions played no great role. In contrast, modern organic syntheses often involve numerous transition metal-catalyzed steps. Main-group compounds, such as BuLi, MeMgBr, or $NaBH_4$, tend to act in stoichiometric quantity as reagents, while the more expensive transition metals, typically complexes of Pd, Rh, or Ru, tend to be used as catalysts and therefore in much lower amounts, for example, 0.1–5 mol% (mmol catalyst per 100 mmol substrate).

Some of the catalytic reactions that enjoy the widest use, such as alkene metathesis, have no parallel in traditional organic chemistry. Others are possible by traditional organic procedures, but catalysis considerably enhances the rate, selectivity, or generality, such as the Buchwald–Hartwig amination of aryl halides. Environmental concerns highlighted by the rise of green chemistry have emphasized atom economic catalytic processes that limit waste and energy input. In yet other cases, catalysis provides asymmetric products, as in hydrogenation. Modern regulatory trends require the production of enantiomerically pure drugs and agrochemicals, both to lower the quantities dispersed

The Organometallic Chemistry of the Transition Metals, Sixth Edition.
Robert H. Crabtree.
© 2014 John Wiley & Sons, Inc. Published 2014 by John Wiley & Sons, Inc.

and to avoid undesired effects of the "inactive" enantiomer. Asymmetric catalysts therefore take a major place in the armory of methods that are needed to meet this challenge. Where catalytic methods are still lacking, stoichiometric applications even of transition metals can still be seen, but typically only with inexpensive metals (e.g., Ti, Cu, or Zn). The role of transition metal catalysis in fine chemicals and pharmaceutical production continues to grow.[3]

The art of organic synthesis[4] involves a judicious combination and sequencing of all the steps, including organometallic steps, into a coherent plan that minimizes the risk of unintended outcomes. In the cases discussed here, the key organometallic steps have been isolated from their context, so that we can cover a broad range of reaction types. The following sections cover the reaction classes that have been most widely applied in recent synthetic work. For historical reasons,[5] quite a few reactions carry names of their discoverers or developers.

14.1 CARBON–CARBON COUPLING

The palladium-catalyzed coupling reactions of Section 9.7 have numerous applications both in the lab and on a large scale in industry.[6] The reactions often start with an oxidative addition of RX (X = Hal or OTs) to Pd(0) to generate an R–Pd–X intermediate that is subsequently functionalized. Aryl or vinyl groups are preferred R groups because these R–Pd intermediates resist decomposition by β elimination.

Some of the cases discussed go beyond simple one-step coupling by using combinations of steps to make more elaborate structures. For example,[7] the annulation (ring-forming) reaction of Eq. 14.1 gives **14.5** by a combination of a Mizoroki–Heck coupling and cyclometalation. Initial oxidative addition gives an arylpalladium(II) species, **14.1**, that undergoes insertion into the strained norbornene cosubstrate from the least hindered face to give the alkylpalladium(II) intermediate **14.2**. Alkylpalladium intermediates normally β-eliminate, but not **14.2** because it lacks the needed syn coplanar arrangement of metal and the β-H. H′ in **14.2** is nearly syn to the Pd and might be available for β elimination except that Bredt's rule prohibits C=C bond formation at a bridgehead. The cyclometalation presumably occurs via agostic species **14.3** that undergoes deprotonation by the base present to give **14.4**. Reductive elimination gives the final product, **14.5**.

A key step (Eq. 14.2) in a new synthesis[8] of strychnine involves a C–C coupling in **14.6** to give the pentacyclic **14.9**. Oxidative addition of vinyl iodide **14.6**, together with deprotonation α to the ketone, gives

14.7. The enolate carbon acts as a nucleophile to displace halide from the metal to give **14.8** and reductive elimination completes the cycle.

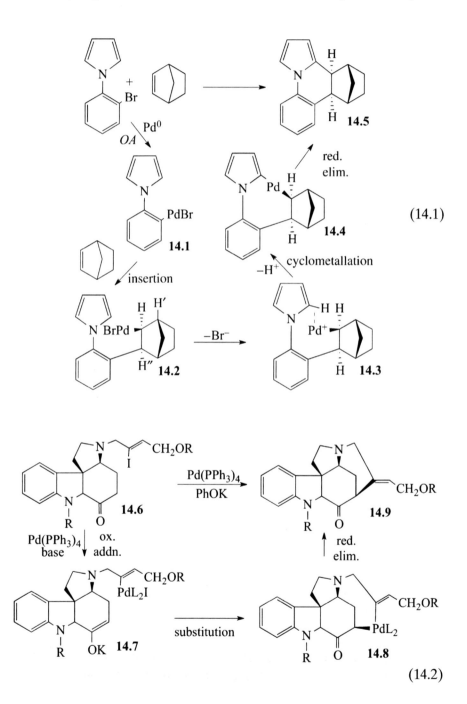

$$(14.1)$$

$$(14.2)$$

Insertion can occur into C—heteroatom multiple bonds—a benzofuran synthesis involving an addition to a nitrile C≡N bond is shown in Eq. 14.3 [L = 2,2'-dipyridyl (dipy)].[9] The catalyst first dissociates to give [Pd(dipy)OH]$^+$, which reacts with ArB(OH)$_2$, as in the Suzuki–Miyaura reaction of Section 9.7, to give a palladium aryl. When the C≡N triple bond inserts into the resulting Pd–Ar bond, the electronegative Ar group ends up on the electropositive CN carbon, as expected from the tendency for new bonds to be formed between partners that differ most in electronegativity. After hydrolysis, ketone **14.10** probably undergoes ring closure by cyclopalladation of the arene ring, directed both by the *ortho/para* activating methoxy groups and by Pd binding to the carbonyl oxygen. The most likely possibility for the ring closure is insertion of the carbonyl C=O into the new aryl–palladium bond to give **14.11**. A standard organic elimination gives the product benzofuran and regenerates the [Pd(dipy)OH]$^+$ catalyst.

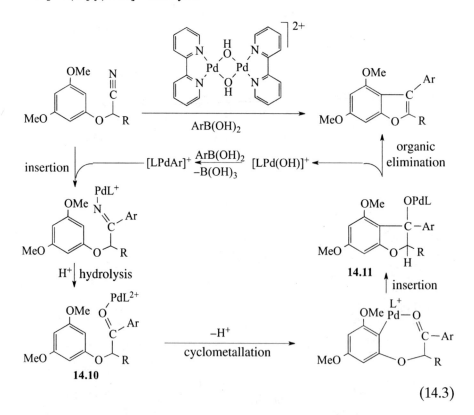

(14.3)

Equation 14.4 shows how a Buchwald–Hartwig amination can be combined with a Mizoroki–Heck reaction in a *tandem* sequence (also termed *cascade* or *domino* reaction) using the same catalyst.[10] In the starting dibromovinyl compound, **14.12**, one C–Br gives rise to an indole ring by attack of the nearby amine in a Pd-catalyzed step. The second C–Br is now available for a Mizoroki–Heck reaction to fix the vinyl substituent at the indole 2-position. 2-Vinyl indoles occur in pharmacologically active materials such as Fluvastatin, a drug for controlling cholesterol. The mechanism of each step is discussed in Section 9.7.

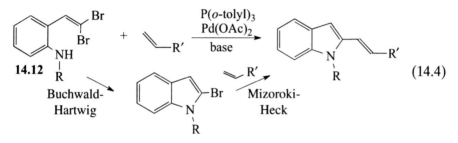

$$(14.4)$$

Rather than starting with a preformed palladium phosphine catalyst, the free phosphine is often combined with Pd(OAc)$_2$. The phosphine is thought to reduce Pd(II) to the Pd(0) state needed for oxidative addition. Another catalyst precursor, Pd$_2$(dba)$_3$, acts as a direct source of Pd(0) by dissociation of the dba (dba = {PhCH=CH}$_2$C=O), but can also give 10- to 200-nm Pd nanoparticles under certain conditions, and conventionally produced Pd$_2$(dba)$_3$ samples can even contain nanoparticles, introducing the possibility of irreproducibility problems. A procedure to make pure [Pd$_2$(dba)$_3$]·CHCl$_3$ is available.[11]

The reactivity of the halide reagent in palladium coupling follows the order I > Br > Cl so that bromides are typically used as a compromise between the higher reactivity of the iodide and the lower cost of the chloride, unless conditions can be found that allow the use of the cheaper but less reactive chlorides.[12]

Oxidative addition of allyl acetates and similar species can occur to Pd(0) to give allylpalladium(II) complexes that are subject to nucleophilic attack by stabilized carbon nucleophiles such as enolates (Section 9.7). In most cases, the nucleophile attacks the CH$_2$ terminus of monosubstituted allyls, so no enantiomeric outcome is possible. Control of regiochernistry to favor the branched product with high enantiocontrol is possible with the Pd complex of Trost's bis-phosphine ligand, **14.13**, one of Trost's modular ligands. Equation 14.5 shows formation of a quaternary carbon with excellent yield and high e.e.[13] Mechanistic work has shown how this catalyst works; the nucleophile hydrogen bonds to one of the amide NH bonds controls the selectivity of the attack.[14] An iridium catalyst prefers to give branched products from monosubstituted allyls (Eq. 14.6) and thus complements the Pd catalyst, which prefers to give linear products in such cases.[15]

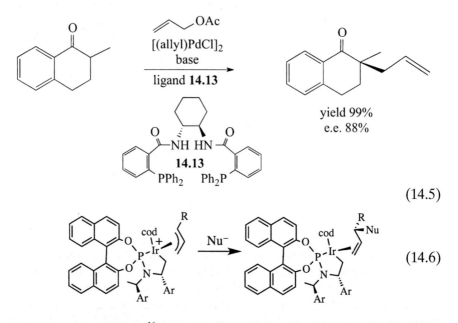

$$(14.5)$$

$$(14.6)$$

Denmark and Wang[16] have shown how hydrosilylation can install a silyl group that can be activated by fluoride via hypervalent intermediate **14.14** so as to act like a boronic acid does in the Suzuki–Miyaura coupling. After vinyl transfer from **14.14** to Pd, reductive elimination with an aryl group from the ArI coreactant gives net addition of Ar–H across a C=C triple bond (Eq. 14.7).

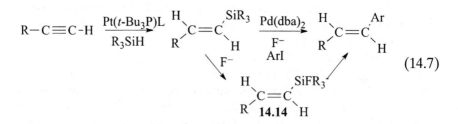

$$(14.7)$$

A wide variety of alternate coupling partners is possible: Coupling of $K[RBF_3]$ with R'OTf to give R–R' can take place with $PdCl_2(dppf)/CsCO_3$, but the reaction appears to go by initial hydrolysis of the B–F bonds.[17] In other Pd-catalyzed reactions, an aryl or vinyl halide couples with an aryl or vinylzinc reagent (Negishi coupling),[18] an aryl or vinyl magnesium reagent (Kumada coupling), an aryl or vinyltin reagent (Stille coupling), or with an aryl or vinylsilicon reagent (Hiyama coupling as in Eq. 14.7). The mechanism in each variant is believed to go by oxidative addition of RBr to give the $Pd(R)(Br)L_n$ species. The reaction partner, R'M, acting as nucleophile, then replaces the bromide to give $Pd(R)(R')L_n$. Reductive elimination provides the product and regenerates the catalyst. Even alkyl halides can react in certain cases (Eq. 14.8),[19] indicating that the coupling steps can be fast enough to beat β elimination. A bulky basic phosphine, such as $P(t\text{-}Bu)_2Me$ in Eq. 14.8, is usually needed for best results in such coupling reactions, where it favors the production of "$Pd(PR_3)$," an intermediate that is extremely active for all the needed steps of the cycle.

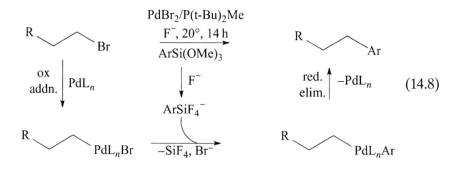

$$(14.8)$$

The wide commercial and synthetic availability of a variety of organoboranes makes the Suzuki–Miyaura coupling (Section 9.7) of aryl or vinyl halide with aryl or vinyl boronic acid $[ArB(OH)_2]$ among the most common coupling variants. For the synthesis of alkynes, the coupling of an aryl or vinyl halide with an alkynyl anion (Sonogashira coupling) is particularly useful. The antifungal Terbinafine (**14.15**) is produced in this way on an industrial scale by Novartis. The route qualifies as green chemistry because it replaces an earlier one using very toxic materials.

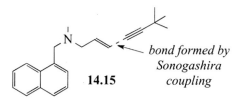

14.15   *bond formed by Sonogashira coupling*

The Claisen rearrangement for C–C bond construction (Eq. 14.9) can be catalyzed both by hard organometallic electrophiles, that bind to the oxygen atom, and by soft organometallic electrophiles that bind to the unsaturated groups. In the latter category, Toste has shown how the Au(I) species $[O\{Au(PPh_3)\}]^+$ retains the chirality of the starting material in the products (Eq. 14.10).

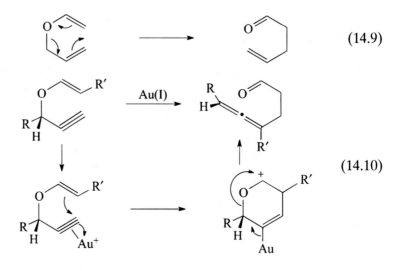

The rapid rise in price of the Pt metals, particularly Rh and Pd, has put increased emphasis on finding good base-metal catalysts for organometallic catalytic reactions. Metals other than Pd that have been successfully employed in coupling include W, Ir, and Mo, with Trost's Mo complex[20] containing ligand **14.12** being highly effective. In the case of Buchwald–Hartwig coupling, a $CuI/HOCH_2CH_2OH$ combination[21] has proved very effective and for aryl–alkyl coupling, and also even so simple a catalyst as $FeCl_3$.[22]

F and CF_3 substituents are valued in pharmaceuticals because they delay the oxidative degradation of drug molecules in the liver, thus prolonging their action. ^{18}F ($t_{1/2} = 110$ min) finds medical use as a preferred isotope for positron emission tomography, in which the fate of an ^{18}F substituted molecule is tracked in the patient by observing the γ ray emission from the ^{18}F; for example, the procedure is useful in imaging metastases from tumors. Specialized Pd coupling procedures have had to be developed for the synthesis of fluoroorganics[23] because fluorine substitution leads to nonstandard chemistry. A Pd complex of Buchwald's bulky *t*BuBrettPhos permits the fluorination of ArOTf with CsF via RE from an $LPd^{II}(Ar)F$ intermediate. An oxidative procedure that goes via a Pd(IV) intermediate favors the RE of Ar–F, normally slow from Pd(II); in this case, the F-containing reagent is an F^+ source, such as $[C_5H_5NF]BF_4$. $CuCF_3$, formed transiently in situ from Cu(I) and Me_3SiCF_3, has also proved a useful reagent for trifluoromethylation of ArI; replacing CsF by Me_3SiCF_3 permits the Pd catalyzed trifluoromethylation of ArBr with Buchwald's catalyst.

14.2 METATHESIS

Like a knight's move in chess, C=C bond metathesis (Section 12.1) can produce surprising outcomes that allow synthetic problems to be approached by unconventional routes, as in Eq. 14.11.[24] At first sight, the starting material and the product seem totally unrelated. A closer look shows how straightforward metathesis steps, carried out with Grubbs second-generation catalyst, lead naturally from starting material to product. Initial ring opening of the strained, and hence more reactive norbornene C=C bond, is followed by ring closure of the resulting ruthenium methylene intermediate with the adjacent vinyl group in a tandem reaction. The ring strain of the starting material provides the necessary driving force to prevent reversal. The initial cycle is presumably carried out with the starting Ru=CHPh catalyst implying that the first turnover of product contains an undesired PhCH= group, but all subsequent cycles go forward with the Ru=CH$_2$ intermediate and give the desired product.

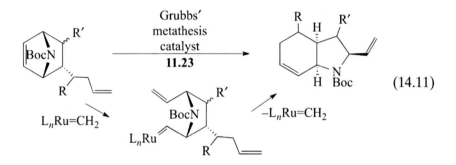

$$(14.11)$$

In a synthesis[25] of the A–E fragment of Ciguatoxin CTX3C (Eq. 14.12), Grubbs' catalyst forms ring A by a ring-closing metathesis (Section 12.1) with the elimination of ethylene, accounting for the loss of two carbons in this step. Incorporation of the required allyl ether and terminal vinyl group into the starting substrate is relatively straightforward, making this a neat solution for forming the A ring.

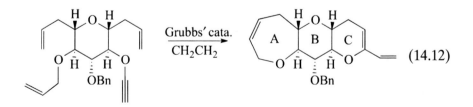

$$(14.12)$$

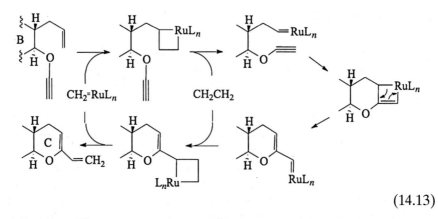

$$(14.13)$$

Ring C and its vinyl substituent is formed concurrently with Grubbs' catalyst but by a ring-closing enyne metathesis reaction.[26] Instead of losing the terminal carbon atoms of the multiple bonds, as in the formation of ring A, the product in the C ring closure has the same number of atoms as the starting substrate. The mechanism of enyne metathesis is not yet fully resolved, but Eq. 14.13 gives a plausible sequence. Instead of a standard metallacyclobutane intermediate (Section 12.1), formed from Ru=C and a C=C double bond, as in the ring A closure, we now have a metallacyclobutene intermediate formed from Ru=C and a C≡C triple bond. This metallacyclobutene ring opens to a new carbene that in turn reacts with ethylene to form the diene portion of the final structure. When the Grubbs catalyst proves insufficiently reactive, Schrock's more reactive Mo catalyst can still be effective, as in a carbafructofuranose synthesis.[27] Cross-metathesis typically gives cis (E) alkenes, but recently developed Mo and Ru catalysts are trans or Z-selective because they contain very bulky OAr ligands that orient both R substituents on the same side of the metalacyclic intermediate, as shown as **14.16** for the Mo catalyst. This provides the cis alkene after extrusion of the alkene (Eq. 14.14).[28] Asymmetric catalysts that give high e.e. are also now available.[29]

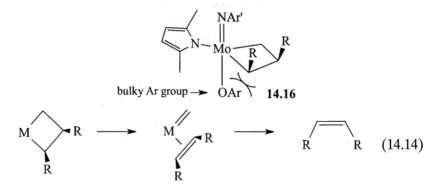

$$(14.14)$$

14.3 CYCLOPROPANATION AND C–H INSERTION

An organic carbene precursor, such as a diazocarbonyl, can transfer the carbene to the metal with subsequent transfer to an alkene to give net cyclopropanation (Eq. 14.15). With a homochiral catalyst, an asymmetric cyclopropanation can occur, as in Doyle's dirhodium(II) carboxamidates (**14.17**).[30] Structure **14.18** shows the substitution pattern of a typical chiral catalyst, with three of the carboxamidates omitted for clarity. The carbene is transferred to the open face of the complex and behaves as a strongly Fischer-type carbene, equivalent to a metal-stabilized carbonium ion. The alkene π system is thought to attack the empty p orbital of the carbene carbon to form the first new C–C bond. The ring is then closed by attack of the newly formed carbonium ion center on the Rh or on the back side of the original carbene carbon in an electrophilic abstraction to form the second C–C bond and regenerate the catalyst.

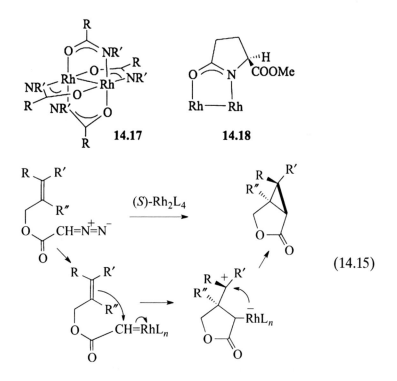

(14.15)

Equation 14.16 shows how carbene insertion into CH bonds is also possible where a CH bond is favorably located. In this case, the less thermodynamically preferred cis geometry of the Me and R groups is nevertheless kinetically preferred, dictated by the transition-state ste-reoelectronic preference. The e.e.s are in the range of 60–80% with the best chiral dirhodium(II) catalysts.[30]

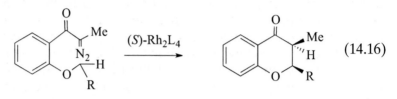

$$(14.16)$$

14.4 HYDROGENATION

In directed hydrogenation, a catalyst first binds to any of a number of different directing groups, such as an alcohol or amide oxygen, that lie on one face of the substrate. Hydrogen addition to the substrate C=C bond then occurs from that same face. The catalyst most often used, $[Ir(cod)py(PCy_3)]^+$, first reacts with hydrogen with loss of cyclooctane to give the $Ir(py)(PCy_3)^+$ fragment. Having only 12e, this can bind the directing group (DG) lone pair, and the substrate C=C bond, and also give oxidative addition of the H_2 without exceeding an 18e configuration. The resulting intermediate $[IrH_2(C=C)(DG)py(PCy_3)]^+$ is then capable of hydrogenating the bound C=C bond. Unlike many catalysts, $[Ir(cod)py(PCy_3)]^+$ is useful even for hydrogenating very hindered C=C groups. In Eq. 14.17, Pd/C adds H_2 to the less hindered side to give the undesired product isomer, while the Ir catalyst adds H_2 from the more hindered side—the top face in Eq. 14.17—via a directing effect of the nearby urea, also on the top face, to give the desired product isomer.[31]

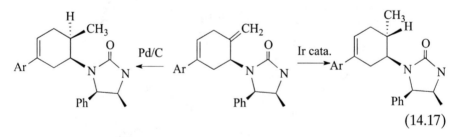

$$(14.17)$$

Kinetic Resolution

Kinetic resolution (KR) in the reduction of racemic mixtures of R and S starting materials requires a chiral catalyst that reacts *very* much faster with one substrate enantiomer. Suppose we have a catalyst that reduces only the R reactant in Eq. 14.18, to give the R product. If the k_{isom} is zero, we will ideally end up with a 50% yield of the S starting material and a 50% yield of the R hydrogenation product. Separation is required, of course, but is often relatively easy since the two compounds are chemically different.

If the k_{isom} rate is very fast, however, we have dynamic KR, or DKR[23], in which the yield can be improved beyond the 50% maximum of simple KR. If the S starting material is constantly interconverting with R, and only R reacts, all the material will ideally go down the R pathway to give a 100% yield of the R hydrogenation product. In the example of Eq. 14.19, the amino group spontaneously switches rapidly between the R and S forms by fast inversion at nitrogen. Two different catalysts were found to give different DKR products, *syn* or *anti*.[32]

$$S_{sat} \xleftarrow[k_S]{H_2, cata.} S_{unsat} \xrightleftharpoons{k_{isom}} R_{unsat} \xrightarrow[k_R]{H_2, cata.} R_{sat} \quad (14.18)$$

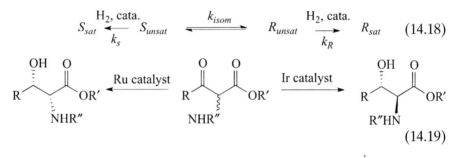

$$(14.19)$$

Asymmetric Hydrogenation

Asymmetric hydrogenation (Section 9.3) of C=C, C=O, and C=N bonds is widely employed with numerous catalysts. The example[33] shown in Eq. 14.20 uses a Noyori catalyst that is believed to operate by an outer-sphere mechanism of Section 9.3 with transfer of H⁻ from the metal to carbonyl carbon and H⁺ from the amino ligand to the carbonyl oxygen, the carbonyl substrate not being directly coordinated to the metal.

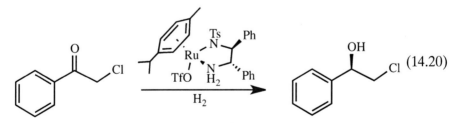

$$(14.20)$$

When the substrate C=C bond is tri- or tetra-substituted, the [(cod)-IrLL′]X series is most useful (Section 9.3); LL′ is typically a homochiral P,N mixed donor chelate, and X is the ′noncoordinating′ BAr^F anion.[34]

In Eq. 14.21, hydrogenation does not proceed as usual because an intermediate vinylrhodium complex is completely trapped to form a new C–C bond.[35] Since the postcoupling intermediate is Rh(III) and unlikely to give oxidative addition, the hydrogenation step most probably occurs via a dihydrogen complex that easily transfers a proton to the basic alkoxide ligand (Section 6.7). This would be followed by reductive elimination of the resulting Rh(III) vinyl hydride to give the final product and regenerate the Rh(I) catalyst.

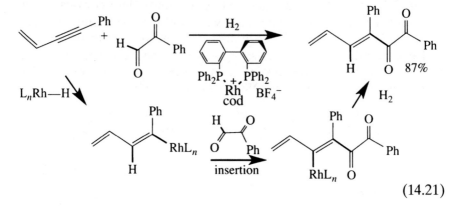

$$(14.21)$$

14.5   CARBONYLATION

Carbonylation comes in many forms, including hydroformylation (Section 9.4) and the Monsanto process (Section 12.3). Three-component carbonylation of an aryl halide, a nucleophile, and CO, catalyzed by palladium (Eq. 14.22), has proved useful for the construction of a wide variety of structures. The reaction starts by oxidative addition of the halide to palladium, followed by CO insertion to give a Pd(II) acyl, leading to nucleophilic abstraction of the acyl by the nucleophile. Buchwald[36] has made Weinreb amides, ArCO–NMe(OMe), in this way. A large bite angle (110°) bis-phosphine, Xantphos (**14.19** in Eq. 14.23), proved essential for high efficiency in this case. (Weinreb amides are useful in ketone synthesis because they reliably react with RLi or RMgBr to give ArCOR.) In another common variant, hydroesterification,[37] an alcohol abstracts the acyl to give an ester.

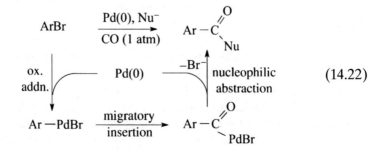

$$(14.22)$$

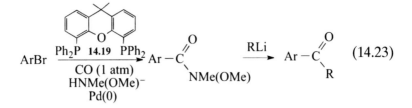

$$\text{ArBr} \xrightarrow[\substack{\text{CO (1 atm)} \\ \text{HNMe(OMe)}^- \\ \text{Pd(0)}}]{\text{Ph}_2\text{P } \mathbf{14.19} \text{ PPh}_2} \text{Ar}-\overset{\overset{O}{\|}}{C}-\text{NMe(OMe)} \xrightarrow{\text{RLi}} \text{Ar}-\overset{\overset{O}{\|}}{C}-\text{R} \quad (14.23)$$

Carbonylation has often required high pressures, involving the use of specialized high-pressure equipment. A goal has therefore been to avoid carbon monoxide or at least to use it at one atmosphere pressure. Coates[38] has applied Cr/Co catalysts in this way for the carbonylation of epoxides to give β-lactones (Eq. 14.24). This is an unusual example in which two transition metal catalysts cooperate. This cooperation is facilitated by the formation of ion pairs in which the two components are already closely associated in solution. The hard Cr(III) salen cation acts as Lewis acid to facilitate nucleophilic attack on the substrate epoxide carbon by the soft Co(CO)$_4$ anion with inversion of stereochemistry at that center. The migratory insertion shown is followed by an attack of the alkoxide on the newly formed carbonyl functionality (nucleophilic abstraction, Section 8.4 and Eq. 8.22 and Eq. 8.23) to close the ring and regenerate the catalyst. As an alkyl containing a β-hydrogen, the intermediate might have been expected to β-eliminate to give the vinyl alcohol and ultimately the ketone. This is a minor pathway probably because the intermediate alkyl is of the type RCo(CO)$_4$. As an 18e species, this must lose CO to generate a 2e site for β elimination, a reaction that is evidently suppressed by the excess CO. In a useful development made possible by a careful mechanistic study, double CO insertion to give an acid anhydride has been shown with a related salt as catalyst in which the Lewis acid role is taken by an aluminum(III) porphyrin cation (Eq. 14.25).[39]

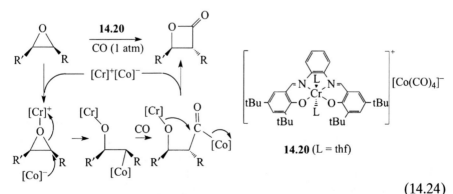

$$(14.24)$$

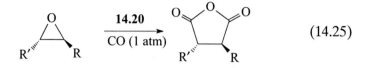

$$(14.25)$$

Traditional alkene hydroesterification involves a three-component, Pd-catalyzed reaction of alkene $RCH=CH_2$, CO, and alcohol $R'OH$ to give RCH_2CH_2COOR'. An intermediate palladium hydride undergoes alkene insertion, then CO insertion, followed by nucleophilic abstraction of the acyl group to give the product ester and regenerate the catalyst. The reaction has been relatively little used in synthesis, perhaps because of the inconvenience of having high pressures of toxic CO. To avoid CO, formates have been introduced to hydroesterify an alkene via addition of the formate ROOC–H bond across the alkene C=C bond, but this did not prove practical until the introduction[40] of a pyridine auxiliary to provide chelate assistance for the required CH activation step. Only the predominant linear product is shown in Eq. 14.26 and Eq. 14.27. When catalyzed by $Ru_3(CO)_{12}$, the resulting method has proved useful in a spirastrellolide A synthesis.[41]

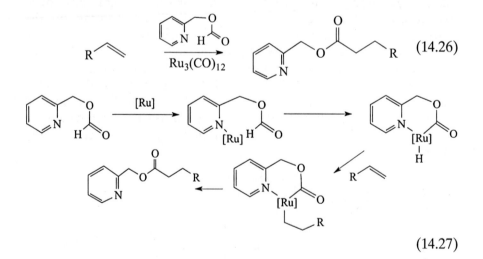

$$(14.26)$$

$$(14.27)$$

The Pauson–Khand reaction forms cyclopentenones from three groups, a C=C, a C≡C, and a CO molecule (Eq. 14.28). Originally, stoichiometric and based on Co, the Rh-catalyzed version is now widely adopted. Equation 14.29 shows the formation of the carbon skeleton of guanacastepene A, a novel antibiotic candidate.[42]

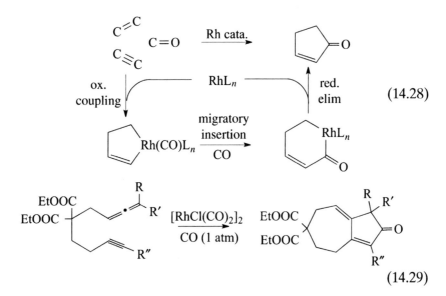

$$(14.28)$$

$$(14.29)$$

In a carbonylation reaction applied to polymer synthesis, a number of cationic Pd(II) complexes, such as $[Pd(dipy)Me(CO)][BAr^F_4]$, convert ethylene—CO mixtures to a perfectly alternating copolymer $(-CO-CH_2-CH_2-)_n$ that allows for easy subsequent functionalization of the carbonyl group. The mechanism involves alternating insertions of CO and ethylene, to account for which, the alkyl must prefer to insert CO and the acyl must prefer to insert ethylene. We have already seen that multiple insertion of CO is not favored (Section 7.2), but multiple insertion of ethylene is seen for Zr(IV) in cases where there is no CO to compete (Section 12.2). As expected, if alternation is to occur, the reaction barrier for the CO insertion into Pd-alkyl must be the lowest (calculated as 15 kcal/mol), followed by ethylene into a Pd-acyl (17 kcal/mol), followed by ethylene into a Pd-alkyl (19 kcal/mol).

14.6   OXIDATION

Organometallic species have traditionally been low valent and reducing rather than high valent and oxidizing, so they are normally involved in catalytic reduction reactions. Oxidation catalysis is now rising in importance and much remains to be discovered. Wacker chemistry is a genuine case where an oxidation proceeds via traditional low-valent organometallic intermediates. The mechanism (Section 8.3) involves nucleophilic attack on a vinyl group to give selective oxidation to a methyl ketone. Equation 14.30 shows its application in one step of a synthesis of tarchonanthuslactone.[43]

$$\text{OH} \quad \xrightarrow[\text{THF--H}_2\text{O}]{\text{PdCl}_2,\ \text{CuCl},\ \text{O}_2,} \quad \text{O} \quad \text{OH} \qquad (14.30)$$

Common transition metal oxidation catalysts are coordination compounds in a hard N- or O-donor ligand environment. Although technically not organometallic compounds, their reactions often involve organometallic intermediates.

Oxidations require a primary oxidant in stoichiometric quantity to reoxidize the catalyst back to its oxidized form at the end of each cycle. The most desirable primary oxidants are cheap and atom economic. For example, O_2 is available from air, H_2O_2 is cheap and both form water as by-product; tBuOOH is formed indirectly from air by reaction with isobutane. O_2 is a 4-electron oxidant not easily compatible with the usual 1e or 2e redox changes in metal complexes, and tBuOOH is particularly prone to give radical pathways that usually lead to low selectivity.

Stahl[44] has a series of Pd catalysts that use air as primary oxidant. When the Pd(0) catalyst is oxidized to Pd(II), the O_2 from air is reduced to give H_2O_2, which can either act as a 2e oxidant in a later cycle or undergo metal-catalyzed disproportionation to $H_2O + 0.5O_2$. The initial O_2 oxidation of Pd(0) seems to occur via an η^2-peroxo Pd(II) intermediate.

For example, Pd(OAc)$_2$/pyridine in aerated toluene at 80° can oxidize alcohols R_2CHOH to R_2C=O via β elimination of an intermediate alkoxide followed by air oxidation of the resulting Pd hydride back to the initial Pd(II). This last step may go via Eq. 9.27 or by O_2 insertion into the Pd–H bond followed by protonation to give H_2O_2.

$$L_n\text{Pd}-\text{H} \xrightarrow{\text{O}_2} L_n\text{Pd}-\text{O}^{\text{O}-\text{H}} \qquad (14.31)$$

With the natural product (−)-sparteine as chelating N-donor ligand, asymmetric Pd catalysis can be achieved in this way via kinetic resolution of PhCH(OH)Me with k_{rel} ratios up to 25. Bäckvall[45] obtained oxidative 1,4-addition of nucleophiles to dienes with good control of stereochemistry in the product. The hydroquinone/benzoquinone pair is the redox partner that couples the Pd catalyst with air as primary oxidant.

$$\xrightarrow[\text{O}_2]{\text{Pd(II), AcOH}} \quad \text{AcO} \cdots \text{OAc} \qquad (14.32)$$

Ferreira and Stolz[46] have reported a Pd-catalyzed oxidative ring-closing reaction (Eq. 14.33) with O_2 as primary oxidant. This probably goes via CH activation on the indole ring, followed by insertion of the CC double bond and β elimination; reoxidation of the resulting palladium hydride regenerates the catalyst. Site selectivity of the reaction is influenced by chelate control and electronic effects.[47]

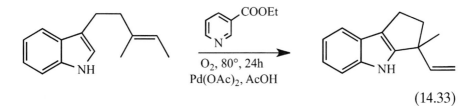

$$(14.33)$$

Catalytic epoxidation of C=C bonds is possible using aqueous H_2O_2 as primary oxidant[48] and, for example, the high-valent organometallic $MeReO_3$ as catalyst. H_2O_2 is one of the cheapest oxidants, making this reaction suitable for industrial applications.

Catalytic dihydroxylation of C–C bonds is possible with OsO_4 as catalyst and aqueous H_2O_2 or Et_3NO as primary oxidant. The intermediacy of the cyclic osmate ester shown in Eq. 14.34 accounts for the *syn* addition. Asymmetric versions of the reaction are possible, as shown in Eq. 14.35, where the asymmetric ligand is the plant alkaloid **14.21**.

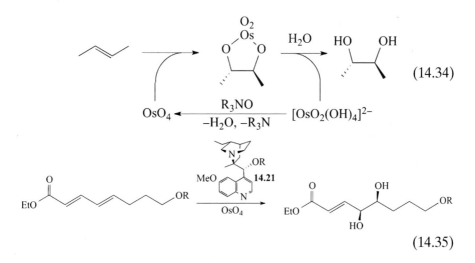

$$(14.34)$$

$$(14.35)$$

14.7 C–H ACTIVATION

Main-group examples of C–H activation, such as arene mercuration, are long known, but tend to involve stoichiometric reagents, not catalysts, and many use metals that are now avoided on toxicity grounds (Hg, Tl, and Pb). Catalytic reactions involving transition metal organometallic activation and functionalization of C–H bonds (Section 12.4) are beginning to move into the applications phase and are likely to become much more common in synthesis.[47,49] Innate selectivity can sometimes permit functionalization of one out of the many

C–H bonds that are present in any given case, but more often chelate control is needed to achieve the desired selectivity.[50] This is the case for the Rh-catalyzed carbene insertions into C–H bonds discussed in Section 14.3, as well as the Murai reaction, in which cyclometallation is followed by alkene insertion, for example, Eq. 14.36. The original catalyst, $RuH_2(CO)(PPh_3)_2$, requires high temperatures, but the more reactive $Ru(H_2)H_2(CO)(PCy_3)_2$ operates at room temperature.[51] In each case, reduction of Ru(II) to Ru(0) by H_2 transfer to the alkene in an initial hydrogenation step is required to attain the Ru(0) state probably needed for the reaction. Equation 14.37 shows that the alkyne, having undergone syn insertion, gives a cis arrangement of methyl and silyl groups in the product, and that the aromatic C–H is the preferred site of cyclometallation.

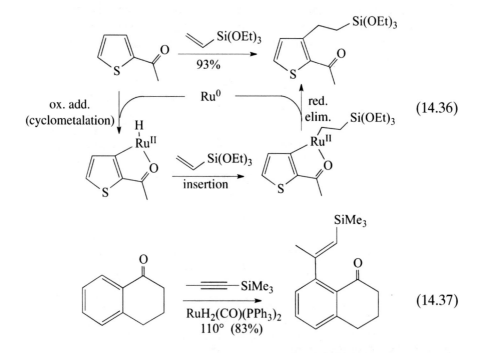

(14.36)

(14.37)

In the reaction of Eq. 14.38,[52] benzene can be functionalized with an alkene, catalyzed by $Ir(acac)_3$. Because Ir–H insertion into the alkene is predominantly anti-Markownikoff, the linear alkyl tends to be obtained, for example, n-Pr not i-Pr from propylene. In the traditional Friedel–Crafts reaction, in contrast, an acid catalyst converts the alkene to a carbonium ion that attacks the arene. This produces branched alkyl substituents, for example, i-Pr not n-Pr from propylene, because carbonium ions are more stable in the order tertiary > secondary > primary.

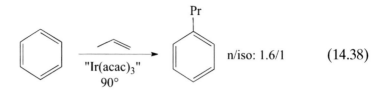

$$n/iso:\ 1.6/1 \qquad (14.38)$$

Ru$_3$(CO)$_{12}$ catalyzes a Murai-like reaction in which an acyl group is introduced β to nitrogen in a heteroarene. Cyclometalation in clusters is particularly easy, and instead of simply forming the Ru alkyl, as in the Murai case, an additional CO insertion step occurs, to produce an α-acyl heterocycle (Eq. 14.39).[53]

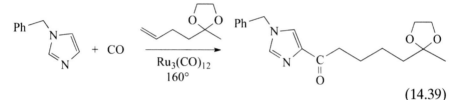

$$(14.39)$$

The arylboron reagents needed for the Suzuki–Miyaura coupling (Section 14.1) are conventionally formed from aryl halides, ArX, via reaction of ArLi or ArMgX with B(OMe)$_3$ to give ArB(OH)$_2$ on hydrolysis. An alternative involves direct reaction of a di-pinacolboron reagent (pin)B-B(pin) with an arene, followed by oxidative hydrolysis. Regiochemistry can be controlled using the 1,3-disubstituted arenes shown in Eq. 14.40 as substrate.[54] Even alkane C–H bonds can be borylated under the right conditions.[55]

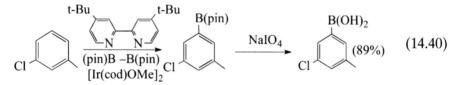

$$(89\%) \qquad (14.40)$$

Palladium(II) acetate readily gives metallation or cyclometallation in a variety of cases. Sanford[56] has made this into a useful catalytic organic synthetic reaction by introducing functionalizing groups, X, that cleave the Pd–Ar bond to give a wide variety of products, ArX. For example, in a series of I(III) oxidants, PhI(OAc)$_2$ gives ArOAc (Eq. 14.41), Ph$_2$I$^+$ gives ArPh, and PhICl$_2$ gives ArCl. These oxidants are effectively donors of X$^+$, a 2e oxidant capable of converting Pd(II) to Pd(IV)–X while releasing PhI and X$^-$. In contrast with the Pd(0)/Pd(II) cycle commonly proposed in palladium chemistry, she has also built a good case for a Pd(II)/Pd(IV) catalytic cycle (Eq. 14.42 and Eq. 14.43), but, as an alternative to Pd(IV), the intermediacy of binuclear

Pd(III)–Pd(III) species has also been documented.[57] Extension to CH bond amination has proved possible with PhI=NR as oxidant and Cu(I) complexes as catalyst.[58]

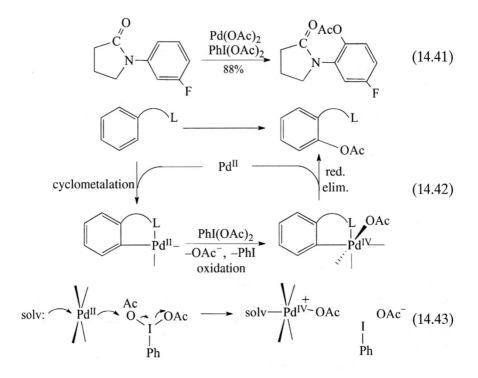

$$(14.41)$$

$$(14.42)$$

$$(14.43)$$

Numerous catalysts are able to hydroxylate C–H bonds, but few have been applied to functionality-rich, complex organic compounds. White[59] reported an iron catalyst, **14.22**, that uses the benign, inexpensive oxidant, hydrogen peroxide, as the ultimate source of the oxygen atom. Depending on the specific case, the remarkably high selectivity is ascribed to a combination of a number of causes. These include the reactive C–H bond being either inherently more reactive than any other or more physically accessible to the catalyst. The catalyst can also be attracted to a specific location by binding to a preexisting functional group within the reactant, thus attacking only a nearby C–H bond. Equation 14.44 shows the application of this procedure to the Chinese antimalarial compound artemisinin, extracted from a widely distributed shrub, *Artemisia annua*, and used in herbal form for millennia in Chinese traditional medicine. This complex reactant has numerous C–H bonds and a delicate peroxide functional group, yet it is efficiently converted to a single product. This implies that the catalyst has high selectivity even for a complex molecule, but predictability for other cases has required more detailed study.

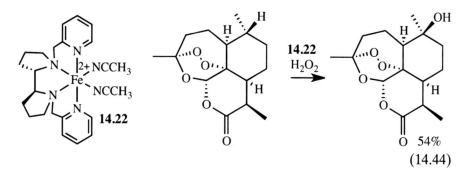

$$(14.44)$$

Cross coupling of R_1–H and R_2–H under oxidative conditions to give R_1–R_2 is an area of rising interest. For instance, an arene ArH can couple with an alkene $RCH=CH_2$ to give $RCH=CHAr$, a Heck-type product, but now made avoiding ArBr as reactant and thus also the waste formation that accompanies the classical procedure.[60]

14.8 CLICK CHEMISTRY

We often need to covalently connect two molecular fragments together in a reliable way whatever the situation. Sharpless emphasized the need for rapid, reliable, and general reactions—click chemistry—that can do this with high yield at room temperature and with essentially complete generality. The Cu(I)-catalyzed, regioselective cycloaddition of azides with alkynes has proved useful for a wide variety of such cases including fixing molecules onto surfaces and in drug discovery and protein chemistry. A proposed mechanism[61] deduced from isotope labeling of the Cu is shown in Eq. 14.45. The thermal cycloaddition also occurs, but more slowly and to give an undesired mixture of the 1,4- and 1,5-triazole regioisomers.

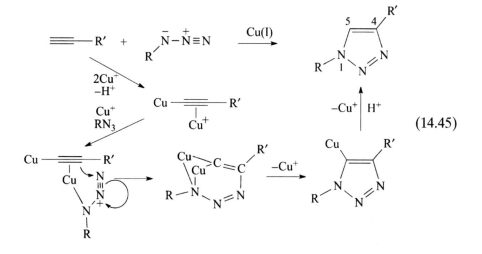

$$(14.45)$$

The catalytic applications of organometallic chemistry to organic synthesis are expanding so rapidly at present that we can expect to continue to see many new reactions and novel combinations of established reactions in the future.

- Organometallic reactions have completely changed the way organic synthesis is planned and performed.
- Green chemistry ideas will lead to increased emphasis on catalysis in synthesis.

REFERENCES

1. See A. J. Birch, *Notes Rec. Roy. Soc. London*, **47**, 277, 1993.

2. See M. E. Bowden and T. Benfey, *Robert Burns Woodward and the Art of Organic Synthesis*, Beckman Center for the History of Chemistry, Philadelphia, 1992.

3. C. A. Busacca, D. R. Fandrick, J. J. Song, and C. H. Senanayake, *Adv. Synth. Catal.*, **353**, 1825, 2011.

4. P. Wyatt and S. Warren, *Organic Synthesis: Strategy and Control*, Wiley, Hoboken, NJ, 2007.

5. C. Drahl, *Chem. Eng. News.*, May 17, 2010, p. 31.

6. J. Magano and J. R. Dunetz, *Chem. Rev.*, **111**, 2177, 2011.

7. D. G. Hulcoop and M. Lautens, *Org. Lett.*, **9**, 1761, 2007.

8. H. J. Zhang, J. Boonsombat, and A. Padwa, *Org. Lett.*, **9**, 279, 2007.

9. B. W. Zhao and X. Y. Lu, *Org. Lett.*, **8**, 5987, 2006.

10. A. Fayol, Y. Q. Fang, and M. Lautens, *Org. Lett.*, **8**, 4203, 2006.

11. S. S. Zalesskiy and V. P. Ananikov, *Organometallics*, **31**, 2302, 2012.

12. F. Littke and G. C. Fu, *Angew Chem Int Ed.*, **41**, 4176, 2002.

13. M. Trost, M. R. Machacek, and A. Aponick, *Acct. Chem. Res.*, **39**, 747, 2006.

14. C. P. Butts, E. Filali, G. C. Lloyd-Jones, P. -O. Norrby, D. A. Sale, and Y. Schramm, *J. Amer. Chem. Soc.*, **131**, 9945, 2009.

15. S. T. Madrahimov and J. F. Hartwig, *J. Amer. Chem. Soc.*, **134**, 8136, 2012.

16. S. E. Denmark and Z. Wang, *Org. Lett.*, **3**, 1073, 2001.

17. M. Butters, J. N. Harvey, J. Jover, A. J. J. Lennox, G. C. Lloyd-Jones, and P. M. Murray, *Angew. Chem. Int. Ed.*, **49**, 5156, 2010.

18. E.-I. Negishi, Z. Huang, G. Wang, S. Mohan, C. Wang, and H. Hattori, *Acc. Chem. Res.*, **41**, 1474, 2008.

19. J.-Y. Lee and G. C. Fu., *J. Am. Chem. Soc.*, **125**, 5616, 2003; L. Ackermann, A. R. Kapdi, and C. Schulzke, *Org. Lett.*, **12**, 2298, 2010.

20. B. M. Trost and K. Dogra, *Org. Lett.*, **9**, 861, 2007.

21. F. Y. Kwong, A. Klapars, and S. L. Buchwald, *Org. Lett.*, **8**, 459, 2002.

22. W. M. Czaplik, M. Mayer, and A. Jacobi von Wangelin, *Angew. Chem. Int. Ed.*, **48**, 607, 2009.

23. Z Jin, G. B. Hammond, and B. Xu, *Aldrichimica Acta*, **45**, 67, 2012.

24. Z. Q. Liu and J. D. Rainier, *Org. Lett.*, **8**, 459, 2006.

25. J. S. Clark, J. Conroy, and A. J. Blake, *Org. Lett.*, **8**, 2091, 2007.

26. C. S. Poulsen, R. Madsen, *Synthesis*, **2003**, 1; S. V. Maifeld and D. Lee, *Chem. Eur. J.*, **11**, 6118, 2005; S. T. Diver, *Coord. Chem. Rev.*, **251**, 671, 2007.

27. A. Seepersaud and Y. Al-Abed, *Org. Lett.*, **1**, 1463, 1999.

28. S. J. Meek, R. V. O'Brien, R. V. Llaveria, R. R. Schrock and A. H. Hoveyda, *Nature*, **471**, 461, 2011; P. Teo and R. H. Grubbs, *Organometallics*, **29**, 6045, 2010; L. E. Rosebrugh, M. B. Herbert, V. M. Marx, B. K. Keitz, and R. H. Grubbs, *J. Am. Chem. Soc.*, **135**, 1276, 2013.

29. S. J. Malcolmson, S. J. Meek, E. S. Sattely, R. R. Schrock, and A. H. Hoveyda, *Nature*, **456**, 933, 2008.

30. M. P. Doyle, R. Duffy, M. Ratnikov, and L. Zhou, *Chem. Rev.*, **110**, 704, 2010.

31. Z. L. Song and R. P. Hsung, *Org. Lett.*, **9**, 2199, 2007.

32. E. Vedejs and M. Jure, *Angew Chem. Int. Ed.*, **44**, 3974, 2005.

33. T. Ohkuma, K. Tsutsumi, N. Utsumi, R. Noyori, and K. Murata, *Org. Lett.*, **9**, 255, 2007.

34. S. J. Roseblade and A. Pfaltz, *Acc. Chem. Res.*, **40**, 1402, 2007.

35. C. W. Cho and M. J. Krische, *Org. Lett.*, **8**, 3873, 2006.

36. J. R. Martinelli, D. M. M. Freckmann, and S. L. Buchwald, *Org. Lett.*, **8**, 4843, 2006.

37. Y. Tsuji and T. Fujihara, *J. Synth. Org. Chem. Jap.*, **69**, 1375, 2011.

38. J. W. Kramer, E. B. Lobkovsky, and G. W. Coates, *Org. Lett.*, **8**, 3709, 2006.

39. J. M. Rowley, E. B. Lobkovsky, and G. W. Coates, *J. Am. Chem. Soc.*, **129**, 4948, 2007.

40. S. Ko, Y. Na, and S. Chang, *J. Am. Chem. Soc.*, **124**, 750, 2002.

41. Y. Pan and J. K. De Brabander, *Synlett*, **853**, 2006.

42. K. M. Brummond and D. Gao, *Org. Lett.*, **5**, 3491, 2003.

43. A. Sabitha, K. Sudhakar, N. M. Reddy, M. Rajkumar, and J. S. Yadav, *Tet. Lett.*, **46**, 6567, 2005.

44. A. N. Campbell and S. S. Stahl, *Acct. Chem. Res.*, **45**, 851, 2012.

45. Y. Deng, Andreas K. Å. Persson, and J. -E. Bäckvall, *Chem. Eur. J.*, **18**, 11498, 2012.

46. E. M. Ferreira and B. M. Stoltz, *J. Am. Chem. Soc.*, **125**, 9578, 2003.

47. S. R. Neufeldt; M. S. Sanford, *Acc. Chem. Res.*, **45**, 936, 2012.

48. B. S. Lane and K. Burgess, *Chem. Rev.*, **103**, 2457, 2003.

49. Dyker ed., *Handbook of C=H Transformations*, Wiley-VCH, Weiheim, 2005.

50. T. Brückl, R. D. Baxter, Y Ishihara, and P. S. Baran, *Acct. Chem. Res.*, **45**, 826, 2012.

51. Y. Guari, A. Castellanos, S. Sabo-Etienne, and B. Chaudret, *J. Mol. Catal. A*, **212**, 77, 2004.

52. J. Oxgaard, G. Bhalla, R. A. Periana, and W. A. Goddard, *Organometallics*, **25**, 1618, 2006.

53. F. Kakiuchi and N. Chatani, *Adv. Synth. Catal.*, **345**, 1077, 2003.

54. J. M. Murphy, C. C. Tzschucke, and J. F. Hartwig, *Org. Lett.*, **9**, 757, 2007.

55. J. F. Hartwig, K. S. Cook, M. Hapke, C. D. Incarvito, Y. B. Fan, C. E. Webster, and M. B. Hall, *J. Am. Chem. Soc.*, **127**, 2538, 2005.

56. T. W. Lyons and M. S. Sanford, *Chem. Rev.*, **110**, 1147, 2010.

57. D. C. Powers and T. Ritter, *Acc. Chem. Res.*, **45**, 840, 2012.

58. R. T. Gephart, D. L. Huang, M. J. B. Aguila, G. Schmidt, A. Shahu, and T. H. Warren, *Angew. Chem. -Int. Ed.*, **51**, 6488, 2012.

59. M. S. Chen and M. C. White, *Science*, **327**, 566, 2010, and **335**, 807, 2012.

60. C. S. Yeung and V. M. Dong, *Chem. Rev.*, **111**, 1215, 2011.

61. B. T. Worrell, J. A. Malik, and V. V. Fokin, *Science*, **340**, 457, 2013.

PROBLEMS

14.1. The reaction shown in Eq. 14.46 occurs via a combination of Pd-catalyzed steps. Trace out the course of the reaction by identifying the reaction steps. Explain why the two products shown are formed.

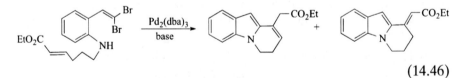

$$(14.46)$$

14.2. Show a plausible detailed mechanism for the Murai reaction of Eq. 14.37, showing the intermediates in full.

14.3. Suggest a mechanism for the transformation of Eq. 14.47. (See *Org. Lett.*, **10**, 2657, 2008.)

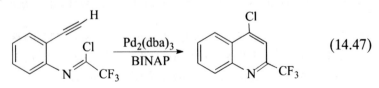

$$(14.47)$$

14.4. Suggest a mechanism for the transformation of Eq. 14.48 and account for the regiochemistry. (See *Org. Lett.*, **10**, 2541, 2008.)

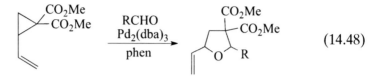

$$(14.48)$$

14.5. Grubbs' catalyst, shown below, tends to decompose by a standard organometallic reaction for R = H but is stable in the case of R = Me. What is the reaction and why does the change of substitution affect the outcome. (See *Org. Lett.*, **10**, 2693, 2008.)

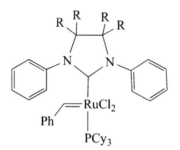

14.6. Suggest a mechanism for the transformation of Eq. 14.49. (See *Org. Lett.*, **10**, 2777, 2008.)

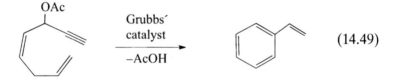

$$(14.49)$$

14.7. Suggest a mechanism for the transformation of Eq. 14.50. (See *Org. Lett.*, **10**, 2829, 2008.)

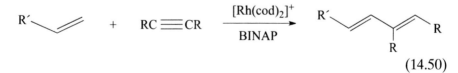

$$(14.50)$$

14.8. Decarbonylation of PhCHO occurs stoichiometrically with $RhCl(PPh_3)_3$ to give PhH and $RhCl(CO)(PPh_3)_3$. Suggest a mechanism for the transformation and a reason for the reaction not being catalytic. The complex, $RhCl(dppe)_2$, in contrast, does give catalytic decarbonylation, although only at 120°C. Why does the dppe complex permit catalysis (dppe = $Ph_2PCH_2CH_2PPh_2$)?

14.9. Suggest a mechanism for the transformation of Eq. 14.51 and account for the stereochemistry. How could you test your mechanism experimentally? (See *Org. Lett.*, **10**, 3351, 2008.)

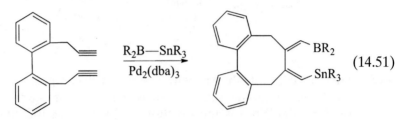

$$(14.51)$$

14.10. Suggest a mechanism for the transformation of Eq. 14.52. (See *Org. Lett.*, **10**, 3367, 2008.)

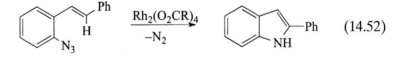

$$(14.52)$$

15

PARAMAGNETIC AND HIGH OXIDATION-STATE COMPLEXES

Diamagnetic complexes have dominated the discussion up to this point because they are easiest to study and are known in the greatest number. With the present increasing interest in nonprecious metals and metals in biology (Chapter 16), paramagnetism is much more commonly encountered. The paramagnetism of these predominantly first-row transition metals reflects their propensity to undergo one-electron redox processes that give odd-electron d^n configurations and their lower ligand field Δ splitting that makes high-spin paramagnetic complexes possible for even d^n configurations. In most of these complexes, we move away from 18e "closed-shell" configurations into "open-shell" territory where at least one orbital is only half filled. The f-block metals (Section 15.4) are often paramagnetic because of partial occupation of the deep-lying f orbitals that are not split by the ligand field, so there are no alternative spin states to consider for any specific metal and OS combination.

Low oxidation states have also dominated the previous discussion because they favor binding soft, π-acceptor ligands (CO, C_2H_4, etc.) that are most typical of organometallic chemistry. If we avoid these ligands and restrict the coordination sphere to alkyl, aryl, H, and Cp, however, high oxidation states appear in the resulting compounds. We look at

The Organometallic Chemistry of the Transition Metals, Sixth Edition.
Robert H. Crabtree.
© 2014 John Wiley & Sons, Inc. Published 2014 by John Wiley & Sons, Inc.

polyalkyls, such as WMe_6 in Section 15.2 and cyclopentadienyls such as $Cp*ReMe_4$ in Section 15.3. Finally, in polyhydrides such as $[ReH_9]^{2-}$, we see the highest coordination numbers with the smallest ligand, hydride (Section 15.3); like polyalkyls, these are also often d^0 and diamagnetic.

The maximum oxidation state possible for any transition element is the group number, N, because only N valence electrons are available for ionization or for forming covalent bonds. The resulting d^0 compounds are normally diamagnetic. Re in group 7 and Os in group 8 are the last elements that are able to attain their theoretical maximum oxidation states (e.g., ReF_7 and OsO_4); Ir and Pt only reach M(VIII) in IrO_4.[1] or M(VI) in PtF_6, and gold shows its highest oxidation state, Au(V), in $[AuF_6]^-$. It is therefore not surprising that most of the organometallic complexes having an oxidation state in excess of 4 come from the elements Ta, W, Re, Os, and Ir. Common for the earlier elements [e.g., Ti(IV), Ta(V)], d^0 oxidation states are rare for the later ones, and, when they do occur, we may expect to find them stronger oxidants. Just as the study of low-valent organotransition metal complexes led to the development of methods for the selective reduction of organic compounds, we can anticipate that high oxidation-state chemistry will lead to better methods of oxidation. The higher oxidation states in general are more stable for the third-row transition metals (Section 2.7). We will see that this is also true for organometallic compounds.

As we saw in Section 2.2, the 18e rule is most likely to be obeyed by low-valent diamagnetic complexes. In this chapter, we will find many examples of stable species with electron counts less than 18e, but this is especially true of polyalkyls, some of which are paramagnetic. One reason is that an alkyl ligand occupies much space around the metal in exchange for a modest contribution to the electron count. Second, the high ∂^+ character of the metal leads to a contraction in its covalent radius because the metal electrons are contracted by the positive charge. This only leads to a slight decrease in the M–L bond lengths because the ligands acquire ∂^- character expanding their covalent radii. An increase in the ligand size and a decrease in the metal size makes it more difficult to fit a given number of ligands around a metal in the high-valent case. The low apparent electron count in such species as $MeReO_3$ may be augmented somewhat by contributions from the ligand (O, Cl, NR, etc.) lone pairs. Agostic interactions with the alkyl C–H bonds are probably not widespread in d^0 and high-valent complexes because this interaction usually needs some back donation from the metal (Section 3.4). This means that electron counting in these species is often ambiguous. High-valent Cp complexes are more likely to be conventional 18e species because Cp contributes many more electrons to the metal in proportion to the space it occupies than do alkyl groups. Polyhydrides are almost always 18e, as we might expect for such a small, tightly-bound ligand.

Oxidation of organometallic compounds typically leads to decomposition, but in an increasing number of cases, useful high oxidation state products have been obtained. The ligands must resist oxidative decomposition to survive the reaction, and in some cases, it is the ligand that is reversibly oxidized rather than the metal.[2] Such ligands are considered redox-active or noninnocent. Oxidation can either be carried out electrochemically or with chemical oxidants, but choosing the right oxidant requires care.[3]

15.1 MAGNETISM AND SPIN STATES

Diamagnetic materials are weakly repelled by a magnetic field gradient while paramagnetic ones are attracted. From the weight change of a sample in the presence or absence of a magnetic field gradient, or by an NMR method (Evans method: ref. 3 in Chapter 10), one can measure the magnetic moment of a complex. This is related to the number of unpaired electrons on the central metal. Specialist texts[4] cover a number of possible complicating factors that can affect the interpretation, such as spin coupling in metal clusters and orbital contributions in third-row ($5d$) transition metals. Table 15.1 shows the situation in the absence of such complications, where the measured magnetic moment in Bohr magnetons gives the number of unpaired electrons. This number is often indicated by the spin quantum number, S, which is simply half the number of unpaired electrons. The multiplicity (singlet, doublet, triplet, etc.) is also used as shown in the table.

TABLE 15.1 Terms for Discussing Magnetism

Spin Quantum Number, S	Number of Unpaired Electrons	Multiplicity	Magnetic Moment (Bohr Magnetons)[a]
0	0	Singlet	0
½	1	Doublet	1.73
1	2	Triplet	2.83
³⁄₂	3	Quartet	3.87
2	4	Pentet	4.90
⁵⁄₂	5	Sextet	5.92

[a]Ideal value—the magnetic moment is also affected by orbital contributions and magnetic coupling in metal clusters, effects ignored here.

The possible S values of a mononuclear complex depend on the d^n configuration. The d^0 and d^{10} cases are necessarily diamagnetic ($S = 0$), having no unpaired electrons. In contrast, d^1 and d^9 are necessarily paramagnetic with one unpaired electron ($\uparrow$, $S = ½$). The d^3, d^5, and d^7 odd-electron configurations are necessarily paramagnetic but may have different accessible spin states depending on how the spins are paired (e.g., ($\uparrow\uparrow\uparrow$, $S = 3/2$) or ($\uparrow\uparrow\downarrow$, $S = ½$) for d^3). Even-electron d^2, d^4, d^6, and d^8 cases may be diamagnetic or paramagnetic depending on spin pairing (e.g., ($\uparrow\uparrow$, $S = 1$) or ($\uparrow\downarrow$, $S = 0$) for d^2).

Spin States

Spin states are isomeric forms with distinct energies, structures, and reactivities. A complex normally exists in its stablest state, but which spin state that is depends on the geometry, ligand set, and consequent d orbital splitting. As we fill these orbitals, the potential exists for alternative spin states, depending on how the electrons distribute themselves. Instead of the idealized octahedral splitting pattern of three d_π below two d_σ orbitals of Chapter 1, giving the high-spin/low-spin alternative spin states of Fig. 1.2, we now have to deal instead with more realistic splitting patterns of low-symmetry organometallic complexes.[5]

As discussed by Poli and Harvey,[6] a simple picture, based on the ionic model, starts from the coordination number, represented in what follows by the symbol m, as given by Eq. 15.1 for the complex $[MX_aL_b]^{c+}$. Of the nine valence orbitals of the metal, we expect to find m orbitals in the M–L σ^* group (Fig. 15.1a). Of these m orbitals, four are the single s and the three p orbitals, so $(m - 4)$ is the number of d orbitals in this M–L σ^* group. For the octahedral case, we have $(6 - 4)$, or two d orbitals, in agreement with the presence of just two d_σ orbitals in the familiar "three below two" octahedral crystal field pattern. We can usually avoid further consideration of these $(m - 4)$ orbitals because electrons rarely go into M–L σ^* antibonding orbitals in organometallic complexes, although this is not uncommon in Werner complexes with their generally lower Δ values. In the middle set of orbitals, in a dotted box in Fig. 15.1a, we find $(9 - m)$ d orbitals, which are either nonbonding or involved in π back bonding. For the familiar octahedral case, we have $(9 - 6)$ or three orbitals, corresponding with the familiar d_π set. Below these orbitals, we have m M–L σ-bonding levels. The electron count of the complex will be $(2m + n)$; for the familiar d^6 octahedral case, this will be $(2 \times 6 + 6)$, or 18 electrons.

$$CN = m = a + b \qquad (15.1)$$

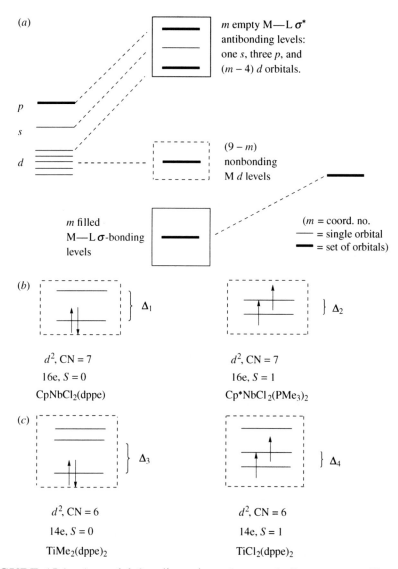

FIGURE 15.1 A model for discussion of open-shell organometallic compounds (dppe = $Ph_2CH_2CH_2PPh_2$). (*a*) Number of nonbonding levels (dotted box) depends on the coordination number, *m*. The number of electrons, *n*, available to fill these levels depends on the d^n configuration. (*b*,*c*) For six- and seven-coordinate species, such as the ones shown, two spin states are possible, $S = 0$ and $S = 1$. Thick lines denote sets of orbitals.

$$\text{Number of M--L antibonding } d \text{ orditals} = (m - 4) \qquad (15.2)$$

$$\text{Number of M--L nonbonding } d \text{ orbitals} = (9 - m) \qquad (15.3)$$

To find the possible alternative spin states from the d^n configuration, we look for different ways to distribute these n electrons among $(9 - m)$ orbitals. To take the d^2 case, typical coordination numbers are 6 and 7. The examples of Fig. 15.1b and 15.1c show how the L_2X-type Cp ligand contributes three to the coordination number. Small changes in the ligand set can be sufficient to alter the energies of the d orbitals so that the magnetism changes from one spin state to the other. If the energies of the two states are close enough, there can even be a spin equilibrium between the two, as for $S = 0$ and 1 spin states of $[(C_5H_4Me)$-$NbCl_2(PEt_3)_2]$, but this is rare.

The relative energies of the spin states is decided by the relative sizes of the electron pairing energy and the HOMO–LUMO splitting, Δ. A large electron pairing energy (PE) favors the $S = 1$ state because it makes it difficult to spin-pair two electrons in the same orbital where e–e repulsion is high. A large Δ favors $S = 0$ because it is now hard to convert $S = 0$ to $S = 1$ because the resulting electron promotion now requires more energy. In Fig. 15.1b and 15.1c, Δ_1 is larger than Δ_2 and Δ_3 is larger than Δ_4, as expected on the basis of this argument.

The value of Δ depends on the geometry, ligands and metal. The geometry therefore often changes to a larger or smaller extent with spin state change. A large change occurs for d^8 16e $NiX_2(PR_3)_2$ where the $S = 0$ complexes are square planar and the $S = 1$ are tetrahedral. The Δ often increases as we move from $3d$ Ni to $4d$ and $5d$ Pd and Pt, so that the heavy analogs, $PdX_2(PR_3)_2$ and $PtX_2(PR_3)_2$, are always square planar with $S = 0$.

Any π bonding also strongly alters Δ by the mechanism of Fig. 1.9 and Fig. 1.10 if different orbitals are differently affected. In $[Cp^*Mo(PMe_3)_2(PPh_2)]$, for example (Fig. 15.2), there is one π-bonding lone pair on the phosphide ligand that raises one of the three nonbonding d levels appropriate for this six-coordinate system. The result is a diamagnetic $S = 0$ state for this d^4 case. If the ligand has two π-bonding lone pairs, as in the chloro analog $[Cp^*Mo(PMe_3)_2Cl]$, however, the two d-orbitals now affected by π bonding are both raised in energy, resulting in an $S = 1$ state.

Influence of Spin State Changes on Kinetics and Thermodynamics

Often, one spin state may be very reactive, the other not. Where alternate spin states are possible, there may be a change of spin state in a

(a)

(b)

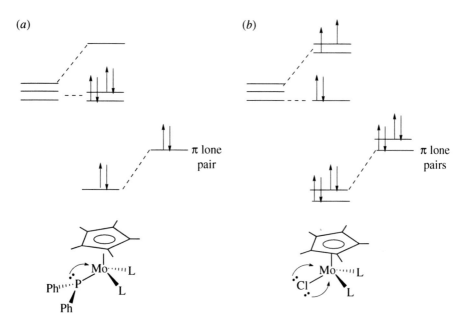

FIGURE 15.2 (a) A single π-donor lone pair of PPh$_2$ splits the d orbitals so that the four d electrons prefer to occupy the two lower levels leading to an $S = 0$ state. (b) The pair of π-donor lone pairs of Cl split the d orbitals so that the four d electrons now prefer to occupy the three lower levels as shown, leading to an $S = 1$ state. The two unpaired electrons are parallel according to Hund's rule.

reaction.[7] A molecule in one spin state could undergo a spin change to give a reactive form if the latter is close enough in energy; the energy cost of the spin state change would merely contribute to the reaction barrier. Such a case is illustrated in Fig. 15.3a for the reaction of **A** to give **B** in a case where we have a ground spin state with a high reaction barrier and an excited spin state with a low barrier. If the spin state change were very fast, the system could take the path **A** → 1 → 2 → 3 → **B**. If the spin change could not occur rapidly enough to happen during the reaction, however, we would have to go via the pathway **A** → **A*** → 2 → **B*** → **B** (where **A*** and **B*** are the excited spin states of reactant and product). In either case, the reaction would still be faster than going via point 4, which would be the case if there were no alternate spin states available (as is often the case in conventional low-valent organometallic chemistry). This implies that organometallic species with alternate spin states can be more kinetically labile than typical 18e complexes.

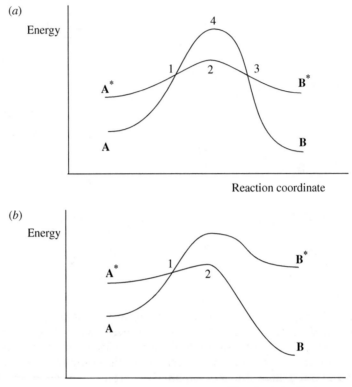

FIGURE 15.3 Reactivity patterns for species with alternate spin states. (*a*) The kinetics of a reaction can be accelerated if a more reactive accessible excited spin state exists with a lower net barrier for the reaction. We assume that spin change is fast. (*b*) The thermodynamics of a reaction can be affected if the product has a spin state different from that of the reagent. In this case, the reaction is unfavorable in the starting spin state but favorable if the system crosses to the other spin state. The star refers to the excited (less stable) spin state in each case.

In a system with alternate spin states, a change of spin state may occur during the reaction.[7] As shown in Fig. 15.3*b*, this can play a role in the thermodynamics of the reaction. Assume the reagent spin state, **A**, leads to an excited spin state of the product, **B***; this can even be an endothermic, unfavorable process, as shown here. If this reaction pathway intersects the corresponding curve for the other spin state, crossover is expected to give not **B*** but **B**. The path is now **A** → 1 → 2 → **B**, and the reaction is only thermodynamically favorable thanks to the accessibility of the alternate spin state.

If the unsaturated product of ligand loss is stabilized by this mechanism, the M–L bond strength will be lower than if no such stabilization occurred because the bond strength is defined as the energy difference between L_nM–L and ground state L_nM + L. Indeed, exceptionally low M–CO bond energies of 10–15 cal/mol have been reported for a series of compounds where this effect applies.[8]

Examples of spin state control of reaction rates have been given by Harvey et al.[7] For example, the slow addition of H_2 to Schrock's $[W\{N(CH_2CH_2NSiMe_3)_3\}H]$ is "spin-blocked" with a high barrier owing to the difficulty of crossing between reactant triplet and product singlet surfaces. In contrast, addition of CO to Theopold's [TpCo(CO)] is fast because the triplet and singlet surfaces cross at an early stage of reaction and therefore at low energy.

3d versus 4d and 5d Metals

First-row (3d) transition metals are the most likely to be paramagnetic with a <18e structure. Later metal analogs often adopt a different, often 18e, structure. For example, in the $CpMCl_2$ series (M = Cr, Mo, and W), **15.1** lacks M–M bonds, and each 15e Cr is $S = 3/2$. In contrast, the Mo and W analogs **15.2** and **15.3** are both 18e, $S = 0$ with M–M bonds. Similarly, the 3d metals may have a lower coordination number in their compounds. For example, **15.1** reacts with dppe to give $S = 3/2$, 15e, **15.4**, having a monodentate dppe, but with **15.2** to give $S = 1/2$, 17e, **15.5**.

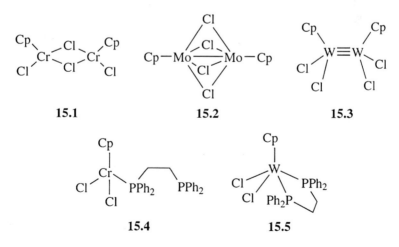

NMR Spectroscopy

Some paramagnetic complexes can give interpretable ^1H NMR spectra, although the signals can appear from −400 to +400 ppm, a much wider range than is usual for diamagnetic complexes, and assignment of the NMR spectrum is more difficult.[9]

15.2   POLYALKYLS AND POLYHYDRIDES

Group 4

The bright yellow crystals of homoleptic (i.e., containing only one type of ligand) $TiMe_4$ decompose above $\sim 0°C$ to methane, but adducts with hard ligands, such as NMe_3, tmeda, or PMe_3, are more thermally stable. The Grignard-like reactivity of the Ti(IV) alkyls implicates a ∂^- carbon, consistent with the electronegativity difference between C (2.5) and Ti (1.5). On going to the right and descending the periodic table from Ti to the heavy platinum metals, the electronegativity increases from 1.5 to about 2.2, and the M–C bond becomes much less polar. This makes the metal less positive and the alkyls less negative in the later metals.

$Ti(CH_2Ph)_4$ has a Ti–C_α–C_β angle of only 84–86° (Fig. 15.4), suggesting that the C_β carbon of the aromatic ring interacts with the metal. The soft ligand CO does not form a stable carbonyl with d^0 $Ti(CH_2Ph)_4$ (**15.6**), although initial formation of a CO adduct has been proposed on the pathway to the final product, $Ti(COCH_2Ph)_2(CH_2Ph)_2$. In contrast to the low thermal stability and high air and acid sensitivity of these alkyls, bulky complexes such as **15.7** are unusually stable, thanks

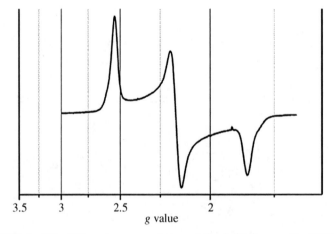

FIGURE 15.4   The X-band epr spectrum of the Ir(IV) complex **15.10** (p. 426). Having four different ligands around Ir provides a rhombic symmetry consistent with the resonance pattern seen here. Source: From Brewster et al., 2011 [30]. Reproduced with permission of the American Chemical Society.

to steric protection of the metal. **15.7** is even stable enough to melt at 234°C. The Zr and Hf alkyls are less well studied but behave rather similarly to their Ti analogs.

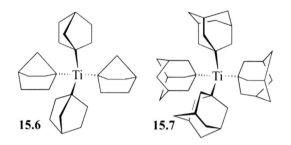

15.6   **15.7**

Group 5

Even though vanadium has a stable (V) oxidation state, the only alkyls so far discovered are the dark paramagnetic d^1 VR_4 species, such as the green-black benzyl complex. The 1-norbornyl is the most stable, decomposing only slowly at 100°C. Tantalum, the third-row element, gives stable alkyls, such as $TaMe_5$, which forms a dmpe adduct. $TaMe_5$ is trigonal bipyramidal, but attempts to make bulkier TaR_5 complexes always lead to α elimination to give carbenes. As we go to the right in the transition series, the differences between the first-, second-, and third-row elements become more marked. An example is the increasing reluctance of the first- and even second-row elements to give d^0 alkyls, a feature that first appears in group 5 and becomes dominant in groups 6 and 7.

Group 6

A dark red Cr(IV) alkyl $[Cr(CH_2SiMe_3)_4]$ is known, but Cr(III) is the common oxidation state, as in the orange $Li_3[CrPh_6]$. WMe_6 was the first homoleptic alkyl of group 6 to have the maximum oxidation state allowed for the group. It can decompose explosively at room temperature but can also give the reactions shown in Eq. 15.4 and Eq. 5.5.

$$WMe_6 \xrightarrow{\quad O_2 \quad} W(OMe)_6 \tag{15.4}$$

$$WMe_6 \xrightarrow{\quad CO \quad} W(CO)_6 + Me_2CO \tag{15.5}$$

Group 7

Only one Mn(IV) alkyl is known, the green $Mn(1\text{-norbornyl})_4$, but rhenium has an extensive series of high oxidation-state alkyls (Eq. 15.6), consistent with the generally higher stability of third-row versus first-row metals in high oxidation states. The higher electronegativity of Re compared with W may help make the Re alkyls generally more stable to air, acids, and attack by nucleophiles. $ReOMe_4$ also fails to react with the Lewis bases that usually give complexes with the polyalkyls of the earlier metals.

$$Cl_4Re\!\equiv\!ReCl_4 \xrightarrow{\text{MeLi}} Me_4Re\!\equiv\!ReMe_4$$
$$\text{red}$$

$$\downarrow O_2$$

$$ReOCl_4 \xrightarrow{\text{MeLi}} \underset{\text{carmine}}{ReOMe_4} \xrightarrow{\text{AlMe}_3} \underset{\text{green}}{ReMe_6} \xrightarrow{O_2} ReOMe_4 \xrightarrow{\text{NO}} cis\text{-}ReO_2Me_3 \xrightarrow{O_2} ReO_3Me$$

$$(15.6)$$

$ZnNp_2$ ($Np = t\text{-}BuCH_2$) and $ReOCl_3(PPh_3)_2$ give the unusual dirhenium dioxo tetraalkyl of Eq. 15.7 with a 2.6Å Re–Re bond.

$$ReOCl_3(PPh_3)_2 \xrightarrow{\text{ZrNp}_2} \begin{array}{c} \text{Me} \quad \text{O} \quad \text{Me} \\ \diagdown \; \diagdown \; \diagup \\ O\!=\!Re\!\!-\!\!Re\!=\!O \\ \diagup \; \diagup \; \diagdown \\ \text{Me} \quad \text{O} \quad \text{Me} \end{array} \qquad (15.7)$$

Groups 8–10

Purple Fe(IV) and brown Co(IV) norbornyls are known, but most alkyls of these groups are M(II) or M(III), such as the yellow $Li_2[FeMe_4]$ or fac-$[RhMe_3(PMe_3)_3]$. Co(III) alkyls are mentioned in connection with coenzyme B_{12} chemistry (Section 16.2).

Ni and Pd alkyls include the golden-yellow $Li_2[NiMe_4]$ or $PdMe_2(bipy)$. In many organic synthetic applications of Pd, formation of a Pd(IV) alkyl had to be postulated, but isolable examples were only found much later.[10] The first aryl, $PdCl_3(C_6F_5)(bipy)$ (1975), and the first alkyl, $PdIMe_3(bipy)$ (1986) (Eq. 15.8), both made use of the stabilizing N-donor bipy group and the exceptionally strong M–C_6F_5 and M–Me bonds.

$$(15.8)$$

Of all polyalkyls, the longest known are the octahedral Pt(IV) species related to the orange complex $[Me_3Pt(\mu^3\text{-}I)]_4$ (**15.8**, some Me groups omitted for clarity), described by Pope and Peachey as early as 1907–

1909. Some of its reactions (Eqs. 15.10; L = NH$_3$, en, py, PMe$_3$) illustrate the water stability of these alkyls consistent with the high electronegativity of the late metal and the strong M–L bonding in the third row metals.

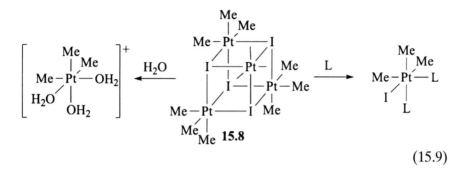

$$(15.9)$$

Group 11

Cu and Ag give only M(I) alkyls, such as the bright yellow and explosive [CuMe]$_n$, but Au forms compounds from Au(I) to (III), such as [Au(C$_6$F$_5$)$_4$]$^-$. The lithium cuprates, Li[CuMe$_2$], are important reagents in organic synthesis, acting as more selective nucleophiles than LiMe itself, but much more active ones than CuMe. Anionic complexes of type [MR$_n$]$^{m-}$ as a class are termed "ate" complexes, from the -ate termination to names such as cuprate.

Catalysis by High-Valent Oxo Complexes

Toste showed that Re(V) oxo complexes can hydrosilylate organic carbonyl groups[11] via a novel mechanism involving [2 + 2] Si–H addition across the Re=O bond to give HRe–OSiR$_3$ in the initial step.[12] The hydrogenation of alkynes to alkenes is also possible, with H$_2$ as reductant using MoO$_2$Cl$_2$, ReIO$_2$(PPh$_3$)$_2$, and CH$_3$ReO$_3$ (MTO) as catalyst precursors.[13] MTO also catalyses a number of oxidations with hydrogen peroxide as primary oxidant. RC≡CH gives RCOOH, RC≡CR yields (RCO)$_2$, and alkenes form epoxides.

Carbenes and Carbynes

Many early-metal Schrock carbenes and carbynes, best seen as d^0 species (Chapter 11), prefer hard ligand sets, as in Eq. 15.10. Chelating amines, often deprotonated, have proved very useful for favoring high oxidation states of early metals;[14] the π lone pair of a deprotonated R$_2$N ligand makes it a π donor, appropriate for a d^0 metal. In Eq. 15.10, the W(IV) starting material has such a high tendency to achieve W(VI) that it dehydrogenates and rearranges ethylene to extrude H_2 to give an ethylidyne (X_3) ligand, and also favors the $M^+ \equiv C{-}O^-$ (X_3, **4.3**) bonding mode of CO that facilitates nucleophilic attack on R'I.

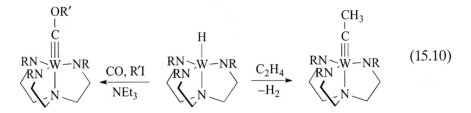

$$(15.10)$$

Polyhydrides

Polyhydrides[15] have H:M ratios exceeding 3, as in $MH_4(PR_3)_3$ (M = Fe, Ru, and Os). Hydrogen is not as electronegative as carbon, and so the metal in a polyhydride is not as oxidized as in a polyalkyl. Polyhydrides therefore retain more of the properties of low-valent complexes than do polyalkyls. For example, many of them are 18e, and relatively soft ligands (PR_3 or Cp in the vast majority of cases) are required to stabilize them. Their high formal oxidation state may only be apparent because they sometimes contain H_2 ligands.[16] For example, $IrH_5(PR_3)_2$ (R = C_6H_{11}) is a classical Ir(V) hydride, but protonation[17] gives Ir(III) $[IrH_2(H_2)_2(PR_3)_2]^+$ (Eq. 15.12), not Ir(VII) $[IrH_6(PR_3)_2]^+$; as a 2e L ligand, H_2 leaves the oxidation state unchanged. The bulky $P(C_6H_{11})_3$ ligand provides steric protection for the relatively labile bis-dihydrogen ligand set, and the "noncoordinating" BF_4 anion remains reliably outer sphere.

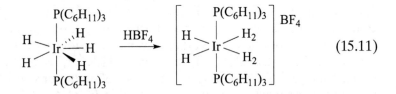

$$(15.11)$$

Although $Re^{VII}H_7(dppe)$ is classical, $ReH_7(P\{p\text{-tolyl}\}_3)_2$ has a $ReH_5(H_2)L_2$ structure with a stretched H–H distance (1.357 Å vs. the usual 0.8–1.0 Å for a standard H_2 complex), making the oxidation state ambiguous because the structure lies between Re(V) and (VII). As these examples show, polyhydrides often have coordination numbers in excess of 6, a consequence of the small size of the hydride ligand.

Almost all polyhydrides are fluxional and the hydrides show coupling in the ^1H NMR spectrum to any phosphines present. The number of hydrides present (n) can be predicted with some confidence from the 18e rule, but a useful experimental method involves counting the multiplicity ($n + 1$) of the ^{31}P NMR peak, after the phosphine ligand protons have been selectively decoupled (Section 10.4), leaving only the coupling to the hydrides.

15.3 CYCLOPENTADIENYL COMPLEXES

The Cp and especially the Cp* ligands are very effective at stabilizing high oxidation states and paramagnetic complexes. While some Cp complexes can be polymeric and difficult to characterize, the Cp* analogs are often soluble and well behaved. High oxidation-state halo complexes are well known, for example, Cp_2TiCl_2, Cp_2NbCl_3, Cp_2TaCl_3, and $[Cp_2MoCl_2]^+$. A route to oxo and halo species is the oxidation of the cyclopentadienyl carbonyls or the metallocenes. The $[CpMO]_4$ complexes, of which the earliest (1960) was Fischer's $[CpCrO]_4$, have the cubane structure (**15.9**).

$$CpV(CO)_4 \xrightarrow[\text{or } Cl_2]{HBr/O_2} CpVOX_2 \qquad (15.12)$$

$$Cp_2Cr \xrightarrow{O_2} \qquad (15.13)$$

15.9

Reaction of carbonyls in CH_2Cl_2 with air or with PCl_5 can give oxo and chloro complexes as in the conversion of $[CpMo(CO)_3]_2$ to $CpMoO_2Cl$ and $CpMo(CO)_3Me$ to $CpMoCl_4$, respectively.

Rhenium

Rhenium has an extensive organometallic oxo chemistry. The early elements are so oxophilic that organometallic groups are unlikely to survive, when lower valent species are oxidized or hydrolyzed. Re is the last element, as we go to the right in the periodic table, for which the M=O bond is still very stable. Herrmann and and Kuehn[18] have shown how to make a whole series of oxo complexes of MeRe and Cp*Re fragments. The Re=O vibrations show up very strongly in the IR spectrum, as in the case of the yellow $Cp*ReO_3$ with bands at 878 and

909 cm^{-1}; indeed, the IR data is an essential item for the characterization of all these complexes. A number of $L_nRe{=}O$ species were originally misidentified as L_nRe for lack of an IR spectrum, providing a useful warning against omitting this measurement.

$$Cp^*Re(CO)_3 \xrightarrow[\text{or } H_2O_2]{O_2/h\nu} Cp^*ReO_3 \qquad (15.14)$$

Partial reduction of Cp^*ReO_3 with Me_3SiCl/PPh_3 gives Cp^*ReCl_4 that on reaction with $SnMe_4$ gives $Cp^*ReMeCl_3$, a compound that is very unusual in having low- and high-spin forms in fast equilibrium leading to very large temperature-dependent ^1H NMR shifts. For example, the broad Re*Me* signal for Cp^*ReCl_3Me in $CDCl_3$ shifts from 13.5δ at $-50°C$ to 36.5δ at $+50°C$.

Other Metals

Maitlis has described a number of Ir(V) and Rh(V) alkyls, such as Cp^*IrMe_4. The strong donor environment of the Ir(III) complex of Eq 15.15 facilitates reversible electrochemical oxidation or chemical oxidation with $[Ru(dipy)_3]^{3+}$ to the corresponding Ir(IV) species **15.10** that gives characteristic epr spectra (Fig 15.4).[19]

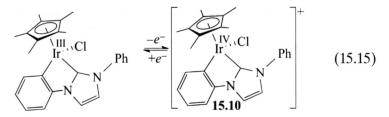

$$(15.15)$$

Bullock and coworkers see reversible dissociation of the W–W bonded dinuclear complex to give the reactive 17e paramagnetic monomer, $CpW(NHC)(CO)_2$.[20] $M(\eta^3\text{-allyl})_4$ complexes also exist for Zr, Nb, Ta, Mo, and W.

15.4 *f*-BLOCK COMPLEXES

The *f* block[21] consists of the 4*f* metals, La–Lu, and the 5*f* metals, Ac–Lr. The common terms *lanthanide* and *actinide* derive from the names of the first elements of each series, and the symbol Ln, not assigned to any particular element, designates the lanthanides as a class; the older term, *rare earths*, is sometimes encountered. The actinides are radioactive, and only Th and U are sufficiently stable to be readily handled outside high-level radiochemical facilities (^{238}U, $t_{1/2} = 4.5 \times 10^9$ years; ^{232}Th, $t_{1/2} = 1.4 \times 10^{10}$ years). Even though they have no *f* electrons, scandium (Sc) and yttrium (Y) in group 3 are also traditionally considered with the *f*-block elements because of their rather similar chemistry.

Unlike the *d* electrons of the *d* block, 4*f* electrons were traditionally considered unavailable for bonding, and where that still holds, we see no equivalent of ligand field effects or of the 18e rule. Instead, the complexes tend to be predominantly ionic without electronic preferences for particular geometries—indeed, irregular geometries are common. The metals become sterically saturated rather than electronically saturated upon ligand binding. If a ligand set does not completely saturate the metal sterically, oligomeric or polymeric structures can form via suitable bridging groups. In such a case, a larger ligand would be needed to prevent bridging and provide a monomeric structure. This accounts for the key role in these elements for ligands having easily adjusted steric bulk. The high tendency to bridge also makes ligand redistribution very fast.

The absence of ligand field effects makes the magnetism of an *f*-block complex identical to that of the parent ion. In the *d* block, a d^2 complex such as Cp_2WCl_2 is typically diamagnetic as a result of *d*-orbital splitting—in contrast, $5f^2$ Cp_2UCl_2 has two unpaired electrons.

Variable valence is a key feature of the *d*-block elements—in contrast, the 4*f* elements generally prefer the tripositive state. Table 15.2 shows the atomic electron configurations of the 4*f* elements, together with the configurations of their common oxidation states. The preference for an unfilled, a half-filled, or a filled *f* shell, helps account for easy access to some non-M(III) states, Ce(IV), Eu(II), Tb(IV), and Yb(II). The f^0 and f^{14} oxidation states being diamagnetic, standard ^{1}H and ^{13}C NMR data can be obtained, greatly facilitating the identification of the complexes involved. Even in other cases, line broadening is relatively small, with the paramagnetic Pr(III), Sm(II), Sm(III), and Eu(III) cases giving the most easily observable spectra. No doubt for this reason, La(III), Ce(IV), Yb(II), and Lu(III)–together with diamagnetic Sc(III) and Y(III) from group 3—are among the most intensively studied states.

The trend in radius, shown for the M(III) ion in Table 15.2, is the result of the increasing number of protons in the nucleus causing the electron shells to contract in the *lanthanide contraction*; the *f* electrons added are deep-lying and inefficient at screening the nuclear charge. In most of chemistry, when we move from one element to the next, the changes in atomic size and preferred valency are abrupt. Here, in contrast, the radius varies smoothly and the M(III) valence state remains preferred, so we have nice control over the M–L bond length. As this

TABLE 15.2   Lanthanide Electronic Configurations and Ion Radii[a]

Element	Atom Config.	M(II) Config.	M(III) Config.	M(IV) Config.	Radius M(III) (Å)
Lanthanum, La	$4f^0 5d^1 6s^2$	[b]	$4f^0$		1.16
Cerium, La	$4f^2 5d^0 6s^2$	[b]	$4f^1$	$4f^0$	1.14
Praseodymium, Pr	$4f^3 5d^0 6s^2$	[b]	$4f^2$	$4f^1$	1.13
Neodymium, Nd	$4f^4 5d^0 6s^2$	$4f^4$	$4f^3$		1.11
Promethium, Pm	$4f^5 5d^0 6s^2$		$4f^4$		1.09
Samarium, Sm	$4f^6 5d^0 6s^2$	$4f^6$	$4f^5$		1.08
Europium, Eu	$4f^7 5d^0 6s^2$	$4f^7$	$4f^6$		1.07
Gadolinium, Gd	$4f^7 5d^1 6s^2$	[b]	$4f^7$		1.05
Terbium, Tb	$4f^8 5d^0 6s^2$	[b]	$4f^8$	$4f^7$	1.04
Dysprosium, Dy	$4f^9 5d^0 6s^2$	$4f^{10}$	$4f^9$		1.03
Holmium, Ho	$4f^{10} 5d^0 6s^2$	[b]	$4f^{10}$		1.02
Erbium, Er	$4f^{11} 5d^0 6s^2$	[b]	$4f^{11}$		1.00
Thulium, Tm	$4f^{12} 5d^0 6s^2$	$4f^{13}$	$4f^{12}$		0.99
Ytterbium, Yb	$4f^{13} 5d^0 6s^2$	$4f^{14}$	$4f^{13}$		0.99
Lutetium, Lu	$4f^{14} 5d^1 6s^2$	[b]	$4f^{14}$		0.98

[a]Oxidation state exists whenever configuration is shown.
[b]Oxidation state only very recently recognized in organometallic derivatives of these elements as having configuration $4f^{(n-1)}5d^1$.

varies, the effective steric size of the ligands gradually varies because the effective ligand cone angle (Section 4.2) increases as the ligand gets closer to the metal. This lanthanide contraction from La–Lu helps account for the fact that the third-row d-block metals, Hf–Hg, which come just after the lanthanides in the periodic table, have a smaller increment in atomic radius over the second row d-block than would be expected by extrapolation of the radius change between the first- and second-row d-block metals. This is illustrated by the metallic radius (Å) trends for some triads–Ti, 1.47, Zr, 1.60, Hf, 1.59; Cr, 1.29, Mo, 1.40, W, 1.41; Ni, 1.25, Pd, 1.37, Pt, 1.39 Å.

As the ionic radius changes, the preferred coordination number can change. For the aqua ions $[Ln(H_2O)_n]^{3+}$, n is 9 for the larger ions, L–Eu, and 8 for the smaller ions, Tb–Lu. For Gd^{3+}, $n = 8$ and $n = 9$ ions have about the same energies. The later lanthanide ions, being smaller, have a slightly greater Lewis acidity.

Consistent with the low Pauling electronegativities of the $4f$ elements (1.0–1.25), ionic bonding plays a greater role in their chemistry than in the d block. The f electrons are low lying in the ions and complexes and do not participate to any great extent in bonding, as shown by the magnetic moments and the color being practically the same in the free ion and in the complexes. The UV–visible *f–f* transitions responsible for the color are very sharp because the deep-lying *f* electrons are isolated from the effects of ligand binding or solvation. These transitions are also involved in the strong luminescence often seen for lanthanide compounds, as in the red Eu-based phosphor in traditional color TVs and Nd-based YAG lasers. Promotion of an *f* electron to the *d* level results in a UV transition. Since the 5*d* levels of lanthanides *are* affected by the ligands, this *f* → *d* band is broad, and the wavelength does depend on the nature of the complex. For example, in [{µ^5-C$_5$H$_3$(SiMe$_3$)$_2$}$_3$Ce], the *f* → *d* band is shifted to such an extent in energy that it appears in the visible range at 17,650 cm^{-1} compared to 49,740 cm^{-1} in the gas-phase UV spectrum of the bare Ce^{3+} ion.

Among the 5*f* elements, we look at Th, with its strongly preferred 5*f*^{0} Th(IV) state, and U with 5*f*^{3} U(III), 5*f*^{2} (IV), 5 *f*^{1} (V), and 5*f*^{0} (VI) states all accessible. In the actinides, the complexes have somewhat more covalency in their bonding than do the 4*f* elements, in line with their higher electronegativities (U, 1.38), and in the case of reduced states of U, a significant tendency to back bond. The 5*f* level is somewhat more available for bonding than is 4*f* in the lanthanides.

Lanthanide Organometallic Chemistry

The chemistry of Ln(III) broadly resembles that of the early *d*-block elements in their highest oxidation states except that the lanthanide complexes are paramagnetic for all configurations from 4*f*^{1} to 4*f*^{13}. The larger size of the Ln(III) ions versus Ti(IV)–Hf(IV) favors higher coordination number for the *f* block.

As oxophilic, hard Lewis acids, Ln^{3+} prefer O donors but Marks' series of bond energies for Cp$_2^*$Sm−X illustrates the bonding preferences are not quite as clear-cut as hard/soft ideas would have it:

$$\text{Cl} > \text{C} \equiv \text{CPh} > \text{Br} > \text{O}(t\text{Bu}) > \text{S}(n\text{Pr}) > \text{I} > \text{H} > \text{NMe}_2 > \text{PEt}_2$$

Simple alkyls, typically formed from LiR and LnCl$_3$, are possible when R is β-elimination resistant, such as in the [LnMe$_6$]$^{3-}$ series of *ate* (anionic) complexes. As 12e(La) to 26e(Lu) complexes, these illustrate the failure of the 18e rule for lanthanides. Bulky alkyls are necessary if bridging is to be avoided, as in the triangular three-coordinate series [Ln{CH(SiMe$_3$)$_2$}$_3$]. β Elimination has a lower driving force in the *f* than in the *d* block because the M–H/M–C bond energy difference is less favorable to M–H. Indeed, α-alkyl elimination, not generally seen in the *d* block, is common here for the same reason.

Cyclopentadienyls[17] have attracted most attention as ligands because they are capable of ionic bonding and can be sterically tuned by varying the substituents. The ionic model is most appropriate for this case because the Cp electrons stay largely on the ligand, but the metal–ligand bond strength can still be very high as a result of the 3+ charge on the metal. The pronounced oxophilicity leads to the formation of a THF complex that only desolvates above 200°C (Eq. 15.16).

$$LnCl_3 + NaCp \xrightarrow{\text{THF}} Cp_3Ln(THF) \xrightarrow[\text{sublime}]{>200°} Cp_3Ln \quad (15.16)$$

The solid state structures form an ordered series. A strictly monomeric structure is only seen for $(\eta^5\text{-Cp})_3Yb$, where steric saturation is precisely attained without the need for bridging. All the other cases involve some degree of Cp bridging between metals. The ions smaller than Yb, Lu, and Sc have $[(\eta^5\text{-Cp})_2M]^+$ units bridged in an infinite chain by $\eta^1\text{-Cp}^-$ groups. The ions larger than Yb have a $(\eta^5\text{-Cp})_3M$ structure with space available for bridges to adjacent Cp_3M units.

Bis-cyclopentadienyl complexes are also seen; Eq. 15.17 shows how Cp_2^*Y can form an adduct with LiCl that is only cleaved by sublimation at 285°C. The monobridged structure of the product contrasts with the bis-bridged $[Cp_2Y(\mu\text{-Cl})_2YCp_2]$ as a result of the lower steric effect of Cp versus Cp*.

$$YCl_3 + LiCp^* \xrightarrow{\text{THF}} Cp^*_2Y \overset{Cl}{\underset{Cl}{\diagdown\diagup}} Li(THF)_2 \xrightarrow[\text{sublime}]{285°} Cp^*_2Y \overset{Cl}{\diagup} \underset{Cl}{\overset{}{Y}}Cp^*_2$$

$$(15.17)$$

The lanthanides are also very fluorophilic, so fluoroborate is far from being noncoordinating (**15.11**), as it is in late d-block chemistry. Methyl groups are also able to bridge, as in $[Cp_2Lu(\mu\text{-Me})_2AlMe_2]$. Their oxophilicity also makes $4f$ and $5f$ organometallics very water and air unstable, resembling early d-block metals in this respect. Cp rings can be connected to give an *ansa* system (Latin = handle), of which two examples are shown in **15.12** and **15.13**.

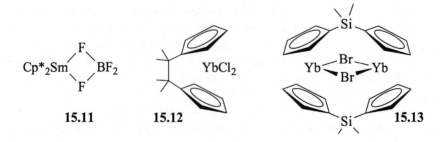

Cp*$_2$Sm $\overset{F}{\underset{F}{\diagdown\diagup}}$ BF$_2$

15.11   **15.12**   **15.13**

For many years, no Cp_3^*Ln compounds were ever seen, and it was assumed that Cp* was just too large. Only the reaction of Eq. 15.18, with its high driving force, permits the formation of the tris species. The tetraene takes one electron from each of two Sm(II) units to give two Sm(III) complexes. Detailed study of the tris complex showed that the Sm–C bond lengths (av. 2.82Å) are longer than usual (2.75 Å) as a result of steric crowding forcing the Cp* ligands to retreat from the metal. As might be expected, one of the Cp* groups easily departs, as in Eq. 15.19.[22]

$$Cp*_2Sm + \; \bigcirc \bigcirc \; \longrightarrow \; Cp*_3Sm + \; \bigcirc\bigcirc\text{—}SmCp* \quad (15.18)$$

$$Cp*_3Sm + H_2 \; \longrightarrow \; Cp*_2Sm \overset{H}{\underset{H}{\diagup\!\!\diagdown}} SmCp*_2 + Cp*H \quad (15.19)$$

Cp_2^*Sm is a reduced Sm(II) organolanthanide with a strongly bent structure quite unlike that of ferrocene. One possible reason is that this predominantly ionic system has no special geometric preference, and the bent arrangement generates a dipole that interacts favorably with neighboring dipoles in the crystal.

It reacts reversibly with N_2 to give a bridging μ^2-N_2 complex (Eq. 15.20), where the N_2 has been reduced to $[N_2]^{2-}$ and the metals have become Sm(III).[23]

$$Cp*_2\overset{II}{Sm} + N_2 \; \rightleftharpoons \; Cp*_2\overset{III}{Sm} \overset{N}{\underset{N}{\diagup\!\!\diagdown}} \overset{III}{Sm}Cp*_2 \quad (15.20)$$

Soft ligands like CO bind very weakly to $4f$ elements: For example, Cp_2^*Eu and CO are in equilibrium with $Cp_2^*Eu(CO)$.[19] For Cp_2^*Yb, the equilibrium includes both and $Cp_2^*Yb(CO)$ and $Cp_2^*Yb(CO)_2$. Crystal structures not being useful here, the IR spectral ν(CO) data were interpreted by comparison with the spectra predicted from DFT calculations.[24] These suggest that CO in $Cp_2^*Eu(CO)$ is conventionally C bound, but that for Yb, the adducts contain O-bound *isocarbonyls*: $Cp_2^*Yb(OC)$ and $Cp_2^*Yb(OC)_2$. This shows both the power of modern computational chemistry, as well as the very high oxophilicity of the $4f$ metal. The bonding between Cp_2^*Ln and CO is largely dipole–dipole in character, and the change from carbonyl to isocarbonyl from Eu to Yb is attributed to larger electron–electron repulsions with the more electron-rich carbon end of the CO in $4f^{14}$ Yb(II) versus $4f^7$ Eu(II). The weak adduct between Cp_2^*Yb and another soft ligand, MeC≡CMe has been isolated and even characterized by X-ray crystallography, but the resulting Yb–C distance, 2.85Å, is rather long compared to 2.66Å for the Yb–C distances to the Cp carbons. Isonitriles, RNC, do bind well to Ln(III), as in $Cp_3Ln(CNPh)$, but only because RNC is a substantial σ donor; back donation is minimal, as shown by the increase in $v(NC)$ of 60–70 cm^{-1} on binding to Ln(III), compared to the decrease seen in complexes like $Cp_2W(NCPh)$.

Since lanthanides cannot back-donate effectively, at least in the M(III) state, any carbene ligands have to be stabilized by their own substituents instead of the metal, and such species as **15.14** have thus been prepared.[25]

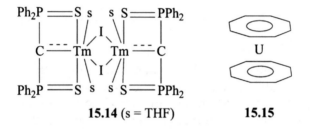

15.14 (s = THF)   **15.15**

The M(II) oxidation state has traditionally been seen only for Nd, Sm, Eu, Dy, Tm, and Yb, with Eu(II) and Yb(II) being stabilized by the resulting half-filled f shell (Table 15.2). This limitation has now been lifted by the synthesis of a series of [K(crypt)][Cp'_3M^{II}] derivatives ($Cp' = C_5H_4SiMe_3$, crypt = 2,2,2-cryptand) for all the f-block, except the radioactive Sm(II), by reduction of the M(III) analog by the very strong reducing agent, KC_8. Still more unexpected, DFT and UV-vis data suggest that the ion configurations include an electron in the 5d shell, for example, Ho(II) and Er(II) configurations are not the expected $4f^{11}$ and $4f^{12}$, but now $4f^{10}5d^1$ and $4f^{11}5d^1$.[26] Once again, we have an example of a long-held view overthrown by experiment.

Actinide Organometallic Chemistry

Most complexes of the actinides involve hard ligands,[21] and their organometallics, such as the trigonal U(III) alkyl, [$U(CH(SiMe_3)_2)_3$], are typically very air and water sensitive. For the sterically small methyl group, steric saturation is achieved by polyalkylation, as illustrated by the eight-coordinate $UMe_4(Ph_2PCH_2CH_2PPh_2)_2$.

Uranocene[27] ($U(cot)_2$, **15.15**) shows how the higher radius and charge of U^{4+} relative to the lanthanides allows stabilization of the planar, aromatic, 10π-electron cyclooctatetraene dianion (cot^{2-}). This pyrophoric 22e compound also shows the failure of the 18e rule in the $5f$ elements.

Cyclopentadienyls are again widely used as spectator ligands and their complexes show extensive catalytic applications.[28] Equation 15.21 shows how a thorium alkyl is hydrogenolyzed by H_2, a reaction step required in catalytic hydrogenation.

$$Cp_2^*Th\left(\begin{array}{c} CH_2 \\ SiMe_3 \end{array}\right)_2 + H_2 \longrightarrow [Cp_2^*ThH_2]_2 + SiMe_4 \qquad (15.21)$$

Carbonyls are somewhat more stable in the $5f$ series. $(Me_3SiC_5H_4)_3U-(CO)$ has a relatively low $\nu(CO)$ value of 1976 cm^{-1}, but it easily loses CO. The more basic $(C_5Me_4H)_3U$ gave $(C_5Me_4H)_3U(CO)$ quantitatively with the surprisingly low $\nu(CO)$ of 1880 cm^{-1}, suggesting strong U–CO π back bonding.[29]

- Paramagnetic organometallics, including f-block species, are hard to study but offer a largely untapped resource for future development.
- Steric saturation, not electron count, decides f-block structures. Ionic bonding dominates and back bonding occurs only for the 5f elements, and then often only weakly.

REFERENCES

1. Y. Gong, M. Zhou, M. Kaupp, and S. Riedel, *Angew. Chem. Int. Ed.*, **48**, 7879, 2009.
2. O. R. Luca and R. H. Crabtree, *Chem. Soc. Rev.*, **42**, 1440, 2013.
3. N. G. Connelly and W. E. Geiger, *Chem. Rev.*, **96**, 877, 1996; A. R. Parent, R. H. Crabtree, and G. W. Brudvig, *Chem. Soc. Rev.*, **42**, 2247, 2013.
4. B. N. Figgis and M. A. Hitchmar, *Ligand Field Theory and Its Applications*, Wiley, New York, 2000.
5. S. Shaik, H. Hirao, and D. Kumar, *Acc. Chem. Res.*, **40**, 532, 2007.
6. R. Poli and J. N. Harvey, *Chem. Soc. Rev.*, **32**, 1, 2003.

7. J. N. Harvey, R. Poli, and K. M. Smith, *Coord. Chem. Rev.*, **238**, 347, 2003; H. Hirao, L. Que, W. Nam, and S. Shaik, *Chem. Eur. J.*, **14**, 1740, 2008 and references cited.

8. J. L. Detrich, O. M. Reinaud, A. L. Rheingold, and K. H. Theopold, *J. Am. Chem. Soc.*, **117**, 11745, 1995.

9. P. Roquette, A. Maronna, M. Reinmuth, E. Kaifer, M. Enders, and H. -J. Himmel, *Inorg. Chem.*, **50**, 1942, 2011.

10. L. M. Xu, B. J. Li, Z. Yang, and Z. J. Shi, *Chem. Soc. Rev.*, **39**, 712, 2010.

11. K. A. Nolin, R. W. Ahn, Y. Kobayashi, J. J. Kennedy-Smith, and F. D. Toste, *Chem. Eur. J.*, **16**, 9555, 2010.

12. J. R. Krumper, M. D. Pluth, R. G. Bergman, and F. D. Toste, *J. Am. Chem. Soc.*, **129**, 14684, 2007.

13. P. M. Reis, P. J. Costa, C. Carlos, J. A. Romao, J. A. Fernandes, M. J. Calhorda, and R. Royo, *Dalton Trans.*, **2008**, 1727.

14. R. R. Schrock, *Acct. Chem. Res.*, **30**, 9, 1997.

15. T. J. Hebden, K. I. Goldberg, D. M. Heinekey, X. W. Zhang, T. J. Emge, A. S. Goldman, and K. Krogh-Jespersen, *Inorg. Chem.*, **49**, 1733, 2010.

16. G. J. Kubas, *Chem. Rev.*, **107**, 4152, 2007.

17. E. Kirillov, S. Kahlal, T. Roisnel, T. Georgelin, J. Y. Saillard, and J. F. Carpentier, *Organometallics*, **27**, 387, 2008.

18. A. M. J. Rost, W. A. Herrmann, and F. E. Kuehn, *Tet. Lett.*, **48**, 1775, 2007 and references cited.

19. T. P. Brewster, J. D. Blakemore, N. D. Schley, C. D. Incarvito, N. Hazari, G. W. Brudvig, and R. H. Crabtree, *Organometallics*, **30**, 965, 2011.

20. E. F. van der Eide, T. Liu, D. M. Camaioni, E. D. Walter, and R. M. Bullock, *Organometallics*, **31**, 1775, 2012.

21. S. A. Cotton, *Lanthanide and Actinide Chemistry*, Wiley, New York, 2006.

22. W. J. Evans and B. L. Davis, *Chem. Rev.*, **102**, 2119, 2002.

23. E. A. MacLachlan and M. D. Fryzuk, *Organometallics*, **25**, 1530, 2006.

24. L. Maron, L. Perrin, O. Eisenstein, and R. A. Andersen, *J. Am. Chem. Soc.*, **124**, 5614, 2002.

25. T. Cantat, F. Jaroschik, L. Ricard, P. Le Floch, F. Nief, and N. Mézailles, *Organometallics*, **25**, 1329, 2006.

26. M. C. Cassani, D. J. Duncalf, and M. F. Lappert, *J. Am. Chem. Soc.*, **120**, 12958, 1998. M. R. MacDonald, J. E. Bates, J. W. Ziller, F. Furche, and W. J. Evans, *J. Am. Chem. Soc.*, **135**, 9857, 2013.

27. J. C. Berthet, P. Thuery, and M. Ephritikhine, *Organometallics*, **27**, 1664, 2008.

28. M. Ephritikhine, *Organometallics*, **32**, 2464, 2013.

29. M. D. Conejo, J. S. Parry, E. Carmona, M. Schultz, J. G. Brennann, S. M. Beshouri, R. A. Andersen, R. D. Rogers, S. Coles, and M. Hursthouse, *Chem. Eur. J.*, **5**, 3000, 1999.

30. T. P. Brewster, J. D. Blakemore, N. D. Schley, C. D. Incarvito, N. Hazari, G. W. Brudvig, and R. H. Crabtree, *Organometallics*, **30**, 965, 2011.

PROBLEMS

15.1. Suggest reasons why $Ti(CH_2Ph)_4$ does not form a stable CO adduct.

15.2. Given that an unstable CO adduct of $Ti(CH_2Ph)_4$ is an intermediate on the way to forming $Ti(COCH_2Ph)_2(CH_2Ph)_2$, suggest reasons why this adduct might be especially reactive.

15.3. Why do you think V gives only VR_4 as the highest oxidation-state alkyl, but Ta can give TaR_5?

15.4. Suggest a possible mechanism for Eq. 15.10.

15.5. The ethylenes in $Mo(C_2H_4)_2(PR_3)_4$ are mutually trans. What do you think the orientation of their C–C bonds would be with respect to one another? (Draw this looking down the principal axis of the molecule.)

15.6. Why are alkene polyhydrides so rare? Why is $Re(cod)H_3(PR_3)_2$ an exception, given that its stereochemistry is pentagonal bipyramidal, with the phosphines axial?

15.7. What values of the spin quantum number S are theoretically possible for: $CpCrLX_2$, $CpMnL_2X_2$, $CpFeLX_2$, and $CpCoLX_2$?

15.8. Cp_2^*LuH reacts with C_6H_6 to give $[(Cp_2^*Lu)_2C_6H_4]$. What structure do you predict for this compound?

15.9. What spin states are in principle possible for (a) d^6 octahedral, (b) f^2 8-coordinate, and (c) d^3 octahedral complexes?

16

BIOORGANOMETALLIC CHEMISTRY

Chemistry continues to be influenced by biology as a result of advances in our understanding of the chemical basis of life. Both organic and inorganic[1] structures have long been known to be essential actors in living things. Only with coenzyme B_{12} (Section 16.2) did it become clear that organometallic species also occur in biology, both as stable species and as reaction intermediates. Nature uses organometallic chemistry sparingly, but the examples we see today may be relics of early life forms, which had to live on simple molecules, such as H_2, CO, and CH_4, and may have made more extensive use of organometallic chemistry.[2] In the reducing environment of the early Earth and of anaerobic environments today, low oxidation states and soft ligand sets would be expected to dominate, but once photosynthesis had done its work and the atmosphere became oxidizing, higher oxidation states and harder ligand sets then became dominant, but some organisms, such as anaerobic bacteria, still retain some of the old biochemical pathways. These can involve organometallic structures, and use of the term *bioorganometallic chemistry* dates from 1986.[3]

The topics covered here have an organometallic connection. Coenzyme B_{12} has M–C or M–H bonds, and the active site cluster in nitrogen fixation has a carbon atom at its heart. The nickel enzymes go

The Organometallic Chemistry of the Transition Metals, Sixth Edition.
Robert H. Crabtree.

via M–H (H$_2$ase) or M–C (CODH) intermediates. Organometallic pharmaceuticals are beginning to see the light. First, however, we review the basic aspects of biochemistry as they apply to proteins, where transition metals have their greatest impact on biology.

16.1 INTRODUCTION

The main catalysts of biology, *enzymes*, can be soluble, or bound to a membrane, or even part of an enzyme complex, in which case they act as a cog in a larger piece of biochemical machinery. Biochemical reactions have to be kept under strict control—they must only happen as they are required, where they are required. One way of doing this is to employ reactions that can only proceed under enzymatic catalysis. The organism now only has to turn these enzymes on and off to control its biochemistry.

Proteins

Most enzymes are *proteins*; that is, they are made up of one or more polypeptide chains having the structure shown in **16.1**. The value of n usually ranges from 20 to 100, and there may be several separate polypeptide chains or *subunits* in each enzyme. Sometimes two or more proteins must associate to give the active enzyme. The monomers from which protein polymers are built up are the amino acids, RC*H(NH$_2$)-COOH, always having an L configuration at C*. More than 20 different amino acids are commonly found in proteins, each having a different R group (Table 16.1). The sequence of the R groups in the protein chain is its *primary structure*. Each enzyme has its own specific sequence, which often differs in minor ways from one species to another. Such chains with similar sequences are said to be *homologous*. In spite of minor sequence differences, the chains can fold in the same way in all cases to give an active enzyme. The sequence is the main factor that decides the way in which the chain will fold, and the R groups also provide the chemical functional groups that enable the protein to perform its function. The problem of predicting the folding pattern of a polypeptide (usually found by X-ray diffraction or NMR) from its primary sequence is still unsolved. Two types of *secondary structure* are common, the rodlike α helix and the flat β sheet. In each case, the folding is decided by the patterns of many hydrogen bonds formed between N–H groups of one peptide bond and CO groups of another. *Tertiary structure* refers to the pattern of secondary structural elements—how helices, sheets, and loops are combined in any subunit. Finally,

TABLE 16.1 Common Amino Acids

Name	Symbol	R	Remarks
Glycine	Gly	H	Nonpolar R group
Alanine	Ala	Me	Nonpolar R group
Valine	Val	i-Pr	Nonpolar R group
Leucine	Leu	i-PrCH$_2$	Nonpolar R group
Phenylalanine	Phe	PhCH$_2$	Nonpolar R group
Glutamic acid	Glu	$^-$O$_2$CCH$_2$CH$_2$	Anionic R group, binds M ions
Aspartic acid	Asp	$^-$O$_2$CCH$_2$	Anionic R group, binds M ions
Lysine	Lys	$^+$H$_3$N(CH$_2$)$_4$	Cationic R group[a]
Arginine	Arg	$^+$H$_2$N=C(NH$_2$) NH(CH$_2$)$_3$	Cationic R group[a]
Tyrosine	Tyr	HO(C$_6$H$_4$)CH$_2$	Polar but unionized, binds M ions
Serine	Ser	HOCH$_2$	Polar but unionized[a]
Threonine	Thr	MeCH(OH)	Polar but unionized
Asparagine	Asn	H$_2$NOCCH$_2$	Polar but unionized
Methionine	Met	MeSCH$_2$CH$_2$	Soft nucleophile, binds M ions
Cysteine	Cys	HSCH$_2$	Binds M ions[b]
Histidine	His	C$_3$N$_2$H$_4$CH$_2$	Binds M ions[c]

Note: Predominant protonation states at pH 7 are given.
[a]These residues occasionally bind metal ions.
[b]Also links polypeptide chains via an –CH$_2$S–SCH$_2$– group.
[c]Via imidazole head group.

quaternary structure refers to the way the subunits pack together. Greek letters are used to designate subunit structure; for example, an $(\alpha\beta)_6$ structure is one in which two different chains α and β form a heterodimer, which, in turn, associates into a hexamer in the native form of the protein.

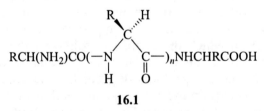

16.1

Certain nonpolar R groups tend to prefer the interior of the structure. Others are hydrophilic and prefer the surface. Some are sufficiently acidic or basic so as to be deprotonated or protonated at physiological pH (generally close to pH 7); these provide a positive or negative charge at the surface of the protein. Among other functions, histidine may act as a nucleophile to attack the substrate of the enzyme or to ligate any metal ions present. Similarly, cysteine may hold chains together by formation of a disulfide link (RS–SR) with a cysteine in another chain or can bind a metal ion as a thiolate complex (RS–ML$_n$). Any nonpolypeptide component of the protein required for activity (e.g., a metal ion, or an organic molecule) is called a *cofactor*. Sometimes, two or more closely related protein conformations are possible. Which is adopted may depend on whether the substrate for the protein or the required cofactors are bound. Such a "conformational change" may turn the enzyme on or off or otherwise modify its properties. Proteins can lose the conformation required for activity on heating, or on addition of urea (which breaks up the H-bond network) or salts, or if we move out of the pH range in which the native conformation is stable. This leads to denatured, inactive protein, which in certain cases can refold correctly when the favorable conditions of temperature, ionic strength, and pH are reestablished.

Metalloenzymes

More than half of all enzymes have metal ions in their structure; these are *metalloenzymes*. In most cases, the metals are essential to the action of the enzyme and are often at the active site where the substrate for the biochemical reaction is bound. All organisms require certain "trace elements" for growth. Some of these trace elements are the metal ions that the organism incorporates into its metalloenzymes. Of the inorganic elements, the following have been found to be essential for some species of plant or animal: Mg, V, Cr, Mn, Fe, Co, Ni, Cu, Zn, Mo, B, Si, Se, F, Br(?), and I. New elements are added to the list from time to time — titanium[4] is a potential future candidate for inclusion, for example. In addition, Na, K, Ca, phosphate, sulfate, and chloride are required in bulk rather than trace amounts. Metal ions also play an important role in nucleic acid chemistry. The biochemistry of these elements is termed *bioinorganic* chemistry.[1]

Modeling

In addition to purely biochemical work, bioinorganic chemists also try to elicit the chemical principles that are at work in biological systems. Two such areas are structural and functional modeling. In structural modeling, the goal is to prepare a small molecule, such as a metal complex, that can be structurally and spectroscopically characterized for comparison with the results of physical measurements on the biological system. This can help determine the structure, oxidation state, or spin state of a metal cofactor. A small molecule complex can often reproduce many important physical properties of the target. Becoming more common is functional modeling, where the goal is to reproduce some chemical property of the target in a small molecule complex and so try to understand what features of the structure promote the chemistry. Typical properties include the redox potential of a metal center or its catalytic activity. Functional models *with the correct metal and ligand set* that reproduce the catalytic activity of the target system are still rare. Many so-called models use the "wrong" metal or ligands, and so provide less relevant information.

Molecular Recognition

A key principle of biochemistry is the recognition of one biomolecule or substructure by another. A substrate binds with its specific enzyme, or a hormone with its receptor protein, or a drug with its receptor, as a result of complementarity between the two fragments with regard to shape, surface charges, and hydrogen bonding. This accounts for the astonishing specificity of biology; for example, only one enantiomer of a compound may be accepted by an enzyme, and only the human, but not the monkey, version of a given protein may be recognized by a suitable antibody (specific binding protein).

   If a protein selectively recognizes and stabilizes the transition state for a reaction by hydrogen bonding or ionic interactions, then the reaction will be accelerated by catalysis because it now becomes easier to reach the transition state. The transition state must be stabilized more than the substrate or product so that the low and high points in the energy profile of Fig. 16.1 become closer in energy—the flatter the energy profile, the faster the reaction. An enzyme that hydrolyzes an ester RCOOMe as substrate should recognize the transition state **16.2** for the attack of water on the ester. Such an enzyme may bind a transition state analog, such as the phosphate **16.3** much more tightly than it binds the starting ester RCOOMe and inhibit the enzyme (poison the catalyst). Drugs are often selective inhibitors of specific target enzymes through which they exert their physiological effects.

$$
\begin{array}{cc}
\underset{\underset{\text{OMe}}{|}}{\overset{\overset{\text{OH}}{|}}{R-C-O^-}} & \underset{\underset{\text{OMe}}{|}}{\overset{\overset{O}{\parallel}}{R-P-O^-}} \\
\textbf{16.2} & \textbf{16.3}
\end{array}
$$

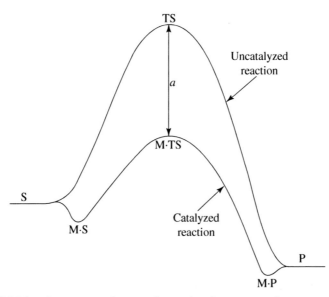

FIGURE 16.1   An enzyme lowers the activation energy for a reaction, often by binding the transition state (TS) for the reaction more tightly than the substrate (S) or product (P). The binding energy for the TS is represented as *a* in this plot of energy versus reaction coordinate.

Coenzymes

Just as a set of reactions may require a given cocatalyst, sometimes a set of enzymes require a given *coenzyme*. For example, coenzyme B_{12}, a Co-containing cofactor, is required for activity in a number of "B_{12} dependent" enzymes.

Protein Structure

The structures of proteins are generally obtained by crystallography. The structural data cannot reveal the oxidation state of any metal present, and for this, we normally need to compare the UV–visible or EPR spectra of the protein with those of model compounds.[5] If the natural enzyme has a metal such as Zn^{2+} that gives uninformative electronic spectra or is EPR silent, it is sometimes possible to replace it with an unnatural but more spectroscopically informative metal, such as Co^{2+}.

Many interesting metalloproteins are not yet crystallographically characterized, but it is always possible to use X-ray spectroscopy even in the absence of suitable crystals. For example, the fine structure on the X-ray absorption edge (EXAFS)[5] for the metal may reveal the number of ligand atoms, their distance, and whether they are first (N,O) or second row (S). The X-ray photon expels a photoelectron from the metal; if it has a certain minimum photon energy required to ionize electrons from a given shell (say, the $2s$), an absorption edge appears at this energy in the X-ray absorption spectrum. As we go to slightly higher X-ray photon energies, the photoelectron leaves the metal atom with a certain small residual energy because of the slight excess energy of the X-ray photon relative to the absorption edge of the metal. The wavelength of the photoelectron depends on the amount of excess energy of the X-ray photon. The backscattering of the electron from the ligands around the metal is also wavelength dependent and affects the probability for absorption of the X ray. Crudely speaking, the ligand atom may backscatter the photoelectron wave in such a way as to give a constructive or destructive interference and so raise or lower the probability of the electron leaving the vicinity of the metal; the probability of absorption of the X-ray photon will be raised or lowered in consequence. Interpretation of EXAFS data is not entirely straightforward and is considerably helped by making measurements on model complexes. Normally, the M–L distance(s) can be extracted to an accuracy of ± 0.002 Å, but the number of ligands of a given type is much less well determined (error: ± 1). The edge position in the X-ray spectrum (X-ray absorption near edge structure, XANES) has become a general method for determining the metal's oxidation state.

In resonance Raman spectroscopy,[5] if the incident radiation is near a UV-vis absorption feature of the metal ion, the Raman scattering involving bonds in the immediate vicinity of the metal is greatly enhanced. This selectivity for the active site region is very useful in bioinorganic studies because the key absorptions are not buried under the multitude of absorptions from the rest of the protein. For iron proteins, Mössbauer measurements[6] can help determine oxidation state and help distinguish 4- from 5- and 6-coordinate metals and hard from soft ligand environments. Computational data can assist the interpretation of both Mössbauer and X-ray spectroscopic data.

16.2 COENZYME B$_{12}$

The story begins with the observation, made early in the twentieth century, that raw beef liver cures an otherwise uniformly fatal disease, pernicious anemia.[4] The active component was finally crystallized in 1948, and in 1965, Dorothy Hodgkin[6] determined the structure **16.4** crystallographically. This showed that the cobalt(III) form of the molecule is an octahedral complex with a *corrin*, a 15-membered 4-nitrogen ring L$_3$X ligand, occupying the equatorial plane. Connected to the corrin is a side chain — the nucleotide loop — terminating in a benzimidazole, which binds as an axial ligand in free B$_{12}$. The benzimidazole can dissociate when B$_{12}$ binds to its site in the appropriate enzymes for which it is a cofactor, in which case it may be replaced by a His imidazole group from the enzyme. The sixth, active site of cyanocobalamin is occupied by cyanide that comes from the isolation procedure. In the cell, a number of other ligands are present, including water (in aquacobalamin or B$_{12a}$), or methyl (in methylcobalamin), or adenosyl groups (**16.5**). Other than B$_{12a}$, all these species have a Co–C bond, the first M–C bonds recognized in biology.

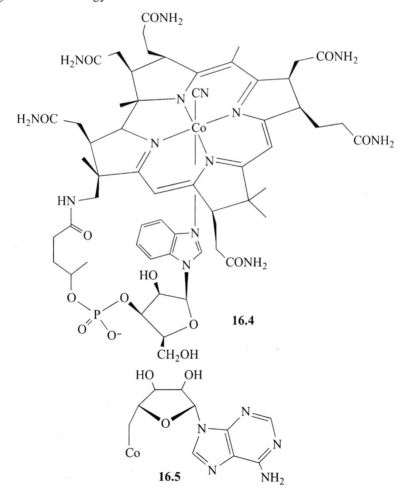

The coenzyme acts in concert with a variety of enzymes to catalyze reactions of three main types. In the first, two substituents on adjacent carbon atoms, –X and –H, are permuted in the *isomerase* or *mutase reaction*. The generalized process is shown in Eq. 16.1, and specific examples are given in Eq. 16.2 and Eq. 16.4. CoA has nothing to do with cobalt, but is the biochemical symbol for coenzyme A, a thiol that activates carboxylic acids by forming a reactive thioester.

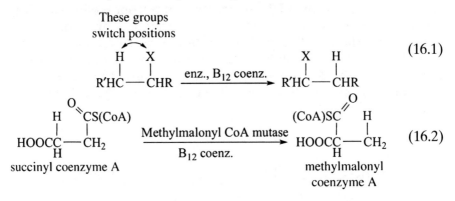

$$(16.1)$$

$$(16.2)$$

In the second general type, methylcobalamin methylates a substrate, as in the conversion of homocysteine to methionine.

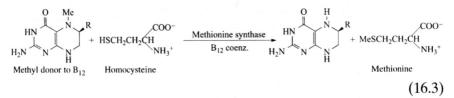

$$(16.3)$$

Finally, B_{12} is also involved as a component of some ribonucleotide reductases that convert the ribose ring of the ribonucleotides that go to make RNA to the deoxyribose ring of the deoxyribonucleotides that go to make DNA. The schematic reaction is shown in Eq. 16.4.

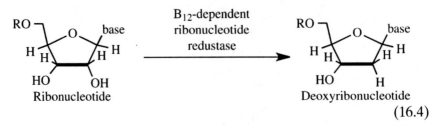

$$(16.4)$$

The coenzyme is required only in small amounts; 2–5 mg is present in the average human, for example, and one of the first signs of deficiency is anemia, the failure to form sufficient red blood cells.

This anemia is not treated successfully by the methods that work for the usual iron-deficiency form of anemia, hence the term "pernicious" anemia.

B$_{12a}$ is easily reducible, first to B$_{12r}$ and then to B$_{12s}$ (r = reduced; s = superreduced). Physical studies showed that B$_{12r}$ contains five-coordinate Co(II), and by comparison with model compounds, B$_{12s}$ was shown to contain four-coordinate Co(I). The B$_{12s}$ state turns out to be one of the most powerful nucleophiles known, reacting rapidly with MeI, or the natural Me$^+$ donor, N^5-methyl tetrahydrofolate, to give methylcobalamin that can in turn transfer the Me group to various substrates as in Eq. 16.3.

Model Studies

Is this chemistry unique to the natural system, or is it a general property of cobalt in a 5-nitrogen ligand environment? At the time that the original model studies were carried out (1960s), it was believed that transition metal alkyls were stable only with very strong field ligands, such as CO or PPh$_3$. This problem was better understood by studying model systems. Early studies revealed that the simple ligand dimethylglyoxime (dmgH) **16.6** gives a series of Co(III) complexes (called *cobaloximes*) **16.7** that have much in common with the natural system. Two [dmg]$^-$ ligands model the corrin, a pyridine models the axial base, and the sixth position can be an alkyl group or water. It was found that these alkyls are stable when the equatorial ligand had some, but not too much, electron delocalization. Neither fully saturated ligands nor the more extensively delocalized porphyrin system, common in other metalloenzymes, allow cobalt to form alkyls easily, but dmg and corrin are both suitable. The second unexpected point was that the longer-chain alkyls, such as -Et or -adenosyl, do not β-eliminate easily. We can now see that this is because the equatorial ligand prevents a vacant site from being formed cis to the alkyl in this 18e system. Such a site would be needed for β elimination to take place by a concerted mechanism (Section 7.5).

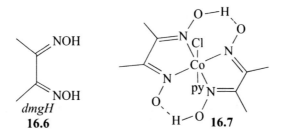

The nature of the B_{12r} and B_{12s} states was made clearer from the behavior of the corresponding Co(II) and Co(I) reduced states of the model cobaloxime. Like B_{12s}, the Co(I) form, $[Co(dmg)_2py]^-$, proved to be a supernucleophile, reacting very fast with MeI to give $[MeCo(dmg)_2py]$ (Eq. 16.5), and the Co(II) form bound water in the sixth site.

$$[Co(dmg)_2 py]^- + MeI \rightarrow [MeCo(dmg)_2 py] + I^- \qquad (16.5)$$

Homolytic Mechanisms

The mechanism of the isomerase reactions starts with reversible Co(III)–C bond homolysis to generate the 17e Co(II) "radical," B_{12r}, and the adenosyl (Ad) radical, $AdCH_2$. This carbon radical abstracts a hydrogen atom from the substrate, QH, to give $AdCH_3$, and the substrate radical, which undergoes a 1,2 shift of the X group (see Eq. 16.1), followed by H atom transfer from $AdCH_3$ to give the final product (Fig. 16.2).

This mechanism implies that the Co–C bond in the coenzyme is not particularly strong because it requires the Co–C bond to be spontaneously homolyzing at ambient temperatures at a rate fast enough to account for the rapid turnover seen for the B_{12}-dependent enzymes ($\sim 10^2$ s^{-1}). Halpern[7] estimated Co–C bond strengths—defined by Eq. 16.8—in B_{12} models by two methods. The first involves measuring the equilibrium constant for Eq. 16.6. From the ΔH and ΔS values, and given the known heats of formation of $PhCH=CH_2$ and $PhCH \cdot -CH_3$, the ΔH and ΔS for Eq. 16.8 can be deduced.

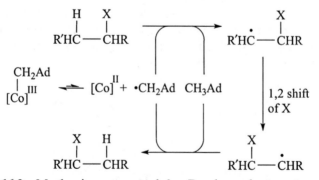

$$\Delta H = 22.1 \text{ kcal/mol \{measured\}} \qquad (16.6)$$

FIGURE 16.2 Mechanism proposed for B_{12} dependent mutase reactions. Ad = adenosyl.

$$H_2C\diagdown Ph + 0.5H_2 \rightleftharpoons H_2C\diagdown Ph$$

$$\Delta H = -2.2 \text{ kcal/mol \{calc'd\}}$$

(16.7)

$$\overset{Ph}{\underset{H}{H_3C}}\diagdown \overset{III}{Co}(dmg)_2py \rightleftharpoons H_2C\diagdown Ph + \overset{II}{Co}(dmg)_2py$$

$$\Delta H = 19.9 \text{ kcal/mol \{calc'd\}}$$

(16.8)

Although Eq. 16.6 looks like a β elimination of the sort that we said should be prevented by the lack of a 2e cis vacancy at the metal, the reaction in fact goes by a pathway that does not require a vacancy: Co–C bond homolysis, followed by H atom abstraction from the resulting carbon radical by the Co(II) (Eq. 16.9).

$$\overset{Ph}{\underset{H}{H_3C}}\diagdown \overset{III}{Co}(dmg)_2py \rightleftharpoons \overset{H_3C\ Ph}{\underset{H}{\diagup\!\!\!\backslash\cdot}} + \overset{II}{Co}(dmg)_2py$$

$$H_2C\diagdown Ph$$

(16.9)

$$\overset{II}{Co}(dmg)_2py + 0.5H_2 \longleftarrow H-\overset{III}{Co}(dmg)_2py$$

Halpern's second method was to trap the R· intermediate from Co–R homolysis with Co(II)aq as $[Co(OH_2)_5R]^{2+}$, the $\Delta H^{\ddagger}$ for this homolysis being a measure of the Co–C bond strength. The answer by this kinetic method turns out to be 22 kcal/mol, very close to the previously determined Co–C bond strength of ~20 kcal/mol. The extra ~2 kcal probably represents the activation energy for the homolysis. Applying the same method to coenzyme B$_{12}$ itself gives a figure of 28.6 kcal/mol for the Co–CH$_2$R bond strength. This figure is too high to account for the rate of turnover of the B$_{12}$-dependent enzymes because the rate of the homolysis of such a strong bond would be much slower than 10^2 s^{-1}. The strong Co–C bond is needed so that the coenzyme does not liberate a radical until required to do so. When the coenzyme binds to the B$_{12}$-dependent enzyme, part of the binding energy to the enzyme is probably used to deform the coordination sphere of B$_{12}$ so that the Co–C bond is weakened, and when the substrate also binds, the bond may be further weakened so that it can now homolyze at the appropriate rate.

Halpern[8] also looked at the rearrangement step itself by making the proposed substrate-derived radical independently in the absence of metal by the action of Bu$_3$SnH on the corresponding halide. For the methylmalonyl mutase reaction, the rate of rearrangement, 2.5 s^{-1}, is only modestly slower than the 10^2 s^{-1} turnover rate for the enzyme. This small difference may arise from the radical being bound to the enzyme, where it is held in a conformation that favors the rearrangement. If so, the only role of the Co–C bond is to reversibly homolyze—the rest is standard organic chemistry.

Bioalkylation and Biodealkylation

Methylcobalamin is important in biological methylation, itself of great importance in gene regulation and even in cancer.[9] In some cases, it has been found that Hg(II) in the sea can be methylated by bacteria to give MeHg$^+$. Being water-soluble, this species can be absorbed by shellfish, which can then become toxic to humans.[8] Mercury is naturally present in small quantities in seawater, but the concentration can rise by pollution. A notorious episode involving numerous fatalities occurred at Minimata in Japan, where abnormally high amounts of mercury were released into the bay as a result of industrial activity.

Certain bacteria have a pair of enzymes, organomercury lyase and mercuric ion reductase, that detoxify organomercury species via the processes shown in Eq. 16.10–Eq. 16.13. The lyase cleaves the R–Hg bond (Eq. 16.10), and the reductase reduces the resulting Hg(II) ion to the relatively less toxic Hg(0) (Eq. 16.11) that is then lost by evaporation. The retention of configuration observed in the lyase reduction of Z-2-butenylmercury chloride and the failure of radical probes to give a radical rearrangement led to the proposal that the reaction goes by an S$_E$2 mechanism in which a cysteine SH group of the reduced protein cleaves the bond (Eq. 16.12; enz = lyase). The reduction of the Hg^{2+} to Hg(0) is believed to go via initial handover of the Hg^{2+} to the reductase with formation of a new dithiolate that loses disulfide (Eq. 16.13; enz$'$ = reductase).

$$\text{R—Hg—Cl} \xrightarrow[\text{lyase}]{\text{organomercury}} \text{RH} + \text{Hg}^{2+} + \text{Cl}^- \qquad (16.10)$$

$$\text{Hg}^{2+} \xrightarrow[\text{reductase}]{\text{Hg}^{2+}} \text{Hg(0)} \qquad (16.11)$$

$$\begin{array}{c}\text{R—Hg—S–enz}\\ \diagup \frown \\ \text{H—S–enz}\end{array} \xrightarrow{-\text{RH}} \begin{array}{c}\text{S—Hg—S–enz}\\ |\\ \text{enz}\end{array} \qquad (16.12)$$

$$\begin{array}{c}\text{S—Hg—S–enz}'\\ |\\ \text{enz}'\end{array} \longrightarrow \text{Hg(0)} + \begin{array}{c}\text{S—S–enz}'\\ |\\ \text{enz}'\end{array} \qquad (16.13)$$

In the absence of Hg(II), the transcription and synthesis of these Hg detoxification enzymes is inhibited by a regulatory protein, merR, that binds to a specific location in the *mer* operon, the section of DNA coding for Hg resistance. When Hg(II) is present, it binds to three Cys residues of the merR protein. This causes a conformational change in both the protein and in the DNA to which it is bound that leads to transcription of the lyase and reductase. In this way, the lyase and reductase are only produced when required.

Arsenic is another toxic element that can cause problems. It is present in groundwater in various locations, such as Bangladesh, where it can accumulate in rice. Rice is particularly affected because it grows in stagnant water, unlike grains, which grow in open fields that receive pure rainwater. In the early nineteenth century, certain green wallpapers contained copper arsenite (Scheele's green) as a dyestuff. In damp conditions, molds, such as *Scopulariopsis brevicaulis*, are able to convert the arsenic to the very toxic $AsMe_3$ by a B_{12}-dependent methylation pathway; many were sickened before the problem was recognized. It has even been argued that in 1821, Napoleon was accidentally poisoned in this way, when he was held at St. Helena by the British; others have blamed the British or a member of his French entourage for deliberately poisoning him,[10a] but the mainstream view is that he died of stomach cancer. [10b]

- Coenzyme B_{12}, the best-established organometallic cofactor in biology, provides a source of carbon-based radicals as well as a methylation reagent.

16.3   NITROGEN FIXATION

Farming communities since antiquity have known that the presence of certain plants encourages the growth of crops.[11] The beneficent action of a fertility goddess associated with the plant was a colorful explanation developed in early times to account for this phenomenon. The truth is only slightly less remarkable: the roots of these plants are infected by soil bacteria, that "fix" atmospheric N_2 to NH_3, by means of a metalloenzyme, nitrogenase (N_2ase), once provided by the plant with the necessary energy input. The resulting ammonia not only fertilizes the host plant, but also escapes into the surroundings, where crop growth is stimulated. Before the advent of fertilizers, almost all the nitrogen required for nutrition was obtained by biological nitrogen fixation—now, much of it comes from the Haber process by Eq. 16.14.

$$N_2 + 3H_2 \xrightarrow[\text{Fe catalyst}]{\text{heterogeneous}} 2NH_3 \qquad (16.14)$$

As early as 1930, it was realized that molybdenum was implicated in the common MoFe type of N_2ase: iron and magnesium are also required. Although alernative nitrogenases also exist that contain no Mo, but instead either V and Fe or Fe alone, the MoFe N_2ase is by far the best understood and is referred to later in the text unless otherwise stated. The only N-containing product normally released by the enzyme is ammonia and never any potential intermediates, such as hydrazine; H_2 is also released from proton reduction. The enzyme, like many organometallic complexes, is air sensitive, and CO and NO are strong inhibitors. These presumably coordinate to the N_2 binding site, a low-valent Fe–Mo cluster, FeMo-co (Fe–Mo cofactor). Other substrates are efficiently reduced: C_2H_2, but only to C_2H_4; MeNC to MeH and $MeNH_2$; and azide ion to N_2 and NH_3. Acetylene reduction is the standard assay for the enzyme, which meant that VFe N_2ase at first escaped detection because it reduces C_2H_2 all the way to C_2H_6.

The Mo enzyme has two components: (1) the Fe protein (molecular weight 57 kDa or 57,000 Da), which contains 4 Fe and 4 S; and (2) the MoFe protein (220 kDa, $\alpha_2\beta_2$ subunits), which contains both metals (2 Mo and 30 Fe). Each also contains S^{2-} ions ($\sim$one per iron), which act as bridging ligands for the metals. The MoFe protein's "P clusters" are Fe_8S_7 clusters that consist of a double Fe_4S_4 cubane sharing one sulfide. The N_2 binding site, the FeMo-co cluster, can be extracted as a soluble protein-free molecule containing 1 Mo, 7 Fe, 9 S^{2-}, and one homocitrate bound to Mo. Protein-free, extracted FeMo-co was known to restore N_2 reducing activity to the apoenzyme—inactive N_2ase that lacks FeMo-co—but no crystal structure of FeMo-co proved possible, and no synthetic model complex was found that could reconstitute the apoenzyme and restore activity.

The crystal structure of the entire enzyme has been central in clearing up some of the mysteries surrounding the system. FeMo-co proves to be a double cubane linked by three sulfide ions (Fig. 16.3). The Mo

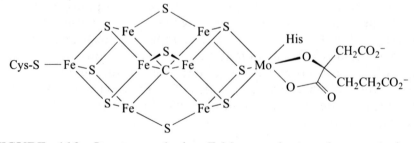

FIGURE 16.3 Structure of the FeMo-co of *Azotobacter vinelandii* nitrogenase.

being six-coordinate made it less likely to be the N_2 binding site even though model studies had for many years concentrated on this element. The probable noninvolvement of the Mo in binding N_2 illustrates one hazard of bioinorganic model chemistry: The data on the biological system may undergo a reinterpretation that alters the significance or relevance of earlier model studies. Systematic mutation of the residues surrounding FeMo-co currently points to the most likely N_2 binding site being the waist region of the cluster, a region that provides four Fe atoms in a rectangular array.[12]

An early state of the refinement, in which the central point of the cluster was taken to be vacant, suggested that six Fe atoms of the cofactor had the unrealistically low coordination number of 3, but subsequent work has put a carbon atom at the center of the cluster, making it unambiguously organometallic.[13]

The FeMo-co cluster does not form by self-assembly but requires biosynthesis on an external template prior to incorporation into the MoFe protein. The P-cluster is synthesized by fusion of two $[Fe_4S_4]$ clusters within the MoFe protein. The organism has thus gone to considerable trouble to make these clusters, otherwise unknown in biology.[14]

The isolated enzyme reduces N_2 and the other substrates if $Na_2S_2O_4$ is provided as an abiological source of the electrons required by Eq. 16.15. Even though the overall process of Eq. 16.15 is exergonic under physiological conditions, adenosine $5'$–triphosphate (ATP) is also needed by the Fe protein to provide energy to overcome the kinetic barrier to N_2 reduction. The Fe protein accepts electrons from the external reducing agent and passes them on to the MoFe protein initially via the P-cluster and finally to FeMo-co. In the absence of N_2, N_2ase acts as a hydrogenase in reducing protons to H_2; indeed, some H_2 is always formed even in the presence of N_2.

$$N_2 + 8H^+ + 8e^- \xrightarrow{N_2ase} 2NH_3 + H_2 \qquad (16.15)$$

Once FeMo-co is liberated from the enzyme, the cluster loses the ability to reduce N_2, so close cooperation must be required in the holoenzyme (= apoenzyme + cofactors) between the cofactor and the polypeptide chain. Similarly, CO is normally an inhibitor of N_2ase, but if valine-70 of the α chain is mutated to alanine or glycine, thus replacing an iPr group by a less bulky Me or H group, CO is now reducible to a mixture of CH_4, C_2H_6, C_2H_4, C_3H_6, and C_3H_8, a process reminiscent of the Fischer–Tropsch reaction (Section 12.3).[15] If the same Val is instead mutated to Ile, thus replacing an iPr group by the bulkier iBu group, only H^+ can now enter the site to give H_2—all other substrates are excluded. Since αVal-70 is adjacent to the waist region of FeMo-co, this is strong evidence for the substrate binding site being located in that region.

Dinitrogen and N_2 Complexes

Dinitrogen is so inert that it reacts with only a very few reagents under the mild conditions employed by nitrogenase. Elemental Li and Mg reduce it stoichiometrically to give nitrides. N_2 also reacts with a number of reduced metal complexes to give N_2 complexes, more than 500 of which are now known, many containing Fe or Mo. In most cases, the N_2 is bound end-on, as in **16.8**. N_2 is isoelectronic with CO, so a comparison between the two ligands is useful. CO has a filled σ-lone pair orbital located on carbon, with which it forms a σ bond to the metal, and an empty π* orbital for receiving back bonding. N_2 also has a filled σ lone pair, but it lies at lower energy than the corresponding orbital in CO, because N is more electronegative than C, and so N_2 is a weaker σ donor. Although the empty π* orbital of N_2 is lower in energy than the CO π* and thus more energetically accessible, it is equally distributed over N^1 and N^2, and therefore the M–N π* overlap is smaller than for M–CO, where the π* is predominantly C-based. The result is that N_2 binds metals very much less efficiently than CO. Of the two M–N_2 interactions, π back donation is the most important for stability, and only strongly π-basic metals bind N_2. Because the two ends of N_2 are the same, the molecule can relatively easily act as a bridging ligand between two metals (**16.9**). If back donation dominates, the terminal N of M–N_2 can be protonated, reducing the N_2 to give a M=N–NH$_2$ complex. The two forms **16.10** and **16.11**, shown below, are resonance contributors to the real structure.

$$M-\overset{1}{N}\equiv\overset{2}{N}$$ $$M-N-N-M$$

terminal   bridging

16.8   **16.9**

$$M-N\equiv N-M$$ $$M=N-N=M$$

16.10   **16.11**

The first recognized dinitrogen complex, $[Ru(NH_3)_5(N_2)]^{2+}$, was isolated as early as 1965 during the attempted synthesis of $[Ru(NH_3)_6]^{2+}$ from $RuCl_3$ and hydrazine. This illustrates how important it can be to avoid throwing out a reaction that has not worked as expected. Terminal M–N_2 complexes have N–N distances only slightly different (1.05–1.16 Å) from that of free N_2 (1.1 Å) as well as a strong IR absorption due to the N–N stretch at 1920–2150 cm^{-1}. Free N_2 is inactive in the IR, but binding to the metal polarizes the molecule (see Section 2.6), with N^1 becoming ∂^+ and N^2 ∂^-. This not only makes the N–N stretch IR active, but also chemically activates the N_2 molecule.

Common preparative routes are reduction of a phosphine-substituted metal halide in the presence of N_2 (Eq. 16.16)[16] and displacement of a labile ligand by N_2.[17]

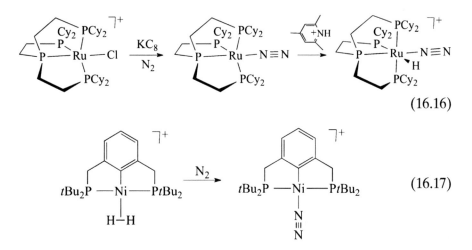

$$(16.16)$$

$$(16.17)$$

As seen in Eq. 16.17, N_2 can often displace η^2-H_2; if this were the substrate-binding step in the catalytic cycle, it would explain why N_2ase always produces at least one mole of H_2 per mole of N_2 reduced.

Reactions of N_2 Complexes

Only the most basic N_2 complexes, notably the bis-dinitrogen Mo and W complexes, can be protonated, as shown in the classic work of Chatt.[18] According to the exact conditions, various N_2H_x complexes are obtained, and even, in some cases, free NH_3 and N_2H_4 (Eq. 16.18 and Eq. 16.19).[19] As strongly reduced Mo(0) and W(0) complexes, the metal can apparently supply the six electrons required by the N_2, when the metals are oxidized during the process. In breaking strong bonds, such as in N_2, we need to compensate for the loss by creating strong bonds at the same time. In Eq. 16.18, the loss of the N≡N triple bond is compensated by the formation of two N–H bonds and a metal nitrogen multiple bond.

$$N\equiv N \xrightarrow{H^+ + e^-} HN\equiv N$$

$$HN=NH \xrightarrow{H^+ + e^-} HN=NH_2 \xrightarrow{H^+ + e^-} H_2N-NH_2 \xrightarrow{-NH_3}$$

A mechanism

$$H^+ + e^- \xrightarrow{-NH_3} NH_2 \xrightarrow{H^+ + e^-} NH_3$$

D mechanism

$$H^+ + e^- \searrow N-NH_2 \xrightarrow{H^+ + e^-} N-NH_3 \xrightarrow[-NH_3]{H^+ + e^-} NH$$

$$H^+ + e^- \nearrow NH_2 \xrightarrow{H^+ + e^-} NH_3$$

FIGURE 16.4 Two proposals for N_2 reduction. The distal *D* mechanism appears to apply to terminal $M-N_2$ complexes, while the alternating *A* mechanism may apply N_2ase itself, where the binding involves a cluster. All of these fragments are coordinated either to a single metal complex or a cluster as in N_2ase itself.

$$(dpe)_2W(N_2)_2 \xrightarrow[-N_2]{2HCl} Cl_2(dpe)_2W=N-NH_2 \xrightarrow{base} Cl(dpe)_2W=N\overset{--}{-}NH \quad (16.18)$$

$$N_2H_4 \xleftarrow[M=W]{H_2SO_4} \quad \xrightarrow[M=Mo]{H_2SO_4} NH_3 \quad (16.19)$$

Two competing types of mechanism have been proposed for N_2 reduction. Each involves additions of protons and electrons to coordinated N_2 with formation of N–H bonds and reduction of the $N\equiv N$ bond order from three to zero. The *D* mechanism calls for reduction of the distal N first, followed by reduction of the proximal N in an $M-N_2$ complex; the *A* mechanism calls for alternation of reduction steps between the two nitrogens (Fig. 16.4). The *D* type Chatt cycle is based on studies of the chemistry of terminal $M-N_2$ complexes. Work on trapped intermediates in the enzyme supports an *A* mechanism, however,[20] so work on terminal $M-N_2$ model compounds may have been misleading in this case where a cluster binding site is involved.

In Schrock's[21] Cp*Me$_3$M=N–N=MMe$_3$Cp* (M = Mo or W), the back donation is so strong that the N_2 is now effectively reduced to a hydrazide tetraanion, as shown by the N–N distance of 1.235 Å (Mo). Ammonia is formed with lutidine hydrochloride as proton source and Zn/Hg as reductant. Dinitrogen can also be reduced to ammonia at room temperature and 1 atm with the molybdenum catalyst $LMo(N_2)$, where L is the bulky trianionic tripodal triamide [{3,5-(2,4,6-i-Pr$_3$C$_6$H$_2$)$_2$ C$_6$H$_3$NCH$_2$CH$_2$}$_3$N]. Addition of a lutidine salt as proton source, and decamethyl chromocene as reductant, gave four catalytic turnovers. The N_2 is reduced at a sterically protected, single molybdenum center that cycles from Mo(III) through Mo(VI).

Since the binding site for N_2 in the enzyme seems to be a rectangular array of four Fe atoms in the waist region of FeMo-co, perhaps the most relevant model system is Holland's four-iron complex. On reduction with the powerful reductant, KC_8, N_2 can be split into two coordinated nitrides as shown in Eq. 16.20.[22] In the Haber process, N_2 is believed to be split into two coordinated nitrides bound to the surface of the Fe catalyst, so there may be a mechanistic similarity with N_2ase.

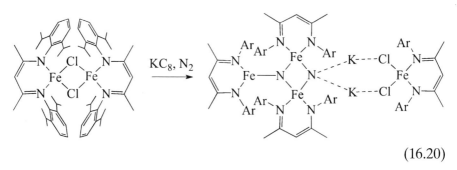

$$(16.20)$$

Fe–S Clusters

The other surprise in the N_2ase structure, apart from the FeMo-co structure, is the nature of the P clusters.[23] To understand this result, we must briefly look at iron–sulfur proteins. Although not strictly organometallic, they do have a soft S-donor environment. Indeed, S donors may be considered as the biological analogs of the P donors that are so common in standard organometallic chemistry. Structures **16.12, 16.13, 16.14**, and **16.15** show some main cluster types that had been recognized in these proteins.[24] In each case, the RS groups represent the cysteine residues by which the metal or metal cluster is bound to the protein chain. Where there is more than one iron atom, S^{2-} ions bridge the metals. The ferredoxin proteins contain Fe_4S_4 or Fe_2S_2 cores, which can be extruded intact from the enzyme by the addition of suitable thiols that can chelate the metal, to give a fully characterizable complex. The metal-free enzyme (the apoenzyme) can then be made active once again simply by adding Fe^{2+} and S^{2-}. These clusters therefore self-assemble; that is, they can form in solution on mixing the components (apoenzyme + metal ions or, for the model compounds, ligands + metal ions) under the correct conditions. This contrasts with FeMo-co, which as yet cannot be formed either from the apoenzyme and metal ions or in models from ligands and metal ions. Multiple genes are present in nitrogen-fixing organisms to direct the inorganic synthesis of the FeMo-co cluster.

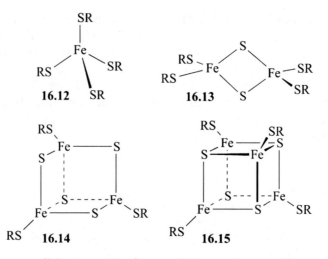

16.12   **16.13**

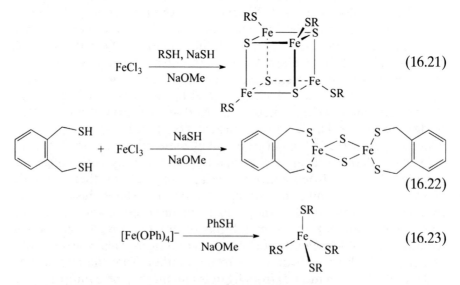

16.14   **16.15**

It has been possible to synthesize model complexes with core geometries similar to those present in the natural Fe–S clusters. Some examples are shown in Eq. 16.21–Eq. 16.23. Normally, adding an oxidizing
metal like Fe^{3+} to RSH simply leads to oxidation to RSSR, and so the
choice of reaction conditions is critical. Millar and Koch have shown
that metathesis of the phenoxide via Eq. 16.23 gives $[Fe(SPh)_4]^-$, an
apparently very simple Fe(III) compound, but one that long resisted
attempts to make it. In spite of being soft ligands, working with S-donors
is hard because of their high bridging tendency.

$$FeCl_3 \xrightarrow[\text{NaOMe}]{\text{RSH, NaSH}} \qquad \qquad (16.21)$$

$$\xrightarrow[\text{NaOMe}]{\text{NaSH}} \qquad \qquad (16.22)$$

$$[Fe(OPh)_4]^- \xrightarrow[\text{NaOMe}]{\text{PhSH}} \qquad \qquad (16.23)$$

The oxidation states present in the natural systems can be determined by comparison of the spectral properties of the natural system

in its oxidized and reduced states with those of the synthetic models; the latter can be prepared in almost any desired oxidation state by electrochemical means. The results show that the monoiron systems indeed shuttle between Fe(II) and Fe(III) as expected. The diiron enzymes are $(Fe^{III})_2$ in the oxidized state, and $(Fe^{II})(Fe^{III})$ in the reduced state. The mixed-valence species have trapped valencies rather than being delocalized. There is also a superreduced state, $(Fe^{II})_2$, which is probably not important in vivo. The four-iron proteins shuttle between $(Fe^{II})_3(Fe^{III})$ and $(Fe^{II})_2(Fe^{III})_2$, such as in the ferredoxins (Fd). One class of four-iron protein has an unusually high oxidation potential (HIPIP, or high potential iron protein) because the system shuttles between $(Fe^{II})_2(Fe^{III})_2$ and $(Fe^{II})(Fe^{III})_3$.

$$
\begin{array}{ccc}
Fe^{II}_3Fe^{III} \rightleftharpoons & Fe^{II}_2Fe^{III}_2 \rightleftharpoons & Fe^{II}Fe^{III}_3 \\
Fd_{red} & Fd_{ox} & Fd_{superox} \\
HIPIP_{superred} & HIPIP_{red} & HIPIP_{ox}
\end{array} \qquad (16.24)
$$

The N_2ase crystal structure, apart from showing FeMo-co, also revealed the structure of the P clusters (16.16), which consist of a pair of Fe_4S_4 cubanes joined by a corner S^{2-} ion and by two cysteine thiolates. This unique structure is presumably required to adjust the potential of the P cluster to make it a suitable electron donor to FeMo-co.

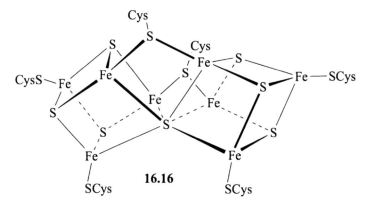

16.16

16.4   NICKEL ENZYMES

Urease is famous in enzymology as the first enzyme to be purified and crystallized (1926).[25] At the time, enzymes were widely viewed as being too ill-defined for detailed chemical study, but James Sumner (1887–1955) argued that its crystalline character meant that urease was a definite single substance. The fact that he could not find any cofactors led him to the conclusion that polypeptides could have catalytic activity on their own. The existence of two essential Ni^{2+} ions per mole of urease was not proved until 1975, so Sumner's conclusion is correct to the extent that cofactors are not *always* required for catalytic activity, but we now know that urease is not a valid example. Nickel was only recognized as a significant catalytic element in metalloenzymes in the 1980s. In three of these, hydrogenase (H_2ase), CO dehydrogenase (CODH), and MeCoM reductase (MCMR), organometallic structures are involved.

Archaea

This group of microorganisms, including the methanogens, the thermoacidophiles, and the halobacteria, are sufficiently different from all other forms of life that they are assigned to their own kingdom, the archaea.[26] The name indicates that they are very early organisms in an evolutionary sense. One of the signs of their antiquity is the fact that many archaea can live on the simple gases, such as H_2 and CO or CO_2, both as energy and carbon source, and on N_2 via nitrogen fixation as nitrogen source. Higher organisms have much more sophisticated nutritional requirements, but few, if any, other life forms must have existed when the archaea evolved, and they therefore had literally to live on air and water. A life form that can synthesize all its carbon constituents from CO_2 is an *autotroph* (from the Greek *autos* "self" and *trophē* "nourishment"); one that requires other C_1 compounds, such as methane or methanol, is *a methylotroph*.

The archaea are very rich in Ni enzymes and coenzymes, and this element is well suited to bring about the initial steps in the anaerobic biochemical utilization of H_2, CO, CH_4, and other C_1 compounds. For H_2ase and CODH, the pathways involve active site organonickel cluster chemistry that is only just beginning to be understood in detail.[27]

CO Dehydrogenase

CODH[28] can bring about two reactions (e.g., Eq. 16.26 and Eq. 16.28) of particular organometallic interest: the reduction of atmospheric CO_2 to CO (CODH reaction, Eq. 16.26) and acetyl coenzyme A synthesis (ACS reaction, Eq. 16.28) from CO, a CH_3 group possibly taken from a corrinoid iron–sulfur protein (denoted CoFeSP in the equation), and coenzyme A, a thiol. These are analogous to reactions we have seen earlier: the water–gas shift reaction (Eq. 16.25) and the Monsanto acetic acid process (Eq. 16.27).

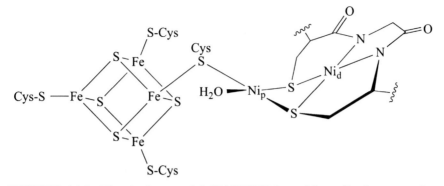

FIGURE 16.5 The A cluster of ACS/CODH from *Moorella thermoacetica*.

The enzyme contains two metal clusters, denoted A and C. CODH activity occurs in the C cluster, a $NiFe_3S_4$ cubane unit capable of reversible CO_2 reduction. ACS activity occurs at a very unusual trinuclear active site in the A cluster (Fig. 16.5). An Fe_4S_4 cubane is bridged by a cysteine sulfur to a four-coordinate Ni that is in turn bridged through two cysteine residues to a square-planar Ni(II) site, also ligated by two deprotonated peptide nitrogens from the peptide backbone. The square plane of Ni_p is completed by a water.[27]

$$CO + H_2O \rightleftharpoons CO_2 + H_2 \tag{16.25}$$

$$CO + H_2O \rightleftharpoons CO_2 + 2H^+ + 2e^- \tag{16.26}$$

$$MeOH + CO \rightarrow MeCOOH \tag{16.27}$$

$$Me\text{-}CoFeSP + CoA + CO \rightarrow MeCO(CoA) + CoFeSP \tag{16.28}$$

In a proposed mechanism of CO oxidation,[28b] an Fe–OH nucleophilically attacks an adjacent Ni(II) carbonyl to form a Ni–COOFe intermediate that releases CO_2. The ACS reaction is proposed to go via a CO insertion into a Ni–Me bond to form a Ni–COMe group. The acetyl then undergoes nucleophilic abstraction by the CoA-SH thiolate to form the CoA–SCOCH$_3$, acetyl CoA.[29]

Methanogenesis

Bacterial methane formation in the digestive system of cattle has gained attention in connection with the resulting global warming gas emission because 10^9 tons of CH_4 are released annually in this way and methane is a much more potent greenhouse gas than CO_2.[30] Methanogens reduce

CO$_2$ to CH$_4$ and extract the resulting free energy via the Wolfe cycle.[31] In the last step, methylcoenzyme M, **16.17**, is hydrogenolyzed to methane by a thiol cofactor, coenzyme B, HS–CoB, catalyzed by the Ni enzyme, methylCoM reductase, MCR.

CH$_3$S⟍⟋SO$_3^-$ + CoB—SH $\xrightleftharpoons{\text{enz.}}$ CoB—S—S⟍SO$_3^-$ + CH$_4$
16.17

$$(16.29)$$

Factor F$_{430}$ (**16.18**), a coenzyme bound within MCR, catalyzes Eq. 16.29. Binding of methyl CoM to the Ni(I) form of F$_{430}$ may lead to release of a transient methyl radical that is immediately quenched by H atom transfer from the adjacent coenzyme B (CoB) HS–HTP thiol cofactor to give methane. The resulting thiol radical may abstract the CoM thiolate from Ni to regenerate the Ni(I) form, as well as give the observed CoM–S–S–CoB heterodisulfide coproduct (Eq. 16.30 and Eq. 16.31).[32] In a truly remarkable C–H activation of methane, methanogenesis has been shown to be reversible, that is, labeled methane can incorporate back into methyl coenzyme M under mild conditions.[33]

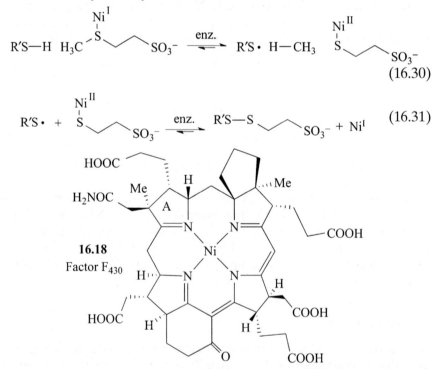

R′S—H H$_3$C–S$^{Ni^{I}}$⟍⟋SO$_3^-$ $\xrightleftharpoons{\text{enz.}}$ R′S• H—CH$_3$ S$^{Ni^{II}}$⟍SO$_3^-$

$$(16.30)$$

R′S• + S$^{Ni^{II}}$⟍SO$_3^-$ $\xrightleftharpoons{\text{enz.}}$ R′S—S⟍SO$_3^-$ + Ni^{I}

$$(16.31)$$

16.18
Factor F$_{430}$

One of the characteristic features of Ni is its aptitude for coordination geometry changes. Unlike d^6 ions that are reliably octahedral, Ni ions can adopt a variety of 4, 5, and 6-coordinate geometries, a property that greatly puzzled early investigators when they obtained yellow, green, and blue compounds from the same ligand. This flexibility may be of importance in F430, where binding of CoB–SH to the enzyme induces a conformational change that has been suggested to involve a change of coordination geometry at Ni.[34]

Hydrogenases

By bringing about Eq. 16.32, hydrogenases[36] allow certain bacteria to thrive on H_2 as energy source, and others to get rid of excess electrons by combining them with protons from H_2O for release as H_2.[35] The nickel-containing [NiFe] hydrogenases are the largest class, but iron-only [FeFe] H_2ases, as well as a cluster-free form, the [Fe] H_2ases[37] also exist. The number of metal ions present varies with the species studied, but the minimum cofactor composition for the [NiFe] or [FeFe] types is one Ni–Fe or Fe–Fe and one Fe_4S_4 cluster per enzyme (Eq. 16.32).

$$H_2 \rightleftharpoons 2H^+ + 2e^- \qquad (16.32)$$

All three H_2ase classes have organometallic active-site clusters, as shown by X-ray crystallography and IR spectroscopy.[38] The [NiFe] protein active-site cluster from *Desulfovibrio gigas* is shown as **16.19**, and the [FeFe] protein's H cluster from *Clostridium pasteurianum* is shown as **16.20**. The active [NiFe] site **16.19** has a nickel tetrathiolate center bridged to a low-spin dicyanoiron(II) carbonyl group—the latter was then an unprecedented ligand set in biology. The bridging oxo or hydroxo group, X, is removed as H_2O on incubation under H_2 for some hours, leading to conversion of the inactive enzyme to the active form. Structure **16.20** has two Fe(CO)(CN) groups bridged both by a CO and by a 2-azapropane-1,3-dithiolate, thus positioning a pendant NH group in the vicinity of the active site. This NH group is believed to act as a local base by deprotonating an intermediate H_2 complex in a key step of the mechanism. One iron has a labile ligand, thought to be water, where the H_2 presumably binds. Theoretical work[39] supports heterolytic splitting of such an intermediate, where the H^+ may move to an internal base, such as the azathiolate N lone pair. As part of an interesting speculation on the origin of life, iron sulfide, dissolved at deep-sea vents by CO, is proposed to give **16.21**, a complex that became incorporated into early proteins to give the first hydrogenases. In any event, **16.21** is a useful synthetic precursor to a series of complexes, such as **16.22**, that resemble the hydrogenase site.[35] The Fe–Fe distance of 2.5Å in **16.22** is consistent with the metal–metal bonding required by the EAN rule.

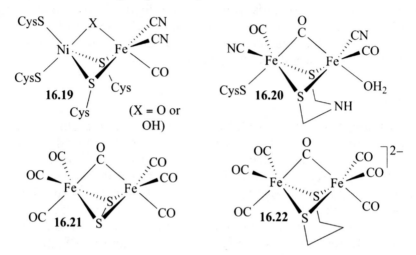

The inactive Ni(III) state having a bridging X group (X = O or OH) seems to form part of a mechanism for protecting the enzyme against exposure to air. The active form involves Ni(II) and more reduced states. Hydrogen activation by the enzyme is heterolytic because D_2 exchanges with solvent protons by Eq. 16.33; dihydrogen complexes are known to catalyze similar reactions (Section 3.4).

$$D_2 + ROH \underset{}{\overset{enz.}{\rightleftharpoons}} HD + ROD \qquad (16.33)$$

Although all three types of hydrogenase have common features: a redox-inactive low-spin, five- or six-coordinate Fe(II) bound to CO or CN, the [Fe] hydrogenase seems to operate by a different mechanism. In the [NiFe] and [FeFe] hydrogenases, electrons from Eq. 16.32 flow through the protein's redox-active clusters to a distant electron acceptor. In the [Fe] hydrogenase, the redox cofactor, methenyl-H_4MPT^+, shown in Eq. 16.34, is nearby. Rather than being involved in accepting an electron, this cofactor accepts a hydride, equivalent to $H^+ + 2e^-$, from the H_2. The structure of the [Fe] hydrogenase[38c,40] active site iron is shown in Eq. 16.34. Once again, we see an organometallic ligand, the acyl group, that provides the high trans effect ligand trans to the active site that favors hydride transfer to the H_4MPT^+ cofactor (when two high trans effect ligands are mutually trans, each accumulates a substantial negative charge). This resembles the mechanism of the Dobereiner catalyst (Eq. 9.15), where hydride transfer occurs to a heteroarenium ion.

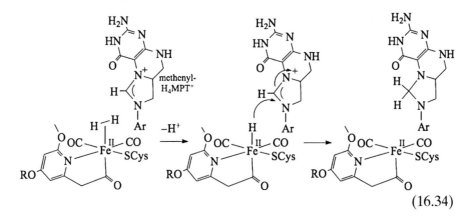

(16.34)

Electrocatalytic water reduction to H_2 has proved possible with a number of Ni catalysts,[41] **16.23** being the most active, having a turnover frequency of 10^5 s^{-1}.[42] The pendant nitrogens probably have a key role in binding protons and transferring one to the reduced nickel and another to the resulting nickel hydride intermediate to form $L_4Ni(H_2)$. If so, a similar proton management function may be fulfilled by the pendant nitrogen in **16.20**.

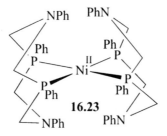

16.23

16.5   BIOMEDICAL AND BIOCATALYTIC APPLICATIONS

As a physician strongly interested in chemistry, Paul Ehrlich (1854–1915),[44] who won the 1908 Nobel Prize for his work as the founder of chemotherapy, is celebrated for his 1906 prediction that therapeutic compounds would be created "in the chemist's retort."[43] His most important discovery—the application of the polymeric organoarsenical, Salvarsan, as the first antisyphilitic—caused an international sensation and led to his being besieged by thousands of desperate sufferers. In spite of this early success, organometallic compounds are at present only just beginning to receive renewed attention in pharmacology and medicine.[45]

Platinum Drugs

Work from 1965 identified cisplatin, cis-[PtCl$_2$(NH$_3$)$_2$], as an anticancer drug that targets DNA by forming intrastrand cross links and thus inhibits cell growth; cancer cells divide faster than normal ones and so are more susceptible to the drug. Its relatively high toxicity means that improved drugs were soon sought—two commercial successes are oxaliplatin (**16.24**) and carboplatin (**16.25**).[46] Pt drugs are now very widely used in cancer chemotherapy, with an estimated 50–70% of all patients receiving them at some point. Organometallics such as **16.26** are also being investigated in this context.[47]

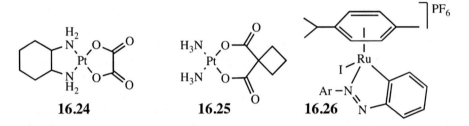

Technetium Imaging Agents

Technetium-99m is a nuclear excited state that decays ($t_{1/2} = 6$h) to the weakly radioactive ^{99}Tc ($t_{1/2} = 2.1 \times 10^5$ y) with emission of 140.5 keV γ rays. These can be detected by a γ ray camera to give images of patients' organs or tumors, as now occurs in millions of diagnostic procedures annually. Numerous organs can be targeted for imaging depending on the ligand environment of the Tc ion. For example, **16.27** and [^{99m}Tc(CN{CMe$_2$OMe})$_6$]$^+$ both image the heart.

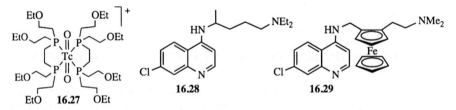

Organometallic Drugs

Of the many research-level agents under study, the antimalarial ferroquine, **16.29**, seems the likely to be the first to reach the clinic.[48] The malarial parasite is thought to affect half a million people annually, and its toll is exacerbated by the increasing level of drug resistance to established drugs. Ferroquine builds on such a drug, chloroquine, **16.28**, by incorporating a ferrocene group that foils the parasite's resistance mechanism. Organometallic drugs also have promise for other tropical diseases.[49]

Organometallic-Enzyme Constructs

Cellular processes require orthogonal catalysts, that is, ones that can function unaffected by all the other cell components. Organometallic catalysts often fail to act in concert with enzymes because of mutual inactivation. A Cp*Ir(chelate)Cl transfer hydrogenation catalyst has now been successfully incorporated into the protein, streptavidin, as an artificial transfer hydrogenase in order to protect it from deactivation in cooperative catalysis with monoamine oxidases.[50]

- Biology uses organometallic chemistry sparingly but always in reactions that are difficult to bring about by conventional means.

REFERENCES

1. I. Bertini, H. B. Gray, E. Stiefel, and J. Valentine, *Biological Inorganic Chemistry: Structure & Reactivity*, University Science Books, Sausalito, CA, 2007; R. R. Crichton, *Biological Inorganic Chemistry*, Elsevier, Amsterdam, 2012.
2. C. Huber, F. Kraus, M. Hanzlik, W. Eisenreich, and G. Wächtershäuser, *Chem. Eur. J.*, **18**, 2063, 2012.
3. J. Halpern, *Pure Appl. Chem.*, **58**, 575, 1986.
4. T. Toraya, *Chem. Rev.*, **103**, 2095, 2003; R. Banerjee, *Chem. Rev.*, **103**, 2083, 2003; R. G. Matthews, *Met. Ions Life Sci.*, **6**, 53, 2009.
5. R. Chang, *Physical Chemistry for the Biosciences*, University Science, Sausalito, CA, 2005.
6. See J. A. K. Howard, *Nature Rev. Molec. Cell Biol.*, **4**, 891, 2003.
7. M. P. Jensen, D. M. Zinkl, and J. Halpern, *Inorg. Chem.*, **38**, 2386, 1999 and references cited.
8. D. W. Boening, *Chemosphere*, **40**, 1, 2000.
9. T. G. Chasteen and R. Bentley, *Chem. Rev.*, **103**, 1, 2003.
10. (a) D. Jones, *New Sci.*, **101**, Oct 14, 1982; (b) http://news.bbc.co.uk/2/hi/health/3913213.stm.
11. B. M. Hoffman, D. Lukoyanov, D. R. Dean, and L. C. Seefeldt, *Acc. Chem. Res.*, **46**, 587, 2013.
12. L. C. Seefeldt, B. M. Hoffman, and D. R. Dean, *Annu. Rev. Biochem.*, **78**, 701, 2009.

13. (a) T. Spatzal, M. Aksoyoglu, L. M. Zhang, S. L. A. Andrade, E. Schleicher, S. Weber, D. C. Rees, and O. Einsle, *Science*, **334**, 940, 2011; K. M. Lancaster, M. Roemelt, P. Ettenhuber, Y. Hu, M. W. Ribbe, F. Neese, U. Bergmann, and S. DeBeer, *Science*, **334**, 974, 2011. (b) Inclusion of N_2ase in the first edition of this book was criticized by a reviewer on the grounds that it was not an organometallic structure.

14. Y. Hu and M. W. Ribbe, *Microbiol. Molec. Biol. Rev.*, **75**, 664, 2011; J. A. Wiig, Y. Hu, C. C. Lee, and M. W. Ribbe, *Science*, **337**, 1672, 2012.

15. Z.-Y. Yang, D. R. Dean, and L. C. Seefeldt, *J. Biol. Chem.*, **286**, 19417, 2011.

16. R. Gilbert-Wilson, L. D. Field, S. B. Colbran, and M. M. Bhadbhade, *Inorg. Chem.*, **52**, 3043, 2013.

17. S. J. Connelly, A. C. Zimmerman, W. Kaminsky, and D. M. Heinekey, *Chem. Eur. J.*, **18**, 15932, 2012.

18. J. Chatt, J. R. Dilworth, and R. L. Richards, *Chem. Rev.*, **78**, 589, 1978.

19. K. Arashiba, K. Sasaki, S. Kuriyama, Y. Miyake, H. Nakanishi, and Y. Nishibayashi *Organometallics*, **31**, 2035, 2012.

20. D. Lukoyanov, SA Dikanov, Z.–Y. Yang, B. M. Barney, R. I. Samoilova, K. V. Narasimhulu, D. R. Dean, L. C. Seefeldt, and B. M. Hoffman, *J. Am. Chem. Soc.*, **133**, 11655, 2011.

21. R. R. Schrock, *Acct. Chem. Res.*, **38**, 955, 2005.

22. M. M. Rodriguez, E. Bill, W. W. Brennessel, and P. L. Holland, *Science*, **334**, 780, 2011.

23. P. V. Rao and R. H. Holm, *Chem. Rev.*, **104**, 527, 2004.

24. D. C. Johnson, D. R. Dean, A. D. Smith, and M. K. Johnson, *Ann. Rev. Biochem.*, **74**, 247, 2005.

25. A. Sigel, H. Sigel, and R. K. O. Sigel (eds.) *Nickel and Its Surprising Impact in Nature*, Wiley Hoboken, NJ, 2007; S. W. Ragsdale, *J. Biol. Chem*, **284**, 18571, 2009.

26. S. Y. Kato (ed.), *Archaea: Structure, Habitats, and Ecological Significance*, Nova Science Publishers, New York, 2011.

27. P. A. Lindahl, *J. Inorg. Biochem.*, **106**, 172, 2012.

28. (a) D. J. Evans, *Coord. Chem. Rev.*, **249**, 1582, 2005; (b) J. -H. Jeoung and H. Dobbek, *Science*, **318**, 1461, 2007; (c) Y. Kung and C. L. Drennan, *Curr. Opin. Chem. Biol.*, **15**, 276, 2011.

29. S. Gencic, K. Kelly, S. Ghebreamlak, E. C. Duin, and D. A. Grahame, *Biochemistry*, **52**, 1705, 2013 and references cited.

30. A. van Amstel, *J. Integr. Environ. Sci.*, **9**, 5, 2012.

31. R. K. Thauer, *Proc. Nat. Acad. USA*, **109**, 15084, 2012.

32. S. L. Chen, M. R. A. Blomberg, and P. E. M. Siegbahn, *Chem. Eur. J.*, **18**, 6309, 2012.

33. S. Scheller, M. Goenrich, R. Boecher, R. K. Thauer, and B. Jaun, *Nature*, **465**, 606, 2010.

34. M. Zimmer and R. H. Crabtree, *J. Amer. Chem. Soc.*, **112**, 1062, 1990; S. Ebner, B. Jaun, M. Goenrich, R. K. Thauer, and J. Harmer, *J. Amer. Chem. Soc.*, **132**, 567, 2010.

35. D. M. Heinekey, *J. Organometal. Chem.*, **694**, 2671, 2009.

36. R. K. Thauer, A. K. Kaster, M. Goenrich, M. Schick, T. Hiromoto, and S. Shima, *Ann. Rev. Biochem.*, **79**, 507, 2010.

37. J. A. Wright, P. J. Turrell, and C. J. Pickett, *Organometallics*, **29**, 6146, 2010.

38. (a) S. Shima and R. K. Thauer, *Chem. Record*, **7**, 37, 2007; (b) D. J. Evans and C. J. Pickett, *Chem. Soc. Rev.*, **32**, 268, 2003; (c) S. Shima, O. Pilak, S. Vogt, M. Schick, M. S. Stagni, W. Meyer-Klaucke, E. Warkentin, R. K. Thauer, and U. Ermler, *Science*, **321**, 5888, 2008.

39. P. E. M. Siegbahn, J. W. Tye, and M. B. Hall, *Chem. Rev.*, **107**, 4414, 2007.

40. B. Hu, D. Chen, and X. Hu, *Chem. Eur. J.*, **18**, 11528, 2012.

41. Z. Han, W. R. McNamara, M. -S. Eum, P. L. Holland, and R. Eisenberg, *Angew. Chem., Int. Ed.*, **51**, 1667, 2012.

42. M. L. Helm, M. P. Stewart, R. M. Bullock, M. Rakowski DuBois, and D. L. DuBois, *Science*, **333**, 863, 2011.

43. E. A. Hillard and G. Jaouen, *Organometallics*, **30**, 20, 2011.

44. See A. Piro, A. Tagarelli, and G. Tagarelli, *Int. Rev. Immunol.*, **27**, 1, 2008.

45. E. A. Hillard and G. Jaouen, *Organometallics*, **30**, 20, 2011.

46. Y. Jung and S. J. Lippard, *Chem. Rev.*, **106**, 1387, 2006.

47. G. Sava, A. Bergamoa, and P. J. Dyson, *Dalton Trans.*, **40**, 9069, 2011.

48. M. Navarro, W. Castro, and C. Biot, *Organometallics*, **31**, 5715, 2012.

49. A. Martínez, T. Carreon, E. Iniguez, A. Anzellotti, A. Sánchez, M. Tyan, A. Sattler, L. Herrera, R. A. Maldonado, and R. A. Sánchez-Delgado, *J. Med. Chem.*, **55**, 3867, 2012.

50. V. Köhler, Y. M. Wilson, M. Dürrenberger, D. Ghislieri, E. Churakova, T. Quinto, L. Knörr, D. Häussinger, F. Hollmann, N. J. Turner, and T. R. Ward, *Nature Chem.*, **5**, 93, 2013.

PROBLEMS

16.1. Why do you think Nature uses first-row transition metals in most of the transition metalloenzymes?

16.2. The oxidation states found in the metal centers we have been discussing in this chapter, Fe(II), Fe(III), Ni(III), and Co(III), are often higher than those usually present in organometallic species we discussed in Chapters 1–14. Why do you think this is so?

16.3. Those mononuclear N_2 complexes, which have the lowest N–N stretching frequency in the IR, are in general also the complexes in which N_2 is most easily protonated. Explain.

16.4. Would you expect the following R groups to dissociate more or less readily as R· from cobaloxime than does ·CH$_2$Ph: –CH$_3$, –CF$_3$, –CPh$_2$H? Explain.

16.5. Many N$_2$ complexes protonate. In the case of ReCl(N$_2$)(PMe$_2$Ph)$_4$, the protonated form HReCl(N$_2$)(PMe$_2$Ph)$_4^+$ (A) is relatively stable. What might happen to the N–N stretching frequency on protonation? Most N$_2$ complexes simply lose N$_2$ on protonation. Given that a complex of type A is the intermediate, explain why N$_2$ is lost.

16.6. If a CODH enzyme were found to incorporate ^{18}O from ^{18}OH$_2$ into the CO reactant, how could we explain this outcome?

16.7. In CODH, the CO that is subject to nucleophilic attack seems to be bound to nickel, which could in principle be Ni(0), Ni(I), or Ni(II). Discuss the relative suitability of these oxidation states for promoting the reaction.

APPENDIX A: USEFUL TEXTS ON ALLIED TOPICS

Bioinorganic Chemistry

S. J. Lippard and J. Berg, *Principles of Bioinorganic Chemistry*, University Science Books, Mill Valley, CA, 1994.

E. Ochiai, *Bioinorganic Chemistry*, Elsevier, Amsterdam, 2008.

R. M. Roat-Malone, *Bioinorganic Chemistry*, Wiley, Hoboken, 2007.

Electrochemistry

E. Gileadi, *Electrode Kinetics for Chemists, Chemical Engineers and Materials Scientists*, Wiley-VCH, 1993.

Encyclopedias

R. A. Scott (ed.), *Encyclopedia of Inorganic Chemistry*, Wiley, New York, 2012.

G. Wilkinson et al. (eds.), *Comprehensive Organometallic Chemistry*, Pergamon, Oxford, 1982, 1987, 1995, 2006.

Homogeneous Catalysis

B. Cornils and W. A. Herrmann, *Applied Homogeneous Catalysis with Organometallic Compounds: A Comprehensive Handbook*, Wiley-VCH, Weinheim, 2002.

The Organometallic Chemistry of the Transition Metals, Sixth Edition.
Robert H. Crabtree.
© 2014 John Wiley & Sons, Inc. Published 2014 by John Wiley & Sons, Inc.

G. Duka, *Homogeneous Catalysis with Metal Complexes*, Springer, Berlin Heidelberg, CA, 2012.

P. W. N. M. van Leeuwen, *Homogeneous Catalysis*, Kluwer, Dordrecht, 2004.

Green Chemistry

P. J. Dunn et al. (eds.), *Green Chemistry in the Pharmaceutical Industry*, Wiley-VCH, Weinheim, 2010.

P. T. Anastas and J. C. Warner (eds.), *Green Chemistry, Theory and Practice*, Oxford U. Press, 1998.

Inorganic Chemistry

F. A. Cotton, W. M. Bochmann, and C. A. Murillo, *Advanced Inorganic Chemistry*, 6th ed., Wiley, New York, 1998.

Kinetics and Mechanism

E. V. Anslyn and D. A. Dougherty, *Modern Physical Organic Chemistry*, University Science, Books, Sausalito, CA, 2007.

J. E. Espenson, *Chemical Kinetics and Reaction Mechanisms*, McGraw-Hill, New York, 1981.

R. B. Jordan, *Reaction Mechanisms of Inorganic and Organometallic Systems*, Oxford University Press, Oxford, 1991.

G. B. Marin, *Kinetics of Chemical Reactions*, Wiley-VCH, Weinheim, 2011.

R. G. Wilkins, *Kinetics and Mechanism of Reactions of Transition Metal Complexes*, 2nd ed., VCH, Weinheim, 1991.

NMR and Physical Methods

A. Bakac (ed.), *Physical Inorganic Chemistry: Reactions, Processes, and Applications*, Wiley, Hoboken, 2010.

R. A. Scott and C. M. Lukehart, *Applications of Physical Methods in Inorganic & Bioinorganic Chemistry*, Wiley, Hoboken, NJ, 2007.

Organic Chemistry, Organometallics in

L. S. Hegedus, *Transition Metals in the Synthesis of Complex Organic Molecules*, University Science Books, Sausalito, CA, 1999.

M. Schlosser, *Organometallics in Synthesis*, Wiley, New York, 2002.

Organometallic Chemistry

J. F. Hartwig, *Organotransition Metal Chemistry: From Bonding to Catalysis*, University Science Books, Sausalito, CA, 2009.

Photochemistry

G. L. Geoffroy and M. S. Wrighton, *Organometallic Photochemistry*, Academic, New York, 1979. (See also Anslyn and Dougherty 2007.)

Preparative Techniques

D. F. Shriver, *The Handling of Air-Sensitive Compounds*, McGraw-Hill, New York, 1969.

APPENDIX B: MAJOR REACTION TYPES AND HINTS ON PROBLEM SOLVING

Alphabetical List of Reaction Types and Where to Find Them in the Text

Reaction Type	Section Number
Abstraction by E^+	8.5
Alkene–carbene cycloaddition	12.1
Association of E^+	6.5, 8.5, 11.1
Association of L	4.5
Association of $X\bullet$	4.6, 6.4
Asymmetric reactions	9.3, 14.3, 14.4
Binuclear oxidative addition	6.1
Binuclear reductive elimination	6.6
Carbene–alkene cycloaddition	12.1
Carbonylation	14.5
C–H activation	12.4, 14.7
Click chemistry	14.8
Coupling	9.7, 14.1
Deprotonation	8.4
Dissociation of E^+	8.4
Dissociation of L	4.4
Eliminations and insertions	7.1–7.4, 9.1–9.6, 12.2

The Organometallic Chemistry of the Transition Metals, Sixth Edition.
Robert H. Crabtree.
© 2014 John Wiley & Sons, Inc. Published 2014 by John Wiley & Sons, Inc.

(*Continued*)

Reaction Type	Section Number
Green and energy chemistry	12.5, 12.6
Hydroformylation	9.4
Hydrogenation	9.3, 14.4
Hydrosilylation	9.6
Isomerization	9.2
Ligand substitution	4.4–4.7
Metalacyclobutane cleavage	12.1
Metathesis	12.1, 14.2
Nucleophilic abstraction of X^+	8.4
Oxidation	9.8, 14.6
Oxidative coupling	6.8
Photochemical dissociation of L or X_2	4.7, 12.4
Polymerization	12.2
Reductive fragmentation	6.8
Single-electron transfer	8.6
α Elimination	7.5
β Elimination	7.5
γ Elimination	7.5
∂ Elimination	7.5
σ-Bond metathesis	3.4, 6.7
σ-CAM	6.7

Hints on Problem Solving

Questions are based on *standard* ideas, structures, and reaction steps, principally OA, RE, insertions, eliminations, nucleophilic and electrophilic additions, and abstractions but also the steps mentioned above. If asked to provide a mechanism for a given organic transformation, try to work backward: what bond in the organic product could be formed by an RE, for example? What organometallic intermediate could have given such an RE, and how could it have been formed? If you have come up with a dead end or have a need for nonstandard reaction steps, try a different approach; for example, do you know anything about the reverse of the process provided in the question? If so, the mechanism of the forward process will be identical by microscopic reversibility arguments, although the steps would then have to be taken in the reverse order.

Δ(CN) → (OS) ↓	−2	−1	0	1	2
−2	Red. Elim. {−2} [6.5, 14.4] Deprotonatn. {0} [8.3]	Nucl. Abs. of X⁺ {0} [8.3]	Metalacyclobutane Clvg. {2} [12.1] Red. Clvg. {2} [6.7]		
−1		Binucl. Red. Elim. {−1} [6.5]			
0		Dissoc of L. {−2} [4.3, photochem., 4.6] Dissoc or Abstrn of E⁺ {0} [6.5, 8.3]	Substn. of L. {0} [4.3–4.7] Insertn. & Elim. {0} [7.1–3, 9.1] SET {±1} [8.6] Ox. Cplg. {−2} [6.7]	Assoc of L. {2} [4.4] Alpha & Beta Elim. {+2} [7.4]	
1				Binucl. Ox Addn {1} [6.3] Assocn. of X• {1} [4.3, 6.3]	
2		Carbene–Alkene Cycloaddn. {−2} [12.1] Ox. Cplg. {−2} [6.7, 14.4]	Assoc of E⁺ incl. Protonation {0} [6.4, 8.4, 11.1]	Ox. Addn {2} [6.1–6.4, 12.4] Gamma, Delta Elim. {+2} [7.4]	

SOLUTIONS TO PROBLEMS

CHAPTER 1

1.1. 4 (if you thought 2, you perhaps missed structures such as $[PtL_4]^{2+}$ $[PtCl_4]^{2-}$).

1.2. Assume octahedral ligand field, high spin: Zn(II), d^{10}, dia; Cu(II), d^9, para; Cr(II), d^4, para; Cr(III), d^3, para; Mn(II), d^5, para; and Co(II), d^7, para.

1.3. The first diphosphine ligand gives a favorable five-membered ring on chelation, while the second gives an unfavorable four-membered ring. The second lone pair of water repels and destabilizes the d_π electrons. Ammonia has no second lone pair.

1.4. (i) $[PtCl_4]^{2-}$ + tu, 1 equiv, which must give $[Pt(tu)Cl_3]^-$; (ii) NH_3, which replaces the Cl trans to the high trans effect tu ligand.

1.5. The Ti complex is a hard acid, so the order is N > P > C (hard base best); the W complex is a soft acid, so C > P > N (soft base best).

1.6. The tetrahedral structure with a two-below-three orbital pattern will be paramagnetic because in a d^8 ion the lower set of two

The Organometallic Chemistry of the Transition Metals, Sixth Edition.
Robert H. Crabtree.
© 2014 John Wiley & Sons, Inc. Published 2014 by John Wiley & Sons, Inc.

orbitals will take four electrons, leaving four for the upper set of three orbitals; two of these must go in with parallel spin, so there will be two unpaired electrons.

1.7. Measure $\nu(CO)$, the better donors will cause greater lowering because they will cause a greater charge buildup on the metal, which will lead to increased $M(d_\pi) \rightarrow CO(\pi^*)$ back donation and a lower C–O bond order.

1.8. The d orbitals are stabilized by the higher nuclear charge, and so back donation (required to form a strong M–CO bond) is reduced. Cu(I) rather than Cu(II) would be best because it would be a stronger π donor.

1.9. Reduced complexes will easily lose electrons to O_2 in an oxidation reaction but will not tend to bind a π donor such as H_2O.

1.10. Assume an octahedral three-below-two splitting pattern, then $MnCp_2$ has five unpaired electrons, one in each of the five orbitals; the Cp* analogue has 4e paired up in the lower pair of orbitals and one unpaired electron in the next higher orbital; Cp* has the higher ligand field because it causes spin pairing.

1.11. The apical sp^3 nitrogens have tetrahedral geometry, meaning the bonds to the adjacent CH_2 groups diverge by only ~109°. The pyridines thus naturally adopt a cis arrangement in which they are 90° apart with less strain than would be the case if they were to occupy trans sites 180° apart.

CHAPTER 2

2.1. The first three are 16e, Pt(II), d^8, then 20e, Ni(II), d^8, 18e, Ru(II), d^6; 18e, Re(VII), d^0; 18e, Ir(V), d^4; 10e, Ta(V), d^0; 16e, Ti(IV), d^0, 14e, Re(VII), d^0.

2.2. $[\{(CO)_3Re\}(\mu_3\text{-}Cl)]_4$. A triply bridging Cl^- in a cubane structure allows each Cl^- to donate 5 electrons (6e ionic model).

2.3. $(\eta^6\text{-}PhC_6H_5)Cr(CO)_3$, with a π-bound arene ring.

2.4. Ti(0) if both ligands are considered as being 4e L_2, but Ti(II) if one is considered as being X_2 and bound via the two N atoms in the MeN–CH=CH–NMe dianionic form, and Ti(IV) if both are considered as being in the X_2 form.

2.5. The same values should be obtained as in answer 2.1.

2.6. M–M counts one for each metal. This rule allows the Os compound to reach 18e. The Rh compound has a tetrahedron of mutually bonded Rh atoms for a total of six Rh–Rh bonds and so is also 18e.

2.7. 8e C for $H_3C^+ \leftarrow :NH_3$ (three X Iigands, one L, and a positive charge) and 8e for $H_2C \leftarrow :CO$ (two X ligands and one L).

2.8. Counting only one lone pair gives an 18e count in both cases.

2.9. 2e either way. A σ-acid metal favors the η^1 form in which the important bonding interaction is L → M σ donation, and a π-basic metal favors the η^2 form where back donation into the C=O π* is the most important interaction. η^1 binding should favor nucleophilic attack.

2.10. $Cp_2W(CO)_2$ with one η^3, and one η^5 Cp gives an 18e count. If each triphos is κ^2, we get a 16e count, which is appropriate for Pd(II), and this is the true structure; a $\kappa^2 - \kappa^3$ structure would be 18e and cannot be ruled out, but an $\kappa^3 - \kappa^3$ would be 20e and is unlikely.

2.11. The left-hand complex has six L-type ligands, so we have 18e, d^6, W(0); the right-hand complex has five L and two X ligands, so we have 18e, d^4, W(II).

CHAPTER 3

3.1. Protonation of the Pt or oxidative addition of HCl gives a Pt–H into which the acetylene inserts.

3.2. $M-CF_2-Me$ (σ-acceptor bonds α to the metal, specially C–F, strongly stabilize an alkyl).

3.3. Oxidative addition of MeCl, followed by reaction of the product with LiMe, which acts as a Me^- donor and replaces the Ir–Cl by Ir–Me.

3.4. Bent, 18e, no π bonding between O lone pairs and filled M d_π.

3.5. 18e in all cases; both structures have the same electron count because (H_2) is a 2e L ligand and $(H)_2$ consists of two 1e X ligands, so no change. Both structures are in fact classical with terminal hydrides only.

3.6. If X or Y have lone pairs, they may complete for binding. Y–H–M is usually not competitive with lone-pair binding as in H–Y–M.

3.7. It is easier to reduce a more oxidized complex.

3.8. 17e (or 18 if M-M bonded), Ru(III), d^5; 18e, Cr(0), d^6; 12e, W(VI), d^0.

3.9. Initial formation of Ir–(i-Pr) with RMgX acting as source of R$^-$ to replace the Cl$^-$ initially bound to Ir. The alkyl then β-eliminates to give propene as the other product.

3.10. Insertion of the alkene into the M–H bond to give M–CHMe(Et), followed by β elimination to give MeCH=CHMe; insertion requires prior binding of the alkene and so does not happen in the 18e case.

3.11. Hydricity involves production of charged species, so the energy needed will strongly depend on the polarity of the solvent, unlike the case for bond dissociation energy where neutral fragments are formed.

CHAPTER 4

4.1. (a) Halide dissociation is bad for two reasons. The product is 16e and cationic, while for proton dissociation, the product is 18e and anionic; 16e species are less favorable and cations are less well stabilized by the π-acceptor CO groups than anions. (b) Solvent is likely to bind to M only in the 16e cation.

4.2. The NO can bend to accommodate the incoming ligand.

4.3. The more ∂^+ the CO carbon, the easier the reaction, so the order is: $[Mn(CO)_6]^+ > Mo(CO)_3(NO)_2 > Mo(CO)_6 > Mo(CO)_4(dpe) > Mo(CO)_2(dpe)_2 > Mo(CO)_5^{2-}$. [This order is decided by (1) cations > neutrals > anions, and (2) within each class, complexes with the better π-acceptor ligands > complexes with less good π-acceptor ligands.]

4.4. The $\nu(CO)$ lowering in the IR or easier oxidation as measured electrochemically; both disfavor reaction.

4.5. $Fe(CO)_5$ and $Fe(CO)_4L$ are Fe(0) d^8; all others are Fe(-I), d^9. $CpMn(CO)_2(L)$ are Mn(I), d^6; all others are Mn(II), d^5.

4.6. NR_3 lacks significant π-acid character and so avoids M(0), but NF_3 should bind better thanks to its N–F σ* orbital, which should be polarized toward the metal and could act as π acceptor; this resembles the cases of CH_3 versus CF_3, where the same applies.

4.7. As a highly reduced metal, Ni(0) prefers π-acceptor ligands such as $P(OMe)_3$. PMe_3 as a poor π acceptor causes the electron density on the metal to rise so much that the NiL_3 fragment is a poor σ acceptor.

4.8. D, A, D, D, A, A because we expect D for 18e, A for 16e and 17e species.

4.9. Eighteen electron structures (or 16e where appropriate) can be achieved as follows: η^6-Ph of BPh_4; η^3 and η^5-Ind groups; $[Me_3Pt(\mu\text{-}I)]_4$; nonadjacent C=C bonds of cot must bind η^2 to each of the two $PtCl_2$ groups; μ-Cl required in a dinuclear complex.

4.10. (a) Labilization of the CO trans to L gives ML_6; (b) preferential labilization of CO by CO would give $(L\text{–}L)M(CO)_4$ or $(L\text{–}L)_2M(CO)_2$.

4.11. Six positive ionic charges on the complex rules it out because the metal would not retain enough π-donor power to bind NO. Very few complexes exceed a net ionic charge of ±2.

4.12. Protonation at the metal (always allowed even for 18e complexes) should introduce a cationic charge that should not only weaken M–CO bonding but also put a high-trans-effect H ligand on the metal. In a D mechanism, a weaker M–CO bond {higher $\nu(CO)$} should lead to faster substitution.

4.13. Extrapolation suggests a very high figure, 2270 cm^{-1} or above, implying the presence of a very weakly bound CO and that the compound would be very hard to make.

4.14. One factor must be the lack of back donation for NR_3, but the short M–N and N–R bonds relative to M–P and P–R may lead to a significant increase in steric size. For the pentacarbonyl, the lack of back donation is not a problem because there are so many good π-acceptor COs present, and the steric problem is minimal because the COs are so small.

4.15. Steric factors are relevant. Arene C–H bonds are most susceptible but the distal ring has i-Pr groups protecting the *ortho* positions. The one accessible aryl C–H on the vicinal aryl ring would be hard to metallate because the arene would have to rotate such that the bulky distal arene would clash with the t-Bu groups. Only the t-Bu groups are plausible candidates for cyclometallation.

4.16. Abnormals have no all-neutral formal charge structure in the free carbene (**4.19**, **4.21** and **4.23** deprotonated at ring position 3).

CHAPTER 5

5.1. Cl$^-$ dissociation, alkyne binding, rearrangement to the vinylidene, nucleophilic attack on the vinylidene by OH_2, rearrangement to a $PhCH_2COIr$ intermediate, from which α elimination gives the product.

5.2. Nucleophilic attack on a halide or tosylate (the latter may be better because the halide may dehydrohalogenate) $2L_nM^- + TsO-CH_2CH_2OTs$. ^{13}C NMR should show two equivalent carbons with coupling to two directly attached H, and coupling to $2n$ L and 2 M nuclei (if these have $I \neq 0$).

5.3. Oxidative coupling of two alkynes to give the metallole, followed by CO insertion and reductive elimination. The dienone should be a good pentahapto ligand.

5.4. From $Cp_2MoClMe$ by abstraction of Cl^- with Ag^+ in the presence of ethylene. C–C should be parallel to Mo–Me for the best back donation because the back-bonding orbital lies in the plane shown in Fig. 5.6. NMR should show inequivalent CH_2 groups, one close to the methyl and one far from this group.

5.5. We expect more LX_2 character (see **5.15**) as L becomes more donor, so C_2C_3 should shorten.

5.6. The allyl mechanism of Fig. 9.2b to give $[(1,5\text{-cod})IrCl]_2$ then displacement of the cod by the phosphite. 1,5-Cod is less stable because it lacks the conjugated system of the 1,3-isomer. The formation of two strong M–P bonds provides the driving force.

5.7. Two optical isomers are possible: the 2-carbon of propene has four different substituents: CH_3, H, CH_2, and Cl_3Pt.

5.8. There are three unpaired electrons for octahedral high spin d^7 Co(II).

5.9. The first complex is the 18e species, $[(\eta^6\text{-indane})IrL_2]^+$ formed by hydrogenation of the C=C bond by the IrH_2 group, and the second is $[(\eta^5\text{-indenyl}) IrHL_2]^+$, formed by oxidative addition of an indane C–H bond, β elimination, then loss of H_2 from the metal and oxidative addition of an indane C–H bond. Substitution only of the arene complex by CO is possible because loss of arene is easier than loss of the Cp-like η^5-indenyl (see Section 5.7).

5.10. The propargyl contributes 3e. The 153° angle must be a compromise between the 180° angle that best accommodates the sp C2 carbon and the bending required for good M–C bonding to C1 and C3.

CHAPTER 6

6.1. **A** reacts by S_N2, **B** by a radical route. i-PrI is an excellent substrate for radical reactions and $MeOSO_2Me$ for S_N2 (see Sections 6.3 and 6.4).

6.2. Assuming steric effects are not important, only the bond strengths change, so these are in the order M–Me < M–Ph < M–H < M–SiR$_3$, favoring silane addition and disfavoring methane addition.

6.3. True oxidative addition is more likely for electron-releasing ligands, better π-donor third-row elements, and better π-donor reduced forms. Dewar–Chatt binding is favored for a weak π-donor site that binds H$_2$ as a molecule.

6.4. For HCl, the steps must be: (1) oxidative addition of HCl; (2a) a second oxidative addition of HCl followed by reductive elimination of H$_2$ and binding of Cl$^-$ or (2b) electrophilic abstraction of H$^-$ by H$^+$ and coordination of the second Cl$^-$ to the empty site so formed. In either case, H$_2$ is also formed. For t-BuCl: (1) SET to give ·PtClL and t-Bu·. t-Bu· may abstract H· from a second molecule of t-BuCl to give Me$_2$C=CH$_2$ and Cl·. In the final step, Cl· adds to PtClL$_2$· to give the product. A Pt(t-Bu) intermediate is also possible from OA, but less likely (M–t-Bu is very rare).

6.5. Oxidative coupling to give the metallacycle followed by β elimination to give L$_n$M(H)(CH$_2$CH$_2$CH=CH$_2$), followed by reductive elimination of 1-butene.

6.6. C > D > B > A. The ν(CO) frequencies increase in the reverse order and lower ν(CO) correlates with a more reduced metal and so faster oxidative addition. After oxidative addition the frequencies should rise because oxidation of the metal should reduce its π basicity.

6.7. Reductive elimination of MeH and PhH are thermodynamically favored relative to reductive elimination of HCl.

6.8. Oxidative addition is not possible for d^0 species, so σ-bond metathesis must be implicated in the first step, probably via formation of H$_2$ complex, which is allowed in a 12e species. PMe$_3$ then displaces H$_2$ from intermediate MH$_2$ species; this process is repeated to give the final product. The final H$_2$ is not lost because W(PMe$_3$)$_6$ is a rather unstable species, for the same reasons we saw for the Ni(0) analog in Problem 7 of Chapter 4.

6.9. The two Hs must be cis in the products. If we run the hydride rearrangement step under D$_2$, D incorporation into products will be seen if H$_2$ is lost.

6.10. PhCN has an unusually unhindered C–C bond and formation of an intermediate η^2-arene complex may help bring the metal close to the C–C bond. Finally, M–CN is unusually strong for a M–C bond because of the π-bonding possible with this CO analog, enhancing the driving force for the OA of the C–C bond.

6.11. Insertion into D_2C–O bond; then β elimination.

CHAPTER 7

7.1. (a) Migratory insertion should give the acyl [CpRu(CO)(COMe)(PPh$_3$)]; (b) insertion into M–H should give the allyl product; (c) attack at an 18e complex is allowed for SO$_2$ (see Section 7.4), so [CpFe(CO)$_2$(MeSO$_2$)] is formed; (d) no reaction is expected because the M–CF$_3$ bond is too strong.

7.2. Cyclometallation of the amine with loss of HCl gives **A**, followed by insertion of the cyclopropene to give **C** or oxidative addition of the strained C–C single bond of the cyclopropene followed by rearrangement to give **D**. Cyclometallation of the amine is not possible for PhNMe$_2$ because of the wrong ring size in this case.

7.3. α Elimination of M–CH$_3$ leaves M=CH$_2$ groups that couple to give H$_2$C=CH$_2$.

7.4. (1) RNC must bind, undergo migratory insertion, and the resulting imine undergo another insertion with the second hydride. (2) Migratory insertion twice over gives a bis-acyl that in its carbenoid canonical form (**7.2**) couples to give the new double bond. (3) Migratory insertion once, followed by alkyl migration from the metal to the carbene carbon in the carbenoid resonance form of the cyclic acyl. (4) Insertion to give MPh(O$_2$CPh) is probably followed by a cyclometallation by a σ-bond metathesis pathway with loss of PhH.

7.5. Oxidative addition of MeI is followed by reductive elimination. The possibility of binuclear reductive elimination is suggested from the label crossover data.

7.6. Ethylene displaces the agostic C–H to give MEt(C$_2$H$_4$). Insertions of ethylene gives an agostic butyl with no α elimination of the growing chain. The process is repeated. The presence of an agostic C–H points to a weakly π-donor metal, which is unable to carry out a β elimination. In the Rh system, neutral Rh(I) is a better π donor, and so β elimination is fast in the first-formed butyl complex.

7.7. Possibilities are $-CH_2-CMe(OMe)_2$ or $-CH_2-CMePh_2$. For C–C bond breaking, we need a strained cyclopropyl or cyclobutyl ring system as in $-CH_2-CMe(CH_2CH_2)$ or $-CH_2-CMe(CH_2CH_2CH_2)$.

7.8. More strongly ligating solvents, more electron-withdrawing ligands, and a poorer π-basic metal will all favor the insertion product. The solvent stabilizes the product, and the ligands and metal make the CO more ∂^+ at carbon and so more reactive.

7.9. Cyclometallation and RE of the cyclopropane should give PtHClL$_2$; the phosphine must cyclometallate in the $-CH_2Nb$ case, which would release CH$_3$Nb and leave a cyclometalated Pd complex.

7.10. The α-CH is β to the second metal, M_2, in a Me–M$_1$–M$_2$ cluster.

7.11. If insertion is first order, L always traps the intermediate acyl, so the rate-determining step is k_1. Increased steric bulk, Lewis acids, and oxidation should all enhance k_1 and speed the overall reaction. If second order, L rarely traps the intermediate acyl, so the rate-determining step is k_2. Increased steric bulk might slow k_2 and slow the overall reaction. Lewis acids and oxidation could both enhance k_2 by making the complex more electrophilic and speed the overall reaction. In an intermediate case, increased steric bulk might speed or slow the rate, while Lewis acids and oxidation should both enhance the rate.

CHAPTER 8

8.1. The rules of Section 8.3 predict attack at (**8.14**) ethylene, (**8.15**) the terminal position of the cyclohexadienyl, and (**8.16**) the butadiene.

8.2. (1) Protonation gives MeH and CpFeL(CO)Cl, (2) SET and nucleophilic abstraction gives MeCl and CpFeL(CO)Cl, (3) electrophilic abstraction gives MeHgCl, and (4) protonation gives MeH and CpL(CO)Fe(thf)$^+$.

8.3. Reduction of Pd(II) to Pd(0) by nucleophilic attack of the amine on the diene complex is followed by oxidative addition of PhI and then insertion of the diene into the Pd–Ph bond to give a Pd(II) allyl. This can either β-eliminate to give the free diene or undergo nucleophilic attack by the amine to give the allylic amine.

8.4. The high $\nu(CO)$ arises from the 2+ charge from the resulting weak π back donation and means that the CO carbon is very electrophilic in character and very sensitive to nucleophilic attack.

8.5. The arene is activated for nucleophilic attack by coordination because the $Cr(CO)_3$ group is so electron withdrawing. The product should be $[(\eta^6\text{-PhOMe})Cr(CO)_3]$ after loss of chloride.

8.6. The H^- group abstracted should be anti to the metal, but in β elimination, expected for a 16e complex, the metal abstracts the syn H.

8.7. We need to make the metal a better σ acid and π base, use a noncoordinating anion, sterically protect the site to prevent dimerization or binding of a solvent C–H bond, and use a poor donor solvent to prevent displacement.

8.8. Nucleophilic attack of MeOH to give the 2-methoxy-5-cyclooctene-1-yl complex is followed by a PR_3-induced β elimination to give **8.18** and the hydride. The 1,4-diene might also be formed.

8.9. Nucleophilic attack of Me^- to give a vinyl complex is followed by electrophilic abstraction of the vinyl with I_2. E and Z isomers of $Me(I)C{=}C(Me)Et$.

8.10. In Eq. 8.26, the metal oxidation state is unchanged during the reaction, but in Eq. 8.24 the Pd(II) is reduced by 2e to Pd(0), accounting for the oxidation of the substrate; Cu(II) helps the reoxidation to Pd(II) required for catalysis.

CHAPTER 9

9.1. Driven by the aromatic stabilization in the product, isomerization should bring all three double bonds together in the right-hand ring to give a phenol, **9.32**.

9.2. Dissociation of L, required for activity, is unlikely for triphos because of chelation, but Cl^- abstraction by BF_3 or Tl^+ opens the required site.

9.3. The initial terminal cyanation step should be followed by isomerization of the remaining internal C=C group to the terminal position and so should give the 1,5-dinitrile as the final product.

9.4. Successive H transfers to the ring are followed by oxidative addition of H_2 and further H transfers. The first H transfer to the arene will be difficult because the aromatic stabilization will be disrupted (this is why arene hydrogenation is hard); this should be easier with naphthalene, where the aromatic stabilization is lower per ring and we only disrupt one ring, at least at first.

9.5. Oxidative addition of the aldehyde C–H bond to Rh is followed by C=C insertion into the M–H to give a metallacycle; this gives the product shown after reductive elimination. Oxidative addition of the strained C–O bond is followed by β elimination and reductive elimination to give the enol that tautomerizes to acetone.

9.6. The first and second are thermodynamically unfavorable unless we find reagents to accept the H_2 or O_2, respectively. The third reaction is favorable, but it will be difficult to prevent overoxidation because the MeOH is usually much more reactive than MeH.

9.7. $H_2[PtCl_6]$ (i.e., an acid, not a hydride).

9.8. Insertion into the M–Si rather than the M–H bond would give M–CR=CHSiR$_3$, and β elimination can now give the unsaturated product. This β elimination produces an MH$_2$ species that could hydrogenate some alkyne to alkene or vinylsilane to alkylsilane.

9.9. Oxidative coupling, followed by β elimination and reductive elimination. For mechanistic support, if the β elimination were suppressed by avoiding β-H substituents, the metallacycle might be isolable.

9.10. Oxidative addition of H_2 is possible after the arene slips to the η^4 form. The substrate can displace the arene to give M(CO)$_3$(diene) H$_2$. We have to assume that the diene adopts a s-cis LX$_2$ form (**5.15**) so that the observed product can be formed by two successive reductive eliminations to place H atoms at the termini of the diene chain. The cis product reflects the conformation of the bound diene, and the monoene is a much poorer ligand in this system and so does not bind and is therefore not reduced.

CHAPTER 10

10.1. The cis form has a doublet of quartets in the hydride region because of the presence of three P nuclei cis to each H and one P trans to H. The trans form has a quintet because of the presence of four P nuclei cis to each H. Using the HD complex will give a 1:1:1 triplet from H coupling to the $I = 1$ D nucleus and after dividing J(H,D) by six to adjust for the lower γ value of the D isotope, we get the J(H,H), which is not observed in the dihydride because equivalent Hs do not couple.

10.2. MH$_3$ and MH(H$_2$) are the most likely. T_1(min) data and ^1J(H,D) in the H$_2$D complexes would be useful. The trihydride should have a long T_1 and a low J(H,D) (see Section 10.7).

10.3. One Ind could be η^3, in which case we should see two distinct sets of Ind resonances. If the two rings were rapidly fluxional, cooling the sample should lead to decoalescence, making the static structure obvious.

10.4. X-ray crystallography would be best and NMR spectroscopy, at low T if need be, should be adequate to make the distinction.

10.5. 31 s^{-1}; $(2500 \times \pi)/\sqrt{2}$ s^{-1}.

10.6. (1) c, a; (2) b, d; (3) d; (4) d; (5) d; (6) b.

10.7. Using Eq. 10.12 gives an angle close to 120°, consistent with a TBP structure with the COs equatorial.

10.8. The CO bond order falls when bridging as μ_2 and falls even further when bridged as μ_3.

10.9. 6-Coordination is expected in both cases, and so loss of Cl$^-$ is necessary to produce an η^2 form; the conductivity should be high for the ionic species, and the IR of the two acetate binding modes are also different. Comparison of the IR with literature examples would be needed to distinguish the two cases.

10.10. If the plane of the pyridine ring is orthogonal to the square plane (from steric effects), we expect diastereotopy of the phosphine methyls because the methyl group of the pyridine breaks the plane of symmetry of the complex.

10.11. The O–H bond of water has a big dμ/dr on vibrating and thus absorbs IR extremely strongly, but Raman uses visible light to which water is transparent, and the O–H bond also has a very small polarization change on vibrating and thus gives no significant Raman bands to swamp out the Raman bands of the dissolved sample.

CHAPTER 11

11.1. The Cp$_2$TiCl$_2$ initially methylates to give Cp$_2$TiMeCl. A second mole of AlMe$_3$ may deprotonate the TiMe group to give MeH and the observed product. Other pathways are possible.

11.2. Initial intramolecular metalacycle formation, presumably with initial reversible CO loss, with metathesis-like cleavage leads to the product. The reaction requires CO dissociation to make a site for the alkene so should be suppressed by excess CO.

11.3. 1,2-Insertion would be followed by alpha elimination in both cases.

11.4. (a) $Ph_3P=CH_2$ has strong Schrock-like character, judging from the strongly nucleophilic character of the methylene group. This is consistent with Fig. 11.1 because C is more electronegative than P. (b) O is more electronegative than C, so $Re=O$ should be more nucleophilic than $Re=CH_2$.

11.5. Initial metathesis of the substrate C=C bond gives $MeCH=CR(OR)$ and a C=W carbene intermediate. This forms a metalacycle with the nearby alkyne and metathesis-like steps lead to product.

11.6. The CH_2, group lines up with the Cp–M–Cp direction to benefit from back donation from W. The two extra electrons of the anion would have to go into the CH_2 p orbital. The CH_2 orientation would be at right angles to that in cation to minimize repulsion between the two filled orbitals. If the extra 2e went into the metal, we would have a 20e complex which would probably undergo insertion to form $[Cp_2WEt]^-$.

11.7. The carbene is a neutral ligand with a lone pair, while Ph is an anionic ligand with a lone pair. Once deprotonated, L-type ligand **11.32** becomes X-type ligand **11.33**.

11.8. See JACS, **133**, 10700, 2012.

CHAPTER 12

12.1. The reverse process should go by the reverse mechanism, which implies (see Fig. 12.6) that MeI will oxidatively add to Pt(II) to give $trans$-$[MePtCl_4I]^{2-}$.

12.2. The Fischer carbene formed on metathesis is stable.

12.3. Cyclometallation of a PMe group in preference to a PPh group is very unusual; perhaps the RLi deprotonates PMe, the CH_2^- group of which then binds to the metal.

12.4. As an 18e species, an η^1-CO_2 adduct is expected; for the indenyl case, slip could generate a site to allow η^2-OCO binding; the 18e complex could only plausibly react by H^- abstraction from the metal by CO_2, which would produce an η^1-OCHO complex. The Re anion is probably the best case because of the negative charge (after all, CO_2 reacts easily with OH^-).

12.5. Cyclometallation of the $ArCH_3$ group followed by CO insertion.

12.6. Loss of PhH by reductive elimination, binding of substrate via the isonitrile C, cyclometallation of the $ArCH_3$ group, migratory insertion involving the isonitrile, isomerization, and reductive elimination of the product.

12.7. Transfer of *endo*-Et to the metal, rotation of Cp, migration of Et back to a different point on the Cp ring, a 1,3 shift on the exo face to bring an H into the endo position from which H transfer to the metal is possible.

12.8. Reductive elimination to form a cyclopropane that immediately oxidatively adds back to the metal.

12.9. Binding of formate as η^1-OCHO, followed by β elimination to deliver H^- to the metal and release CO_2. This can be a good synthetic route to hydrides. Two tests could be NaOOCD and Na acetate.

12.10. CO_2 insertion into the terminal M–C bond to give an η^4-OCOCH$_2$CHCHCH$_2$ carboxylato-allyl complex. Oxidation then leads to the coupling of the allyls by binuclear reductive elimination.

12.11. Oxidative addition of Si–H, followed by coordination and insertion of the alkyne into M–H or M–Si, followed by reductive elimination.

12.12. The intermediate acyl could be hydrogenated; if so, with D_2, one would get MeCD$_2$OH. The methanol could undergo CH activation to Rh–CH$_2$OH, which might undergo RE with the Me–Rh; if so, one would get MeCH$_2$OH.

12.13. Deprotonated OAc^- ion must attack the acyl rhodium intermediate to form the anhydride, so basic conditions should increase the rate by increasing $[OAc^-]$.

12.14. OA/RE is not allowed for d^0 metals. β Elimination is slow for d^0 metals, but olefin insertion is fast, as in olefin polymerization (Section 12.2).

CHAPTER 13

13.1. Any bridging CO complex with L_nM isolobal with CH, for example, $Cp_2Ni_2(CO)$. This might be formed from $NiCp_2$ and CO.

13.2–3. (1) 48e, 3 M–M bonds; (2) 50e, 2 M–M bonds; (3) 52e, 1 M–M bond. The S's are counted as vertex atoms—they retain their lone pair as shown by easy methylation.

13.4. This 60e cluster **13.36** is 2e short of the 62e system expected; Wade's rules give 14 skeletal electrons appropriate for an octahedron counting each of the EtC carbons as vertices because each $Co(CO)_2$ contributes 1e, the two bridging COs contribute 2e, and each EtC contributes 3e.

13.5. **13.37** is isolobal with tetrahedrane, **13.38** with cyclopropane.

13.6. The Fe_4 species is 60e and should be tetrahedral. Four $Fe(CO)_3$ groups are likely, which leaves a single CO, which might be bridging; but we cannot tell from counting electrons. The Ni_5 structure is 76e, and so a square pyramid with one Ni–Ni bond opened up is most likely. The 36e Cr_2 system is expected to have no M–M bond but be held together by the bridging phosphine.

13.7. Two $W\equiv C$ bonds bind to Pt in the cluster just as two alkynes should bind to Pt in the alkyne complex, so $n = 2$. On an 18e rule picture, the alkynes are 4e donors. The unsaturated ligands are orthogonal so that each $X\equiv C$ bond ($X\equiv W$ or C) can back-bond to a different set of d_π orbitals.

13.8. The most symmetric structure for **13.40** is a square pyramid with Fe at the apex and four B's at the base; $(\eta^4\text{-}C_4H_4)Fe(CO)_3$ is the carbon analog.

13.9. Elements to the left of C are electron deficient; elements to the right are electron rich. As long as electron-deficient elements dominate a structure, a cluster product can be formed.

13.10. An $\eta^2\text{-}\mu\text{-}CH_2CO$ complex with the ligand bridging two Os atoms that have lost their direct M–M bond.

CHAPTER 14

14.1. Oxidative addition of the *endo* vinyl C-Br to Pd(0), probably steered by Pd precoordination to the $C=C(CO_2Et)$ alkene, is followed by a Buchwald–Hartwig amination sequence. Oxidative addition of the *exo* vinyl C–Br to Pd(0) is then followed by a Mizoroki–Heck sequence with β elimination in two alternative directions.

14.2. Ru must bind to the ketone O, cyclometallate at the adjacent ring, and undergo insertion of the alkyne into the resulting Ru—aryl bond. Reductive elimination with the Ru–H acquired at the cyclometallation step, completes the process.

14.3. Oxidative addition of the vinyl C–Cl to Pd(0) must be followed by the insertion of the alkyne into the Pd–C bond. Reductive elimination with the Pd–Cl acquired at the oxidative addition step completes the process. Such a Cl–Pd–C reductive elimination to Cl–C is relatively rare.

14.4. Precoordination of Pd(0) to the vinyl group may facilitate subsequent oxidative addition of the strained cyclopropyl C–C bond. The bond adjacent to the $C(CO_2Me)_2$ group cannot be chosen for steric reasons, therefore electronic effects must predominate. An M–C bond is stronger if the carbon bears electronegative substituents, as here. Insertion of the aldehyde C=O group must occur, followed by reductive elimination. The regiochemistry seen suggests the insertion may occur into the $Pd-C(CO_2Me)_2$ bond with the stabilized malonate anion attacking the C end of the C=O bond. C–C bond formation then requires reductive elimination.

14.5. The catalyst may decompose by a cyclometalation. This requires the Ph group to rotate such that it becomes coplanar with the azole ring. This is possible for R=H, but when R=Me, a prohibitive steric clash occurs.

14.6. The enyne metathesis pathway of Eq. 14.13 is most plausible.

14.7. An alkene–alkyne oxidative coupling to give a metalacyclopentene, could be followed by β elimination and reductive elimination.

14.8. The $RhClL_3$ complex can easily lose an L to give stable $RhCl(CO)L_2$, but $RhCl(L-L)_2$ cannot so easily lose an L because of the chelate effect; presumably Cl^- is now lost instead. The appropriate intermediate is $[Rh(CO)(L-L)_2]^+$. This should lose CO much more easily than $RhCl(CO)L_2$ because it is five-coordinate and has a positive charge, discouraging back donation.

14.9. Alkyne–alkyne oxidative coupling leads to a metalacyclopentadiene (metallole). Oxidative addition of R_2BSnR_3 is then followed by reductive elimination. This accounts for the *endo–endo* arrangement of the vinyl groups. Presumably, if the R_2BSnR_3 were omitted, the metallole might be isolated.

14.10. If the azide loses N_2, it can give rise to a Rh–nitrene intermediate. By analogy with carbene insertions, a nitrene insertion into the adjacent CH would give the observed product.

CHAPTER 15

15.1–2. The metal is d^0, and therefore CO does not bind well enough to give a stable complex, but weak binding is possible and the absence of back donation increases the electrophilic character of CO carbon and speeds up migratory insertion in the weakly bound form.

15.3. The third-row element prefers the higher oxidation state and has longer M–C bonds, allowing a greater number of R groups to fit around the metal.

15.4. Ethylene insertion into W–H could be followed by a double alpha elimination of the H, followed by RE of H_2. CO insertion into the H to give an eta-2 formyl could be followed by alkylation at O and deprotonation at the alpha CH.

15.5. The two alkenes are orthogonal to allow the metal to back-donate efficiently to both alkenes by using different sets of d_π orbitals.

15.6. Alkene hydrogenation normally occurs in the presence of many hydride ligands. The stereochemistry of the Re compound makes the (C=C) groups of the bound alkene orthogonal to the M–H bonds and prevents insertion.

15.7. Cr, $S = 1/2$ and 3/2; Mn, 0; Fe, 1/2; Co, 0.

15.8. $(Cp^*)_2Lu$ groups at 1 and 4 positions on benzene ring to avoid steric clash.

15.9. d^6 Oct, $S = 0, 1$ or $2; f^2, S = 1; d^3$ Oct, $S = 1/2$ or 3/2.

CHAPTER 16

16.1. These are the most abundant metals in the biosphere.

16.2. Most organisms live in an oxidizing environment and proteins have mostly hard ligands.

16.3. A low $\nu(N_2)$ implies strong back donation, which also means that the terminal N will also have a large ∂^- charge and therefore be readily protonated.

16.4. The stability of radicals R· is measured by the R–H bond strength, which is the ΔH for splitting the bond into R· and H·. For these species, this goes in the order $HCN > CF_3H > CH_4 > PhCH_3 > Ph_2CH_2$. C–H bonds to sp carbons are always unusually strong because of the high s character, while Ph groups weaken C–H bonds by delocalizing the unpaired electron in the resulting radical. This is the reverse of the order of case of loss of R·.

16.5. Protonation lowers the electron density on Re and reduces the back donation to N_2, resulting in an increase in $\nu(N_2)$ and weaker M–N_2 binding, making the N_2 more easily lost.

16.6. This would need reversal of the proposed nucleophilic attack on CO by OH⁻. In order to reverse the reaction while maintaining the label on the carbon, however, the proton of the Ni–COOH group has to switch from the labeled O to the normal O before the reversal step.

16.7. CO binds best to Ni(0) but strong back donation would tend to minimize nucleophilic attack. Ni(II) might be too weakly back-donating to bind CO but if it did, nucleophilic attack would be favored. Ni(I) is midway in properties.

INDEX

Note: Page numbers in **bold** indicate main entries; Greek letters appear at the end of the index.

A vs. D substitution mechanisms 115–122

Abbreviations xiii

Acetic acid process, Monsanto 333–334, 458

Acetylides 73

Acid with noncoordinating anion, use of 173

Actinide complexes 158, 426

Activation of ligands 61–63, **332–342**, *see also* Ligand

Actor ligands 33

Acyl complexes 78, 84

Adamantyl complexes 73

ADMET (acyclic diene metathesis) 319

Agostic species 74–75, 89, 91, 167, 278, 298–300

Alcohol activation catalysis 344

Alcohols, as reducing agents 85, 138, 250

Alkane activation 336
  C–C bond cleavage in 342
  dehydrogenation, homogeneous catalysis of 339
  metathesis, homogeneous catalysis of 340

Alkene
  coupling 82
  hydroboration catalysis 246
  hydroformylation, homogeneous catalysis of 242
  hydrogenation, homogeneous catalysis of 233, 394–396
  hydrosilation, homogeneous catalysis of 246
  isomerization, homogeneous catalysis of 231

The Organometallic Chemistry of the Transition Metals, Sixth Edition.
Robert H. Crabtree.
© 2014 John Wiley & Sons, Inc. Published 2014 by John Wiley & Sons, Inc.

Alkene (*cont'd*)
  metathesis, homogeneous catalysis
    of 317, 391–392
  polymerization, homogeneous
    catalysis of 319, 323, **324–329**
Alkene complexes 134–138
  bonding models 135–136
  masked carbonium ion character
    138
  nucleophilic addition to 205,
    209
  strain in 136–137
  synthesis, reactions 137
Alkoxides 85
Alkylidene complexes 293, 300,
    see also Carbenes
Alkyls and aryls, organometallic
    69–79
  agostic 74–75
  bond strengths 92–93
  bridging 81
  bulky, special stability of 73, **76**
  d⁰ 72, 76, 81, 84, 91
  decomposition pathways of
    72–73, 76–77
  electrophilic abstraction of 219
  fluoro-, 76, 80
  homolysis of 446
  main group 69–71, 81
  metalacycles 82, 298, 301
  polarity of M–C bond in 70–71
  preparation of 77–79
  stability of 72–74
  as stabilized carbonium ions
    70–71
Alkynes 302
  complexes, and bonding in **139**
  coupling of 180
  hydration of 215
  hydrosilylation 388
  two vs. four electron ligands 139
Allenyl complexes 143
Allyl complexes 140–143
  bonding in 140
  fluxionality 141
  NMR of 141
  syn and anti groups in 140–141

Alpha elimination 103, **198**, 298,
    304, 421, 429
Ambidentate ligands 31–33
Ambiguity
  in catalysis, homogeneous vs.
    heterogeneous 225
  in oxidation states 47, 292,
    302–303
Amido (-NR₂) complexes 85
Amino acids 437
Ammines (NH₃ complexes) 6–8
Anion, noncoordinating **128**, 395,
    424, 430
Antimalarial drug 3
Antitumor drug 10
Apoenzyme 455
Aqua ions 4, 48
Aquacobalamin 443
Archaea (microorganisms) 458
Arene complexes 154–156
  from diene 155
  nucleophilic addition to 209–210
Arene hydrogenation 226, 233, 242
Aromaticity 143, 155–158
Aryl complexes 71, 77, 79,
    83–84, 94
Associative substitution 120–122
Asymmetric catalysis 226, 230, 231,
    236–239, 240, 242
  alkene hydrogenation 236, 395
  in organic synthesis 383–384,
    392–393, 395, 400–401
Atom economy 333, 343, 400

Back bonding 23–26, 81, 89–91,
    144, 146, 149–50, 158, 204, 216,
    291–293, 296, 303, 313, 414
  in CO complexes 98–102
  evidence for 25
  in PR₃ complexes 109–110
  in sigma complexes 30–31, 90
"Barf" anion 83, 128, 395, 399
Benzyl complexes 143
Beta-elimination 72–75, 77, 85, 87,
    198
  of alkyl 137, 198, 429
Bioalkylation and dealkylation 448

Biofuels 345
Bioinorganic chemistry 3, 436–462
Biomedical applications 463
Biomethylation reactions 444
Bioorganometallic chemistry
    436–464
Biosynthesis, of methane 459
Bismuth donor ligands 33
Bite angle of chelate ligands **112**,
    173, 244, 396
Bond strengths, organometallic
    92–94, 167, *see also specific*
    ligands
Bonding models
    for alkene complexes 134–136
    for alkyne complexes 139
    for allyl complexes 140
    for carbene complexes 291, 300
    for CO and its complexes
        98–102
    for complexation in general
        19–20
    for cyclopentadienyl complexes
        147–150
    for diene complexes 144–145
    for metallocenes 149–150
    for paramagnetic organometallics
        414
    for phosphine complexes
        109–110
    reactivity rules based on 70. 88
Borane clusters 358
Boryl ligand 302
Bridging ligands 5, 42–43
    electron counting in 43, 46–47
    μ-symbol for 5
Buchwald–Hartwig reaction 249,
    386, 390
Bulky groups, stabilization from 73,
    76, 167

CF₃ group 78–79
C₆F₅ group 76
Carbene complexes 207, **296–310**,
    432
    agostic 298, 308
    in alkene metathesis 317–324

bonding in 291, 300
bridging in 305–306
fluxionality in 301
Fischer vs. Schrock type 290
insertion into C–H bonds 393
IR spectra 300
NMR of 293–295, 300
Carbide clusters 82
Carbon dioxide, activation of 332
Carbon-hydrogen bond cleavage
    336–342
Carbon monoxide, *see also*
        Carbonyls
    activation of 332
    double insertion of, apparent 192
    electronic structure of 98–102
    polarization on binding 99
Carbon monoxide dehydrogenase
    458
Carbonate complex 8
Carbonyls, metal 16, 25, 64–65,
    98–105, 125–127, 459, 461
    bond strengths (M–CO) 93
    bonding in 98–102
    bridging 104
    cluster 105, 242, 281, **353–364**,
        449–462
    containing hydrides 86–87
    d⁰ 101
    first row, structures 41, 65
    migratory insertion involving
        187–192
    infrared spectra of 64–65, 99,
        101, 166, **276–9**
    nucleophilic attack on 208
    photochemical substitution of
        124–127
    preparation 102
    removal of CO from 103
    substitution in 119–124
Carbyne ligand 302–306, 423
Catalysis, homogeneous 1–3,
    224–251
    acetic acid process 333
    acid, hidden, 253
    alkene metathesis 317, 320, 391
    asymmetric **236–241**, 393–395

Catalysis, homogeneous (*cont'd*)
  carbonylation 396
  C–C coupling 248, 384
  C–H activation and
    functionalization 230, 251, 336,
    393, 401
  CO_2 reduction 334
  cooperative 253
  enzymatic Ch. 16
  hydration of alkynes 215
  hydrocyanation 245
  hydroformylation 242
  hydrogen borrowing 343
  hydrogenation 233–242, 394
  hydrosilylation 246, 388
  isomerization **231–233**, 235, 239,
    244, 330
  isotope exchange 90
  kinetic competence 229
  living 319, 375
  organic applications 383–404
  oxidative 225, **250–251**, 399
  polymerization of alkenes 324
  supported 372
  tests for homogeneity of 242
  thermodynamics 226
  Wacker process 212
  water gas shift 332
  water splitting 251
  yield, conversion and selectivity in
    228
Catalytic cycles, general features of
  224–229
Chain theory of complexation 6–7
Chatt cycle of N_2 reduction
  453–454
Chauvin mechanism of alkene
  metathesis 320
C–H bond activation 167, 251,
  336–342, 401–404, 460
Chelate
  definition 5
  trans-spanning 177
  wide bite angle 173
Chelate effect,
Chemotherapy 463

Chromocene 150
CIDNP method 172
Click chemistry 405
Clusters, metal 82, 306, **354–364**
  in biology 449–462
  descriptors (closo, nido, etc.) 360
  electron counting in 355–362
CO complexes, *see* Carbonyls
CO dehydrogenase 458
CO stretching frequencies 25, 65,
  101–105
Cobaloximes 445
Coenzyme A 444
Coenzyme B 460
Coenzyme B_{12} 442
Coenzyme M 460
Coenzymes 441
Complex and complexation
  chiral 8
  definition 4
  effects of complexation 61–63
  changing metal 63
  high spin and low spin 12–14
  with lone pair donor ligands
    29–31
  net ionic charge, effect of 51, 65
  optical activity 8
  with π-bonding pair as donor
    29–31
  with σ-bonding pair as donor
    29–31
Computational methods 110, 112,
  156, 214, 229, **283–284**
Cone angle 110, 116
  of Cp ligands 147
  of PR_3 110–112
Coordination complexes 4–11
Coordination geometries, common
  57
Coordination number 49, 57–58
Coordinatively inert and labile
  complexes 14, 120, 122
Corrin ring system 443
Cossee mechanism, for alkene
  polymerization 326
Counter ions, choice of 128

Counting electrons **40–51**, 292, 312
 ionic vs. covalent models for
  40–43
 in metal clusters 355–362
Coupling, to form C–C bonds
  384–390
Covalent and ionic models for
  electron counting 40–43
Cross metathesis 318–319
 selectivity in 322
Crossover experiment 178
 double crossover experiment 320
Crystal field theory, stabilization
  energy 11–19
 in photochemical substitution
  124–126
 splittings for various geometries
  13, 15, 18, 117
Crystallography 128, 154–156, **279**
Cyanocobalamin 443
Cycloheptatriene and -trienyl
  complexes 156
Cyclometalation **78–79**, 266, 378,
  384, 402–403
Cyclooctatetraene, complexes
  formed from 157–158
Cyclopentadienyl complexes 45, 94,
  147–153, 293, 312, 425–426
 analogues of, with Cp-like ligands
  153
 bonding in 147–150
 electrophilic addition to 218–219
 fluxionality of 268
 pentamethyl (Cp*) 150–152, 293,
  426
Cyclopropanation 393

d^n configurations 12, 17, 19, 33
d^0 configuration, special properties
  of 17, 24, 27, 33, 48–49, 52–53,
  56–59, 64, 72–78, 81, 84, 86, 91,
  101, 159, 179, 304, 310, 326,
  412–414, 420–421
d^2 configuration, special properties
  of 28, 30, 33, 53, 57, 58, 64,
  144, 310

d^3 configuration, special properties
  of 14–15, 33, 53, 57
d^4 configuration, special properties
  of 17, 33, 53, 55, 57, 312
d^5 configuration, special properties
  of 14, 17, 33, 52–53, 57
d^6 configuration, special properties
  of 12–27, 33, 45, 53, 55, 57
d^7 configuration, special properties
  of 14–15, 33, 57
d^8 configuration, special properties
  of 17–18, 33, **48–49**, 52–53, 57,
  86
d^9 configuration, special properties
  of 17, 33
d^{10} configuration, special properties
  of 17, 33, 58–59
d-orbital energies, crystal field
  behavior 12
 effect of oxidation state changes
  28
 effect of changing the metal
  18–19
Density Functional Theory (DFT)
  283
Dewar–Chatt bonding model
  135–136, 144, 264, 291
Dialkylamido ligands (NR_2) 84–86.
Diamagnetism 13, 17
Diastereotopy 261
Diene
 complexes of 142, **144–146**, 159
 bonding in 145
 metathesis 319
 nucleophilic addition to 145
 s-trans binding mode of
  145–146
Dihydrogen bond 94
Dihydrogen complexes 30, **89–91**,
  424
 bioinorganic aspects 462
 H . . . H distance from J(H,D)
  91
 stretched 91
Dinitrogen (N_2) complexes 452
 IR spectra 453

Dioxygen (O₂)
  insertion of, into M–H   400
  reactions involving   400
Directing effects, in alkene
    hydrogenation   235
Disproportionation   127, 252, 335,
    400
Dissociative substitution   115–120
Dodecahedral geometry   57, 58
Double insertion, of CO, apparent
    192
Drugs, organometallic   464
Dynamic kinetic resolution (DKR)
    395

Effective atomic number (EAN)
    rule, in clusters   355
Eight coordination   57
Eighteen electron rule   **40–50**,
    411–413, 425
  ionic/covalent conventions for   40
  limitations of   48
Electrochemical methods   88, 251,
    281, 345, 371, 413, 426, 457
Electron counting   40–50, 355–362
  different conventions for   40–50
  of reagents   50
Electron paramagnetic resonance
    (EPR)   420
Electronegativity   22, 27, 63, 428
Electroneutrality   27
Electrophilic addition and
    abstraction   216–221
  single electron transfer pathways
    in   219–220
Eliminations, α, β, γ, and δ   72–75,
    77, **198**, 298, 304
Energy chemistry   344–346
Entropy of activation   167, 169
Enzymes   226, 237, 250, **439**
Epoxidation, catalytic   401
Ethynyls, *see* Acetylides
EXAFS   442

Factor F₄₃₀   460
fac- vs. mer-stereochemistry   34, 119
f-block metals   57, 195, 329, 411, **426**

FeMo-co, in nitrogen fixation   450
Ferredoxin proteins   455
Ferrocene (FeCp₂)   54, **147–150**, 464
Ferromagnetism   13
Fischer carbene   290–298, 365, 393
Five coordination   42, 57, **115–119**,
    121, 175–176, 234
Fluoro complexes (M–F)   85
Fluoroalkyls   76, 80
Fluxionality   59, 118, 148, 260, 265,
    268, 270, 301, 424
Formation constants   6, 11
Formyl complexes   104
Four coordination   9, 17, 57, 93, 121,
    312, 445
Free radicals, *see* Radicals
Frontier orbitals (HOMO and
    LUMO)   26, 99, 144
Fullerene complexes   155–156

Geometries, typical for specific d^n
    configurations   57
Green chemistry   3, 56, 128, 215,
    224, 317, **343–344**
Green–Davies–Mingos rules
    209–211
Green's MLX nomenclature   43
Grubb's catalyst, for alkene
    metathesis   318, 323, 391

Halocarbons, as ligands   128
Hapticity changes in π complexes
    140, 147, 155
Haptomers   155
Hard and soft ligands   10
Heck reaction   249, 384, 405
Heterolytic activation of H₂   90
Hieber's hydride (H₂Fe(CO)₄)   86
High field and low field ligands   16
High spin and low spin complexes
    12
HOMO and LUMO   26, 99, 119
Homoleptic complexes   106–108,
    139, 420
Hydrides, metal   86–89, 424, 463
  acidity of   89–90
  bond strengths of   92–93

bridging in   46, 89, 356
characterization   86–87
crystallography   86, 279
H atom transfer in   88
IR spectra of   91
kinetic vs. thermodynamic
  protonation   90
NMR spectra of   86, 91
nonclassical structures in   90
photochemical substitution of
  124–126
preparation and characterization
  86–87
reactivity   63, 87–88
Hydroboration, catalysis   246
Hydrocyanation, catalysis of   245
Hydrogenases   461
Hydroformylation, catalysis of   242
Hydrogen bonding   21, 94
Hydrogenases   458, **461**
Hydrogenation, catalysis of   233
Hydrosilylation, catalysis of   246
Hydrozirconation   193
Hypervalency   21

Indenyl complexes   122
Inert vs. labile complexes   14
Infrared spectroscopy   9, 64–65, 75,
  276, *see also specific ligands*
  of agostic alkyl complexes   75
  of carbenes   300
  of carbonyls   64–65, 99, 101, 166,
    276–279
  of hydrides and H_2 complexes
    86, 91
  of isonitriles   106
  isotope labeling in   279
  of metal oxos   312
  of N_2 complexes   453
  of NHCs   307
  of nitrosyls   107
  of thiocarbonyls   106
Insertion   78–80, **185–198**
  1,1 vs. 1,2 types   185–186
  apparent   191
  alternating, of ethylene/CO   197
  in catalysis   224–248

of CO into M–H   190
comparison of M–H vs. M–R   195
coplanarity requirement in 1,2
  case   193
double, of CO   192
enhanced rate with Lewis acid
  190
enhanced rate by oxidation   190
involving alkenes   192–194, 249,
  324–329
involving alkynes   194
involving dienes   196
involving carbon dioxide   197,
  198, 333
involving carbonyls (migratory
  insertion)   185–192, 333, 459
involving fluoroalkenes   138
involving isonitriles   192
involving M–R   192–197
involving O_2 196, 198, 250
involving radicals   196
involving SO_2 186, 197
Lewis acid promoters for   190
mechanism of   187–189
of M–H vs. M–R   195
multiple   192, 325–330
oxidation as promoter for   190
in polymerization   324–329
regiochemistry of, M–H/alkene
  193
syn vs. anti   194
Inter- vs. intramolecular reaction,
  test for   178
Interchange mechanism of
  substitution   122
Inversion of normal reactivity in
  ligands (umpolung)   209
Ion pairing   191
Ionic and covalent models, e
  counting and   40–50
Iron–sulfur proteins   455
Isolobal analogy   364
Isomerase reaction   444
Isomers, linkage and optical   7–8
Isonitriles (RNC)   105
Isotope labeling   166, 198, 200, 279,
  285

Jahn–Teller distortion   15

Karplus relation   171
Kinetic isotope effect   285
Kinetic vs. thermodynamic products
   90
Kinetic resolution   394
Kinetics   116, 120–122, 164–170,
   187–188, 416–417
   of CO insertion   187–188
   of substitution   116, 120
Kumada coupling   388

L vs. X_2 binding   **135**, 292, 313
Lanthanide complexes   429–432
Lanthanide contraction   29
Ligand field theory   19, 41, 58–60
Ligands
   bulky   73, **76**, 85, 104, 110–111,
      167
   bridging   5
   definition   4
   effects of complexation   60
   electron counting for   40–50
   binding geometry like excited
      state   145
   hard vs. soft   10–11
   high and low field   16
   polarization of on binding   61, 101
   π-bonding, π -acid, π -donor   16,
      23–26, 99–101
Linkage isomers   7
Living catalysts   319
Low and high spin forms   12

Magic numbers, in nanoclusters   369
Magnetic moment   428
Magnetic properties of complexes
   17, 148, 150, 153
Main group compounds   21–23
Manganocene ($MnCp_2$)   150
Mass spectroscopy   285
Materials   371–378
   bulk   372
   electronic   375
   MOFs   373
   NLOs   376

OLEDs   377
organometallic polymers   374
POPs   373
porous   373
sensors   378
mer- vs. fac-stereochemistry   34, 119
Metal-to-ligand charge transfer   126
Metal–metal bonds   42, **354–370**
   homolysis   127, 426
   multiple   363
Metal organic frameworks (MOFs)
   373
Metalabenzenes   158
Metalaboranes   361
Metalacarboxylic acid (M–COOH)
   333
Metalacycles, metal   82, 158, 180,
   298, 301
Metalacyclopropane bonding model
   135–136
Metallocenes (MCp_2)   150
   bent   150
   bonding in   149–150
   in polymer synthesis   324
   polymers containing   374
Metalloenzymes   439
Metalloles   158
Metals, Earth-abundant (cheap)   3
Metathesis, alkene   301, 309, 317–323
   Chauvin mechanism for   320
Methane oxidation, catalytic   338
Methanogenesis   459
Microscopic reversibility   175, 473
Migratory insertion   **185–192**
Mizoroki–Heck reaction   249, 384,
   405
MLX nomenclature   43
MO model for ligand binding, see
   Bonding model
Model studies, bioinorganic   445
Molecular electronics   375
Molecular recognition   440
Molecular wires   375
Mond, Ludwig, discovery of $Ni(CO)_4$
   98
Monsanto acetic acid process   333
Murai reaction   402

N₂, *see* Dinitrogen
N-Heterocyclic carbene (NHC)
113–115, **306–310**
abnormal (mesoionic) NHC   115
detachment from metal by RE
307
Nanoparticles   368–371
Neutron diffraction   87, 91, 280
Nickel enzymes   457
Nickelocene (NiCp₂)   150
Nine coordination   59
Nineteen electron configuration
122–4, 127, 220, 375
Nitride complexes   452
Nitrogen fixation   449
Nitrogenase   449
NO complexes (linear and bent)
106–108, 122
IR stretching frequencies of   107
Noble gas configuration   40–42
Nonclassical hydrides (H₂ complex)
90
Noncoordinating anions   **128**, 395,
424, 430
Nonlinear optical materials (NLOs)
376
Noyori catalyst   395
Nuclear magnetic resonance
spectroscopy   **260–276**
of alkene complexes   136
CIDNP effects in   172
coupling in   86, 91, 424
of dihydrogen complexes   91
of hydride complexes   86, 91
NOE effects in   272
of paramagnetic compounds   282,
413, 419
stereochemical information from
171
Nucleophilic abstraction   207–216
Nucleophilic addition   101, 136, 138,
140, 207–215, 297–298
on alkynes   215
on CO by Et₃NO   208
effect of metal on tendency for
57, 60
on isonitriles   208

ligand hapticity changes caused by
205
rules for predicting products in
209–211

O₂, *see* Dioxygen
Octahedral geometry   4–5, 59
Odd-electron organometallics   17
Odd vs. even dⁿ configurations   17
OLED   2, 377
Oligomerization, catalysis of   324
Open shell systems   411
Orbitals
d, role in M–L bonding   11–21
f, role in f block   411
π*, role in M–L bonding   23–25
σ*, role in M–L bonding   30
σ*, role in oxidative addition
166
Organic light emitting diodes
(OLEDs)   2, 377
Organoaluminum species   70
Organosilicon reagents   246
Organozinc reagents   69, 388
Outer sphere reactions   197, 240
Oxidase reactions, organometallic
250
Oxidation, accelerating substitution
by   122
Oxidation state   45–48, **51**, 64
ambiguities in assigning   47, 54,
292, 302, 303, 424
complexes of unusually high
420–426
and dⁿ configurations   49
limitation on maximum and
minimum   56, 179, 412, 421
variation of ligand type with
32–33
Oxidative addition   77–79,
163–173
of alkane C–H bonds   340
binuclear   164, 172
concerted mechanism   166–168
ionic mechanism   172–173
radical mechanism   170–172
S_N2 mechanism   168–170

Oxidative coupling 180–181
Oxo complexes (M=O) 251, 300,
    310–312, 425, 429
    IR spectra 312
Oxo wall 311
Oxophilic character 84, 431
Oxygen donor ligands, *see*
    Alkoxides; Dioxygen; Oxo
    complexes

Palladium (II)
    promotion of nucleophilic attack
        by 212–216
    substitution 121
Para hydrogen induced polarization
    (PHIP) 275
Paramagnetic organometallics,
    bonding model 414
Paramagnetism 11, 411–424,
    426–432
Pauson–Khand reaction 398
Pentadienyl complexes 153
Pentamethylcyclopentadienyl (Cp*),
    special features of 150, 152
Perfluoro ligands 79
Periodic table xvi
Periodic trends 28, 427
Phosphide (PR$_2$) ligand 85
Phosphine ligands (PR$_3$) 109–112
Photochemistry 124–127
Piano stools 147
Pincer ligands 56, 79, 113, 253, 312,
    339–340
Platinum (II), substitution 121
Platinum drugs 464
Polar organometallics 70
Polarity of M–C bonds 71
Polarization of ligands 61, 99, 453
Polyene complexes 158, 159
    stability to dissociation 159
Polyhydrides 424
Polymerization, alkene, catalysis of
    324
Polymers
    organic 324–326
    organometallic 374

Pressure, effect on reaction rates
    126
Problem solving, hints for 38, 473
Propargyl complexes 143
Proteins 437
Proton-coupled electron transfer
    (PCET) 251
Protonation 46, 61
    kinetic vs. thermodynamic 90

Radicals
    chain vs. nonchain reactions of
        organic 170–172
    clock reactions of 172, 251
    mechanistic pathways involving
        organic **170–172**, 194, 196,
        219
    metal-centered 123, 170–172
    ligand-centered 35, 283
    solvents appropriate for reactions
        involving organic 172
Radioactivity 426, 432
Raman spectroscopy, resonance
    442
RCM (ring-closing metathesis)
    319
Reactivity of alkyls, factors
    governing 70–71
Real charge on atoms 64
Reduction, accelerating substitution
    by 123
Reductive elimination 76, 127, 163,
    173–178, 239, 246, 249, 307
    binuclear 179
    C–O, C–N bond formation in
        179, 249, 386
    kinetics and mechanism
        175–178
Reductive fragmentation 180–181
Regiochemistry
    in hydroformylation 242–244
    of nucleophilic attack of π ligand
        209–211
Relaxation in NMR work on metal
    complexes 264–265, 272–276
Rh(I), substitution 120

ROM (ring-opening metathesis) 319

ROMP (ring-opening metathesis of polymerization) 319, 323

Rubber, synthetic 331

Saturation, coordinative 72

Schrock carbene 290–293, 298–301, *see also* Carbene complexes

Schrock catalyst (for alkene metathesis) 318

sd^n model 21–23

Sensors 378

Seven coordination 57, 415

Seventeen electron configuration 41–42, 49, 122, 419, 426, 446

Shell Higher Olefins Process (SHOP) 324

Shilov chemistry (alkane reactions) 337

Sigma bond metathesis 179–180

Sigma complexes 89–92, *see also* σ-Complexes

Silyl complexes (SiR$_3$) 77, 84

Single electron transfer 219

Single molecule imaging 286

Single site catalyst 325

Six coordination 4–9, 57

Sixteen electron species, d^8 metals preferring 49, 120

intermediates 107, 115

Skeletal electron pair theory (Wade's rules) 358–363

Slip, of π ligands 122, 167

Soft vs. hard ligands 10

Solar cell 346

Solvents (and other weakly bound ligands) 121, 127–128

Spectator vs. actor ligands 33

Spin saturation transfer 271

Spin state changes 413

effect on reaction rates 418

Splitting, crystal field and ligand field 11–16

Square planar geometry 5, 9, 17–18, 49–50, 58–60, 76, 120, 166, 176–177, 416, 459

distorted 167

typical metals that adopt 49

Square pyramidal **17–18**, 59, 117, 169, 268

Stability, of alkyls 70–75

of polyene and polyenyls 159

Stereochemistry

of 1,2-insertion 194

determining 260–268, 276–279

of electrophilic attack on an alkyl 219

fac vs. *mer* 34–35

of hydrogenation 233–234

at metal 101, 117, 121

of migratory insertion 189

of nucleophilic attack on a ligand 209–211

of substitution 117, 121

Stereoscopic representation, of molecules 156

Steric effects 73, **76**, 85, 104, 110–111, 167, 299–300, 307, 342, 421, 424

Steric saturation 427–430

Strained hydrocarbons, enhanced binding and reactivity of 136–137

Substitution **115–129**

associative 120

dissociative 115

effect of pressure 126

kinetics of 116, 120–122

ligand rearrangement in 122

mechanism 116, 120–122

photochemical 124

radical mediated 124

redox catalysis of 122–124

stereochemistry of 5, 117, 121

Subunits (of enzymes) 437

Supramolecular effects 94

Sulfur dioxide, insertion reactions involving 197

Supported organometallic chemistry, on polymer 251–252

Surface organometallic chemistry 252–253

Suzuki–Miyaura coupling 386, 388
Symbiotic and antisymbiotic effects
   63

T- vs. Y-geometry 117–118
Technetium imaging agents 464
Tetrahedral enforcer ligand, Tp as
   154
Thiocarbonyl complexes (CS) 106
Thiolate (SR) 85, 448, 458,
   460–461
Three coordination 57, 175–177,
   234, 248, 429
Titanocene dichloride (Cp$_2$TiCl$_2$)
   45, 59, **150–151**
Tolman electronic and steric
   parameters
   for NHCs 307
   for PR$_3$ 110–112
Trace elements in biology 439
trans effect 9
   rationale 117–121
   use in synthesis 10
trans influence 10
Transfer hydrogenation **241**, 286,
   465
Transition state analogue 440
Transmetalation 78
Tricapped trigonal prism 58
Trigonal bipyramidal geometry 58,
   117–121, 167, 176, 268, 356
Trigonal prismatic geometry 55, 58,
   74
Trimethylenemethane as ligand
   146
Trimethylsilylmethyl complexes
   77
Tris(pyrazolyl)borates 154
Tungsten hexamethyl 73, 91
Turnover limiting step 228
Twenty electron species 122
Two coordination 57
2-electron, 3-center bond 30

Unsaturation, coordinative 75
Uranocene 433
UV-visible spectroscopy 285, 429,
   432

Vacant site, definition 72, 75
Vanadium, alternative nitrogenase
   containing 450
Vanadocene (Cp$_2$V) 150
Vinyl complexes 81, 84
   isomerization 84
   synthesis 81
   η^2-form 84
Vinylidene 139, 295

Wacker Process 212–215
Wade's rules (for clusters) 358–363
Water, as ligand 4
Water gas shift reaction 332
Water oxidation catalysis 251
Werner complexes 4–9

X-ray crystallography 86–87, 279
   of diene complexes 144
   of fullerene complexes 156
   of hydrides and H$_2$ complexes
      86–87
   of PR$_3$ complexes 118

Y- vs. T-geometry 117–118

Zeise's salt 134
Zeolites 373
Zero electron ligands and reagents
   21, **47**, 50, 138, 216–217
Ziegler–Natta polymerization
   catalysis 326

Δ, in crystal field and ligand field
   models 12–20
   effect of metal on 16

π-Acid (π-acceptor) ligand 19–25
   CO as 98–105
   PR$_3$ as 109–112
π-Donor ligand 26–27
   alkoxide as 85
   amide as 85
   halide as 94

σ-CAM 336
σ-Complexes 30–31, 75, **89–92**
   as reaction intermediates 166

Lightning Source UK Ltd.
Milton Keynes UK
UKOW04n1653020414

229304UK00001B/1/P